Embryonic Stem Cells

METHODS IN MOLECULAR BIOLOGY™

John M. Walker, SERIES EDITOR

METHODS IN MOLECULAR BIOLOGY™

Embryonic Stem Cells

Methods and Protocols

Edited by

Kursad Turksen

Ottawa Health Research Institute, Ottawa, Ontario, Canada

Humana Press ✳ Totowa, New Jersey

© 2002 Humana Press Inc.
999 Riverview Drive, Suite 208
Totowa, New Jersey 07512

humanapress.com

This publication is printed on acid-free paper. ∞
ANSI Z39.48-1984 (American Standards Institute)
Permanence of Paper for Printed Library Materials.

Production Editor: Diana Mezzina

Cover design by Patricia F. Cleary.

For additional copies, pricing for bulk purchases, and/or information about other Humana titles, contact Humana
at the above address or at any of the following numbers: Tel.: 973-256-1699; Fax: 973-256-8341; E-mail:
humana@humanapr.com; or visit our Website: www.humanapress.com

Printed in the United States of America. 10 9 8 7 6 5 4 3 2 1

Library of Congress Cataloging in Publication Data

Embryonic Stem Cells: methods and protocols / edited by Kursad Turksen.
    p. cm. -- (Methods in molecular biology ; v. 185)
Includes bibliographical references and index.
ISBN 0-89603-881-5 (alk. paper)
 1. Embryonic Stem Cells--Laboratory manuals. I. Turksen, Kursad. II. Series.

QH440.5 .E43 2002
612'.0181--dc21

                                                        2001026459

Preface

It is fair to say that embryonic stem (ES) cells have taken their place beside the human genome project as one of the most discussed biomedical issues of the day. It also seems certain that as this millennium unfolds we will see an increase in scientific and ethical debate about their potential utility in society.

On the scientific front, it is clear that work on ES cells has already generated new possibilities and stimulated development of new strategies for increasing our understanding of cell lineages and differentiation. It is not naïve to think that, within a decade or so, our overall understanding of stem cell biology will be as revolutionized as it was when the pioneering hemopoietic stem cell studies of Till and McCulloch in Toronto captured our imaginations in 1961. With it will come better methods for ES and lineage-specific stem cell identification, maintenance, and controlled fate selection. Clearly, ES cell models are already providing opportunities for the establishment of limitless sources of specific cell populations. In recognition of the growing excitement and potential of ES cells as models for both the advancement of basic science and future clinical applications, I felt it timely to edit this collection of protocols *(Embryonic Stem Cells)* in which forefront investigators would provide detailed methods for use of ES cells to study various lineages and tissue types.

We are pleased to provide *Embryonic Stem Cells*: *Methods and Protocols,* a broad-scaled work of 35 chapters containing step-by-step protocols suitable for use by both experienced investigators and novices in various ES cell technologies. In the first section of the volume, there are chapters with detailed protocols for ES cell isolation, maintenance, modulation of gene expression, and studies of ES cell cycle and apoptosis. *Embryonic Stem Cells* also includes chapters with protocols for the use of ES cells to generate diverse cell and tissue types, including blood, endothelium, adipocytes, skeletal muscle, cardiac muscle, neurons, osteoclasts, melanocytes, keratinocytes, and hair follicle cells. The second part of the volume contains a series of cutting edge techniques that have already been shown to have, or will soon have, tremendous utility with ES cells and their differentiated progeny. These chapters include the use of cDNA arrays in gene expression analysis, phage display antibody libraries to generate antibodies against very rare antigens, and phage display libraries to identify and characterize protein and protein interactions, to name a few. Collectively, these protocols should prove a useful resource not only to those who are using or wish to use ES cells to study fate choices and specific lineages, but also to those interested in cell and developmental biology more generally. We hope that this book will also serve as a catalyst spurring others to use ES cells for lineages not yet being widely studied with this model and to develop new methodologies that would contribute to both the fundamental understanding of stem cells and their potential utility.

Embryonic Stem Cells would not have materialized at all had the contributors not recognized the special value of disseminating their protocols and hard-won expertise. I am extremely grateful to them for their commitment, dedication, and promptness with submissions! I am also grateful to Dr. John Walker for having faith in and supporting me throughout this project. I wish also to acknowledge the great support provided by many at Humana Press, specifically Elyse O'Grady, Craig Adams, Diana Mezzina, and Tom Lanigan. A special thank you goes to my dedicated coworker, Tammy-Claire Troy, who, with her infectious optimism and tireless commitment, became a crucial factor in the editing and completion of the volume.

I am grateful to N. Urfe, P. Kael, and M. Chambers for their unintentional "awesome" contributions.

Finally, I hope that the volume will achieve the intent that I had originally imagined: that it will prove a volume with something for both experts and novices alike, that it will serve as a launching point for further developments in stem cells, and that we will all-too-soon wish to expand and update it with other emerging concepts, insights and methods!

Kursad Turksen

Contents

Contributors

GERNOT ACHATZ • *Department of Genetics, University of Salzburg, Hellbrunnerstrasse Salzburg, Australia*

HELMUT ACKER • *Institute of Neurophysiology, University of Cologne, Koln, Germany*

DAVID M. ADELMAN • *Abramson Research Institute, Department of Cancer Biology, University of Pennsylvania Cancer Center, Philadelphia, PA*

ANDRÉ-PATRICK ARRIGO • *Laboratoire du Stress Oxydant, Chaperons et Apoptose, Center de Genetique Moleculaire et Cellulaire, University Claude Bernard Lyon-I, Villeurbanne, France*

JANE E. AUBIN • *Department of Anatomy and Cell Biology, University of Toronto, Toronto, Ontario, Canada*

RAJESHWAR B. AWATRAMANI • *Department of Genetics, Harvard Medical School, Boston, MA*

ANDREW BAIRD • *Selective Genetics Inc., San Diego, CA*

KATHRIN BANACH • *Institute of Neurophysiology, University of Cologne, Koln, Germany*

VICTORIA L. BAUTCH • *Department of Biology, The University of North Carolina at Chapel Hill, Chapel Hill, NC*

KENNETH R. BOHELER • *In Vitro Differentiation Group, Institute of Plant Genetics and Crop Plant Research, Gatersleben, Germany*

ERIC W. BRUNSKILL • *Division of Developmental Biology, Children's Hospital Medical Center, Cincinnati, OH*

MICHAEL A. BURG • *Selective Genetics Inc., San Diego, CA*

G. ANTONIO CANDELIERE • *Department of Anatomy and Cell Biology, University of Toronto, Toronto, Ontario, Canada*

FRANCESCO CECCONI • *Department of Biology, University of Rome Tor Vergata, Roma, Italy*

J. RICHARD CHAILLET • *Department of Pediatrics University of Pittsburgh, School of Medicine, Children's Hospital of Pittsburgh, PA*

GRACE CHUNG • *Division of Protein and Nucleic Acid Chemistry, Cambridge, Medical Research Council Laboratory of Molecular Biology, UK*

RETO CRAMERI • *Swiss Institute of Allergy and Asthma Research, Davos, Switzerland*

CHRISTIAN DANI • *Institute of Signaling, Developmental Biology, and Cancer Research, Centre de Biochimie, Nice, France*

CHARLES DECRAENE • *CEA Service de Genomique Fontionnelle, Batiment Genopole, Evry, France*

SUSAN M. DYMECKI • *Department of Genetics, Harvard Medical School, Boston, MA*

TAKUMI ERA • *Howard Hughes Medical Institute, University of California, Los Angeles, CA*

YONG FAN • *Department of Pediatrics University of Pittsburgh, School of Medicine, Children's Hospital of Pittsburgh, PA*

LOREN J. FIELD • *Herman B. Wells Center for Pediatric Research, James Whitcombe Riley Hospital for Children, Indianapolis, IN*

BERND K. FLEISHMANN • *Institute of Neurophysiology, University of Cologne, Koln, Germany*

THOMAS FLOSS • *GSF-Institute of Mammalian Genetics, Neuherberg, Germany*

STUART T. FRASER • *Department of Molecular Genetics, Faculty of Medicine, Kyoto University, Sakyo-ku, Kyoto, Japan*

ANDRÉ GALARNEAU • *Department of Biochemistry, University of Montréal, Québec, Canada*

MARINA GERTSENSTEIN • *Samuel Lunenfeld Research Institute, Mount Sinai Hospital, Toronto, Ontario, Canada*

R. DANIEL GIETZ • *Department of Human Genetics, University of Manitoba, Winnipeg, Manitoba, Canada*

PETER GRUSS • *Department of Molecular Cell Biology, Max-Planck-Institute of Biophysical Chemistry, Göttingen, Germany*

KAOMEI GUAN • *In Vitro Differentiation Group, Institute of Plant Genetics and Crop Plant Research, Gatersleben, Germany*

SHIN-ICHI HAYASHI • *Department of Immunology, School of Life Science, Faculty of Medicine, Tottori University, Yonago, Japan*

JÜRGEN HESCHELER • *Institute of Neurophysiology, University of Cologne, Koln, Germany*

THORSTEN HOEVEL • *Department of Cell Analytics, Roche Pharmaceutical Research, Roche Diagnostics GmbH, Penzberg, Germany*

KRISTEN JENSEN-PERGAKES • *Selective Genetics Inc., San Diego, CA*

LUDMILA JIRMANOVA • *Laboratoire de Biologie Moleculaire de Cellulaire de I'Ecole Normale Superieure de Lyon, Lyon, France*

TOMOYUKI KANEKO • *Department of Pathology and Cell Regulation, Kyoto Prefectural University of Medicine, Kyoto, Japan*

HIROSHI KAWASAKI • *Department of Medical Embryology and Neurobiology, Institute for Frontier Medical Sciences, Kyoto University*

MANFRED KUBBIES • *Department of Cell Analytics, Roche Pharmaceutical Research, Roche Diagnostics GmbH, Penzberg, Germany*

TAKAHIRO KUNISADA • *Department of Hygiene, Faculty of Medicine, Gifu University, Gifu, Japan*

HÉLÈNE LAPILLONE • *Laboratoire de Biologie Moleculaire de Cellulaire de I'Ecole Normale Superieure de Lyon, Lyon, France*

DAVID LAROCCA • *Selective Genetics Inc., San Diego, CA*

MENG LI • *Center for Genome Research, University of Edinburgh, Edinburgh, UK*

FINA LIU • *INSERM, Hõpïtal Edouard Herriot, Lyon, France*

JUN LU • *Department of Laboratory Medicine and Joint Program in Transfusion Medicine, Children's Hospital, Harvard Medical School, Boston, MA*

CORRINNE LOBE • *Cancer Research Division, Sunnybrook and Women's College Health Science Center, Toronto, Ontario, Canada*

PATRICK MEHLEN • *Laboratoire Différenciation et Apoptose, CNRS, Université Claude Bernard Lyon-I, France*

JOHN D. MCNEISH • *Genetic Technologies, Pfizer Global Research and Development, Groton, CT*

STEPHEN W. MICHNICK • *Department of Biochemistry, University of Montréal, Québec, Canada*

KENJI MIZUSEKI • *Department of Medical Embryology and Neurobiology, Institute for Frontier Medical Sciences, Kyoto University*

TAHMINA MUJTABA • *Department of Neurobiology and Anatomy, University of Utah Medical School, Salt Lake City, UT*

ANDRAS NAGY • *Samuel Lunenfeld Research Institute, Mount Sinai Hospital, Toronto, Ontario, Canada*

SATOMI NISHIKAWA • *Department of Molecular Genetics, Faculty of Medicine, Kyoto University, Sakyo-ku, Kyoto, Japan*

SHIN-ICHI NISHIKAWA • *Department of Molecular Genetics, Faculty of Medicine, Kyoto University, Sakyo-ku, Kyoto, Japan*

MINETARO OGAWA • *Department of Molecular Genetics, Faculty of Medicine, Kyoto University, Sakyo-ku, Kyoto, Japan*

MASAHITO OYAMADA • *Department of Pathology and Cell Regulation, Kyoto Prefectural University of Medicine, Kyoto, Japan*

YUMIKO OYAMADA • *Department of Pathology and Cell Regulation, Kyoto Prefectural University of Medicine, Kyoto, Japan*

KISHORE B.S. PASUMARTHI • *Herman B. Wells Center for Pediatric Research, James Whitcomb Riley Hospital for Children, Indianapolis, IN*

GENEVIÈVE PIÉTU • *CEA Service de Genomique Fontionnelle, Batiment Genopole, Evry, France*

S. STEVEN POTTER • *Division of Developmental Biology, Children's Hospital Medical Center, Cincinnati, OH*

GABRIELE PROETZEL • *Deltagen Inc., Menlo Park, CA*

TERENCE H. RABBITTS • *Division of Protein and Nucleic Acid Chemistry, Medical Research Council Laboratory of Molecular Biology, Cambridge, UK*

MAHENDRA S. RAO • *Department of Neurobiology and Anatomy, University of Utah Medical School, Salt Lake City, UT*

INGRID REMY • *Department of Biochemistry, University of Montréal, Québec, Canada*

CLAUDIO RHYNER • *Swiss Institute of Allergy and Asthma Research, Davos, Switzerland*

MARSHA L. ROACH • *Genetic Technologies, Pfizer Global Research and Development, Groton, CT*

CAROLYN I. RODRIGUEZ • *Department of Genetics, Harvard Medical School, Boston, MA*

JACQUES SAMARUT • *Laboratoire de Biologie Moleculaire de Cellulaire de l'Ecole Normale Superieure de Lyon, Lyon, France*

YOSHIKI SASAI • *Department of Medical Embryology and Neurobiology, Institute for Frontier Medical Sciences, Kyoto University*

HEINRICH SAUER • *Institute of Neurophysiology, University of Cologne, Koln, Germany*

PIERRE SAVATIER • *Laboratoire de Biologie Moleculaire de Cellulaire de I'Ecole Normale Superieure de Lyon, Lyon, France*

M. CELESTE SIMON • *Abramson Research Institute, Department of Cancer Biology, University of Pennsylvania Cancer Center, Philadelphia, PA*

STEVEN R. SLOAN • *Department of Laboratory Medicine and Joint Program in Transfusion Medicine, Children's Hospital, Harvard Medical School, Boston, MA*

TETSURO TAKAMATSU • *Department of Pathology and Cell Regulation, Kyoto Prefectural University of Medicine, Kyoto, Japan*

TAMMY-CLAIRE TROY • *Ottawa Health Research Institute, Ottawa, Ontario, Canada*

ERIC TSE • *Medical Research Council Laboratory of Molecular Biology, Division of Protein and Nucleic Acid Chemistry, Cambridge, UK*

KURSAD TURKSEN • *Ottawa Health Research Institute, Ottawa, Ontario, Canada*

M. TODD VALERIUS • *Department of Molecular Cell Biology, Harvard University, Cambridge, MA*

LUIGI VITELLI • *Laboratoire de Biologie Moleculaire de Cellulaire de I'Ecole Normale Superieure de Lyon, Lyon, France*

MARIA WARTENBURG • *Institute of Neurophysiology, University of Cologne, Koln, Germany*

MICHAEL WEICHEL • *Swiss Institute of Allergy and Asthma Research, Davos, Switzerland*

MICHAEL V. WILES • *Deltagen Inc., Menlo Park, CA*

OWEN N. WITTE • *Howard Hughes Medical Institute, University of California, Los Angeles, CA*

ANNA M. WOBUS • *In Vitro Differentiation Group, Institute of Plant Genetics and Crop Plant Research, Gatersleben, Germany*

ANNA M. WOBUS • *In Vitro Differentiation Group, Institute of Plant Genetics and Crop Plant Research, Gatersleben, Germany*

STEPHANE WONG • *Howard Hughes Medical Institute, University of California, Los Angeles, CA*

ROBIN A.WOODS • *Department of Biology, University of Winnipeg, Winnipeg, Manitoba, Canada*

WOLFGANG WURST • *Clinical Neurogenetics, Max-Planck Institute of Psychiatry, Munich, Germany*

TOSHIYUKI YAMANE • *Department of Immunology, School of Life Science, Faculty of Medicine, Tottori University, Yonago, Japan*

HUANG-TIAN YANG • *In Vitro Differentiation Group, Institute of Plant Genetics and Crop Plant Research, Gatersleben, Germany*

Color Plates

Color plates 1–16 appear as an insert following p. 254.

1

Methods for the Isolation and Maintenance of Murine Embryonic Stem Cells

Marsha L. Roach and John D. McNeish

1. Introduction

Embryonic stem (ES) cells were first isolated in the 1980s by several independent groups. *(1–4)*. These investigators recognized the pluripotential nature of ES cells to differentiate into cell types of all three primary germ lineages. Gossler et al. *(5)* described the ability and advantages of using ES cells to produce transgenic animals *(5)*. The next year, Thomas and Capecchi reported the ability to alter the genome of the ES cells by homologous recombination *(6)*. Smithies and colleagues later demonstrated that ES cells, modified by gene targeting when reintroduced into blastocysts, could transmit the genetic modifications through the germline *(7)*. Today, genetic modification of the murine genome by ES cell technology is a seminal approach to understanding the function of mammalian genes in vivo. ES cells have been reported for other mammalian species (i.e., hamster, rat, mink, pig, and cow), however, only murine ES cells have successfully transmitted the ES cell genome through the germline. Recently, interest in stem cell technology has intensified with the reporting of the isolation of primate and human ES cells *(8–11)*.

ES cells are isolated from the inner cell mass (ICM) of the blastocyst stage embryo and, if maintained in optimal conditions, will continue to grow indefinitely in an undifferentiated diploid state. ES cells are sensitive to pH changes, overcrowding, and temperature changes, making it imperative to care for these cells daily. ES cells that are not cared for properly will spontaneously differentiate, even in the presence of feeder layers and leukemia inhibitory factor (LIF). In addition, healthy cells growing in log phase are critical for optimal transformation efficiency in gene targeting experiments.

Targeted murine ES cells have little value if they lose the ability to transmit the introduced mutations through the germline of the resulting chimeras. Therefore, it is critical that murine ES cells have a normal 40 XY karyotype. It is standard practice in our laboratory to have complete karyotypic analysis of all targeted ES cells prior to the production of chimeras. The criteria used in our laboratory to qualify an ES cell clone for making chimeras is that at least 50% of the chromosome spreads analyzed must be 40 XY. In our experience, our DBA/1LacJ ES cells *(12)* meet or exceed that criterion

From: *Methods in Molecular Biology, vol. 185: Embryonic Stem Cells: Methods and Protocols*
Edited by: K. Turksen © Humana Press Inc., Totowa, NJ

at least 86% of the time, whereas our 129 strain of ES cells meet or exceed the criteria 45% of the time.

The many opportunities that exist in stem cell biology today, combined with the need to further explore and develop new technologies, makes it necessary to clearly define the process of developing stem cell lines. Therefore, this chapter will present the methods used in our laboratory to develop murine ES cell lines and maintain them in an undifferentiated state.

2. Materials

2.1. Mice for Blastocyst Stage Embryos and Primary Embryonic Fibroblasts

1. DBA1/LacJ, 129/SvJ, and C57BL/6 inbred mice were obtained from Jackson Laboratories.
2. MTK-neo CD1 transgenic mice were obtained from Dr. Colin Stewart for the production of primary embryonic fibroblasts (PEF) for feeder cells.

2.2. Tissue Culture Plastic and Glassware

1. 35-mm Petri dish (Falcon cat. no. 1008).
2. 4-Well multiwell tissue culture dish (Nunc cat. no. 176740).
3. 24-Well multiwell tissue culture dish (Nunc cat. no. 143982).
4. 12-Well multiwell tissue culture dish (Nunc cat. no. 150628).
5. 6-Well multiwell tissue culture dish (Nunc cat. no. 152795).
6. T-25 Flask (Nunc Cat. no. 163371).
7. 100-mm Tissue culture dishes (Falcon cat. no. 3003).
8. 60-mm Tissue culture dishes (Falcon cat. no. 3002).
9. 50-mL SteriFlip filter unit (Millipore cat. no. SCGP00525).
10. 150-mL Stericup filter unit (Millipore cat. no. SCGPU01RE).
11. 250-mL Stericup filter unit (Millipore cat. no. SCGPU02RE).
12. 500-mL Stericup filter unit (Millipore cat. no. SCGPU05RE).
13. Nalgene controlled-rate freezer (VWR cat. no. 55710-200).
14. Bright-Line hemacytometer (improved Neubauer counting chamber) (VWR cat. no. 15170-172).

2.3. Media and Reagents

1. ES cell qualified light mineral oil (Specialty Media cat. no. ES-005-C).
2. M2 Medium (Specialty Media cat. no. MR-015D).
3. KSOM (Specialty Media cat. no. MR-023-D).
4. Knockout™ Dulbecco's Modified Eagle medium (KO-DMEM) (Invitrogen Life Technologies, I-LTI cat. no. 10829-018).
5. ES cell qualified fetal bovine serum (FBS) (I-LTI cat. no. 10439-024).
6. 0.2 mM L-Glutamine (100×) (I-LTI cat. no. 25030-081).
7. 0.1 mM MEM nonessential amino acids (NEAA) (100X) (I-LTI cat. no. 11140-122).
8. 50 U/ml penicillin/50 µg/mL streptomycin (100X) (I-LTI no. 15140-122).
9. 1000 µ/mL ESGRO or LIF (Chemicon cat. no. ESG-1107).
10. 0.1 mM 2-Mercaptoethanol (BME) (Sigma cat. no. M-7522).
11. Dulbecco's phosphate-buffered saline (PBS) (I-LTI cat. no. 14190-144).
12. 0.05% Trypsin EDTA (I-LTI cat. no. 25300-054).
13. 10 µg/mL Mitomycin C (Sigma cat. no. M-0503).
14. 10% Dimethyl sulfoxide (DMSO) (Sigma cat. no. D-2650).

Table 1
Media Protocols for ES Cells and Feeder Cells

Reagents	sDMEM	SCML	G418/Ganc/SCML	G418/SCML	HAT/SCML
KO-DMEM	500 mL	500 mL	500 mL	500 mL	500 mL
FBS	50 mL	90 mL	90 mL	90 mL	90 mL
L-Glutamine	5 mL	6 mL	6 mL	6 mL	6 mL
MEM/NEAA	—	6 mL	6 mL	6 mL	6 mL
BME	4 µL	4 µL	4 µL	4 µL	4 µL
LIF	—	60 µL	60 µL	60 µL	60 µL
Pen/Strept	2.5 mL	3 mL	3 mL	3 mL	3 mL
G418	—	—	2.1–3.6 mL	2.1–3.6 mL	—
Gancyclovir	—	—	2 µM	—	—
HAT	—	—	—	—	6 mL

Store at 4°C until used and discard after 14 d.

15. 175–300 µg/mL G418 (Geneticin™ 50 mg/mL) (I-LTI cat. no. 10131-035).
16. 2 µM/L Gancyclovir (Ganc) (Hoffman-LaRoch—no cat. no.).
17. HAT supplement (100X) 10 mM sodium hypoxanthine, 40 µM aminopterin, and 1.6 mM thymidine (I-LTI cat. no. 31062-011).
18. 0.1% Gelatin in sterile water (Specialty Media cat. no. ES-006-B).
19. Mouse Y-ES system (I-LTI cat. no. 10357-010).
20. Mycoplasma *Plus*™ PCR detection primer set (Stratagene cat. no. 302008).
21. Mycoplasma stain kit (Sigma cat. no. MYC-1).

3. Methods

3.1. Preparation of Media Used for Feeders and ES Cells

1. The list of reagents for the different culture media's used for ES cells and PEFs can be found in **Table 1**. All reagents are combined and filtered through 0.2-µm filter units. ES cells are sensitive to pH change, therefore, when a bottle is about half full, the remaining medium is filtered into a smaller bottle. This practice minimizes the air space in the bottle that causes the pH to raise as air gases and medium reach equilibrium. (*See* **Notes 1–5**).

3.2. Preparation of Feeder Layers from PEF

1. PEFs were isolated from 12–14-d-old transgenic MTK-neo CD1 embryos and frozen as described *(13)*. Frozen vials of PEF cells are thawed by agitation in a 37°C water bath until cell suspension becomes a slurry. Transfer the cell suspension into 49 mL DMEM with serum, L-glutamine, and BME (sDMEM) in the 50-mL tube. Pipet up and down gently and transfer 10 mL cell suspension into each of 5 labeled 100-mm dishes (approx 1.5–2.0 × 10^6 cells/dish). Rotate plates back and forth to distribute cells evenly over entire dish.
2. Incubate 2–3 d and examine for confluence. When approx 80% confluent, remove media and replace with 6 mL mitomycin C (10 µg/mL in sDMEM) and incubate 2–5 h. After treatment, remove mitomycin C solution, wash with 10 mL PBS, then add 10 mL sDMEM. Incubate in sDMEM until ready to use.
3. The day before harvesting blastocysts to develop new ES cell lines, remove media from one 100-mm PEF feeder layer, and rinse with 10 mL PBS. Incubate 2–3 min in 2 mL

trypsin EDTA. Dislodge the PEF cells by tapping the dish against the palm of your hand. When cells release from the dish, add 24 mL sDMEM to neutralize the trypsin and pipet up and down to produce a single-cell suspension (approx $2.5–3.5 \times 10^5$ cells/mL). Transfer 1 mL/well of six 4-well dishes. Incubate overnight. The next day, remove media, wash with 1 mL PBS/well, then add 1 mL (SCML). These 4-well dishes are ready to receive embryos.

3.3. Preparation of Gelatin-Coated Dishes

1. Warm the 0.1% gelatin solution in a 37°C water bath. Transfer enough gelatin solution to cover the bottom of the dish (i.e. 0.5 mL/well for 4 or 24 wells, 1 mL/well for 12 wells, 2 mL/well for 6 wells, 3 mL for 60-mm dishes and 6 mL for 100-mm dishes). Let gelatin solution sit at room temperature for 30 min in a tissue culture hood.
2. Remove the excess gelatin solution and use dishes immediately. Do not allow the gelatin to air-dry.

3.4. Obtaining Blastocyst Stage Embryos

1. Blastocysts can be obtained from super-ovulated or naturally mated females. However, we believe blastocysts are generally more fit from natural matings.
2. For natural matings, place two females per male on Thursday mornings. Check for copulation plugs daily. This is typically done before 10 AM to ensure the identification of all mated females. Separate plugged females and label for blastocysts embryos 3 d later. Set up 10–15 males and 20–30 females this way.
3. On d 3.5 post coitus (p.c.), sacrifice plugged females, and flush blastocyst stage embryos from both uterine horns as described *(14)*. Transfer the embryos through several M2 drops to wash away uterine fluids and debris. Finally, transfer one washed embryo into a 4-well dish with fresh PEF feeder layer in SCML. PEF feeders may be eliminated if you have 1000 U/mL LIF (ESGRO) in the medium.

3.5. Culture of the Blastocyst and Picking of the ICM

1. Observe the embryos daily to monitor fitness, hatching, and attachment to the feeder layer or gelatin-coated plastic. When the embryos have attached, the ICM will become apparent (*see* **Fig. 1**).
2. Using a drawn mouth pipet, tease the ICM away from the rest of the embryo and gently aspirate it into the pipet. Transfer the ICM into one well of a 24-well dish previously prepared with fresh PEF feeders and SCML. If you prefer not to use feeder layers, gelatin coat the wells (*see* **Subheading 3.3., step 1**) and proceed in the same manner as with PEF feeders.

3.6. Isolation of Putative ES Cells from the ICM

1. The ICM should attach to the feeder layer or gelatin-coated dish overnight. The next day, remove the media and wash the cell layer with 0.5 mL PBS/well. Remove the PBS and add four drops of 0.05% trypsin EDTA. Incubate for 1–2 min. Vigorously tap the dish against the palm of your hand to dislodge the cells into suspension. When fully detached, add 2 mL SCML/well and pipet up and down to dissociate cells into a single-cell suspension. Record this as S1:1 p1 (split one to one, passage one) and return the cells to the incubator.
2. Twenty-four hours after splitting, remove the media from each well and replace with 2 mL SCML/well. Examine the cells in each well and record the morphology. Following examination, feed the cells daily by removing the old medium and replacing with 2 mL fresh SCML. Every second or third day, the colonies must be dissociated and the passage

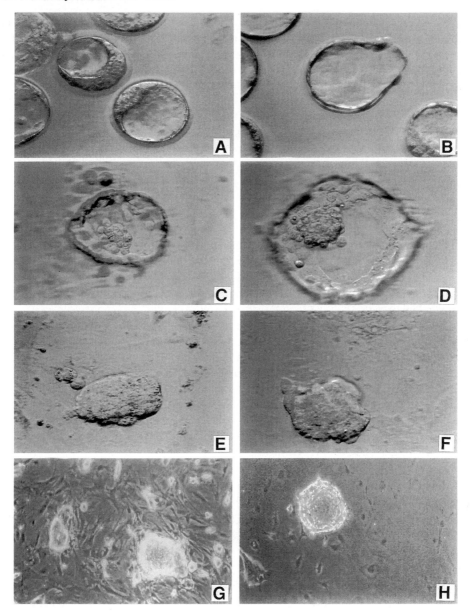

Fig. 1. From blastocyst stage embryos to ES cells. (**A**) Blastocyst stage. (**B**) Blastocyst embryo hatching from the zona pellucida. (**C**) Blastocyst embryo attached to a PEF feeder layer 2 d after hatching—ICM is apparent inside the blastocyst. (**D**) Blastocyst embryo attached to tissue culture plastic without a PEF feeder 2 d after hatching—ICM is apparent inside the blastocyst. (**E**) ICM is distinctive and extends above the the flat trophoblasts and PEF feeders. (**F**) ICM is distinctive and extends above the flat trophoblasts without PEF feeders. (**G**) ES cell colonies on PEF feeders. (**H**) ES cell colony on tissue culture plastic without PEF feeders.

number recorded. Never allow colonies to become larger than 400 μm in diameter. If the colonies are less than 100 μm in diameter, wait another day before dissociating. We believe that keeping the colonies small aids in maintaining pluripotency. Large colonies tend to flatten and differentiate.

3. The new ES cells generally remain in the 24-well dish for 2–3 passages. When the colonies appear to be evenly dispersed over the dish, it is time to move the cell population to a larger 12-well dish. Individual colonies should never be allowed to overgrow, forming a monolayer. Follow the same procedure as in **Subheading 3.6., step 1** above, to trypsinize the cells.

4. When the trypsinized cells are in suspension and no longer attached to the dish, they are ready to be moved to the next size dish. Using a 5-mL pipet, aspirate 3 mL SCML into the pipet. Tilt the 24-well dish and express 2 mL SCML into the well, then immediately aspirate the entire contents of the well into the pipet. Quickly transfer 2 mL of the volume into one well of a previously prepared 12-well dish (PEF feeders or gelatin-coated). With the remaining 1 mL SCML in the pipet, go back and wash the well in the 24-well dish to ensure that all cells have been removed. Then add the remaining 1 mL to the 2 mL cell suspension already in the well of the 12-well dish. Pipet up and down to completely dissociate the cells into a single-cell suspension. Repeat this procedure for each well and make sure to record passage number. Note that, at this stage, only a few embryos will move into the 12-well dish, because many will die at this stage.

5. The next day, examine each well, record morphology, and change the media with 2.5 mL fresh SCML/well. Follow the same media change and dissociating procedures as described in this section, with the exception that the 12-well dish will use 0.5 mL trypsin. Generally, there will be only one S1:1 in the 12-well dish.

6. When there are enough colonies to move to the next sized vessel, transfer to one 100-mm dish. At this point, the cells are typically at passage 5. Prepare a 100-mm dish with 10 mL fresh SCML on a PEF feeder layer or gelatin. Remove the media from the 12-well dish and wash with 1 mL PBS. Remove the PBS and add 0.5 mL trypsin. Incubate for 1–2 min, then dislodge the cells from the dish by tapping the dish against the palm of your hand. Once these cells are dislodged, aspirate 5 mL SCML into a 10-mL pipet. Tilt the 12-well dish and express 2 mL SCML into the well, then quickly aspirate the contents of the well into the pipet. Immediately express 3 mL into the previously prepared 100-mm dish. Return to the 12-well dish and express the remaining 2 mL SCML in the pipet into the well, then quickly aspirate the contents of the well back into the pipet. This is to ensure that you have removed all the cells from the well. Add the last 2 mL to the 100-mm dish and pipet up and down to dissociate the cells into a single-cell suspension. There should be approx $0.5–1.0 \times 10^7$ total cells in the suspension. Incubate overnight.

7. The following day, record morphology and change the media with 15 mL fresh SCML. On the second day after the move into the 100-mm dish, either change the media again or, if the cells are ready, split them 1:2 based on colony size (if colonies are less than 100 µm in diameter, feed that day and wait another day to split).

8. From this point on, the new ES cell population is being expanded and cryopreserved. Therefore, every time the cells are split, part of the cell suspension must be passed for expansion (approx 2×10^6 cells/100-mm dish) and part will be cryopreserved. Pass the cells in a 100-mm dish by removing SCML and washing with 10 mL PBS. Remove the PBS and incubate in 2 mL trypsin for 1–2 min. After incubation, vigorously tap the dish against the palm of your hand to dislodge the ES cells from the dish.

9. Once the cells are completely in suspension, tilt the dish and add 8 mL SCML to wash the cells into a pool at the bottom of the tilted dish. Aspirate the cell suspension into the pipet and transfer into a 15-mL conical tube. In the 15-mL tube, gently aspirate the cells up and down 3–4 times to dissociate into a single-cell suspension. Leave 5 mL of the cell suspension in this tube and transfer the remaining 5 mL cell suspension into another 15-mL tube (one tube is for freezing and one is to maintain cells). Pellet the cells by centrifugation at 110*g* for 5 min.

10. While the cells are in the centrifuge, prepare two 100-mm dishes of fresh PEF feeders by washing the monolayer with PBS and adding 5 mL SCML. (If using a gelatin-coated dish, just add 5 mL SCML to the dish.) After centrifugation, aspirate the supernatant from both tubes, taking care not to disturb the cell pellet. Resuspend the cell pellet from one tube in 10 mL SCML. Count the cells using a Neubauer counting chamber, then transfer 2×10^6 cells/dish into the previously prepared 100-mm dishes with PEF feeders or gelatin and record the passage number (should be around p6). At this stage, there should be enough cells to plate one or two 100-mm dishes. Resuspend the cell pellet in the other 15-mL tube with enough freezing medium to freeze $4-6 \times 10^6$ cells/mL for each cryovial. Transfer 1 mL of cells in freezing media into 1.5-mL cryovials labeled with the name of the cell line, with or without feeders, the passage number, freeze number (F1 in this case), and your initials. Place cryovials of cells into a controlled-rate freezer at $-80°C$ overnight.

11. The next day, transfer the cryovial of cells into long-term freezer storage, in either liquid nitrogen or a $-150°C$ freezer. Record location in freezing log. Next, examine the cells that were passed and record morphology. Change the media by removing the old media and replacing with 15 mL SCML.

12. Once the cells are into the 100-mm dish, the new ES cell line is usually established. Continue to carry the cells for expansion of the line to ensure many vials in cryopreservation. The next split should be S1:6 or S1:8. Freeze 3 or 4 vials, respectively. Aim to freeze $4-6 \times 10^6$ cells/vial in 1 mL freezing medium. We typically accumulate approx 50 vials.

3.7. Characterization of Putative ES Cells

It is necessary to characterize the ES cell lines to determine sex, karyotype, pluripotency, and absence of pathogens. It is preferred to have a male cell line, because XY ES cells can sex convert an XX blastocyst in a chimeric embryo development, and these resulting chimeric males can produce more offspring than females *(15)*. In addition, it is necessary to determine the karyotype of the ES cell lines, because transmission of the ES cell genome through the germline of the chimeras is dependent upon the ES cells having a normal chromosome number *(16)*. Finally, the ability to differentiate into many cell types and the ability to make healthy chimeras is dependent upon the cells being free of pathogens, such as mycoplasma and murine viruses. Therefore, it is necessary to test for mycoplasma contamination and murine antibody production (MAP) testing for antibodies against murine viruses *(17)*.

3.7.1. Sex Determination to Identify XY ES Cell Lines

1. The first step in determining the sex of the novel ES cells is a PCR screen. Pick 6 colonies into individual microfuge tubes that contain 10 μL sterile water. Put the tubes in a $-20°C$ freezer for 10 min. Next, remove the tubes from the freezer, vortex mix for several seconds, and then pulse-spin to collect lysate in the bottom of the tube. Follow the instructions for the Y-ES system to PCR screen for the Y chromosome *(18)*.

2. The next step is to do a full karyotype of all cell lines determined to be male by PCR. Karyotyping can be done according to published protocols *(19,20)* or contracted. We typically contract our ES cell karyotyping. At the time of splitting, $1-1.5 \times 10^6$ cells are transferred into a T25 Flask in 10 mL SCML and cultured overnight. The next day, the medium is removed, and the flask's lid, if filled to the brim with SCML, is closed tightly, and the lid and neck are wrapped in parafilm to prevent leakage. The flasks are packed

and shipped to Coriell Cell Repository (Cytogenetics Laboratory, 401 Haddon Avenue, Camden, New Jersey 08103; phone 1-800-752-3805) for full karyotyping.

3.7.2. Mycoplasma and Murine Viral Contamination Testing

1. To test for mycoplasma contamination, you may do a simple Hoechst stain using the Sigma kit (follow insert instructions) or do a PCR of the supernatant (follow Stratagene insert instructions).
2. To test for murine viral contamination, we send a vial of frozen cells to Charles River Laboratories (252 Ballardvale Street, Wilmington, MA 01887; phone 1-508-658-6000) for MAP testing.

3.7.3. In Vitro Differentiation (IVD)

1. To remove the ES cells from the PEF feeders, aspirate the media from the dish and wash the cell layer with 10 mL PBS. Remove the PBS and add 2 mL trypsin. Immediately take the dish to the microscope and place on the stage. While observing the cells through the eyepieces of the microscope, tap the dish to dislodge the rounded ES cell colonies. As soon as many of the colonies are floating and the feeder layer is still attached, return the dish to the hood and aspirate the colony suspension and transfer into a 15-mL conical tube. Add 8 mL SCML, pipet up and down to dissociate the colonies, then pellet by centrifugation at 110g for 5 min. Resuspend the pelleted cells in 15 mL SCML, plate in a 100-mm tissue culture dish without PEF feeder layer, and incubate overnight. The next day, change the media on the feeder-free ES cells by removing the old media and adding 15 mL SCML.
2. To begin the IVD experiment, change the media and add 15 mL SCML, approx 1–2 h before dissociating the cells. Next, remove the media and wash the cell layer with 10 mL PBS. Remove the PBS, add 2 mL fresh trypsin, and incubate 1–3 min. Check the cells every 30 s for dissociation by tapping the dish against the palm of your hand. When the colonies are completely free-floating, return the dish to the hood, add 8 mL SCML, and pipet up and down until the cells are in a single-cell suspension. Count the cells using a hemocytometer, then pellet the cells by centrifugation at 110g for 5 min.
3. After centrifugation, aspirate the supernatant, taking care not to disturb the cell pellet, then resuspend the cells in 10 mL stem cell medium (without LIF) (SCM). Plate the cells at a concentration of 1–2 × 10^5 cells/mL in a vol 10 mL SCM in a 100-mm bacterial dish. This suspension culture will allow the cells to form cell aggregates called embryoid bodies (EBs).
4. Change the media every 2–3 d by transferring the EBs into a 15-mL conical tube and letting them settle out of suspension into the bottom of the tube. Aspirate the supernatant, add 10 mL fresh SCM, then transfer the EB suspension back into the bacterial dish.
5. After 7–9 d of culture, transfer the EB suspension into a 15-mL conical tube and again allow to them to settle out. Remove the supernatant, add 10 mL PBS, and allow the EBs to settle out. After the EBs have settled to the bottom, again remove the supernatant, add 3 mL of trypsin, and incubate for 3 min at 37°C. Following incubation, add 7 mL SCM to the trypsin solution and pipet up and down vigorously to dissociate the EBs. Pellet the cell suspension by centrifugation at 110g for 5 min. Remove the supernatant and resuspend the cells in 10 mL SCM. Transfer into two 100-mm tissue culture dishes and increase the vol to 12 mL SCM in each dish.
6. Examine for differentiated morphology daily and feed SCM every second day. Many different cell populations should become apparent, including blood islands and contracting myocytes. Additional details of IVD methods can be found in other chapters of this text.

3.7.4. Gene Targeting Ability and Germline Transmission

1. To test for the ability of your ES cells to undergo homologous recombination, a vector of known targeting frequency should be used. Electroporations are carried out as described in **Subheading 3.9.** below.
2. Ultimately, the novel ES cells must be capable of colonizing the germline of chimeric mice. The ES cells can be microinjected into blastocysts or aggregated with morula, according to standard protocols. Producing chimeras with host blastocysts or morula from strains different from the ES cells allows one to use coat color genetics to identify germline transmission of the ES cell genome *(21)*.

3.8. Maintenance of ES Cells

3.8.1. Thawing ES Cells

1. To prepare a fresh 100-mm PEF feeder plate, remove the old media, wash with 10 mL PBS, then add 15 mL SCML. Check the date on the feeder dish and examine to determine that feeder cells are healthy. Primary embryonic fibroblast feeders usually last 7–10 d. Put prepared feeders back into the incubator to equilibrate cells with higher serum concentration. (If you are thawing clones from an electroporation to expand, prepare a well in a 6-well dish.) These clones are 1/2 well of a 24-well dish when frozen.
2. Remove a vial of cells from the –150°C freezer and plunge into 37°C water bath, agitating the vial until the frozen suspension becomes a slurry. Sterilize the vial with 70% ethanol and transfer to a tissue culture hood.
3. Transfer the contents from the vial into the previously prepared PEF feeder plate. Most vials have enough cells to evenly plate a 100-mm dish with colonies (approx $4–6 \times 10^6$). Gently swirl the plate to distribute cells over the entire PEF feeder surface. Label the dish with the cell line, passage number, date, and then return the plate to the incubator.
4. Change the media the next morning, by removing the old media and replace with fresh SCML. Return the dish to the incubator and culture another day. If the cells recovered easily from the freeze–thaw, they should be ready to split approx 48 h after thawing.

3.8.2. Daily Feeding of ES Cells

1. Examine the dish for the condition of ES cell colonies and record observations. It is critical to monitor colony morphology, since this is the only gauge of culture conditions. Healthy ES cell colonies have smooth borders, the cells are tightly packed together so the individual cells are not detectable, and the entire colony has depth, giving a refractile ring around it (*see* **Fig. 1G**).
2. Remove the media from the healthy cells and replace with SCML. Slowly aspirate the media down the side of the dish so that the cell layer is not disturbed. The media volumes for each dish are in **Table 2**.

3.8.3. Subculture of ES Cells

1. Change the media by replacing with 15 mL fresh SCML approx 1–2 h prior to passage and return the dish to the incubator.
2. Examine the dish for colony morphology, density, and size, and prepare feeder plates based on the determined split ratio. Decide the ratio to split the cultures based on the size and distribution of ES cell colonies. An even distribution of colonies averaging in size 200–400 µm in diameter and spaced around 400 µm apart in a 100-mm dish will have

**Table 2
Media Volumes and Cell Counts for ES Cells in Various
Different Tissue Culture and Multiwell Dishes**

Dish Size	Media Volume	Cell Count
4-well dish	0.5 mL	Embyros
24-well dish	1.0 mL	2.0×10^4
12-well dish	2.0 mL	3.0×10^5
6-well dish	5.0 mL	4.0×10^5
35-mm dish	3.0 mL	4.0×10^5
60-mm dish	5.0 mL	6.0×10^5
100-mm dish	15.0 mL	2.0×10^6
4-chamber slide	1.0 mL	1.4×10^5
8-chamber slide	0.5 mL	6.0×10^4

$1.0–1.5 \times 10^7$ total cells. We typically split cultures at ratios from $1:6$ to $1:8$ resulting in approx $1.5–2.0 \times 10^6$ cells to be plated in each new 100-mm tissue culture dish. Splitting ES cells will ensure healthy passage and no overcrowded or undercrowding.

3. Remove the media and wash with 10 mL PBS. Remove the PBS, add 2 mL trypsin (for a 100-mm dish; 0.5 mL/well of a 6- or 12-well dish; 4 drops/well of a 24-well dish), and incubate for 1–2 min, checking the dish every 30 s by tapping the dish against the palm of your hand to dislodge the colonies.
4. Once the cells are no longer attached, add 8 mL SCML to the trypsin cell suspension. Pipet up and down vigorously to dissociate cells. Then plate 2×10^6 cells to each prepared 100-mm dish and cryopreserve the remaining cell suspension.

3.8.4. Freezing ES Cells

1. Transfer the remaining cell suspension (*see* **Subheading 3.8.3., step 4**) into a 15-mL tube and pellet the cells by low-speed centrifugation at $110g$ for 5 min.
2. Remove the supernatant taking care not to disturb pellet. A 100-mm dish will yield enough cells to freeze 4–5 vials (approx $3–6 \times 10^6$ cells/vial).
3. Add 1 mL freezing medium (50% FBS, 40% SCML, and 10% DMSO) for each vial frozen based on cell number. Pipet up and down to dissociate the ES cells and transfer 1 mL cell suspension per cryovial.
4. Put cryovials into a Nalgene controlled-rate freezer box and then put the box into a –80°C freezer. The next day, transfer the vials of frozen ES cells into the –150°C freezer for long-term storage.

3.9. Electroporation of ES Cells for Gene Targeting

3.9.1. ES Cell Preparation

1. Thaw ES cells 4–5 d prior to electroporation. Follow the maintenance protocol in **Subheading 3.8.1.**
2. Approximately 48 h after thawing, the cells should be ready to be split. Prepare two 100-mm feeder dishes with fresh SCML, then follow **Subheading 3.8.3, steps 1–4**. Freeze the cell suspension that is left by following **Subheading 3.8.3., steps 1–4** or pellet for DNA as a control for wild-type. (*See* **Note 6**).
3. Change the media on the ES cells with fresh SCML 1–2 h before electroporation. At the same time, dissociate the PEF cells from two 100-mm dishes and make 5 new dishes. This is done to minimize feeders rescuing ES cells during the selection process.

4. Prepare the ES cells from one of the two dishes made 2 d previously for subculture (*see* **Subheading 3.8.3.**). While the cells are in trypsin, remove the old media from one PEF feeder dish made in **Subheading 3.9.1., step 3**, wash with PBS, and add 15 mL fresh SCML. Dissociate the cells as described in **Subheading 3.8.3.** Transfer 1.0 mL trypsinized cell suspension (approx 2×10^6 cells) into the newly prepared feeder dish, which will be used as a control for selection, and transfer the remaining 8.5 mL ES cell suspension to a 15-mL centrifuge tube (approx $1–1.5 \times 10^7$ cells) for electroporation.
5. To the remaining dish, add 7 mL SCML and pipet up and down. Transfer the cell suspension to another 15-mL centrifuge tube for freezing. Pellet the contents of both tubes by centrifugation at 110g for 5 min. For freezing *see* **Subheading 3.8.4.**
6. Aspirate the supernatant and resuspend the cells to be electroporated in 10 mL SCML. Pellet again as in step 5. This is to ensure that all the trypsin has been removed.

3.9.2. Electroporation of ES Cells

1. Ideal electroporation settings must be determined for each different type of instrument used. We use the BTX ECM-600 electroporator with the following conditions: set volts at 280 V, capacitance at 50 μf, and resistance timing at R8 (360 ohms). Turn BTX unit on and push reset button to clear. Place a 0.4-mm disposable cuvette into the hood taking care not to touch the rim of the lid or the metal sides of the cuvette.
2. Transfer 25 μg DNA into a microfuge tube. Care must be taken when removing the microfuge tube from the container so that sterility is maintained, therefore handle the tubes by the sides and avoid touching the inside of the cap or rim of the tube.
3. Remove the supernatant from the cell pellet in the 15-mL tube, then with a 1-mL pipet add 375 μL SCML to the DNA, and then pipet up and down to thoroughly mix the DNA and SCML. Transfer the SCML/DNA solution to the cell pellet and pipet up and down to ensure a single-cell suspension. Finally, transfer the cell suspension into a 0.4-mm cuvette. Replace the lid on the cuvette to maintain sterility.
4. Place the cuvette into the holding apparatus of the electroporator and make sure there is good contact to the electrodes. Push reset button to clear. To electroporate, press the "automatic charge and pulse" button. When electroporation is complete, record actual voltage and pulse length (time is in milliseconds.) Remove the cuvette from the holder and return to hood.
5. Following electroporation, set the cuvette off to the side to allow the ES cells to recover for approximately 10–15 min. Prepare the feeder dishes. Remove the media from the 4 feeder dishes that were previously prepared in **Subheading 3.9.1., step 3** and add 15 mL fresh SCML to each dish. Also, transfer 12 mL SCML to a 15-mL tube and set aside.
6. Using the transfer pipet that came with the cuvette, aspirate a small volume of SCML from the 15-mL tube to wet the inside of the pipet so that the cells will not stick to the pipet. Now aspirate the electroporated cell suspension into the pipet slowly. Transfer the suspension to the 15-mL tube and repeat to ensure that most of the cells have been transferred to the tube. Using a 10-mL pipet, gently pipet up and down to disperse the cells, then transfer 3 mL cell suspension into each of the 4 new feeder dishes previously prepared (**Subheading 3.9.2., step 5**). It is very important to pipet the newly electroporated ES cells gently to ensure minimal cell damage. Incubate overnight in SCML.
7. The next morning, examine the dishes for colony morphology and cell survival. Record your observations. Remove the old media from the 4 dishes that contain the electroporated ES cells and the one selection control dish, and then replace with selection media. The selection medium used depends on the type of ES cell line and targeting vector used. HAT/SCML is used when the targeting vector restores the hyposanthine phosphoribosyl transferase (HPRT) function in HPRT-deficient ES cells, whereas 6-thioguanine/SCML is used when the targeting vector deletes the HPRT function in an ES cell line. G418-

Gancyclovir/SCML is used for positive–negative selection when the targeting vector contains the neomycin resistance gene and the thymidine kinase (TK) gene. Positive selection selects for cells that are neomycin resistant, whereas negative selection selects for cells that have lost the TK gene during homologous recombination. Since prolonged use of gancyclovir is harmful, we only use it in our medium for the first 4 d of selection. Then, on d 5, we switch to G418/SCML and use this medium throughout the remainder of selection (*see* Media Protocol, **Subheading 3.1.1.**, and **Table 1**).

8. Examine all 5 dishes and record observations daily. Then remove the old media and replace with fresh selection media. Selection generally takes 7–9 d (*see* **Note 7**).

3.9.3. Picking ES Cell Colonies

1. Approximately 7–13 d following electroporation, the ES cell colonies are ready to be picked. Prepare 24-well feeder plates using one 100-mm PEF feeder dishes for each 24-well dish. Wash with 10 mL PBS, then add 2 mL 0.05% trypsin EDTA to each 100-mm dish. Incubate 1–2 min, then check for dissociation. Tap the dish against the palm of your hand to dislodge cells from the dish. If cells are not completely free-floating, incubate for another 30–60 s. When completely dissociated, add 22 mL sDMEM and pipet up and down, then transfer 1 mL to each of the 24 wells. Return the dishes to the incubator until ready to use.

2. When ready to pick colonies, remove the old media from each well of the 24-well feeder dish and replace with fresh selection medium. Prepare a 100-mm bacteriology dish with microdrops of PBS or SCML. These will be used to wash the pipet between picks. Make sure you have sterile drawn pipets to use for picking and a filter on your mouth pipet tubing. This will help ensure the cultures remain free of contamination.

3. Place a dish with selected colonies on the microscope stage and examine it for colonies with the best morphology. Pick colonies that are approx 300 μm in diameter using a drawn mouth pipet. (*see* **Fig. 1G** and **H** and **Note 8**).

4. Transfer the colony to one well in a 24-well dish and blow until bubbles appear in the well. Draw media from the well up and down in the pipet to transfer all ES cells into the well. Wash the pipet in a microdrop of PBS or SCML and pick next colony. We generally pick 48 colonies into two 24-well dishes over a 2- to 3-d period with DBA/1LacJ ES cells. However, with 129 ES cells, it is often better to pick all colonies the same day.

3.9.4. Expanding Picked Colonies into Clonal ES Cell Lines

1. The days after you pick colonies, examine each well for the presence of ES cells. Observe each well to determine the average size of the surviving colonies. When the colonies are nearly 300 μm in diameter, dissociate them. If they are smaller and look fragile, change the media and leave the cells alone until the next day.

2. When the colonies are ready to dissociate (1–2 days after picking), remove the old media from each well. Wash by adding 0.5 mL PBS to each well, remove the PBS, then add 4 drops of trypsin solution per well, and incubate 1–2 min.

3. After incubation, vigorously tap the dish against the palm of your hand to dislodge the cells. Once the cells are completely dissociated, add 2 mL selection medium to each well. The next day, examine each well and record observation. Change the media in each well with 1.5 mL of fresh selection medium.

4. To keep ES cells undifferentiated, they must be dissociated every other day and the media changed daily. Dissociation and media changes may need to be done several times in the 24-well dish before there are enough ES cells to split 1:2 (half for freezing and half for DNA analysis). Not all clones grow at the same rate, therefore each clone must be handled

as a separate cell line. When there are enough colonies (200–400 μm in diameter) to cover the dish, spaced 200–400 μm apart, they are ready to split.

5. Examine each well and mark the colonies that will be dissociated and left in the 24-well dish and the colonies that are ready to split 1:2 (half will be cryopreserved and half transferred into 12-well dishes). Record the clone numbers in the data book. To prepare the 12-well dishes, follow the same protocol as for the 24-well dish (**Subheading 3.9.3., step 2**). One 100-mm dish of PEFs (6–8 × 10⁶ cells) will make two 12-well dishes. After trypsinizing the cells in the 100-mm dish, add 47 mL of sDMEM to the 2.0 mL of trypsin cell suspension. Pipet up and down and transfer 2 mL cell suspension/well into the two 12-well dishes. Let incubate about 1–2 h prior to use.

6. Before splitting the ES cells, change the media in each well of the previously prepared 12-well dishes and replace with 0.5 mL selection medium. Remove the media from the clones in the 24-well dish and wash with PBS. Add 4 drops of trypsin solution to each well and incubate 1–2 min at 37°C. After incubation, vigorously tap the dish against the palm of your hand to dissociate all the cells in the wells. For the wells that are just being dissociated and not split 1:2, add 2 mL selection media to each well. For each well to be split 1:2, aspirate 3 mL of selection medium into a 5-mL pipet. Transfer 1 mL of this medium into one well of the trypsinized 24-well dish and aspirate the entire contents of that well, then transfer to the 12-well dish. Pipet the entire volume up and down several times in the 12-well dish to ensure that all the ES cells are completely dissociated. Then transfer 1.5 mL of the cell suspension into the appropriate prelabeled cryovial, leaving the remaining cell suspension in the 12-well dish to continue growing for DNA analysis. When all the clones are transferred into the 12-well dish, fill each well with selection media to total 3 mL.

7. Pellet the cells in the cryovials by centrifugation at 110*g* for 5 min. Pour off the supernatant and add 0.5 mL freezing medium to each vial. Vigorously shake all the vials and place in a controlled-rate freezer at –80°C. The next day, transfer the vials into a liquid nitrogen freezer or –150°C freezer until the targeted clones are identified.

8. Prepare the ES cells for DNA analysis. Change the media on the ES cells for DNA analysis daily until they are overly confluent. At that point, remove the media from the cells and prepare for the extraction of genomic DNA for analysis by preferred method.

9. Once the targeted clones are identified, thaw those clones as described in **Subheading 3.8.1.**, except transfer the thawed cells into a prepared well of a 12-well dish (remember the frozen clone was half of a 24-well). When the cells are ready to be split, move half of the well into a 100-mm dish on a new feeder in SCML. The remaining half in the 12-well dish can be grown for DNA as described in **Subheading 3.9.4.** We do this routinely to confirm that the ES cells thawed are the targeted line. Once targeting is confirmed, all nontargeted ES cell isolates can be discarded

10. The targeted cells in the 100-mm dish will most likely need to be split 1:1 the first time. With the next split, begin freezing vials. We typically freeze 8–10 vials of targeted ES cells from the first two splits.

11. Once targeting is confirmed, choose 2–3 targeted ES cell lines for karyotyping. Follow **Subheading 3.7., step 2**.

3.10. Preparation of ES Cells for Aggregations or Microinjection into Blastocyst Stage Embryos

3.10.1. Whole Plate Shake-off Method

1. When ready to prepare ES cells for microinjection or aggregation, remove the media and wash with 10 mL PBS by tilting the dish and letting the PBS run down the dish to a pool in the bottom. Remove the PBS wash and add 2 mL trypsin solution. Immediately place

the dish on the microscope stage and tap the dish gently while observing that some of the ES cell colonies will dislodge from the feeders. This process takes approximately 30 s if the trypsin is fresh.

2. When enough ES cell colonies are free from feeders, return the dish to the hood. Tilt the dish so the loose colony suspension pools at the lower edge. Using a 1-mL pipet, aspirate 0.5 mL of the colony suspension in trypsin and transfer to a 1.5-mL microfuge tube. If the cells are to be used for blastocyst microinjection, let the tube set for about 30–60 s to allow the cells to further dissociate from the colonies. Then add 1 mL M2 medium to the microfuge tube to inactivate the trypsin and pipet up and down to completely dissociate the ES cells into a single-cell suspension.

   If the cells are to be used for aggregations, after 15–30 s in trypsin, add 1 mL M2 medium to the microfuge tube to inactivate the trypsin. Pipet up and down several times so the cells are still in small clumps.

3. Allow the cells remaining in the original dish to finish dissociating in the trypsin solution (1–2 min total). When these cells are completely free-floating, aspirate 4 mL SCML into a 5-mL pipet, tilt the dish, and wash the cell population into a pool at the bottom of the tilted dish. Pipet up and down several times to completely dissociate the cells, then transfer 1.5 mL of the cell suspension into a 1.5-mL microfuge tube to freeze for DNA (*see* **Note 9**). Transfer 1 mL cell suspension into the new PEF feeder (1:10 split) prepared in **Subheading 3.10.1., step 1**. This dish will be used to carry the cells for additional microinjection or aggregation. Transfer the remaining 3 mL of cell suspension into a 15-mL conical tube for freezing if this is the first split (follow maintenance protocol in **Subheading 3.8.4., steps 1–4**). Only freeze the first split, then discard the surplus cell suspension thereafter.

4. Place the two microfuge tubes (ES cells to inject and cells to pellet for DNA) into the microfuge and spin for 5 min at 110g. To make sure trypsin is removed, aspirate the supernatant and add 1 mL M2 to the cells for injection and add 1 mL PBS to the cells for DNA. Repeat microfuge to pellet again. Aspirate the supernatant from the cells used for microinjection or aggregation and resuspend the cell pellet in 50 µL M2 by gently pipetting up and down to dissociate the cells. The cells are ready to be injected into blastocysts or aggregated with morula.

4. Notes

1. It is important to emphasize aseptic technique. Always scrub hands before handling dishes with 70% ethanol. Always douse bottles and vials with 70% ethanol before putting into hood. Always flame bottles before opening them. Never reenter a bottle with the same pipet more than once. A good motto for all tissue culture practices is "when in doubt throw it out". It takes several months to generate targeted clones, so aseptic technique cannot be overemphasized.

2. ES cells are very sensitive to pH and temperature changes, as well as overcrowding and undercrowding. Once the cells are thawed, you must be committed to caring for them every day and even over the weekends and holidays. There are no good short cuts for this routine care.

3. ES cells maintain their pH best if the CO_2 concentration is between 5–10%. Therefore, to reduce the risk of fluctuations above or below that range, we keep our incubators set at 6%. If an incubator is not very stable, consider replacing it.

4. The quality of the reagents used in tissue culture of ES cells is also critical. Where possible, purchase products that are qualified for ES cell culture. We found that improving reagent quality has increased our clone survival and targeting frequency.

5. ES cells from different mouse strains react differently to the FBS used in the medium. When working with ES cell lines from different mouse strains, make sure to include all cell lines in your tests of different serum lots. We found this to be necessary even for ES cell qualified serum.

6. Keep in mind that ES cells grow more slowly following freeze–thaw and need time to recover. It is not unusual to dissociate the cells within the same dish or split 1 : 2. When this happens, an extra 2 d should be estimated into the time to thaw prior to electroporation, and use 2 µ gancyclovir for ± selection and only G418 for + selection.

7. Specific selection conditions need to be established for ES cell lines developed from different mouse strains. We use 300 µg G418/mL SCML ES cells from 129 mouse strains, and the 129 ES cell colonies are picked on d 9 and 10. For DBA1/LacJ ES cells, we use 175 µg G418/mL SCML, and ES cell colonies surviving selection are picked as early as d 7 and as late as d 13 following electroporation.

8. Typically, after electroporation and selection, you will notice differentiated cells and dead floating cells. Therefore, you must carefully choose the colonies to pick. Also, after you pick a colony, it sometimes dies. These we call "tried but died" and are probably being rescued from selective pressure by feeder cells in the 100-mm dish. Finally, we avoid large (>450 µm in diameter) "perfect" looking colonies, because of the observation that these colonies may have developed trisomy 8 *(22)*.

9. We store a pellet of ES cells, which were used to produce chimeras from each day of injections or aggregations, just in case there is a problem later and the resultant mice do not demonstrate the introduced genetic modification. This is an excellent control to have available, if the targeted mutation or germline transmission is not successful.

References

1. Evans, M. J. and Kaufman, M. H. (1981) Establishment in culture of pluripotential cells from mouse embryos. *Nature* **292,** 154–156.
2. Axelrod, H. R. (1984) Embryonic stem cell lines derived from blastocysts by a simplified technique. *Dev. Biol.* **101,** 225–228.
3. Wobus, A. M., Holzhausen, H., Jakel, P., and Schneich, J. (1984) Characterization of a pluripotent stem cell line derived from a mouse embryo. *Exp. Cell Res.* **152,** 212–219.
4. Doetschman, T. C., Eistattaer, H., Katz, M., Schmidt, W., and Kemler, R. (1985) The in vitro development of blastocyst derived embryonic stem cell lines: formation of yolk sac, blood islands and myocardium. *J. Embryol. Exp. Morphol.* **87,** 27–45.
5. Gossler, A., Doetschman, T., Korn, R., Serfling, E., and Kemler, R. (1986) Transgenesis by means of blastocyst derived embryonic stem cell lines. *Proc. Natl. Acad. Sci. USA* **83,** 9065–9069.
6. Thomas, K. R. and Capecchi, M. R. (1987) Site-directed mutagenesis by gene targeting in mouse embryo-derived stem cells. *Cell* **51,** 503–512.
7. Koller, B. H., Hageman, L. J., Doetschman, T. C., Hagaman, J. R., Huang, S., Williams, P. J., et al. (1989) Germline transmission of a planned alteration made in the hypoxanthine phosphoribosyltransferase gene by homologous recombination in embryonic stem cells. *Proc. Natl. Acad. Sci. USA* **86,** 8927–8931.
8. Thomson, J. A., Kalishman, J., Golos, T. G., Durning, M., Harris, C. P., Becker, R. A., and Hearn, J. P. (1995) Isolation of a primate embryonic stem cell line. *Proc. Natl. Acad. Sci. USA* **92,** 7844–7848.
9. Thomson, J. A., Itskovitz-Eldor, J., Shapiro, S. S., Waknitz, M. A., Swiergiel, J. J., Marshal, V. S., and Jones, J. M. (1998) Embryonic stem cell lines derived from human blastocysts. *Science* **282,** 1145–1147.

10. Shamblott, M. J., Axelman, J., Wang, S., Bugg, E. M., Littlefield, J. W., Donovan, P. J., et al. (1998) Derivation of pluripotent stem cells from cultured human primordial germ cells. *Proc. Natl. Acad. Sci. USA* **95,** 13,726–13,731.

11. Reubinoff, B. E., Pera, M. F., Fong, C.-Y., Trounson A., and Bongso, A. (2000) Embryonic stem cell lines from human blastocysts: somatic differentiation in vitro. *Nat. Biotechnol.* **18,** 399–404.

12. Roach, M. L., Stock, J. L., Byrum, R., Koller, B. H., and McNeish, J. D. (1995) A new embryonic stem cell line from DBA/1LacJ mice allows genetic modification in a murine model of human inflammation. *Exp. Cell Res.* **221,** 520–525.

13. Robertson, E. J. (1987) *Teratocarcinomas and Embryonic Stem Cells, a Practical Approach.* IRL Press, Eynsham, Oxford. pp. 76–78.

14. Hogan, B., Beddington, R., Costantini, F., and Lacy, E. (1994). *Manipulating the Mouse Embryo, a Laboratory Manual.* CSH Press, Cold Spring Harbor, N.Y. pp. 144–145.

15. Voss, A. K., Thomas, T., and Gruss, P. (1997) Germline chimeras from female ES cells. *Exp. Cell Res.* **230,** 45–49.

16. Longo, L., Grave, A. B., Grosveld, G. F., and Pandolfi, P. P. (1997) The chromosome make-up of mouse ES cells is predictive of somatic and germ cell chimerism. *Transgenic Res.* **6,** 321–328.

17. Rowe, W. P., Hartley, J. W., Estes, J. D., and Huebner, R. J. (1959) Studies on mouse polyoma virus infection. *J. Exp. Med.* **109,** 379–391.

18. Darfler, M. M., Dougherty, C., and Goldsborough, M. D. (1996) The mouse YES system: a novel reagent system for the evaluation of mouse chromosomes. *Focus* **18,** 15–16.

19. Hogan, B., Beddington, R., Costantini, F., and Lacy, E. (1994). *Manipulating the Mouse Embryo, a Laboratory Manual.* CSH Press, Cold Spring Harbor, N.Y. pp. 311–315.

20. Cowell, J. K. (1984) A photographic representation of the variability in the G-banded structure of the chromosomes in the mouse karyotype. *Chromosoma* **89,** 294–320.

21. Wood, S. A., Allen, N. D., Rossant, J., Auerbach, A., and Nagy, A. (1993) Non-injection methods for the production of embryonic stem cell-embryo chimeras. *Nature* **365,** 87–89.

22. Liu, X., Wu, H., Loring, J., Hormuzdi, S., Disteche, C. M., Bornstein, P., and Jaenisch, R. (1997) Trisomy eight in ES cells is a common potential problem in gene targeting and interferes with germline transmission. *Dev. Dyn.* **209,** 85–91.

2

The Use of a Chemically Defined Media for the Analyses of Early Development in ES Cells and Mouse Embryos

Gabriele Proetzel and Michael V. Wiles

1. Introduction

During embryonic development, primitive ectoderm forms three primary germ layers, the mesoderm, the ectoderm, and the endoderm. These germ layers interact forming all the tissues and organs of the developing embryo. The influences controlling the transition of ectoderm to visceral and parietal endoderm in the blastocyst, followed by the formation of mesoderm at gastrulation, are only beginning to be defined. In the mouse, this process occurs between d 3 and 7 post-fertilization, and as such, it is both difficult to monitor and to experimentally influence. With this in mind, many groups have used mouse embryonic stem (ES) cells, and more recently human ES cells, to study the control of germ layer formation and their subsequent differentiation.

The history of ES cell in vitro differentiation began with Tom Doetschman and Anna Wobus *(1,2)* who independently observed that ES cells grown in suspension form clusters of cells referred to as embryoid bodies (EBs). Under these conditions, ES cells rapidly differentiate to many recognizable cell types, including spontaneously beating heart muscle and islands of primitive erythrocytes (blood islands) *(1,2)*. This approach was refined by Michael Wiles and Gordon Keller, who succeeded in both improving the percentage of EBs, which formed mesoderm and hematopoietic cells, and its reproducibility *(3)*. However, the approach was still totally dependent upon the presence of fetal calf serum (FCS) in the media and, more importantly, the "batch" of serum used (i.e., the main influence of differentiation was the presence of unknown factors in fetal bovine serum [FBS]; *see also 4*). These observations spurred the development of a completely chemically defined media (CDM) for use in such experiments. The use of fully defined reagents aimed to make these experiments independent of variations due to serum and/or poorly defined "proteolytic digests" of meat, sheep brains, or other bizarre FCS substitutes. Further, a defined media would facilitate characterizing exactly those influences that control ES cell differentiation and, thus, early mammalian development.

The use of a totally CDM as a media for studying early mammalian development was further inspired by the observations of research groups working with *Xenopus laevis* embryos. The use of the extremely robust *X. laevis* embryo as a research tool

From: *Methods in Molecular Biology, vol. 185: Embryonic Stem Cells: Methods and Protocols*
Edited by: K. Turksen © Humana Press Inc., Totowa, NJ

in the study of early vertebrate development made significant inroads into understanding the mechanisms controlling early germ layer formation. Although the *X. laevis* embryos as an experimental system is conceptually similar to many experiments with mammalian cells, there is one a major difference. Experiments using cells derived from the *Xenopus* blastula are routinely conducted in a defined simple salt solution. In contrast, mammalian cell models (e.g., ES cell invitro differentiation), use a defined media, which was then supplemented with 5–30% serum (FCS). In essence, the *Xenopus* experimenter has total control over the initial environment used to conduct their experiments, while those using serum are embroiled in the complexities of ill-defined FCS batches and their variable constituents. This difference also explains the results obtained with the two systems. For example, when *Xenopus* blastula cells are exposed to the transforming growth factor beta (TGFβ) family member, activin A, mesodermal and neural differentiation is induced *(5-7)*. If however, mouse ES cells are differentiated as EBs in 10% FCS containing media (without leukemia inhibiting factor [LIF]) in the presence of activin A, no striking change in the "spontaneous" pattern is observed.

In 1995, Johansson and Wiles described how ES cell differentiation could be achieved in a completely CDM and that specific growth factors added to this media could directly influence the course of differentiation *(8)*. At this time, the media contained bovine serum albumin (BSA), which although of a very high purity could still be regarded as only one step above supplements containing serum substitutes. The work of M. T. Kane had previously demonstrated that the BSA component of media could be replaced with polyvinylalcohol (PVA) for the culture of rabbit eggs and blastocysts *(9)*. Using this idea, Johansson and Wiles demonstrated that PVA could be used to replace BSA in the original formulation of CDM and that the media could both support ES cell growth and differentiation *(10)*. As such, the ES cell invitro differentiation model could be used to test the effects of exogenously added growth factors in a fully definable environment. The replacement of FCS with CDM or similar completely defined media removes one of the principal undefined influences in the study of ES cell differentiation. In CDM, ES cells are now responsive to many exogenous growth factors and are capable of differentiating to many lineages, including neuronal cells, mesoderm derivatives, including hematopoietic cells, myocytes, and endoderm precursors. Further, recent data have suggested that this media can also support the early development of mouse egg cylinders from premesodermal (E6.0) to a fully expanded egg cylinder expressing markers for mesoderm and hematopoiesis.

The simplicity of the CDM is its strength, however it can also be a major drawback. Many cell lineages can develop in vitro from ES cells during differentiation, however if novel cell types arise in an environment that is not supportive of their specific requirements, the cells may die or at least be severely selected against. As concisely put by Martin Raff, ". . . most mammalian cells constitutively express all of the proteins required to undergo programmed cell death and undergo programmed cell death unless continuously signaled by other cells not to . . ." *(11)*. The basal CDM described here contains only three growth factors, insulin, transferrin, and a very low concentration of LIF. As such, cells grown, or those which arise in this media, have access to a very limited environment in regard to growth factors and signaling molecules. This means that as ES cells differentiate to new cell types, only those cells that continue to receive the appropriate survival signals will flourish, and those that are not sufficiently supported either by CDM, exogenously added factors, or by factors made by the cells themselves, will die. For example, in basal CDM, this effect is evident upon the development of neuroectoderm cells and derivatives from ES cells. After 6–8 days, EB grown in CDM alone do not continue to grow well, and cell death is evident. However, these cultures can be rescued if they are fed, for example, with FCS containing media (M.V.W. unpublished observations) or with CDM containing neuronal survival factors (e.g., nerve growth factor [NGF]). From this, it is also evident that the system lends itself to testing specific growth factors and combinations, acting as an assay system to monitor growth factor regimes supportive of specific cell survival and expansion.

Recently, these ideas have gained a new dimension with the advent of human ES cells. These cells can be used as tools, allowing us to examine and understand the earliest events of human development *(12–14)*. Further, as human ES cells share some of the IVD capabilities of mouse ES cells, they may provide an abundant source of many different (stem) cell types with possible applications to tissue repair, etc. Schuldiner et al. and others have differentiated human ES cells in a serum-free media and defined growth factors allowing the generation of several cell lineages *(15–17)*. Thus, in the near future, it may be possible to tailor cell culture environments leading to the induction and then the selective expansion of medically useful cells.

2. Materials

2.1. Reagents

1. 100X Chemically defined lipid concentrate (Gibco-BRL, Life Technologies, cat. no. 11905-031).
2. 200 mmol/L GlutaMAX -I (Gibco-BRL, Life Technologies, cat. no. 35050-061).
3. Ham's F12 nutrient mixture with GlutaMAX-I (Gibco-BRL, Life Technologies, cat. no. 31765).
4. Insulin (Sigma, cat. no. I2767 powder; alternatively, Gibco-BRL, Life Technologies, cat. no. 13007).
5. Iscoves modified Dulbeccos medium (IMDM) with GlutaMAX-I (Gibco-BRL, Life Technologies, cat. no. 31980).
6. LIF (Chemicon International, cat. no. ESG1107).
7. Monothioglycerol (MTG) (Sigma, cat. no. M6145).
8. PVA (Sigma, cat. no. P8136).
9. Transferrin (Roche Biochemicals, cat. no. 1073982).
10. Trypsin inhibitor (Sigma, cat. no. T6522), made up at 1 mg/mL in serum-free medium.
11. Phosphate-buffered saline (PBS), pH 7.2 (Gibco-BRL, Life Technologies, cat. no. 20012043).

2.2. Schema for the Preparation of Basal CDM from Stock Solutions

Reagent	Concentration of work stock solution	Final concentration
49 mL IMDM	2X	1X
49 mL Ham's F12	2X	1X
1 mL PVA	10% w/v autoclaved in water	0.1% w/v
300 µL MTG	27 µL MTG in 2 mL IMDM/F12	450 µM

The media must be filter-sterilized *before* adding lipids and proteins (including growth factors).

Reagent	Concentration of work stock solution	Final concentration
1 mL Synthetic lipids	100X	1X
10 μL LIF	10 U/μL	1–2 U/mL
0.5 mL Transferrin	30 mg/mL	0.1% w/v
70 μL Insulin	10 mg/mL	7.0 μg/mL

3. Methods
3.1. Differentiation Protocols:

To fully understand the following protocols, it is essential that the Notes given below are read and understood (*see* **Note 1**).

3.1.1. Preparation of ES Cells for Differentiation Protocols (see **Note 2***).*

1. Wash standard ES cell cultures with basal CDM twice, then culture the ES cells for a further 30 min in basal CDM.
2. Trypsinize ES cells and make a single-cell suspension in CDM, centrifuge to pellet the cells.
3. Resuspend approx 3 mL basal CDM containing 1 mg/mL trypsin inhibitor, then centrifuge to pellet the cells.
4. Resuspend the cells in basal CDM and count the cells.

The ES cells are now clear of undefined substances and are now ready for the differentiation studies.

3.1.2. ES Cell Differentiation in Suspension Culture (see **Note 3***)*

1. Seed a single-cell suspension of approx 6000 ES cells onto a 35-mm bacterial grade non-tissue-culture grade dish in 1 mL of CDM.
2. Place the plates within a larger dish and add a few open plates containing water to avoid the drying out of the CDM cultures.
3. Culture for 1 to 8 d and then assess differentiation status.

3.1.3. ES Cell Differentiation in Hanging Drop Culture (see **Fig. 1** *and* **Note 4***)*

1. Dilute ES cells to approx 5–50 cells/20 μL (i.e., 250–2500 cells/mL) in 20 μL CDM ± test factors.
2. Place individual drops of 20 μL CDM plus cells carefully onto the surface of a 35-mm non-tissue-culture grade plate (each drop must remain separate).
3. Place lid on the plate and invert the whole assembly rapidly and keep leveled. The individual drops are now hanging from the top of the plate (*see* **Fig. 1**).
4. Place the plates within a larger dish and add a few open plates containing water to avoid drying out of the cultures.
5. Incubate for 24–48 h and then inspect the EBs.
6. Re-invert the plate and flood it with 1 mL CDM into 20 μL CDM ± test factors. The individual EBs are now floating in the media.
7. Culture in this condition for a further 0–7 d and assess differentiation—(note the EBs will generally remain in suspension during this culture period).

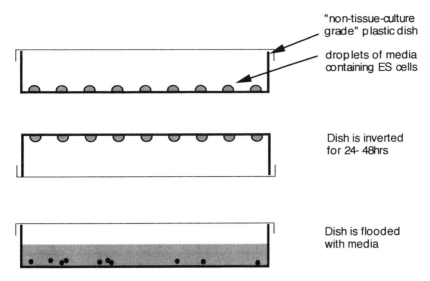

Fig. 1. Outline of hanging drop culture. The hanging drop approach is an efficient and highly controllable method to make a defined number of regular sized EBs. ES cells are placed in drops of 20 µL of media in a non-tissue-culture grade plate. When the plate is inverted, the drops hang, and the ES cells coalesce to form an EB. After 48 h, the plate is re-inverted and flooded with growth media.

3.1.4. ES Cell Differentiation in Hanging Drop Culture Followed by Attachment Culture (see **Note 5**)

1. Differentiate cells using either suspension culture or hanging drop culture.
2. Transfer EBs in CDM into a sterile 1.5-mL Eppendorf tube and allow the EBs to settle out.
3. Carefully remove the majority of CDM and transfer the EBs with a wide bore pipet tip to a standard tissue culture plate.
4. Add tissue culture media containing 5–10% FCS.
5. The EBs will attach and spread in the next 24–48 h.
6. Development assessment.

3.1.5. Culture of Egg Cylinder Embryos in CDM (see **Fig. 2** and **Note 6**)

1. Dissect mouse egg cylinder embryos at E6.0 to E7.5.
2. Transfer egg cylinder into PBS to remove all maternal tissue.
3. Transfer embryos into 20 µL CDM ± test factors.
4. Follow from **step 3** of the ES cell differentiation in hanging drop culture (**Subheading 3.1.3.**).
5. Incubate for 24–48 h.
6. Assess development.

3.2. Assessment of ES Cell Differentiation

The assessment of differentiation by visual inspection of EBs in culture is not very informative, being mainly limited to counting EBs, which are visibly red due

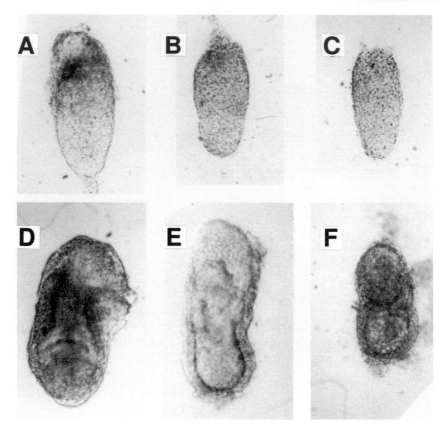

Fig. 2. Egg cylinder e6 and after 30 h in basal CDM. Mouse egg cylinders were dissected out of the decidua at d 6.0 post coitus (p.c.) and grown in hanging drop cultures for 30 h in basal CDM. After 30 h, it is evident that further differentiation has occurred. Additionally, RT-PCR (not shown) detected markers for mesoderm (Brachyury) and hematopoiesis (β-H1 globin).

to hematopoiesis, beating after the formation of cardiac muscle, or judging cell morphology for muscle or neuronal cells after EB attachment and cell outgrowth. A more quantitative approach is to use reverse transcription polymerase chain reaction (RT-PCR) and assess the expression of specific lineage marker genes. For this, we isolated total RNA from the EBs after various time points and treatments. cDNA synthesis used random hexamers as primers. For RT-PCR the approximate amounts of cDNA used was previously assessed using hyposanthine phosporibosyl transferase (IIPRT) as a concentration standard *(4)*. For the experiments described here, we used a Biometra TRIO thermal cycler. PCR regimes were: 96°C for 6 s, 50° or 55°C for 15 s, 72°C for 60 s, for 30 cycles, and finally 72°C for 10 min. PCR products were assessed by gel electrophoreses, Southern blotting, and hybridization (*see* **Fig. 2**).

When ES cells are grown in suspension or in hanging drops in CDM, EBs develop within 48 h. These clusters of cells form by both cell division and cell–cell collision. During the first 24–48 h, there is a rapid decline, as measured by RT-PCR, of Rex-1 and activin βB RNAs, indicative of the loss of the undifferentiated ES cell phenotype. In many experiments, low variable levels of Pax6 mRNA were also detectable in

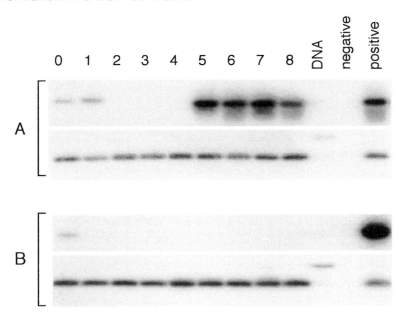

Fig. 3. ES cell differentiation RT-PCR time course for Pax6. Southern blots of RT-PCR analysis of ES cells grown in (**A**) basal CDM and (**B**) CDM plus 2ng/mL BMP-4. Cultures were harvested for RNA from 0–8 d, cDNA was synthesized, and RT-PCR was conducted. HPRT was used as a cDNA loading control (lower panel of each set) and compared with Pax6 (upper panel of each set). In basal CDM, Pax6 expression increases over time, indicative of neuroectoderm formation. When BMP-4 is present, Pax6 expression is not detectable after 24 h of culture. The figure was derived from the linear output of a Phosphor Imager (Molecular Dynamics, Sunnyvale, CA).

undifferentiated cells, however, within 24–48 h of EB formation, Pax6 became undetectable.

Where cultures were maintained in basal CDM, the EBs continue to grow for 6–8 d, although at a slower rate compared with FCS-containing cultures. To monitor the progress of differentiation, a number of genes can be examined. For example, a marker of neuroectoderm formation is Pax6. **Fig. 3** shows that after 5 d of culture, Pax6 mRNA abundance rises rapidly. In contrast, markers for mesoderm are not readily detectable *(8)*.

However, after 7 to 9 d, the physical state of EBs in basal CDM begins to deteriorate with an increase in cell debris, suggesting that the ES-derived differentiated cells are beginning to die. These cultures can be rescued if the EBs are transferred into tissue grade plastic dishes in the presence of FCS. Under these conditions, the EBs will attach, spread, and in general (depending upon the batch of FCS used), produce large lattice works of neuronal cells in 4–10 d. It is conceivable to use specific growth factors or growth factor cocktails instead of FCS.

ES cells in CDM plus BMP-2, 4, or 7 rapidly develop into EBs. Under this regime, the EBs grow more rapidly than in CDM alone. Further, they do not show cell death as observed in 7–9 d basal CDM cultures. Expression of genes related to mesoderm (BMP-2, 4 or 7) and hematopoietic formation (BMP-2 or 4) are readily detectable within 3–4 d *(10)*.

Whole-mount *in situ* hybridization can also be used to derive exact localization information, which can be correlated to defined morphological changes observed during ES cell differentiation *(8)*.

4. Notes

1. When beginning to work with serum-free tissue culture, it is important to appreciate that cells are far less buffered to any toxic substances that may inadvertently be introduced into the culture system. It is, therefore, essential that all reagents used for the media are of the very highest quality and that media preparation is conducted in a perfectly clean manner. With this in mind, we suggest that disposable plasticware be used wherever possible. Further, although significant batch variations in the various chemicals used in the formulation of CDM was not observed, it is recommended that reagent batch tracking records be maintained as part of good laboratory practice.

   The optimal concentration of any new exogenous factors should be assessed empirically, as many novel factors may have variable specific activities depending upon their source and the purification method used to obtain them. Further, it should be noted that many factors, for example members of the TGFβ family, could show dramatically different effects depending upon concentration used.

   In the work described here, cultures were maintained for varying periods of time. In some cases, we returned the EBs to tissue culture plates allowing them to attach and spread. The effects of a number of growth factors have been assessed during ES cell differentiation using CDM. We give an example of data obtained when EB were differentiated in basal CDM and activin A. Interestingly, many other growth factors tested failed to have any striking effect on the parameters monitored, e.g., mesoderm formation *(8)*.

2. For the experiments reported here, we used the 129/Sv-derived ES line CCE *(19)*, similar data were obtained with other 129-derived ES lines, including D3 and E14.1 *(20)*. For routine culture of ES cells, we used Dulbecco's modified Eagle medium (DMEM) supplemented with 15% FCS, 1.5×10^{-4} mL MTG and 1000 u/mL LIF. For all ES cell experiments, cells were adapted to grow off feeders, as the presence of variable numbers of feeders in the differentiation culture would complicate interpretation. As ES cells were maintained in FCS for routine culture, residual growth factors derived from the FCS have to be removed before the initiation of CDM differentiation experiments. To do this, we washed the attached ES cells with CDM twice. The cells were then cultured for a further 30 min. in basal CDM before proceeding. Cells were trypsinized to obtain a single-cell suspension and resuspended in CDM containing a trypsin inhibitor to inactivate any residual trypsin. Residual trypsin will considerably reduce cell viability in subsequent culture. Following trypsin inactivation, cells were pelleted by centrifugation and resuspended in basal CDM without trypsin inhibitor and counted. Cells were now ready for experimental tests.

3. Nontissue-culture grade plastic is used for these experiments, this is to reduce the number of cells adhering to the plate's surface. When using the ES line CCE, approx 10–20 EBs/mL formed after 5 d. Other ES cells lines have different plating efficiencies and, hence, required different cell densities to give a reasonable number of EBs. The approximate density of EBs in the media is crucial, because the density of EBs increases so will any effects of growth factors synthesized by the developing EBs themselves.

4. This is an alternative strategy and is strongly recommended as the approach lends itself to more uniform EB development and exact control of the final EB density (*see* **Fig. 1**). The hanging drop procedure was first described for ES cell differentiation by Anna Wobus *(21)*.

Individual drops of cells in CDM are placed onto the surface of a 35-mm non-tissue-culture grade plate; for example, by using a repeating pipet, approx 20 drops can be easily placed per 35-mm plate. Because of the hydrophobic nature of the non-tissue-culture grade plastic, the drops will not spread, but remain as individual separate drops. The lid is then placed on the plate, and the whole assembly is rapidly inverted and then kept level. The individual drops are now hanging from the top of the plate (*see* **Fig. 1**). This maneuver is simple after a few practice runs.

Following 48 h of culture in a humidified incubator, the majority of the hanging drops will have developed a single uniform EB. We tend to allow 24–48 h to develop the EBs before flooding the dish with 1 mL of media. If the hanging drop cultures are left longer, the rapidly growing cells will exhaust the 20 mL of media in the drop and begin to deteriorate.

5. Cultures were maintained for varying periods of time. In some cases, we returned the EBs to tissue culture plates, allowing them to attach and spread. Although, it must be noted that in basal CDM, EBs attach and spread to a lesser extent when compared to the use of conventional FCS-containing media as the secondary culture media. This suggests that basal CDM is lacking factors necessary for efficient attachment and proliferation, however, these could be added specifically, e.g., collagen.

6. We tested if CDM can support early mouse development in hanging drops. For this, egg cylinder stage embryos (E6.0-E7.0) were grown in basal CDM without added factors. As can be seen in **Fig. 2**, basal CDM cultures are capable of supporting early embryos for at least 48–72 h. Although the developing embryos lose coherent organization, they do continue to develop recognizable structures. Further, the embryos develop mesoderm and hemoglobin-producing cells containing embryonic globin (M.V.W. own observations; *see also* **ref. 22**). A few final closing words concerning the use of CDM or similar media and the interpretation of the data obtained. The developing embryo is a highly dynamic environment, in which the fate of cells is directed by the continually changing environment of growth factors and other influences. Cells interpret these signals using multiple interacting networks of genes, which together, provide a high degree of developmental homeostasis *(18)*. The end result of these dynamic interactions is a fully functional organism. Although the use of CDM represents a gross simplification of any in vivo environment, it is fully defined, and its use in combination with growth factors and ES cell IVD offers a method to dissect and direct the differentiation processes.

Acknowledgments

We thank the Genetic Institute Inc. for the gift of many of the factors used in these studies and Britt Johansson for excellent technical assistance.

References

1. Wobus, A. M., Holzhausen, H., Jakel, P., and Schoneich, J. (1984) Characterization of a pluripotent stem cell line derived from a mouse embryo. *Exp. Cell Res.* **152,** 212–219.
2. Doetschman, T. C., Eistetter, H., Katz, M., Schmidt, W., and Kemler, R. (1985) The in vitro development of blastocyst-derived embryonic stem cell lines: formation of visceral yolk sac, blood islands and myocardium. *J. Embryol. Exp. Morphol.* **87,** 27–45.
3. Wiles, M. V. and Keller, G. (1991) Multiple hematopoietic lineages develop from embryonic stem (ES) cells in culture. *Development* **111,** 259–267.
4. Keller, G., Kennedy, M., Papayannopoulou, T., and Wiles, M. V. (1993) Hematopoietic commitment during embryonic stem (ES) cell differentiation. *Mol. Cell Biol.* **13,** 473–486.

5. Slack, J. M., Darlington, B. G., Gillespie, L. L., Godsave, S. F., Isaacs, H. V., and Paterno, G. D. (1990) Mesoderm induction by fibroblast growth factor in early *Xenopus* development. *Philos. Trans. R. Soc. Lond. Biol. Sci.* **327,** 75–84.

6. Thomsen, G., Woolf, T., Whitman, M., Sokol, S., Vaughan, J., Vale, W., and Melton, D. A. (1990) Activins are expressed early in *Xenopus* embryogenesis and can induce axial mesoderm and anterior structures. *Cell* **63,** 485–493.

7. van den Eijnden-Van Raaij, A. J., van Zoelent, E. J., van Nimmen, K., Koster, C. H., Snoek, G. T., Durston, A. J., and Huylebroeck, D. (1990) Activin-like factor from a Xenopus laevis cell line responsible for mesoderm induction. *Nature* **345,** 732–734.

8. Johansson, B. M. and Wiles, M. V. (1995) Evidence for involvement of activin A and bone morphogenetic protein 4 in mammalian mesoderm and hematopoietic development. *Mol. Cell Biol.* **15,** 141–151.

9. Kane, M. T. (1987) Minimal nutrient requirements for culture of one-cell rabbit embryos. *Biol. Reprod.* **37,** 775–778.

10. Wiles, M. V. and Johansson, B. M. (1999) Embryonic stem cell development in a chemically defined medium. *Exp. Cell Res.* **247,** 241–248.

11. Raff, M. C. (1992) Social controls on cell survival and cell death. *Nature* **356,** 397–400.

12. Thomson, J. A., Itskovitzeldor, J., Shapiro, S. S., Waknitz, M. A., Swiergiel, J. J., Marshall, V. S., and Jones, J. M. (1998) Embryonic stem cell lines derived from human blastocysts. *Science* **282,** 1145–1147.

13. Shamblott, M. J., Axelman, J., Wang, S. P., Bugg, E. M., Littlefield, J. W., Donovan, P. J., et al. (1998) Derivation of pluripotent stem cells horn cultured human primordial germ cells. *Proc. Natl. Acad. Sci. U.S.A.* **95,** 13,726–13,731.

14. Pera, M., Reubinoff, B., and Trounson, A. (2000) Human embryonic stem cells. *J. Cell Sci.* **113,** 5–10.

15. Schuldiner, M., Yanuka, O., Itskovitz-Eldor, J., Melton, D. A., and Beurenisty, N. (2000) Effects of eight growth factors on the differentiation of cells derived from human embryonic stem cells. *Proc. Natl. Acad. Sci. U.S.A.* **97,** 11,307–11,312.

16. Itskovitz-Eldor, J., Schuldiner, M., Karsenti, D., Eden, A., Yanuka, O., Amit, M., et al. (2000) Differentiation of human embryonic stem cells into embryoid bodies compromising the three embryonic germ layers. *Mol. Med.* **6,** 88–95.

17. Reubinoff, B., Pera, M., Fong, C., Trounson, A., and Bongso, A. (2000) Embryonic stem cell lines from human blastocysts: somatic differentiation in vitro. *Nat. Biotechnol.* **18,** 399–404.

18. Waddington, C. H. (1942) Canalization of development and the inheritance of acquired characters. *Nature* **150,** 563–565.

19. Robertson, E., Bradley, A., Kuehn, M., and Evans, M. (1986) Germ-line transmission of genes introduced into cultured pluripotential cells by retroviral vector. *Nature* **323,** 445–448.

20. Fisher, J. P., Hope, S. A., and Hooper, M. L. (1989) Factors influencing the differentiation of embryonal carcinoma and embryo-derived stem cells. *Exp. Cell Res.* **182,** 403–414.

21. Wobus, A. M., Wallukat, G., and Hescheler, J. (1991) Pluripotent mouse embryonic stem cells are able to differentiate into cardiomyocytes expressing chronotropic responses to adrenergic and cholinergic agents and Ca^{2+} channel blockers. *Differentiation* **48,** 173–182.

22. Palis, J., McGrath, K. E., and Kingsley, P. D. (1995) Initiation of hematopoiesis and vasculogenesis in murine yolk sac explants. *Blood* **86,** 156–163.

3

Analysis of the Cell Cycle in Mouse Embryonic Stem Cells

Pierre Savatier, Hélène Lapillonne, Ludmila Jirmanova, Luigi Vitelli, and Jacques Samarut

1. Introduction

The molecular mechanisms underlying self-renewal of pluripotent embryonic stem (ES) cells is still poorly understood. Deciphering these mechanisms is of prime importance for at least two reasons: (1) ES cells derive from, and are closely related to, the pluripotent stem cells of the blastocyst, the founder cells of the whole embryo proper. Hence, they constitute a unique model for studying embryonic development at the time of implantation, when embryos are inacessible to experimental manipulation; and (2) Isolating and manipulating ES cells in species of economic or therapeutic interests is more difficult than in the mouse. It is likely that better defining their growth requirements will lead to major improvements in their culture conditions.

During the past 6 yr, intrinsic features of mouse ES cells regarding the regulation of their growth cycle have been pinpointed. These features may serve not only to understand how the cell cycle machinery of ES cells works, but also to better characterize ES cells isolated from embryos of other species. Hence, a striking feature of mouse ES cells is their unusual cell cycle distribution. The three phases of the cell cycle, G1, S, and G2/M, represent 15, 75, and 10%, respectively, of the total cell cycle, with a G1 phase of approx 1 h. Hence, ES cells reenter the S-phase very shortly after exit from mitosis *(1,2)*. These preliminary observations have paved the way to the analysis of cell cycle control in ES cells, focusing on the regulation of G1→ S transition.

1.1. Retinoblastoma Pathway

The proliferation of mammalian cells is controlled largely during the G1 phase of their growth cycles. The decision to initiate a new round of DNA synthesis is largely dependant upon the phosphorylation and functional inactivation of the retinoblastoma (RB) protein. This phosphorylation is driven by components of the cell cycle apparatus, specifically cyclins and cyclin-dependent kinases (CDKs). Of prime importance are complexes of D-type cyclins (cyclin D1, D2, and D3) and CDK4 or CDK6 *(3)*. Moreover, the cellular machinery that is organized to collect extracellular signals and transduce them via tyrosine kinase receptors and the SOS-RAS-MEKK-MAPK pathway seems to be dedicated largely to driving RB phosphorylation *(4)*. This control

From: *Methods in Molecular Biology, vol. 185: Embryonic Stem Cells: Methods and Protocols*
Edited by: K. Turksen © Humana Press Inc., Totowa, NJ

circuitry appears to be operative in virtually all cell types. In contrast, the control of the ES cell mitotic cycle is likely to be markedly different. First, ES cells seem to lack the CDK4-associated kinase activity that characterizes all RB-dependent cells. They express very low levels of D-type cyclins, as a result of the very poor activity of the respective promoters. This is somewhat surprising as hypophosphorylated RB remains undetectable during the M→G1→S transition, indicating that RB is rapidly rephosphorylated in G1 *(2)*. Secondly, ES cells appear to be resistant to the growth inhibitory effect of the cyclin D:CDK4-specific inhibitor p16[ink4a], further suggesting that RB phosphorylation may not rely on proper CDK4-associated kinase activity in ES cells. Not surprisingly, induction of differentiation restores the expression of all three D-type cyclins, strong CDK4-associated kinase activity, and the sensitivity to the growth-inhibitory activity of p16[ink4a] *(1,2,* and unpublished results), suggesting that differentiating ES cells resume a normal cell cycle control.

Another important aspect of G1 control lies in the regulation of cyclin D1 expression by the Ras→MAPK pathway. Phosphorylated ERKs activate cyclin D1 expression through fos and ets transcription factors *(5)*. In ES cells, inhibition of ERK phosphorylation by wortmannin (an inhibitor of Ras activation) or by PD98059 (an inhibitor of *MEK*) neither inhibits background expression of cyclin D1 nor induces growth retardation. Induction of differentiation up-regulates the steady-state level of cyclin D1, whose expression then becomes sensitive to the inhibitors of the Ras→MEK→ERK cascade (Jirmanova et al., unpublished results). Hence, cyclin D1 expression seems not to be regulated by Ras in ES cells. This regulation is likely to be restored upon differentiation.

Recently, it has been shown that Rb-E2F forms a transcriptional repression complex by recruiting histone deacetylase and SWI/SNF subunits *(6)*. These large complexes are capable of blocking the transcription of cell cycle genes and remodeling chromatin *(7)*. However, it is unclear if these large complexes have a specific role in chromatin organization of ES cells. Thus far, our preliminary immunoprecipitation experiments suggest that HDAC1 binds to the low amount of RB in ES cells. This could be a key aspect in the ES renewing cell cycle that should be investigated.

1.2. p53 Pathway

In somatic cells, cell cycle checkpoints limit DNA damage by preventing DNA replication under conditions that may produce chromosomal abberations. The tumor suppressor p53 is involved in such control as part of a signal transduction pathway that converts signals emanating from DNA damage, ribonucleotide depletion, and other stresses into responses ranging from cell cycle arrest to apoptosis *(8)*. Stress-induced stabilization of nuclear p53 results in the transactivation of downstream target genes encoding, for example, the cyclin-dependent kinase inhibitor p21[cip1/waf1/sdi1] or Mdm2. p21[cip1/waf1/sdi1] inhibits RB phosphorylation, thereby preventing transition from G1 to S *(8)*. ES cells do not undergo cell cycle arrest in response to DNA damage (caused by γ-radiations, UV light, genotoxic agents) or nucleotide depletion *(9,10)*. ES cells express abundant quantities of p53, but the p53-mediated response is inactive because of cytoplasmic sequestration of p53. Moreover, enforced expression of nuclear p53 still fails to induce cell cycle arrest, suggesting that, in addition to its cytoplasmic sequestration, p53 cannot activate the downstream targets required for growth arrest

(9). One of these targets is p21[cip1/waf1/sdi1]. ES cells do not express p21[cip1/waf1/sdi1] *(2)*, suggesting that the p21[cip1/waf1/sdi1] promoter is not responsive to p53 in ES cells. Therefore, it appears that ES cells have a very effective mechanism for rendering them refractory to p53-mediated growth arrest. Induction of differentiation restores the p53-mediated cell cycle arrest response *(9)*.

Taken together, these results suggest fundamental differences in the regulation of cell proliferation in ES cells as compared to somatic cells. Firstly, they suggest that the complex apparatus that operates in most cells with extracellular mitogens, transducing signals through the SOS-RAS-RAF-MEKK-MAPK pathway and that ultimately leads to pRB phosphorylation is not engaged in ES cells. Induction of differentiation would reactivate this mechanism. Secondly, ES cells do not seem to have a p53-mediated checkpoint control. This control would also become operative when differentiation occurs.

In the second part of this chapter, we describe experimental procedures used to synchronize ES cells and to analyze their cell cycle distribution. These procedures have been developed to characterize the growth cycle of mouse ES cells.

2. Materials

1. Feeder-independent ES cell line: CGR8 *(11)*.
2. Complete medium: Glasgow's Modified Eagle's Medium (GMEM) (BioMedia, cat. no. GMEMSPE2052) supplemented with 10% fetal calf serum (FCS) (PAA, cat. no. A15-043), 2 mM L-glutamine (BioMedia, cat. no. GLUN2002012), 1% nonessential amino acid solution (BioMedia, cat. no. AANE0002012), 1 mM sodium pyruvate (BioMedia, cat. no. PYRU0002012), 0.1 mM 2-mercaptoethanol (Sigma, cat. no. M7522), 100 U/mL penicillin, 100 mg/mL streptomycin, and 1000 U/mL human leukemia inhibitory factor (LIF). For LIF preparation and testing (*see* **ref. 12**).
3. 0.25% (w/v) Trypsin in Phosphate-Buffered Saline (PBS).
4. 20 ng/mL Demecolcine (Sigma, cat. no. D6165).
5. 0.1% and 0.2% Gelatin (Sigma, cat. no. G9391) dissolved in H$_2$O.
6. 5 mM BrdU (Sigma, cat. no. B9285) (100× stock solution).
7. 1 mg/mL RNAse dissolved in PBS + 0.13 mM EGTA.
8. PBT: PBS + 0.5% Bovine Serum Albumin (BSA) (Sigma, cat. no. A2153) + 0.5% Tween-20 (Sigma, cat. no. P7949).
9. Anti-BrdU (Becton Dickinson, cat. no. 347583).
10. 100 µg/mL Propidium iodide (Sigma, cat. no. P4170) (100X stock solution).
11. Sterile flasks and Petri dishes: sterile 5- and 10-mL pipets.
12. FACScan (fluorescence-activated cell sorter) (Becton-Dickinson), equipped with a 15-mW 488-nm air-cooled argon-ion laser. Filters used: 530 nm fluorescein isothiocyanate (FITC), 585 nm (propidium iodide). Data aquisition and analysis are performed using CellQuest (Becton-Dickinson) software.

3. Methods
3.1. Synchronization of ES Cells by Mitotic Shake-Off

This protocol is intended to generate large numbers of synchronized ES cells exiting from mitosis, entering into G1, and then into S phase, synchronously.

1. At d 1, trypsinize ES cells and seed at a density of 20–30 million cells in 25 mL complete medium in T160 flasks coated with 0.2% gelatin (gelatin is added to flasks at least 2 h

before seeding the cells. Gelatin is thoroughly removed by aspiration just before seeding the cells). Incubate at 37°C in 7.5% CO_2 (*see* **Note 1**).

2. At d 2, add 50 mL complete medium (removing the exhausted medium is not necessary) and incubate overnight.

3. At d 3, ES cells should form a confluent layer. Check that each flask is confluent. Discard those in which empty spaces are visible, as isolated clumps of cells are likely to detach from the flasks during the shake-off procedure. Then:

   a. Remove the loosely attached cells by preshaking the flasks 5 times by hitting the flasks against the palm of the hand.

   b. Quickly aspirate the medium and replace it with 25 mL complete medium containing 20 ng/mL demecolcine (*see* **Note 2**). Incubate for 3–4 h at 37°C in 7.5% CO_2.

   c. Shake the flasks 5 times by hitting them against the palm of the hand. Collect the medium in 50-mL disposal plastic tubes. From this step on, sterile manipulation is not required.

   d. Spin mitotic cells at 500g for 5 min. Aspirate the medium. Invert the tubes on absorbing paper for 5 min.

   e. Gently resuspend each pellet with 1 mL of prewarmed demecolcine-free medium using a P1000 Gilson pipet. Do not pipet the cells up and down more than required to get a single-cell suspension. Fill the tubes with complete medium.

   f. Spin at 500g for 5 min. Discard the medium. Invert the tubes onto absorbing paper to dry.

   g. Gently resuspend each pellet with 1 mL of prewarmed medium and pool into a single tube. Count the cells. This procedure yields approx 2×10^6 mitotic cells/T160 flask (i.e., approx 1% of the total number of cells).

   h. Prepare a cell suspension containing approx 10^6 mitotic cells/mL. Seed 6-cm dishes (coated with 0.1% gelatin as described in **step 1**) with 5 mL cell suspension. Incubate at 37°C in 7.5% CO_2.

   i. Collect the cells at various time points and analyze them for cell cycle distribution as described in **Subheading 3.2.** Since mitotic cells usually take 4–5 h to attach to the dish, do not aspirate the medium. Any supplements should be added dropwise using 10X stock solutions (*see* **Note 3**).

3.2. Analysis of DNA Content in Synchronized ES Cells

As mitotic ES cells usually take 4–5 h to attach strongly to the Petri dish, the following protocol must be used to prepare a single-cell suspension suitable for flow cytometry:

1. Collect the cells by pipetting up and down approx 10 times with a P1000 Gilson to dissociate the loosely adherent cells. Trypsinization is not required (*see* **Note 4**). Transfer the suspension (>1 million cells) into a conical 15-mL tube.

2. Spin for 5 min at 500g. Discard the medium and wash in PBS. Repeat once.

3. After the last spin, resuspend cells in 100 μL of PBS. Pipet up and down with a P200 Gilson until clumps are no longer visible. Dropwise, add 1 mL of 70% ethanol at –20°C (1 drop/s to avoid formation of clumps of cells). Store the fixed cells at 4°C.

4. To analyze the DNA content, add 10 mL PBS directly to cells in ethanol. Incubate for 5 min at room temperature to allow cells to rehydrate.

5. Spin for 5 min at 500g. Resuspend the pellet in 100 μL of PBS. Add 10 mL PBS. Incubate for 5 min at room temperature.

6. Resuspend the pellet in 100 μL of 1 mg/mL RNase. Incubate for 20 min at room temperature. Store at 4°C if required (<24 h).

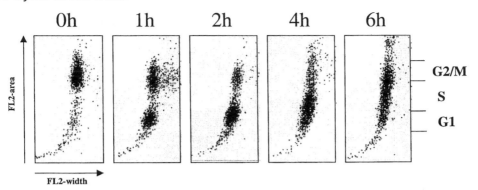

Fig. 1. Analysis of DNA content of ES cells synchronized by mitotic shake-off, determined according to the protocol described in **Subheading 3.2.** FL2-width indicates the size of the cells. FL2-area quantitates fluorescence associated to propidium iodide. 0h, 1h, 2h, 4h, and 6h indicate the time after release from the mitotic block.

7. Add propidium iodide to a final concentration of 1 µg/mL. Incubate for 5 min at room temperature.
8. Analyze fluorescence using conventional setups. The result of a representative synchronization experiment is given in **Fig. 1**.

3.3. Analysis of Cell Cycle Distribution in Exponentially Growing ES Cells

1. Refeed exponentially growing ES cells with complete medium. Incubate for 1 h at 37°C.
2. Add BrdU at a final concentration of 50 µM and incubate for 30 min.
3. Trypsinize the cells and take 5 million cells for analysis of BrdU incorporation.
4. Spin for 5 min at 500g. Discard the medium and wash in PBS. Repeat once.
5. After the last spin, resuspend the pellet of cells in 100 µL of PBS. Dropwise, add 1 mL of 70% ethanol at –20°C (1 drop/s to avoid formation of clumps of cells). Store the fixed cells at 4°C for up to several weeks.
6. Add 10 mL PBS to ethanol-fixed cells. Incubate for 5 min at room temperature to allow cells to rehydrate.
7. Spin for 5 min at 500g. Resuspend the pellet in 200 µL of 2 N HCl and incubate for 20 min at room temperature.
8. Wash 3–4 times in 10 mL PBT.
9. Resuspend the pellet in 100 µL FITC-conjugated antibody raised to BrdU (Becton Dickinson), diluted 1:10 in PBT, and incubate for 30 min at room temperature.
10. Wash 3–4 times in 10 mL PBT.
11. Resuspend the pellet in 100 µL of 1 mg/mL RNase. Incubate 20 min at room temperature. Store at 4°C if required (<24 h).
12. Add propidium iodide to a final concentration of 1 mg/mL. Incubate for 5 min.
13. Analyze fluorescence associated to FITC and to propidium iodide using conventional setups. The result of a representative experiment is given in **Fig. 2**.

4. Notes

1. Check carefully that T-flasks are horizontal in the incubator, as uniformity is essential for recovery of pure populations of mitotic cells.

Savatier et al.

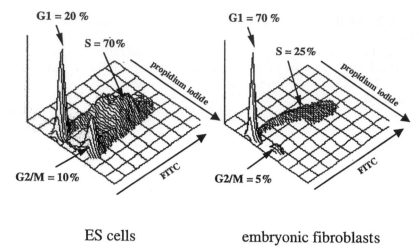

ES cells   embryonic fibroblasts

Fig. 2. Cell cycle distribution of ES cells and mouse embryonic fibroblasts, determined according to the protocol described in **Subheading 3.3.**

2. Do not leave the cells free of medium for more than 1 min, as this will lead to clumps of cells detaching during the shake-off procedure.
3. Following this protocol, one can obtain a population of ES cells containing >95% mitotic cells. Ninety percent of those cells will resume cell cycle progression within 1 h following incubation in demecolcine-free medium. ES cells will start entering the S-phase within 2 h, and the vast majority of them will be replicating their DNA at 4 h post-release from the mitotic block. Note that increasing the incubation time with demecolcine will result in a larger proportion of mitotic cells not being released from the mitotic block.
4. Collecting post-mitotic cells by pipetting up and down may lead to cell damage. Pipetting must be done very gently to avoid this, but for a long enough time to obtain a single-cell suspension suitable for cell cycle analysis. In any case, cell debris will be discarded during FACS analysis.

Acknowledgments

This work was supported by the Association pour la Recherche contre le Cancer (ARC) and comité Interdépartemental de la région Rhône-Alpes de la Ligue Nationale Française contre le Cancer.

References

1. Savatier, P., Huang, S., Szekely, L., Wiman, K. G., and Samarut, J. (1994) Contrasting patterns of retinoblastoma protein expression in mouse embryonic stem cells and embryonic fibroblasts. *Oncogene* **9,** 809–818.
2. Savatier, P., Lapillonne, H., van Grunsven, L. A., Rudkin, B. B., and Samarut, J. (1996) Withdrawal of differentiation inhibitory activity/leukemia inhibitory factor up-regulates D-type cyclins and cyclin-dependent kinase inhibitors in mouse embryonic stem cells. *Oncogene* **12,** 309–322.
3. Weinberg, R. A. W. (1995) The retinoblastoma protein and cell cycle control. *Cell* **81,** 323–330.
4. Bartek, J., Bartkova, J., and Lukas, J. (1997) The retinoblastoma protein pathway in cell cycle control and cancer. *Exp. Cell Res.* **237,** 1–6.

5. Kerkhoff, E. and Rapp, U. R. (1998) Cell cycle targets of Ras/Raf signalling. *Oncogene* **17,** 1457–1462.

6. Zhang, H. S., Gavin, M., Dahiya, A., Postigo, A. A., Ma, D., Luo, R. X., et al. (2000) Exit from G1 and S phase of the cell cycle is regulated by repressor complexes containing HDAC-Rb-hSWI/SNF and Rb-hSWI/SNF. *Cell* **101,** 79–89.

7. Harbour, J. W., Luo, R. X., Dei Santi, A., Postigo, A. A., and Dean, D. C. (1999) Cdk phosphorylation triggers sequential intramolecular interactions that progressively block Rb functions as cells move through G1. *Cell* **98,** 859–869.

8. Hansen, R. and Oren, M. (1997) p53: from inductive signal to cellular effect. *Curr. Opin. Genet. Dev.* **7,** 46–51.

9. Aladjem, M. I., Spike, B. T., Rodewald, L. W., Hope, T. J., Klemm, M., Jaenisch, R., and Wahl, G. M. (1998) ES cells do not activate p53-dependent stress responses and undergo p53-independent apoptosis in response to DNA damage. *Curr. Biol.* **29,** 145–155.

10. Prost, S., Bellamy, C. O., Clarke, A. R., Wyllie, A. H., and Harrison, D. J. (1998). p53-independent DNA repair and cell cycle arrest in embryonic stem cells. *FEBS Lett.* **425,** 499–504.

11. Nichols, J., Zevnik, B., Anastassiadis, K., Niwa, H., Klewe-Nebenius, D., Chambers, I., et al. (1998) Formation of pluripotent stem cells in the mammalian embryo depend on the POU transcription factor Oct4. *Cell* **95,** 379–391.

12. Smith, A. G. (1991) Culture and differentiation of embryonic stem cells. *J. Tiss. Cult. Methods* **13,** 89–94.

4

Murine Embryonic Stem Cells as a Model for Stress Proteins and Apoptosis During Differentiation

André-Patrick Arrigo and Patrick Mehlen

1. Introduction

The early phase of murine embryonic stem (ES) cell differentiation is accompanied with the expression of proteins that are key players in this phenomenon. The function of some these proteins can be analyzed by using the classical "knock-out gene technology", that consists of the inactivation of the endogenous gene that encodes the studied protein. This approach can be considered if the analyzed protein does not play a vital role; in other words, if the cells can still undergo differentiation without expressing the protein in question. To analyze proteins whose expression is essential during early differentiation, an antisense strategy can be used that interferes with the expression of the studied protein. One interesting protein we studied with this technique is the small stress protein Hsp27 *(1)*.

Hsp27 is a member of the so-called heat shock or stress proteins whose synthesis is stimulated when cells are exposed to heat shock or other type of aggression *(1)*. Hsp27 and the other small stress proteins (sHsp) are molecular chaperones that share the property to be transiently expressed during early cell differentiation. This phenomenon has been observed in all the organisms analyzed so far and does not appear to be cell specific *(1)*. This ubiquitous phenomenon usually occurs concomitantly with the differentiation-mediated decrease of cellular proliferation. In contrast, during development, the different sHsp show a tissue-specific pattern of expression *(1)*. For example, Hsp27 is transiently expressed through d 13 to 20 of mouse development, where it accumulates in neurons, brain, and muscles. In young mice, this protein is no more detectable in the brain, but is still present in different muscles.

In undifferentiated mouse ES cells, a small level of constitutively expressed Hsp27 can be observed. In cells committed to differentiation, the Hsp27 level increases and is maximal after 24 h when DNA synthesis is decreased by about 80%. After 72 h of differentiation, the Hsp27 level is drastically decreased, and this protein is almost undetectable. We have analyzed the role of the transient expression of Hsp27 during early differentiation by transfecting ES cells with an expression vector containing the murine or human Hsp27 gene placed either in sense or antisense orientation. ES clones either overexpressing murine or human Hsp27 or expressing reduced levels of

From: *Methods in Molecular Biology, vol. 185: Embryonic Stem Cells: Methods and Protocols*
Edited by: K. Turksen © Humana Press Inc., Totowa, NJ

endogenous Hsp27 were obtained. Maintained undifferentiated, these clones showed similar growth rates. However, when these cells were committed to differentiate, we observed that Hsp27 constitutive overexpression enhanced the differentiation-mediated decreased rate of ES cell proliferation, but had no effect at the level of the morphological changes that characterize ES cell differentiation. In contrast, Hsp27 underexpression, which attenuated cell growth arrest, aborted the differentiation process of ES cells because of their overall death by apoptosis. Hsp27 transient expression appears, therefore, essential for preventing differentiating ES cells from undergoing apoptosis *(2)*. Our results also suggest that the expression of Hsp27 during early differentiation of ES cells is not related to a specific type of cell differentiation. Similar aborting or aberrant differentiation processes were observed when Hsp27 expression was impaired in rat olfactory precursor neurons *(3)* or human leukemic HL-60 cells *(4)*.

In this chapter, we describe the procedure for the generation of murine ES cells that are genetically modified to underexpress or overexpress Hsp27. This procedure has been developed in our laboratory and could be used in any well-established cellular laboratory. In addition, the procedure could be extended to test proteins other than Hsp27, which are expressed during early differentiation of ES cells.

2. Materials

All reagents and materials used in the culture, differentiation, and transfection of ES cells are sterile.

1. Phosphate-buffered saline (PBS) medium: 137 mM NaCl, 2.7 mM KCl, 8 mM Na$_2$HPO$_4$, and 1.5 mM KH$_2$PO$_4$, pH 7.4.
2. Murine ES cells CGR8 (early passages 18–30) were obtained from A. Smith (Center of Genome Research, University of Edinburgh, U.K.) *(5)*.
3. CGR8 cell culture medium: BHK21 medium (Gibco BRL, cat. no. 22100-028) containing 0.05 mM β-mercaptoethanol, 2 mM L-glutamine, 1 mM sodium pyruvate, 100 U/mL penicillin, 0.1 mg/mL streptomycin, 1× minimum essential medium (MEM) (Gibco BRL, cat. no. 31095-060), leukemia inhibitor factor (LIF), and 10% fetal calf serum (FCS) (Gibco BRL, cat. no. 10084-168). Conditioned medium from p10-6R DIA-LIF transfected COS cell line was used as a source of LIF (dilution 1/1000) (*see* **ref. 5**). The cell culture medium was stored at 4°C without serum and used within 5 wk of supplementation. Store FCS at –70°C. The medium containing the serum is made fresh and used within 1 wk of supplementation (storage at 4°C). Geneticin G418 (Gibco BRL, cat. no. 11811-049) must be added to the medium to select transfected clones.
4. Trypsination buffer: 0.4g EDTA, 7.0g NaCl, 0.3g Na$_2$HPO$_4$ (12 H$_2$O), 0.24g KH$_2$PO$_4$, 0.37g KCl, 1.0g D-glucose, and 3.0g Tris-HCl were dissolved in water. Adjust the pH to 7.6 with diluted HCl/NaOH, warm at 37°C, and add 2.5g of trypsin (Gibco BRL, cat. no. 35400-027). The final volume is 1 L. The solution is filtered on 0.22-μm filters before adding 1% chicken serum (Gibco BRL, cat. no. 16110-033). Aliquot the solution and store at –20°C. Thaw just before use and discard the buffer that was not used.
5. Gelatinized flasks are used to allow CGR8 cells adhesion and growth. One h before plating the cells, 10 mL of 0.1% autoclaved gelatin is added to Falcon 100-mm dishes. After gelatin removal, the cells are plated in BHK21 medium.
6. Bacterial grade Petri dishes (Bibby Sterilin, Ltd.) are used when differentiation is induced.

7. Mammalian expression vector psvK3 (Pharmacia), plain or bearing human (pSVHsp27; *see* **ref. 6**) or murine (psvWT; *see* **ref. 7**) HSp27 cDNAs, under the control of the early promoter of simian virus 40 (SV40) virus. HSp27 antisense expression vectors (*see* **ref. 2**) were constructed by subcloning *Eco*RI-*Eco*RI or *Sac*I-*Sac*I DNA fragments containing the entire human or murine Hsp27 coding sequences, respectively, in reverse orientation in the corresponding site of psvK3 plasmid polylinker. These expression vectors were denoted psvant-hhsp27 (human Hsp27) and psvant-mhsp27 (murine Hsp27).

8. pC1-neo plasmid (Promega) encoding the G418 resistance was used for cotransfection and subsequent selection of positive clones.

9. Purified proteins: recombinant human Hsp27 (Stressgen, cat. no. SPP-715) and recombinant murine Hsp27 (also denoted Hsp25) (Stressgen, cat. no. SPP-710). Antibodies: anti-human Hsp27 (Stressgen, cat. no. SPA-800), anti-murine Hsp27 (Stressgen, cat. no. SPA-801), anti-Hsp70 (Stressgen, cat. no. SPA-820), and anti-Hsp90 (Stressgen, cat. no. SPA-830).

10. Optimem medium (Gibco BRL, cat. no. 31985-047).

11. Cell electroporation apparatus (i.e., gene pulser from Bio-rad).

12. [^{3}H]-thymidine (Amersham International).

13. Hemocytometer chamber.

14. Inverted photomicroscope equipped with a phase-contrast equipment (i.e., TMS Nikon).

15. RNase A (Sigma, cat. no. R 4875).

16. Propidium iodide solution: 50 mg/mL of propidium iodide (Sigma, cat. no. P 4170) in PBS.

17. Flow cytometry apparatus (Facscalibur, Becton Dickinson).

18. Trizol reagent (Gibco, cat. no. 15596-026).

19. Dithiothreitol (DTT) (Gibco, BRL, cat. no. 15508-013).

20. Reverse-transcription (RT) medium: 50 mM Tris-HCl, pH 8.3, 10 mM DTT; 1 U/mL Rnasin (Promega, cat. no. N2511), 0.5 mM random hexamers (Pharmacia, cat. no. 27-2166-01), 0.5 mM of each dNTP (Pharmacia, cat. no. 27-2035-01), 75 mM KCl, 5 mM MgCl$_2$, and 20 U Moloney murine leukemia virus (MMLV) reverse transcriptase (Gibco-BRL, cat. no. 28025-013).

21. Polymerase chain reaction (PCR) medium: 20mM Tris-HCl, pH 8.3, 50 mM KCl, 1.5 mM MgCl$_2$, 0.2 mM of each dNTP, 0.2 mM of each primer, and 1 U of *Taq* DNA polymerase (Gibco-BRL, cat. no. 18038-042).

22. Hypoxanthine phosphoribosyl transferase (HPRT)-, collagen IV-, or β-major globin-specific primers are as follows:
    HPRT: 5′ CAC AGG ACT AGA ACA CCT GC 3′ and 5′ GCT GGT GAA AAG GACCTCT 3′.
    collagen IV: 5′ CAA GCA TAG TGG TCC GAG TC 3′ and 5′AGG CAG GTC AAG TTC TAG CG 3′.
    β-major globin: 5′ CTG ACA GAT GCT CTC TTG GG 3′ and 5′CAC AAC CCC AGA AAC AGA CA 3′.

23. Thermal cycler (Perkin Elmer Cetus).

24. Agarose (Pharmacia, cat. no. 17-0554-01).

25. Tris-borate-EDTA buffer for nucleic acids gel electrophoresis: 89 mM Tris-base, 89 mM boric acid, pH 8.0, 2 mM EDTA, pH 80.

26. Nucleic acids precipitation reagents: phenol-chloroform, sodium acetate, ethanol.

27. Formaldehyde (Sigma, cat. no. F 1635).

28. Triton X-100 (Sigma, cat. no. T 9284).

29. Hybond C extra membrane (Amersham, cat. no. RPN 303E).

30. X-Omat AR films (Eastman Kodak Co, cat. no. 1651454).

31. Bioprofil scanning system (Vilber Lourmat).

32. Trypan blue (Sigma, cat. no. T 0776).
33. Ethidium bromide (Sigma, cat. no. E 8751).
35. Phosphorothioate oligodeoxynucleotides for antisense inhibition of endogenous murine Hsp27 expression: the sequence of the antisense oligonucleotide was 5'-TCGGT**CAT** GTTCTTGGCTGGT-3' (the underlined bold letters indicate the position of the corresponding AUG initiation codon). Control oligonucleotide is the reverse antisense sequence to the mentioned antisense oligodeoxyribonucleotide (synthesized in the opposite 5' to 3' orientation compared to antisense oligophosphothioates, so that it contains the same bases in the reverse order as the antisense oligonucleotide). As an additional negative control, an antisense oligonucleotide of vesicular somatistis virus (VSV): 5'-TTGGGATA-ACACTTA-3' should also be used.
36. Permeabilization buffer (100 mM piperazine-N-N'-bis [2-ethanesulfonic acid], 1,4 piperazine diethanesulfonic acid [PIPES], pH 7.4, 137 mM NaCl, 5.6 mM glucose, 2.7 mM KCl, 2.7 mM EGTA, 1 mM ATP, 0.1% bovine serum albumin [BSA]) containing 0.2 U/mL streptolysin-O (Sigma, cat. no. S 5265).
37. Scintillation counter (LS6000 SC; Beckman).
38. Glass microfibre filters (GFC filters; Whatman, cat. no. 1822-025).
39. Detection of immune reaction in western immunoblot using the electrochemiluminescence (ECL) kit from Amersham (cat. no. RPN 2106).

3. Methods

3.1. ES Cell Cultures and Induction of Their Differentiation

The stem cell cultures are handled as follow:
Undifferentiated CGR8 cells are grown on gelatinized (*see* **Subheading 2., item 5**). flasks in BHK21 cell culture medium containing LIF. Culture medium is changed every 2 d adding fresh LIF. Observe under the microscope for the outgrowth of cells. As the cultures proliferate, aliquots of the cells are cryopreserved at early passage (<30) (i.e., in the log phase of growth). CGR8 cultures should be examined in an inverted tissue culture microscope every other day to monitor the growth of the cells, their morphology, and any signs of poor growth, unwanted differentiation, or contamination.
When cells are 80–100% confluent proceed as follows to subculture the cells:

1. Aspirate by pipet the medium from the cell cultures.
2. Wash with PBS.
3. Add 1 mL of trypsin-EDTA solution (*see* **Subheading 2.**) to the culture flasks. Let the flasks stand for 2–3 min at 37°C and monitor under the microscope for the detachment of cells.
4. Remove the trypsin-EDTA solution containing the detached cells from the flask and place it in a conical centrifuge tube. Rinse the flasks with the complete culture medium and add it to the centrifuge tube. Serum is required to inhibit the action of trypsin.
5. Centrifuge the cell suspension at 1500g for 3 min and resuspend the cell pellet in 10 mL of complete culture medium. Count cells and record for their viability using trypan blue staining exclusion assay.
6. Cell cultures are splitted by 1×10^6 cells/100-mm dish.
7. Add the LIF (*see* **Subheading 2.**).

3.2. In Vitro Differentiation of ES Cells

To induce differentiation, CGR8 cells are dissociated by trypsinization and seeded at 3×10^5 cells/mL on bacterial grade Petri dishes in LIF devoid growth medium.

1. Induction of differentiation is performed as follows:
   a. Trypsinize the cells as described previously.
   b. Cells are resuspended in complete BHK-21 medium devoid of LIF and numerated.
   c. Cells are then plated in bacterial grade petri dishes at a concentration of 3×10^6 cells/100-mm dish.
   d. CGR8 cell cultures are examined in an inverted tissue culture microscope every other day to monitor cell differentiation. After 48 h, CGR8 cells begin to form cluster of unadherent cells called "cystic embryoid bodies". After 4 to 8 d, the clusters dissociate and the cells begin to attach to the plate. Channel boundary precursors of the cystic automatic myocardiac cells can then be observed *(8)*.

3.3. Detection of Differentiation Markers and Analysis of Hsp27 Transient Accumulation

ES cell differentiation can be monitored by the detection of specific mRNA markers. In our hands, collagen IV and β-major globin mRNAs are the best candidates that can be detected by RT-PCR analysis respectively 96 and 120 h after LIF removal. RT-PCR analysis of HPRT mRNA was used as a control.

3.3.1. RT-PCR Detection of Differentiation Markers

1. Total RNA from 4 dishes per condition (1.2×10^7 cells from undifferentiated CGR8 cells or CGR8 cells derived from differentiating embryoid bodies) is prepared with TRIzol reagent according to manufacturer instructions.
2. RT-PCR analyses were performed as follows: 1 µg (representing about 7500 cells) of total RNA is denatured for 10 min at 65°C. RT is then performed for 1 h at 37°C in a specific medium (*see* **Subheading 2.**) containing MMLV reverse transcriptase. Reactions were stopped by incubating the mixtures for 5 min at 95°C. For PCRs, 5 µL of the RT reactions were transferred in a PCR medium. HPRT-, collagen IV-, or β-major globin-specific primers are used (*see* **Subheading 2.**). PCRs were carried out with a thermal cycler for 30 cycles with the following regimen: 94°C for 1.5 min, 50°C (for HPRT), or 55°C (for collagen IV and globin β-major) for 1.5 min, and 72°C for 2 min, followed by 72°C for 10 min. Aliquots (10 µL) of each of the PCRs were analyzed by electrophoresis in agarose (2%) in Tris-borate-EDTA buffer.

3.3.2. Monitoring Hsp27 Expression During ES Cell by Northern and Western Blot Analysis

1. Northern blot: RNA (10 eg µg) was analyzed in 1% agarose/formaldehyde gel, transferred to Hybond C extra membrane, and hybridization is performed at 65°C according to the classical method described in Yang et al. *(9)*. The murine Hsp27 probe was a 0.6-kb *Sac*I cDNA fragment of the psvWT plasmid.
2. Western blot: gel electrophoresis and immunoblots using Hsp27 or Hsp70 antisera are performed as already described *(6,10)* and revealed with the ECL kit from Amersham.
3. Autoradiographs from the northern and western blot experiments are recorded onto X-rays films (*see* **Note 2**). A Bioprofil system (*see* **Subheading 2.**) is used for quantification.

3.4. Transfection of CGR8 Cells and Clones Selection

To establish stable transfectants of ES cells that under-express Hsp27, we have used an antisense strategy. Undifferentiated LIF-maintained CGR8 cells in the log phase of growth are transfected using an electroporation technique with vectors encoding the Hsp27 gene, cloned either in sense or antisense orientation, and a gene for selection, such as neomycine phosphotransferase. Stable transfectants are then selected for resistance to neomycine by culture in the presence of G418, a neomycin analog that is active in eukaryotic cells. Hsp27 expression is then analyzed in G418-resistant clones.

Transfection and selection procedures were performed as follows:

3.4.1. Transfection Procedure

1. DNA mixtures made of 10 μg of pC1-neo plasmid and 50 μg of either psvK3, psvant-hhsp27, psvant-mhsp27, psvhsp27, or psvWT vector (*see* **Subheading 2.**) are used to transfect the cells.
2. Linearized DNA (50 μg) from the different DNA mixtures are resuspended in 30 μL of H$_2$O before being added to a Bio-Rad gene pulser cuvette containing 800 μL of Optimem medium and 5×10^6 cells.
3. The mixture is incubated for 10 min at room temperature. Thereafter, the cuvette is placed in the gene pulser apparatus and exposed to an electric shock of 250 V, 500 mF, for 6 ms.
4. Cells are then allowed to recover for 30 min before being plated on gelatinized dishes in BHK-21 medium containing LIF.

3.4.2. Selection of G418-Resistant Clones

1. Forty-eight hours after the electroporation procedure, 250 μg/mL of G418 are added to the culture, and resistant clones are isolated 10 d later.
2. Hsp27 level is analyzed in G418-resistant clones by Western blot analysis.
3. Several independent clones are frozen down.

3.5. DNA Oligonucleotide-Based Antisense Inhibition of Hsp27 Expression

Another possibility to interfere with Hsp27 expression is to take advantage of antisense phosphorothioate-modified oligodeoxynucleotides. This technique is based on the fact that oligonucleotides that have a complementary and reverse sequence to a specific mRNA can hybridize with this messenger and interfere with its translation. This phenomenon may then lead to a decreased level of the polypeptide encoded by the targeted messenger. The antisense oligonucleotide corresponding to the first nucleotides of the coding region of murine Hsp27 gene is used (*see* **Subheading 2.**). The sense oligonucleotide and an antisense oligonucleotide of vesicular somatistis virus (VSV) are also used as negative controls. The oligonucleotides, purified by sodium acetate–ethanol precipitation, were resuspended in sterile double-distilled water.

3.5.1. Introduction of DNA Oligonucleotides into Living Cells

1. Undifferentiated LIF-maintained CGR8 cells (10^6) in the log phase of growth are washed in PBS medium, trypsinized, and resuspended in 1 mL of permeabilization buffer (*see* **Subheading 2.**) containing 0.2 U/mL streptolysin-O and 50 μM (final concentration) of oligonucleotides.
2. Cells are incubated for 5 min at room temperature in permeabilization buffer before adding 5 mL of complete growth medium (containing LIF) to the mixture to stop the permeabilization process.
3. Cells are then spun and resuspended in 3-d-old conditioned growth medium containing 10 μM oligonucleotide *(11)*. After 2 d of incubation, cells are replated in fresh medium and processed for Hsp27 expression analysis.

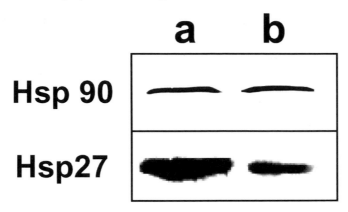

Fig. 1. Reduced Hsp27 level induced by Hsp27 antisense oligonucleotide. Cells are treated with streptolysin-O in the presence (**b**) or not (**a**) of Hsp27 antisense oligonucleotide. Cells are then grown for 2 d in the presence or not of 10 µmol/L of the oligonucleotide before being analyzed. Thereafter, equal amounts of total cellular proteins are analyzed in immunoblots probed with antisera specific for Hsp27 or Hsp90. Note the reduced level of Hsp27 in cells containing Hsp27 antisense oligonucleotide. Hsp27 level is unaltered when control sense or VSV oligonucleotides are analyzed. No alteration of Hsp90 level is induced by Hsp27 antisense oligonucleotide.

4. Cells showing a reduced level of Hsp27 expression are then incubated in LIF-deprived medium to study their behavior during differentiation. An illustration of a decreased expression of Hsp27 is shown in **Fig. 1**.

3.6. Cellular Proliferation and Cell Cycle Analysis

Cellular proliferation is an important parameter to analyze in undifferentiated CGR8 cells expressing different levels of endogenous Hsp27. Moreover, the early phase of the differentiation process is characterized by a rapid decrease in cell proliferation. Cellular proliferation and cycle analysis are then monitored as follows:

1. Undifferentiated cells or cells undergoing differentiation are trypsinized (*see* **Subheading 3.2.**), resuspended in PBS, and counted using an hemocytometer chamber and an inverted photomicroscope equipped with a phase-contrast equipment.
2. Cells can also be labeled for 1 h with 1 µCi [^{3}H]-thymidine to study their ability to synthesize DNA. Cells are labeled for 15 min with 10 µCi/mL [^{3}H] thymidine (88 Ci/mmol) before being washed with PBS and lyzed in 0.5% sodium dodecyl sulfate (SDS). Trichloroacetic acid precipitation is performed and collected on glass microfibre filters. Dried filters are counted on scintillation counter.
3. Cell cycle analysis can be performed essentially as follows: undifferentiated cells are trypsinized and washed twice with PBS. Resuspend 10^5 cells in 100 µL of PBS and add slowly (drop by drop) 2 mL of 70% ethanol kept at –20°C. After this fixation step, the cells can be stored for 1 wk at 4°C. Add 8 mL of PBS and spin down the fixed cells. Wash the cell pellet twice with PBS and resuspend the fixed cells with 1 mL of PBS. Add RNase A (0.1 mg/mL final concentration) and incubate for 30 min at room temperature. Then add 2 µL of a 50 mg/mL solution of propidium iodide (10 µg/mL final concentration). Analyze the cells immediately by flow cytometry (the recording of fluorescence should be positioned on the FL2-A setting). This last technique is difficult to perform with

differentiating ES cells because of their strong tendency to aggregate (formation of embryonic bodies).

3.7. Cell Death Analysis

Cell death can be monitored using different techniques. One technique, which does not discriminate between necrosis and apoptosis, is the staining of dead cells by the vital dye trypan blue Cells (3×10^6 (either undifferentiated or differentiating) are used. At different time points, before or after (e.g., 12, 24, 48, 72, and 120 h) removing the LIF containing medium, cells were trypsinized (*see* **Subheading 3.2.**), washed in PBS, and analyzed for their viability using trypan blue staining exclusion or absence of DNA fragmentation.

1. Trypan blue cell staining demonstrating membrane permeabilization is used to estimate the level of cell death. Add 0.5 mL of trypan blue solution (0.4% in PBS) in a test tube. Add 0.2 mL of cells (10^4 to 5×10^4) resuspended in PBS to the test tube. Incubate at room temperature for 10–15 min. Avoid incubating longer because, with time, viable cells may also be stained by the dye. Transfer a small quantity of the solution to a hemacytometer and count the stained (nonviable cells) and nonstained (viable) cells using an inverted microscope (*see* **Fig. 2A**).
2. To specify apoptosis induction, DNA fragmentation analysis can be performed essentially as follows: cells (3×10^6) are harvested, washed in PBS, and lyzed for 20 min at 4°C in a medium containing 5 mM Tris buffer, pH 7.4, 0.5% Triton X-100, and 20 mM EDTA. After centrifugation at 20,000*g* for 15 min, the supernatants are extracted with phenol-chloroform, and nucleic acids are precipitated in ethanol before being analyzed by gel electrophoresis (1.5% agarose) in Tris-borate buffer. After completion of the electrophoresis run, the agarose gel is incubated for at least 3 h at 37°C in water containing 20 µg/mL of RNase A. This treatment eliminates most of the contaminating RNA from the gel. DNA ladders can then be observed following ethidium bromide staining of the gel *(12)*. An illustration of the anti-apoptotic effect of Hsp27 is presented in **Fig. 2B** (*see also* **Note 1**).

4. Notes

1. There are several new methods that can be used for apoptosis detection. However, the use of caspases cleavage (or cleavage of substrates of caspases) can be tricky if the experiment is performed using the Western immunoblot technique. Indeed, most of the commercially

---

Fig. 2. *(opposite page)* Enhanced apoptosis in cells that contain down-regulated Hsp27 level. **(A)** Analysis of cell viability. Control (control-1) or Hsp27 under-expressing (ant-hsp27-2) cells growing in the presence of LIF are trypsinized and replated (3×10^6 cells/100-mm dish) in a medium devoid of LIF. Cell viability is determined daily by trypan blue exclusion. The percentage of survival cells (% trypan blue negative cells) is calculated by dividing the number of trypan blue negative cells to the total number of cells. Standard deviations are indicated ($n = 3$). **(B)** DNA fragmentation analysis. Control and Hsp27 under-expressing cells are treated as above, and DNA fragmentation is determined prior to 0 h and after 12, 24, or 48 h of LIF withdrawal. M: molecular marker. Note that the down-regulation of Hsp27 induces an early and intense DNA fragmentation in cells grown in the absence of LIF. Reproduced from **ref. 2** by copyright permission of the American Society for Biochemistry and Molecular Biology.

A

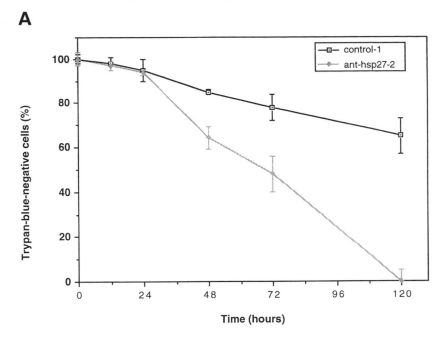

B

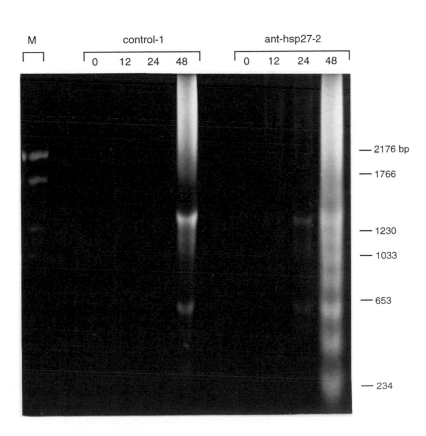

available antibodies specific for caspases (i.e., caspase 3) or caspase substrates will poorly recognize murine proteins. Most of these antibodies work well with human proteins but not with murine proteins. In contrast, the methods that detect caspase cleavage through the use of colorimetric or fluorogenic polypeptides could be used.

2. The level of Hsp27 polypeptide expressed in the different CGR8 cell lines is estimated in immunoblots by comparing the different signals to that of serial dilutions of the purified recombinant protein. Autoradiography must be performed within the range of proportionality of the film.

References

1. Arrigo, A.-P. (1998) Small stress proteins: chaperones that act as regulators of intracellular redox and programmed cell death. *Biol. Chem. Hoppe Seyler* **379,** 19–26.
2. Mehlen, P., Mehlen, A., Godet, J., and Arrigo, A.-P. (1997) Hsp27 as a shift between differentiation and apoptosis in embryonic stem cells. *J. Biol. Chem.* **272,** 31657–31665.
3. Mehlen, P., Coronas V., Ljubic-Thibal V., Ducasse, C., Grangier, L., Jourdan, F., and Arrigo, A.-P. (1999) Small stress protein Hsp27 accumulation in differentiating olfactory neurons counteracts apoptosis. *Cell Death Differ.* **6,** 227–234.
4. Chaufour, S., Mehlen, P., and Arrigo, A.-P. (1996) Transient accumulation, phosphorylation and changes in the oligomerization of Hsp27 during retinoic acid-induced differentiation of HL-60 cells: possible role in the control of cellular growth and differentiation. *Cell Stress Chaperones* **4,** 225–235.
5. Smith, A. G., Heath, J. K., Donalson, D. D., Wong, G. G., Moreau, J., Stahl, M., and Rogers, D. (1988) Inhibition of pluripotential embryonic stem cell differentiation by purified polypeptides. *Nature* **336,** 688–690.
6. Mehlen, P., Preville, X., Chareyron, P., Briolay, J., Klemenz, R., and Arrigo, A.-P. (1995) Constitutive expression of human hsp27, *Drosophila* hsp27 and human α-Bcrystallin confers resistance to tumor necrosis factor- and oxidative stress-induced cytotoxicity in stably transfected murine L929 fibroblasts. *J. Immunol.* **154,** 363–374.
7. Knauf, U., Jakob, U., Engel, K., Buchner, J., and Gaestel, M. (1994) Stress- and mitogen-induced phosphorylation of the small heat shock protein Hsp25 by MAPKAP kinase 2 is not essential for chaperone properties and cellular thermoresistance. *EMBO J.* **13,** 54–60.
8. Wang, R., Clark, R., and Bautch, V. L. (1992) Embryonic stem cell-derived cystic embryonic bodies from vascular channels: an in vitro model of blood vessel development. *Dev.* **114,** 303–316.
9. Yang, H., McLeese, J., Weisbart, M., Dionne, J. L., Lemaire, I., and Aubin, R. A. (1993) Simplified high throughput protocol for northern hybridization. *Nucleic Acids Res.* **21,** 3337–3338.
10. Arrigo, A.-P., Suhan, J., and Welch, W. J. (1988) Dynamic changes in the structure and locale of the mammalian low molecular weight heat shock protein. *Mol. Cell. Biol.* **8,** 5059–5071.
11. Barry, E. L. R., Gesek, F. A., and Friedman, P. A. (1993) Introduction of antisense oligonucleotides into cells by permeabilization with streptolysin O. *BioTechniques* **15,** 1016–1020.
12. Hockenbery, D. M., Oltvai, Z. N., Yin, X. M., Milliman, C. L., and Korsmeyer, S. T. J. (1993) Bcl2 functions in an antioxidant pathway to prevent apoptosis. *Cell* **75,** 241–251.

5

Effects of Altered Gene Expression on ES Cell Differentiation

Yong Fan and J. Richard Chaillet

1. Introduction

Little is known about the molecular mechanisms controlling the earliest cellular differentiation events of the mammalian embryo. Pluripotent embryonic stem (ES) cell lines, derived from cells of the inner cell mass of the blastocyst, are an important experimental system that can be used to study the differentiation of the mammalian embryo into its earliest recognizable tissue lineages. For instance, experiments performed in ES cells demonstrate the essential role of the homeobox-containing gene *Oct3/4* in the totipotent cells of the preimplantation embryo *(1)*. When two alleles of *Oct3/4* are mutated in ES cells, *Oct3/4*-null cells lose their pluripotency and differentiate into only trophoblast cells. In contrast, when *Oct3/4* is overexpressed in ES cells, spontaneous differentiation occurs and a variety of mesoderm-specific and extraembryonic endoderm-specific genes are expressed. Similar results have been obtained from in vivo experiments using *Oct3/4* knock-out mice *(2)*. Mouse preimplantation embryos, with two null alleles of *Oct3/4*, develop into blastocysts containing primary trophoblast giant cells but lacking inner cell mass cells. Taken together, the in vitro and in vivo studies of *Oct3/4* function indicate that *Oct3/4* is one of the key factors regulating the differentiation of totipotent cells of the preimplantation embryo.

Like *Oct3/4*, the X-linked homeobox-containing gene *Pem* is present in the mouse preimplantation embryo. Following implantation, *Pem*'s expression is largely restricted to extraembryonic tissues (trophoblast and extraembryonic endoderm), with limited expression in the embryonic portion of the mouse embryo *(3)*. Interestingly, *Pem* and *Oct3/4* are also both expressed in undifferentiated primordial germ cells (PGCs) in the mouse embryo. When PGCs begin to differentiate into either male or female germ cells at approx 15 d of gestation, both *Pem* and *Oct3/4* expression are rapidly downregulated. Because of the pattern of *Pem* expression in the mouse, and its abundant expression in ES and embryonal carcinoma (EC) cell lines *(4)*, *Pem* could very likely regulate the differentiation of totipotent cells of the mouse embryo. However, in the absence of an abnormal phenotype in *Pem*-null mice, the exact function of *Pem* would be difficult to determine by further in vivo experiments *(5)*.

From: *Methods in Molecular Biology, vol. 185: Embryonic Stem Cells: Methods and Protocols*
Edited by: K. Turksen © Humana Press Inc., Totowa, NJ

Alterations of genes expressed in ES cells can be a powerful method to evaluate the function of specific genes. Therefore, to determine the role of *Pem* in early mouse development, we altered *Pem*'s expression in ES cells and studied their differentiation during their growth as embryoid bodies *(6)*. In many ways, embryoid bodies mimic normal mouse embryonic development, differentiating over the course of a few days into both embryonic and extraembryonic cell types *(7)*. To overexpress *Pem* in D3 ES cells, we generated *Pgk-Pem* ES lines that contain a transgene expressing *Pem* from the mouse phosphoglycerate kinase (*Pgk1*) promoter *(6)*. To reduce *Pem* expression, we used standard targeted mutagenesis techniques to knock out the single *Pem* allele in male D3 ES cells, thereby generating *Pem*$^{-/Y}$ ES lines. We measured the levels of expression of a number of tissue-specific genes to assess the differentiation of *Pgk-Pem* and *Pem*$^{-ly}$ cell lines (in a comparison to normal D3 ES cells). Interestingly, the forced overexpression of *Pem* in *Pgk-Pem* cells blocked ES cell differentiation, whereas ablation of *Pem* gene expression in *Pem*$^{-/Y}$ cells caused defects and/or delays in primitive endoderm differentiation *(6)*. The absence of differentiated cell types in experimentally derived teratomas from *Pgk-Pem* cells substantiated the in vitro results. Moreover, *Pgk-Pem* cells did not differentiate either in vitro or in vivo when mixed with normal ES cells, indicating a cell autonomous role of *Pem* in mouse development. Here, we describe the protocols used to measure the effects of altered *Pem* gene expression on ES cell differentiation. Standard techniques were used for establishing *Pem*-null ES cells, and these are not described here.

2. Materials

All tissue culture reagents are from Gibco-BRL unless noted.

1. Fetal bovine serum (FBS) (cat. no. 26140-079) was heat inactivated at 56°C for 30 min, and stored at –20°C as 37.5 mL aliquots in 50-mL sterile plastic centrifuge tubes (Corning, New York, cat. no. 25330-50).
2. Dulbecco's modified Eagle's medium (DMEM) (high glucose with L-glutamine; cat. no. 11965-092) is stored at 4°C.
3. Leukemia inhibitory factor (LIF) (Chemicon International, cat. no. ESG1107) was purchased as 1×10^7 U/mL in a rubber-capped vial. Use a 1-mL syringe with a needle to push 1 mL of air into the bottle and pull out all the liquid. Aliquots of 100 µL each (1×10^6 U) are stored in sterilized cryotubes at –80°C. Before use, one tube (containing 1×10^6 U) is thawed, and the entire contents is added to 10 mL DMEM with 15% FBS. This stock has a LIF concentration of 1×10^5 unit/mL (100X) and can be stored up to 6 mo at 4°C.
4. Geneticin (G418) (cat. no. 10131-035; with 50 mg/mL active component) is treated as 200X and stored at 4°C.
5. 100 m*M* MEM Sodium Pyruvate solution (100X, cat. no. 11360-070) and nonessential amino acids (100X; Specialty Media, N.J., cat. no. TMS-001-C) are stored at 4°C.
6. L-Glutamine (200 m*M*; 100X; cat. no. 25030-081) and penicillin–streptomycin (100X, 10,000 U each; cat. no. 15140-122) are stored in 15-mL centrifuge tubes (Corning, cat. no. 430052) as 5 mL aliquots at –20°C.
7. 2-Mercaptoethanol (5.5×10^{-2}, used as 500X; cat. no. 21985-023) is stored for up to 1 yr at 4°C.

8. 20X Phosphate-buffered saline (PBS) stock solution: 160 g NaCl, 4 g KCl, 23 g Na_2HPO_4, 4 g KH_2PO_4. Bring to 1 L with dH_2O. Autoclave and store at room temperature. Prior to use, dilute to 1X PBS (pH 7.2) with dH_2O and autoclave. Warm to 37°C before use.

9. Trypsin-EDTA (10X): 0.5% trypsin and 5.0 mM NaEDTA (cat. no. 10131-035) is stored in 10 mL aliquots at –20°C. Prior to usage, one tube is thawed and diluted to 1X with 90 mL of 1X PBS.

10. ES cell culture medium: DMEM containing 15% FBS, 1 mM sodium pyruvate, 2 mM L-glutamine, 1X nonessential amino acid, 1X penicillin–streptomycin, 1000 U/mL LIF, 110 µM 2-mercaptoethanol, and 250 µg/mL G418. For growth of embryoid bodies, do not add LIF.

11. Proteinase K digestion solution: 10 mM Tris-HCl (pH 7.5), 100 mM NaCl, 10 mM Na-EDTA, 0.5% sodium dodecyl sulfate (SDS), 200 µg/mL proteinase K (Roche Molecular Biochemicals, Indianapolis, cat. no. 745723) dissolved in 10 mM Tris-HCl, pH 7.5.

12. Glutaraldehyde fixation solution: 0.1 M phosphate buffer (pH 7.3), 0.2% glutaraldehyde, 5 mM EGTA (pH 8.0), and 2 mM $MgCl_2$.

13. Detergent solution: 0.1 M phosphate buffer (pH 7.3), 2 mM $MgCl_2$, 0.01% sodium deoxycholate, and 0.02% Nonidet P-40 (NP40).

14. Staining solution: 0.1 M phosphate buffer (pH 7.3), 2 mM $MgCl_2$, 0.01% sodium deoxycholate, 0.02% NP40, 5 mM $Fe_3[CN]_4$, 5 mM $Fe_2[CN]_3$. Add 5-bromo-4-chloro-3-indolyl-β-D-galactopyranoside (X-gal) (20 mg/mL stock solution in dimethylformamide) to a final concentration of 1 mg/mL before use.

15. CO_2 incubator (37°C, 5% CO_2).

16. Gene Pulser II Electroporator system from Bio-Rad (cat. no. 165-2105 and 165-2107. Use 0.4-cm Gene Pulser Cuvette, cat. no. 165-2088. Set electroporator at 300 V, 500 µF).

17. 10-cm Tissue culture dishes (Falcon, VWR cat. no. 25382-166), treated with 5 mL of sterilized 0.1% gelatin for 30 min prior to use.

18. Bacteriologic dishes (Valmark, Canada, cat. no. 900).

19. Toluidine blue 1% ready to use solution (Tousimis Research Corporation, MD, USA, cat. no. 4168).

20. Embedding capsules (Tousimis Research Corporation, MD, USA, cat. no. 7004C).

21. Resin embedding kit (Tousimis Research Corporation, MD, USA, cat no. 3112).

3. Methods

3.1. DNA Constructs for the Forced Overexpression of the Mouse Homeobox-Containing Gene Pem

The mouse phosphoglycerate kinase (Pgk1) promoter element was isolated from the pPNT plasmid (*8*) and cloned into the pBluescript II KS (+) vector (Stratagene) as an *Eco*RI (5′)-*Pst*I (3′) fragment. A 6-kb *Bam*HI genomic fragment containing all the coding exons, as well as 3′ noncoding sequences, of the *Pem* gene (exons 3–6) was ligated into the *Bam*HI site of the multiple cloning site of pBluescript, 3′ of the Pgk1 promoter element (**Note 1**). Correct orientation of the overexpression plasmid (designated Pgk-Pem) was verified by sequencing. Prior to transfection into ES cells, the Pgk-Pem plasmid is linearized by digestion with *Sal*I.

3.2. Establishment of Pgk-Pem ES Cell Lines

3.2.1. Maintenance of D3 ES Cells in Culture

1. D3 ES cells are removed from liquid nitrogen storage and quickly thawed in a 37°C water bath.

2. Transfer ES cells into a 15-mL Corning centrifuge tube containing 5 mL of ES cell medium at 37°C.

3. Spin at low speed (350g for 5 min). Remove medium by suction and resuspend the cell pellet in 1 mL of ES cell medium by repeated pipetting (approx 5 times with a 5-mL pipet attached to an electronic pipettor set at medium speed).

4. The cell suspension is transferred to a gelatin-treated 10-cm tissue culture dish containing 9 mL of prewarmed ES cell medium and cultured in a tissue culture incubator (37°C, 5% CO_2). Change the medium every 24 h.

5. Once the ES cells grow to 60–70% confluence, passage them to new tissue culture dishes (3–5 fresh gelatin-treated dishes). To passage ES cells, rinse twice with 5 mL warm 1X PBS, discard the PBS solution, and add 2 mL of 1X trypsin-EDTA. Incubate at 37°C for 3 min and then break up the cell aggregates by repeated pipetting 5–10 times with a 5-mL pipet. Transfer the trypsinized suspension to 15-mL Corning centrifuge tubes with 5 mL warm ES cell medium and pipet up and down an additional 5 to 10 times. Spin at low speed (350g) for 5 min. Remove the medium and resuspend the cell pellets in 6 mL ES cell medium by repeated pipetting. Transfer 1.2 mL (about 3–4 × 10^6 cells) to a new plate containing 9 mL of prewarmed ES cell medium. ES cells must be passaged at least once before transfection.

3.2.2. Transfection

1. Harvest D3 ES cells as described above (**Subheading 3.2.1., step 5**). After centrifuging, resuspend the cell pellet in 5 mL warm ES cell medium. Transfer 0.5 mL to a 15-mL Corning tube containing 9.5 mL 1X PBS. Mix well and count the cell number with a hemacytometer.

2. Transfer approximately 4 × 10^7 ES cells (harvested from 2 plates, each approximately 70–80% confluent) to a 15-mL Corning tube and spin at low speed (350g for 5 min). Resuspended in 0.8 mL 1X PBS.

3. Add 22.5 µg of linearized Pgk-Pem plasmid DNA, together with 2.5 µg of linearized pPNT (containing the bacterial *neo^r* gene under control of the mouse Pgk1 promoter), in 1X PBS to the resuspended ES cells (**Note 2**). Mix well by repeated pipetting 5 times. Transfer to a 0.4-cm Gene Pulser cuvette and leave on ice for 10 min.

4. Prior to electroporation, pipet twice with a 1-mL tissue culture pipet and wipe the cuvette with a Kimwipe to eliminate the moisture. After electroporation, leave the cells on ice for 10 min. Plate the whole volume onto one tissue culture plate containing 10 mL of prewarmed ES cell medium. Incubate at 37°C in a tissue culture incubator (5% CO_2).

3.2.3. Isolation of Pgk-Pem ES Clones

1. After 24 h of electroporation, change the medium to ES cell medium supplemented with 250 µg/mL G418.

2. Change the medium 1 to 2 times/d depending on its acidity (change medium when it turns yellowish-orange). Cell death should be observed after 4 to 5 d of culture.

3. After 10 to 14 d of G418 selection, ES cell colonies of 1–2 mm diameter appear, and they are ready to be isolated and transferred to 24-well dishes. Wash the plate three times with 5 mL prewarmed 1X PBS. Leave the last wash in the plate.

4. Under a dissecting microscope, use a 20 µL pipet tip to scrape a circle around the colony to be transferred. Set the pipettor to 10 µL and pipet the colony into the tip.

5. Transfer the colony into 60 µL 1X typsin-EDTA in a 96-well tissue culture dish. Digest for 5 min at room temperature.

6. Pipet up and down 10 times to disperse the cells and transfer to one well of a 24-well tissue culture dish with 2 mL ES cell medium containing both LIF and G418. Incubate at 37°C (5% CO_2).
7. When the cells are grown to 60–80% confluence, wash twice with 1X PBS, and treat with 200 µL 1X trypsin for 5 min at room temperature (**Note 4**).
8. Set a pipettor at 130 µL and pipet 10 times. Transfer to a sterilized cryotube containing 200 µL ES cell medium.
9. Add 330 µL ES medium containing 20% dimethyl sulfoxide (DMSO) to the cryotube. Mix by tapping several times and transfer the tubes to an insulated styrofoam box. Leave at room temperature for 20 min, followed by 60 min at 4°C, 2 h at –20°C, and overnight at –80°C. Transfer the cryotubes to liquid nitrogen for long-term storage.

3.2.4. Analysis of Isolated G418-Resistant Clones

1. After transferring the 130 µL to a cryotube, add 1 mL of ES cell medium to the well. Culture the remaining cells to confluence to isolate genomic DNA for analysis of transgene integration.
2. To extract genomic DNA, wash once with 0.5 mL 1X PBS. Remove the 1X PBS and add 400 µL proteinase K digestion solution with a pipettor.
3. Pipet up and down 2 to 3 times and transfer the lysate to a 1.5-mL centrifuge tube. Leave at 50°C for 4 h. Mix by vortex mixing briefly every 30 min.
4. DNA is extracted by phenol–chloroform extraction (400 µL), chloroform extraction (400 µL), and ethanol precipitation (800 µL, at –20°C). Spin down DNA at 13,800g in a microcentrifuge for 10 min and resuspend in 200 µL TE buffer after washing once with 70% ethanol.
5. DNA samples are subjected to Southern blot analysis to examine the transgene integration. Twenty-four clones are usually picked and analyzed. Clones with high copy numbers of the *Pgk-Pem* transgene integration are identified, grown in ES cell medium with LIF, and examined for transgene *Pem* expression. This is done by extracting RNA from selected clones and determining the level of Pem expression by Northern blot and ribonuclease protection analyses (**Note 3**).

3.3. Using Growth and Differentiation of ES Cell Embryoid Bodies to Analyze Mutant ES Cells

3.3.1. Grow ES Cell Embryoid Bodies

1. Culture D3 and *Pgk-Pem* ES cells on 10-cm tissue culture dishes to 60–70% confluence.
2. Wash cells twice with 5 mL 1X PBS and treat the cells with 2 mL of 1X trypsin for three min at 37°C.
3. Pipet repeatedly (up and down 10 times) with a 5-mL pipet and transfer to 15-mL Corning centrifuge tubes, each containing 4 mL of ES cell medium. Pipet up and down 10 more times to separate the cells into a single-cell suspension.
4. Spin at low speed (350g for 5 min) and resuspend the pellet in 5 mL of ES cell medium.
5. Count cell numbers and transfer 1×10^7 cells to 15-mL Corning centrifuge tubes. Spin down and resuspend pellet in 10 mL of ES cell medium without LIF to a density of 1×10^6 cells/mL.
6. Aliquot 1 mL to each bacteriologic dish containing 9 mL of ES cell medium without LIF (supplemented with G418 for *Pgk-Pem* cells). Incubate at 37°C, 5% CO_2 for the

desired period of time. In the case of *Pgk-Pem* cell cultures, samples are collected every 4 d, for up to 12 d. Therefore, 3 bacteriologic plates are seeded with 1×10^6 *Pgk-Pem* cells each.

7. Change 5 mL of the appropriate ES cell medium (minus LIF) every other day by slightly tilting the bacteriologic dish and pipetting out 5 mL of medium from the surface. Afterwards, add 5 mL fresh medium.

3.3.2. Examination of Embryoid Body Morphology

Cellular differentiation in embryoid bodies can be monitored in culture using light microscopy. After 7 d, balloon-like cysts (fluid-filled cavities) can be observed in normal D3 ES cell embryoid bodies. However, there is no cavity formation in *Pgk-Pem* embryoid bodies. Abnormal development of embryoid bodies derived from genetically modified ES cells can also be histologically analyzed, using a variety of fixation and embedding procedures. We chose to use a protocol commonly used for electron microscopy, which can preserve the embryoid body morphology.

1. Harvested embryoid bodies are fixed with 2.5% glutaraldehyde for 24 h at 4°C.
2. Wash twice with ice-cold 1X PBS.
3. Dehydrate embryoid bodies through alcohol (50% ethanol, 70% ethanol, 90% ethanol, and 100% ethanol, 1 h at each step).
4. Wash with 100% ethanol one more time for 1 h.
5. Soak embryoid bodies in the solution containing 1/2 100% ethanol and 1/2 freshly prepared embedding resin (following the Tousimis embedding kit instruction) and rotate overnight at room temperature.
6. Change the solution to 100% freshly prepared embedding resin and rotate for an extra 1 h.
7. Transfer the embryoid bodies to embedding capsules and fill the capsules with embedding resin.
8. Incubate at 60°C to let the embedding resin harden.
9. Sections as thick as 1 μm can be made and stained with 1% Toluidine blue.
10. Examine the slides using light microscopy.

3.3.3. Cell Mixing Experiments

To further examine defects in genetically modified ES cells upon embryoid body differentiation, a cell mixing experiment (growth of wild-type ES cells together with mutant ES cells) can be performed. If essential intercellular signals for ES cell differentiation are missing in mutant cell lines, wild-type ES cells mixed with the mutant cells should be able to rescue the differentiation defects in the mutant lines. In contrast, if the mutation affects intracellular signals for differentiation, mutant ES cells (such as *Pgk-Pem*) will behave the same in the presence or absence of wild-type ES cells. To perform the cell mixing experiment, ROSA26 cells, which express the bacterial β-galactosidase gene, are used to distinguish the wild-type ES cells from the mutant cells *(6)*.

1. Grow ROSA26 and Pgk-Pem ES cells as described above.
2. Harvest the cells and dilute to a density of 1×10^6 cells/mL with ES cell medium (minus LIF), then transfer 0.5 mL cells from each cell line to a fresh Corning centrifuge tube (15 mL).

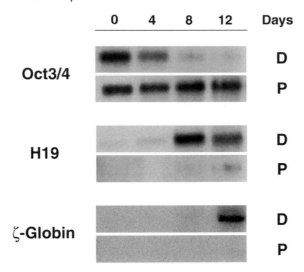

Fig. 1. Tissue-specific gene expression in differentiating embryoid bodies. Expression of *Oct 3/4* and *H19* were determined by Northern blot analysis. Expression of S-globin was determined by a ribonuclease protection assay. The number of days of embryoid body growth are displayed at the top. Undifferentiated ES cells are designated as day O. P, *Pgk-Pem* cells. D, D3 ES Cells.

3. Mix well by pipeting up and down 5 times and transfer to a bacteriologic dish containing 9 mL of ES cell medium.
4. After 8 d of culture, embryoid bodies developing in mixed cultures are transferred to 3-mL glass vials and washed three times with 1 mL ice-cold 1X PBS.
5. Fix embryoid bodies in 0.5 mL 0.2% glutaraldehyde fixation solution for 10 min.
6. Rinse with detergent solution three times at room temperature, allowing 10 min for each rinse.
7. Add staining solution and incubate in dark at 37°C for 3 h or overnight. To maintain moisture, cap the vials loosely, put wet paper towels at the bottom of a plastic container, place the glass vials in the container, and then seal it.
8. After staining, rinse the embryoid bodies with 1X PBS three times and fix with 2.5% glutaraldehyde at 4°C overnight for further histological analysis.

3.3.4. Examination of Tissue-Specific Gene Expression Upon Embryoid Body Differentiation

In addition to analysis of morphologic changes, examination of tissue-specific gene expression can help to identify the formation of specific cellular lineages in the embryoid bodies. Genes such as Oct3/4 (undifferentiated stem cells), H-19 (primitive endodermal derivatives), and ζ-globin (mesoderm) are especially useful markers for the early lineage specification events of embryoid body differentiation. Their expression can be measured by Northern blot analysis, ribonuclease protection assay, or semiquantitative reverse transcription polymerase chain reaction (RT-PCR). We chose Northern blot analysis and ribonuclease protection assay in our studies because these lineage marker genes are relatively highly expressed. Both assays were performed using the protocols described in *Current Protocols in Molecular Biology (9)*. **Fig. 1** is a representative figure for examining tissue-specific gene expression.

3.4. Using Teratomas to Analyze Mutant ES Cells In Vivo

3.4.1. Analysis of the Growth and Differentiation of Mutant ES Cells In Vivo as Teratomas

1. Grow and harvest ES cells as in **Subheading 3.3.3.**
2. Wash ES cell pellets 3 times with 1X PBS and resuspend to a final density of 1×10^6 cells/mL.
3. Inject 0.5 mL (5×10^5 cells) subcutaneously into the flank of NIH Nu/Nu nude mice.
4. Teratomas are harvested and fixed in 10% formulin for histological sections after 4–6 wk.

3.4.2. Teratomas Derived from a Mixture of ROSA26 Wild-Type ES Cells and Mutant Cells

1. Equal numbers of ROSA26 cells and *Pgk-Pem* cells (5×10^5 each) are mixed and washed with 1X PBS 3 times.
2. Resuspend the cell pellet in 1 mL of 1X PBS solution and inject 0.5 mL subcutaneously into the nude mouse.
3. After 6 wk, teratomas are harvested and chopped into small cubes of approx 2 mm to facilitate fixation and X-gal staining.
4. Small cubes of teratoma are fixed and stained for β-galactosidase expression in the same manner described above for embryoid body staining (*see* **Subheading 3.3.2.**).

4. Notes

1. To force expression of the *Pem* gene in undifferentiated ES cells, we obtained higher levels of *Pem* expression from the Pgk1 promoter using the genomic *Pem* sequence rather than a full-length *Pem* cDNA. The genomic sequence contained all the coding exons together with 3′ untranslated sequences. For unknown reasons, the level of transgene *Pem* expression in *Pgk-Pem* ES cells (genomic *Pem* sequences in the transgene) was higher than that in ES cell lines transfected with full-length *Pem* cDNA construct, even though the number of integrated transgene copies was similar.
2. Cotransfection of the *Pgk-Pem* DNA construct with a *neo^r* gene expressed from the Pgk1 promoter was used in establishing *Pgk-Pem* ES cell lines. We did not observe any loss of transgene *Pem* expression even after numerous passages of *Pgk-Pem* cells in vitro. Loss of transgene gene expression after many passages has been occasionally observed in ES cells containing expression constructs that transcribe the gene of interest and a *neo^r* gene as a single transcript. The use of genomic *Pem* sequences in the expression transgene may ensure consistent and long-term expression of both cotransfected (and presumably co-integrated) expression constructs.
3. To accurately measure *Pgk-Pem* expression, it had to be carefully distinguished from endogenous *Pem* transcript. This proved to be somewhat difficult because *Pgk-Pem* contains a *Pem* genomic fragment, whose pre-mRNA is spliced in nearly the same way as the endogenous *Pem* gene. Thus, transgenic and endogenous *Pem* bands on Northern blots were the same size. We were able to distinguish the transgenic and endogenous *Pem* transcripts in a ribonuclease protection assay using a riboprobe constructed from a cDNA containing 5′ UTR sequences (not in *Pgk-Pem*) and coding sequences (in both *Pgk-Pem* and endogenous *Pem* transcripts).
4. When ES cell clones are picked, trypsinized, and expanded in 24-well plates, some wells will contain only a few colonies (<10) and grow very slowly. To increase their growth and also the number of cells per clone stored in liquid nitrogen, the wells with fewer colonies can be trypsinized and replated in the same well. Remove the ES cell medium and wash

once with 1X PBS. After the PBS is removed, add 0.2 mL of 1X trypsin-EDTA for 3 min at 37°C. Disperse the colonies by repeated pipetting (5–10 times) using a pipettor. One milliliter of ES cell medium is then added to the well. After 1–2 d, many colonies will appear, and the well will rapidly become confluent.

References

1. Niwa, H., Miyazaki, J., and Smith, A. G. (2000) Quantitative expression of Oct-3/4 defines differentiation, dedifferentiation or self-renewal of ES cells. *Nat. Genet.* **24,** 372–376.
2. Nichols, J., Zevnik, B., Anastassiadis, K., Niwa, H., Klewe-Nebenius, D., Chambers, I., et al. (1998) Formation of pluripotent stem cells in the mammalian embryo depends on the POU transcription factor Oct4. *Cell* **95,** 379–391.
3. Lin, T. P., Labosky, P. A., Grabel, L. B., Kozak, C. A., Pitman, J. L., Kleeman, J., and MacLeod, C. L. (1994) The Pem homeobox gene is X-linked and exclusively expressed in extraembryonic tissues during early murine development. *Dev. Biol.* **166,** 170–179.
4. Wilkinson, M. F., Kleeman, J., Richards, J., and MacLeod, C. L. (1990) A novel oncofetal gene is expressed in a stage-specific manner in murine embryonic development. *Dev. Biol.* **141,** 451–455.
5. Pitman, J. L., Lin, T. P., Kleeman, J. E., Erickson, G. F., and Macleod, C. L. (1998) Normal reproutive and macrophage function in Pem homeobox gene-deficient mice. *Dev. Biol.* **202,** 196–214.
6. Fan, Y., Melhem, M. F., and Chaillet, J. R. (1999) Forced expression of the homeobox-containing gene Pem blocks differentiation of embryonic stem cells. *Dev. Biol.* **210,** 481–496.
7. Keller, G. M. (1995) In vitro differentiation of embryonic stem cells. *Curr. Opin. Cell Bio.* **7,** 862–869.
8. Tybulewicz, V. L. J., Crawford, C. E., Jackson, P. K., Bronson, R. T., and Mulligan, R. C. (1991) Neonatal lethality and lymphopenia in mice with a homozygous disruption of the c-abl proto-oncogene. *Cell* **65,** 1153–1163.
9. Ausubel, F. M., Brent, R., Kingston, R. E., Moore, D. D., Seidman, J. G., Smith, J. A., and Struhl, K. (eds.) (1994) *Current Protocols in Molecular Biology*, John Wiley & Sons, N.Y. pp. 4.7.1–4.9.16.

6

Hypoxic Gene Regulation in Differentiating ES Cells

David M. Adelman and M. Celeste Simon

1. Introduction

Mammalian development occurs in the hypoxic environment of the uterus *(1,2)*. Initially, the limited oxygen available can adequately diffuse to all the cells of the growing conceptus. However, with continued growth, diffusion becomes less efficient and results in a "physiologic hypoxia" within the embryo *(3)*. Current models suggest this physiologic hypoxia acts as a stimulus to coordinate the development of the cardiovascular system. Once formed, oxygen can be transported throughout the organism to enable aerobic respiration and promote increased ATP production. Unless the increasing energy demands of the embryo are met, further development will be halted, and early lethality will occur.

The hypoxia-inducible factor (HIF) is an obligate heterodimeric transcription factor, comprised of HIF-1α (or HIF-2α) and ARNT (arylhydrocarbon receptor nuclear translocator) subunits. Embryonic stem (ES) cells deficient in either HIF-1α or ARNT are incapable of activating genes normally responsive to low oxygen, and mice derived from these ES cells die at mid-gestation *(4–8)*. In addition to the prominent vascular defects in both animals, *Arnt*[-/-] embryos display hematopoietic *(9)*, placental, and cardiac abnormalities *(10)*. Genes such as vascular endothelial growth factor (VEGF) and erythropoietin (Epo) are required for vascular and erythroid development, respectively, and both are known to be regulated, in part, by hypoxia-responsive elements (HREs) that bind the HIF complex *(3)*. Therefore, accumulating evidence suggests that HIF activity is an essential component of the hypoxic response and critical to this important cardiovascular developmental cue.

ES cells can be differentiated in vitro to form embryoid bodies (EBs), which contain tissues of all three germ cell layers. Due to their three-dimensional structure, EBs exhibit gradients in oxygen levels, even when cultured under normoxic (i.e., atmospheric) conditions *(11)*. A great benefit of in vitro culture is the ability to exacerbate this oxygen gradient by culturing EBs under hypoxic conditions. In doing so, the requirement for low oxygen in the formation of many different tissues can be examined. *Arnt*[-/-] ES cells are an extremely valuable reagent since they do not form HIF complexes and, therefore, do not respond to the inherent hypoxic environment of the EB. Because the *Arnt*[-/-] animals exhibit multiple developmental defects, in vitro differentiation allowed us to establish which phenotypes are primary in nature, rather

From: *Methods in Molecular Biology, vol. 185: Embryonic Stem Cells: Methods and Protocols*
Edited by: K. Turksen © Humana Press Inc., Totowa, NJ

than secondary to developmental delay. We focused our initial efforts on studying the decreased hematopoietic progenitor phenotype within EBs, since they, unlike embryos, do not require a functional placenta or mature cardiovascular system to survive. However, we hope future studies may take advantage of different tissue differentiation protocols to study hypoxia and its requirement for endothelial and cardiac development. In addition, research has recently indicated that trophoblast stem (TS) cells can be cultured and maintained in vitro *(12)*. Much like ES cells, TS cells are nontransformed multipotent progenitors capable of differentiating towards various extra-embryonic and placental cell lineages. We are beginning to analyze these cells for hypoxic requirements as well and, if such requirements exist, their dependence on ARNT activity *(13)*.

In this chapter, we outline our protocol for the in vitro differentiation of ES cells towards hematopoietic progenitor generation and enumeration, with specific reference to hypoxic culture conditions. Our procedure has two steps: (*i*) a primary differentiation to generate progenitors within EBs, followed by (*ii*) disaggregation and secondary plating to enumerate progenitor colony forming units (CFUs). The primary differentiation is performed under both normoxia and hypoxia, whereas the secondary plating is accomplished under solely normoxic conditions. In this way, any progenitor that has formed will have an equal opportunity to survive and proliferate.

2. Materials

All materials and reagents should be kept as sterile as possible, by filtering or autoclaving where appropriate.

2.1. Hypoxic Environment

The hypoxic environment consists of 3% oxygen, 5% carbon dioxide, and 92% nitrogen. This may be obtained with a premixed gas tank fed into a closed environment, such as a nalgene canister with inlet and outlet tubes, or it may also be obtained with the use of a gas mixer. Levels of less than 5% oxygen are necessary to see HIF protein stabilization. Be sure to humidify the gas, either by bubbling the premixed gas through sterile water or using a humidified incubator.

2.2. Media

2.2.1. ES Cell Medium

Totaling 500 mL, the ES cell medium consists of the following:

1. Dulbecco's modified Eagle medium (DMEM) (high glucose, with L-Glut, without Na Pyruvate) (Life Technologies, cat. no. 11965-092), 405 mL.
2. Fetal bovine serum (FBS) (Stem Cell Technologies, cat. no. 06951), 75 mL.
3. Penicillin–streptomycin 100X (Life Technologies, cat. no. 15140-122), 5 mL.
4. Nonessential amino acids 100X (Life Technologies, cat. no. 11140-050), 5 mL.
5. L-Glutamine 100X (Life Technologies, cat. no. 25030-081), 10 mL.
6. β-mercaptoethanol (Sigma, cat. no. M 7522), 4 μL.
7. Leukemia inhibitory factor (LIF); ESGRO, 10^7 U (Chemicon, cat. no. ESG1107), 50 μL; or collected supernatant from Chinese hamster ovary (CHO) cell line transfected with LIF cDNA, 250 μL.
8. Filter sterilize and store at 4°C. Do not use for more than 3 wk.

2.2.2. ES Cell Freeze Medium

Totaling 10 mL, the ES cell freeze medium consists of the following:

1. ES cell media, 8 mL.
2. FBS, 1 mL.
3. Dimethyl sulfoxide (DMSO) (Sigma, cat. no. D 2650), 1 mL.
4. Filter sterilize. Make fresh as needed and keep on ice until use.

2.3. Differentiation Methylcellulose

Totaling 16.0 mL, differentiation methylcellulose consists of the following:

1. 0.9% Base methylcellulose (Stem Cell Technologies, cat. no. H4100), 14.4 mL. Prepare by thawing one bottle (40 mL) at 37°C. Add 50 mL Iscove's MDM (Life Technologies, cat. no. 12440-020), cover with foil, and rock O/N at 4°C to mix. Store at 4°C.
2. 10% FBS (Stem Cell Technologies, cat. no. 06950), 1.6 mL.
3. 5 ng/mL Recombinant murine (rm)IL-3 (R&D Systems, cat. no. 403-ML-010), 5.5 µL.
4. 500 U/mL rmIL-1, 10^6 U/mL stock (R&D, cat. no. 400-ML-005), 8.0 µL.
5. 10 µg/mL Insulin, bovine, in H_2O (pH 2.0), 10 mg/mL stock (Sigma, cat. no. I6634), 16.0 µL.
6. 200 µg/mL Transferrin, human, iron-saturated, 20 mg/ml stock in PBS (Sigma, cat. no. T0665), 160 µL.
7. 1×10^{-4} *M* α-monothioglycerol, dilute 1 : 10 in phosphate-buffered saline (PBS) (Sigma, cat. no. M 6145), 6.2 µL.
8. 25 m*M* HEPES buffer solution (1 *M*) (Life Technologies, cat. no. 15630-080), 400 µL.
9. Add methylcellulose to a 50-mL conical tube. Combine remaining ingredients and filter into methylcellulose through a syringe and 0.2-µm filter. Vortex mix and keep on ice. Make fresh before each use.

2.4. Replating Methylcellulose

1. "Complete" methylcellulose (Stem Cell Technologies, cat. no. 03434), contains 1.0% methylcellulose, 10% FBS, 1% bovine serum albumin (BSA), 10 µg/mL bovine pancreatic insulin, 200 µg/mL human transferrin, 10^{-4} *M* β-mercaptoethanol, 2 m*M* L-glutamine, 50 ng/mL (rm) stem cell factor (SCF), 10 ng/mL rmIL-3, 10 ng/mL recombinant human (rh)IL-6, and 3 U/mL rhEPO.
2. Pokeweed mitogen murine spleen cell-conditioned medium (PWMSCCM) (Stem Cell Technologies, cat. no. 02100). Contains IL-3 and granulocyte macrophage-colony-stimulatory factor (GM-CSF).

Thaw the complete methylcellulose and make 15-mL aliquots in 50-mL conicals to minimize freeze–thaw cycles. For use, thaw briefly at 37°C and then keep on ice. Add 300 µL of filter-sterilized PWMSCCM to 15 mL complete methylcellulose and vortex mix. This mixture can be stored at 4°C and used for up to 2 wk.

2.5. Additional Materials

1. Gelatin, porcine (Sigma, cat. no. G 1890), 0.1% solution in ddH_2O, autoclaved.
2. PBS, without Ca^{2+} or Mg^{2+} (Life Technologies, cat. no. 14190-136).
3. Mitomycin C (Boehringer Mannheim, cat. no. 107 409).
4. Trypsin-EDTA, 0.25% (Life Technologies, cat. no. 25200-056).
5. α-MEM (Life Technologies, cat. no. 12571-048).

6. 3-mL Syringes (Becton Dickenson, cat. no. 309585).
7. 0.2-µm Sterile filter discs (Millipore, cat. no. SLGV025LS).
8. Pasteur pipets (glass), autoclaved (Fisher, cat. no. 13-678-20D).
9. Falcon 2054 capped tubes (Falcon, cat. no. 352054).
10. Culture dishes, specially treated to prevent the growth of anchorage-dependent cells (Stem Cell Technologies, cat. no. 27150).
11. Cell lifters (Costar, cat. no. 3008).
12. 21-gauge needles (Becton Dickenson, cat. no. 305167).
13. 100×25-mm-deep Petri dishes (Fisher, cat. no. 08-757-11).
14. Gridded scoring dish (Stem Cell Technologies, cat. no. 27500).

3. Methods

3.1. Culture of ES Cells

In contrast to the culture of ES cells for generation of chimeric mice, ES cells for in vitro differentiation must be cultured in the absence of feeder cells (MEFs, STOs, SNLs, etc.). To gelatin-adapt ES cells:

3.1.1. Gelatin-Adapting

1. Plate mitomycin-c-treated feeder cells onto a gelatin-coated 6-well plate (add 2 mL of 0.1% gelatin to each well and aspirate after 15 min).
2. Thaw ES cells quickly at 37°C and transfer to 4 mL of prewarmed ES media. Centrifuge at approx 500g for 5 min. Aspirate media, resuspend in fresh media, and plate onto feeders.
3. Once ES cells have reached confluency, passage them at a dilution of no more than 1:6 onto a 0.1% gelatin-treated feeder-free well. Take the cells through at least 2 more passages to completely adapt the cells to a feeder-free environment.
4. Freeze down these gelatin-adapted cells for later use (a single well of a 6-well plate/vial). After trypsinization, resuspend in 0.5 mL freezing media/single 6-well, transfer to a cryovial, and place at –80°C overnight. Transfer to liquid N_2 the following day (*see* **Note 1**).

3.1.2. Thawing ES Cells

1. Thaw gelatin-adapted ES cells into a gelatinized well of a 6-well plate.
2. Once confluent, passage the ES cells no more than 1:6 1 d prior to setting up a differentiation.
3. Change media on the ES cells the morning of the differentiation, at least 1 h prior to use (*see* **Note 2**).

3.2. Primary Differentiation of ES Cells to Form EBs Under Normoxia and Hypoxia

1. Prior to harvesting the ES cells, prepare the methylcellulose media. The primary differentiation should be performed in a cytokine-depleted medium to avoid masking any cell-extrinsic defects. Each culture will require 1.5 mL of methylcellulose cocktail and are usually plated in triplicate per experiment. Therefore, to assess wild-type, heterozygous, and homozygous mutant clones of a given gene, 9 dishes (13.5 mL) are used. As the methylcellulose is very viscous, we prepare an extra 2–3 mL. Therefore, 16 mL would be made for these 9 dishes (amounts should be adjusted accordingly for more or less plates).

2. Add the base methylcellulose to a 50-mL conical tube. Add the remaining ingredients into a 3-mL syringe attached to a 0.22-μm filter and filter sterilize into the base methylcellulose. Vortex mix when everything is added and place on ice until needed (*see* **Notes 3** and **4**).

3. Wash ES cells with PBS and trypsinize (0.5 mL/well) for 5 min. Kill the trypsin with an equal volume of 10% FBS in DMEM solution (filter-sterilized). From this point on, it is important to keep the cells free of LIF. Ensure that you have single-cell suspensions by pipetting with a 200-μL tip on the end of a 1-mL tissue culture pipet. Resuspend the cells in 5 mL total volume with 10% FBS in DMEM in a 15-mL conical and count the cells (hemacytometer, Coulter counter, etc.).

4. For each differentiation, approx 3×10^3 cells are used. This number will vary from genotype to genotype and from one freeze to another. It is best to empirically determine this number for your cells. Aim for about 200–300 EBs/dish. Too few or too many cells can yield a less than perfect differentiation and may adversely affect your results (*see* **Note 5**).

5. Once cells are counted, spin down (approx 500g for 5 min) and resuspend in 10% FBS in DMEM according to the dilution you need (*see* **Note 6**). It is easier to pipet a larger volume than a smaller volume, but avoid over-dilution of the methylcellulose. Aim for about 10 μL of cells/1.5 mL methylcellulose.

6. Add 1.5 mL methylcellulose into each sterile Falcon 2054 capped tubes, using a 5-mL pipet and pipetting very slowly (*see* **Note 7**).

7. Use a P-20 pipetteman to add 3×10^3 of your cells (10 μL or whatever volume is required) to each 2054 tube with methylcellulose. Cap securely when finished and vortex mix. While cultures are settling, label the culture dishes.

8. Using an autoclaved glass pasteur pipet attached to a pipet aid, transfer the methylcellulose plus cells from the 2054 tube into the 35-mm dish. Avoid getting air bubbles in the methylcellulose. Rotate the dish at an angle to spread the methylcellulose when finished.

9. Place 2 dishes (covered) with 1 water dish (uncovered) into a 100×25-mm tissue culture dish, cover, and place in a tissue culture incubator OR hypoxia setup. Allow the differentiation to go for approximately 9 d. Check on the cultures periodically to ensure that no fungal contamination is present and that your plating density is appropriate (not too dense and not too sparse) (*see* **Notes 8** and **9**).

3.3. Disaggregation of EBs and Replating of Progenitors

1. After 9 d in culture, count the number of EBs/dish. This is best accomplished with a microscope under low power (4× objective), using a gridded scoring dish underneath your culture dish (*see* **Note 10**).

2. Add 2 mL of prewarmed PBS to each dish. Use a sterile cell lifter to break up the methylcellulose and get all the EBs in suspension (they may rest on the bottom of the dish). Using a P-1000 with autoclaved tips, pipet the diluted contents into a 15-mL conical tube that is prefilled with 4 mL of PBS. Using a fresh tip, place another 1 mL of PBS into the dish to ensure that you have transferred all the EBs into the tube. Spin down the EBs at approx 500g for 5 min.

3. Aspirate off the diluted methylcellulose. Add 1 mL of trypsin using a P-1000 to ensure good resuspension of the EBs. Place at 37°C for 5–7 min. While incubating, prepare a 3-mL syringe with a 21-gauge needle for each tube (*see* **Note 11**).

4. After 5–7 min, use the syringe/needle to shear the EBs, plunging up and down 4–5 times. When all the tubes are finished, add 1.5 mL 10% FBS in DMEM to inactivate the trypsin, and shear again 4–5 times. These actions are usually enough to give single-cell suspensions, but if whole EBs can still be seen, repeat the shearing. After shearing, there

should be 2.5 mL total volume; however, this can vary depending on your initial accuracy and how many bubbles were introduced during syringing. Add 10% FBS in DMEM to obtain a final vol of 3 mL in each tube. The actual volume is not so important so long as the precise volume is known (*see* **Note 12**).

5. Count the cells, then centrifuge at approx 500g for 5 min. Resuspend in 500 μL α-MEM solution. We add the equivalent number of cells in 50 wild-type normoxic EBs from each sample. This can be determined for each experiment, or a set number may be plated each time (e.g., 2×10^5). The important point is to plate the same number of cells for each sample (*see* **Note 13**).

6. Vortex mix a thawed aliquot of complete methylcellulose (plus PWMSCCM) and add to Falcon 2054 tubes as before. Add the appropriate volume of cells (e.g., 50 μL of 2×10^5 cells) with a P-200, cap securely, and vortex mix. After labeling the 35-mm culture dishes, plate out each culture, using a fresh pasteur pipet for each tube (*see* **Note 14**).

7. Assemble the humidifying chambers as before and incubate under normoxic conditons for 5–7 d. It is not necessary to keep the cultures under hypoxia at this point. Any progenitors that have formed during the differentiation should now proliferate to form a colony. Score cultures when the colonies are recognizable but not yet overgrown, otherwise they can be hard to distinguish from one another (*see* **Note 15**).

3.4. Scoring of CFUs

1. Once colonies have formed, place a culture dish into a gridded scoring dish and view under a microscope. Start with low power (4× objective) and move to a higher power if necessary. Score the types and numbers of hematopoietic colonies present. A good atlas of hematopoietic CFU morphologies in methylcellulose is available from Stem Cell Technologies (cat. no. 28700) (*see* **Note 16**).

2. If the colony morphologies in methylcellulose are hard to distinguish, individual colonies may be picked, cytospun, and stained to assess individual cellular morphology. Use a P-20 pipetman to isolate a single colony and place into 150 μL PBS. Spin the cells onto a slide (e.g., using a Wescor Cytopro Cytocentrifuge at 700 rpm, medium acceleration, for 10 min). Once dry, fix the cells and stain (e.g., Wright-Giemsa stain).

4. Notes

1. With increased passage number, the totipotency of ES cells may decline. It is, therefore, advisable to make multiple freezes of low passage cells, in order to be thawed for each experiment.

2. The differentiation will be optimal if the cells are in log phase of growth. Therefore, it is best to passage each clone the day before an experiment. Changing media the morning after passage ensures that any residual trypsin has been removed.

3. Filtering all reagents into the methylcellulose decreases chances of contamination. Remember: these cultures grow for 9 d, and fungus grows very well in this rich medium. It is also best to add all reagents to the methylcellulose fresh, to avoid cycles of freeze–thaw.

4. Cultures in hypoxia for extended periods of time may become acidic. IMDM has 25 mM HEPES, but we find adding an additional 25 mM HEPES helps greatly. However, the addition of too much HEPES (e.g., 100 mM) will kill your cells. You may want to determine what level of buffering is best for your cell line.

5. These experiments were performed with R1 ES cells. For other ES cell strains and genotypes, it is necessary to empirically determine the number of cells to be plated in the primary differentiation. Too many EBs can lead to a depletion of nutrients in the methylcellulose, and too few EBs may not generate enough progenitors to be quantitative.

Plating efficiency will also be affected by the cells' totipotency and phase of growth, therefore, be certain to set up cultures under similar conditions.

6. A sample calculation would be: 600×10^3 cells (counted) $\times$ 5 mL (total vol) = 3.00×10^6 (total cells from a single well) divided by 3.0×10^3 (amount you want per dish) = 1000 (how many times more cells you have than you need) $\times$ 10 µL (volume of cells to be added) = 10,000 µL (resuspended volume so that you can add 10 µL and be adding 3.00×10^3 cells).

7. It is important to pipet the methylcellulose very slowly, due to its high viscosity. We find that sterile glass tissue culture pipets work well, but are certainly not required.

8. It is important to keep the methylcellulose humidified to ensure an optimal differentiation. We find that uncovered water dishes within the covered 10-cm dishes works well.

9. A 9-d differentiation period was determined to be optimal for the R1 ES cells, from a time course of 5–10 d. Some strains will differentiate faster or slower, and therefore the length of time until the peak of hematopoiesis should be empirically determined.

10. EBs can differ in size, shape, number, and types of cells present. It is best to normalize secondary platings to a definable constant, such as 2×10^5 cells, or the equivalent number of cells in 50 wild-type normoxic EBs, etc.

11. Some published reports suggest that d 9 and later EBs require collagenase treatment for disruption of the Reicharts membrane. We find that trypsin, combined with mechanical shearing, works sufficiently. It is important to let the trypsin continue working after initial shearing, to allow for continued disruption of cellular aggregates within the EB.

12. Be as expedient as possible during the trypsinization steps, as you do not want the fragile progenitors to be exposed to trypsin for longer than necessary. Also, the less bubbles introduced during syringing, the better (less oxidation of cell surface components and a more accurate volume). Try to empty the contents of the syringe against the side of the 15-mL conical.

13. The volume of α-MEM, in which your cells are resuspended, is dependent on the total number of cells you have. As with the primary differentiation, overly diluting the methylcellulose with cells is less than ideal. If your differentiation yields fewer cells, requiring replating of a large percentage of them, resuspend in a smaller volume, such as 250 µL.

14. To let any cell-extrinsic (e.g., cytokine-mediated) defects become known, the primary differentiation uses a cytokine-poor methylcellulose. The EBs will produce cytokines, another reason to ensure your plating efficiency is not too low. However, once a progenitor is formed, the cytokine-rich secondary methylcellulose should allow it to form a scorable colony (unless there is a cell-intrinsic phenotype that prevents this).

15. Colony growth will appear within 2–3 d of disaggregation and replating. However, colony types may not be distinguishable for 5–7 d. After this period, erythrocytes and granulocytes may start to die, making their colony identification difficult. Therefore, try to score colonies within the 5–7 d window.

16. Some strains of ES cells tend to give nonhematopoietic colony growths in the secondary plating, which resemble "secondary EBs." These do not necessarily affect the type or number of hematopoietic CFUs, but can make their scoring more difficult. The addition of Protein-Free Hybridoma Medium II (Life Technologies, cat. no. 12040-093) to the secondary plating (5%) tends to minimize this population.

References

1. Rodesch, F., Simon, P., Donner, C., and Jauniaux, E. (1992) Oxygen measurements in endometrial and trophoblastic tissues during early pregnancy. *Obstet. Gynecol.* **80,** 283–285.

2. Fischer, B. and Bavister, B. D. (1993) Oxygen tension in the oviduct and uterus of rhesus monkeys, hamsters and rabbits. *J. Reprod. Fertil.* **99,** 673–679.
3. Maltepe, E. and Simon, M. C. (1998) Oxygen, genes, and development: an analysis of the role of hypoxic gene regulation during murine vascular development. *J. Mol. Med.* **76,** 391–401.
4. Kozak, K. R., Abbott, B., and Hankinson, O. (1997) ARNT-deficient mice and placental differentiation. *Dev. Biol.* **191,** 297–305.
5. Maltepe, E., Schmidt, J. V., Baunoch, D., Bradfield, C. A., and Simon, M. C. (1997) Abnormal angiogenesis and responses to glucose and oxygen deprivation in mice lacking the protein ARNT. *Nature* **386,** 403–407.
6. Carmeliet, P., Dor, Y., Herbert, J. M., Fukumura, D., Brusselmans, K., Dewerchin, M., et al. (1998) Role of HIF-1alpha in hypoxia-mediated apoptosis, cell proliferation and tumour angiogenesis. *Nature* **394,** 485–490.
7. Iyer, N. V., Kotch, L. E., Agani, F., Leung, S. W., Laughner, E., Wenger, R. H., et al. (1998) Cellular and developmental control of O_2 homeostasis by hypoxia-inducible factor 1 alpha. *Genes Dev.* **12,** 149–162.
8. Ryan, H. E., Lo, J., and Johnson, R. S. (1998) HIF-1a is required for solid tumor formation and embryonic vascularization. *EMBO J.* **17,** 3005–3015.
9. Adelman, D. M., Maltepe, E., and Simon, M. C. (1999) Multilineage embryonic hematopoiesis requires hypoxic ARNT activity. *Genes Dev.* **13,** 2478–2483.
10. Adelman, D. M., Gertsenstein, M., Nagy, A., Simon, M. C., and Maltepe, E. (2000) Placental cell fates are regulated in vivo by HIF-mediated hypoxia responses. *Genes Dev.* **14,** 3191–3203.
11. Gassmann, M., Fandrey, J., Bichet, S., Wartenberg, M., Marti, H. H., Bauer, C., et al. (1996) Oxygen supply and oxygen-dependent gene expression in differentiating embryonic stem cells. *Proc. Natl. Acad. Sci. USA* **93,** 2867–2872.
12. Tanaka, S., Kunath, T., Hadjantonakis, A. K., Nagy, A., and Rossant, J. (1998) Promotion of trophoblast stem cell proliferation by FGF4. *Science* **282,** 2072–2075.
13. Adelman, D. M. and Simon, M. C. (2000) unpublished observations.

7

Regulation of Gap Junction Protein (Connexin) Genes and Function in Differentiating ES Cells

Masahito Oyamada, Yumiko Oyamada, Tomoyuki Kaneko, and Tetsuro Takamatsu

1. Introduction

Gap junctions are specialized cell–cell junctions that directly link the cytoplasm of neighboring cells. They mediate the direct transfer of low molecular weight (<1000 D) metabolites and ions, including second messengers such as cyclic adenosine monophosphate (cAMP), inositol trisphosphate, and Ca^{2+} between adjacent cells. Therefore, gap junctional intercellular communication is considered to play an important role in the control of cell growth, differentiation, morphogenesis, and the maintenance of homeostasis. Gap junctions are composed of oligomeric proteins consisting of 6 subunits called connexins (Cxs) that are coded for by a multigene family *(1,2)*. Thus far, more than 14 different *Cxs* have been cloned in rodent genes. The expression of each *Cx* has organ and cell-type specificity and is developmentally controlled. During the cardiomyocytic differentiation of embryonic stem (ES) cells in vitro, the expression of multiple *Cxs* is differentially regulated, and gap junctional intercellular communication is modulated *(3–5)*.

In this chapter, we describe methods for the analysis of gap junctions during cardiomyocytic differentiation of ES cells in vitro. The analytical methods are divided into two groups, those for the expression of *Cx* genes and proteins and those for functions of gap junctions. The former includes reverse transcription polymerase chain reaction (RT-PCR) analysis for *Cx* genes and immunofluorescence for *Cx* proteins, and the latter includes dye coupling and calcium imaging.

2. Materials

Materials for the hanging drop culture method for cardiomyocytic differentiation of ES cells in vitro are described in detail in Chapter 13.

2.1. RT-PCR Analysis for Cx Expression in Differentiating ES Cells

All the buffers except for Tris-HCl-containing buffers must be made with diethyl pyrocarbonate (DEPC)-treated water.

1. An inverted microscope (e.g., Olympus, IX-70, Tokyo, Japan) equipped with micromanipulators (e.g., Narishige, WR-88, Tokyo, Japan).

From: *Methods in Molecular Biology, vol. 185: Embryonic Stem Cells: Methods and Protocols*
Edited by: K. Turksen © Humana Press Inc., Totowa, NJ

2. Solution D: 4 *M* guanidium thiocyanate, 25 m*M* sodium citrate (pH 7.0), 0.5% sarcosyl, 0.1 *M* 2-mercaptoethanol. Store in the dark for up to 1 mo at room temperature.
3. 2 *M* sodium acetate (pH 4.0).
4. Water-saturated phenol. Store in the dark for up to 1 mo at 4°C.
5. Chloroform.
6. Isopropyl alcohol.
7. 70% Ethanol.
8. 4 *M* LiCl.
9. H$_2$O (autoclaved).
10. Isoamyl alcohol.
11. 14-mL Polypropylene round-bottom centrifuge tubes (Becton Dickinson Labware, Franklin Lakes, NJ, cat. no. 352059).
12. 5X DNase buffer: 200 m*M* Tris-HCl, pH 8.0, 50 m*M* NaCl, 30 m*M* MgCl$_2$.
13. RNase-free DNase I (Boehringer Mannheim, Indianapolis, IN, cat. no. 776785).
14. RNase inhibitor (Boehringer Mannheim, cat. no. 799017).
15. A kit for RT-PCR (e.g., TAKARA RNA PCR Kit ver 2.1; Takara Shuzo, Otsu Shiga, Japan, cat. no. R019A).
    a. Random 9-mer.
    b. Avian myeloblastosis virus reverse transcriptase XL.
    c. MgCl$_2$ solution.
    d. dNTP mixture.
    e. RNase inhibitor.
    f. 10X RNA PCR buffer.
    g. *Taq* DNA polymerase.
    h. RNase-free H$_2$O.
16. PCR primers. Store at –20°C.
    a. *Connexin 40 (Cx40)*, 5′-CCACGGAGAAGAATGTCTTCA-3′ and 5′-TGCTGCTGG CCTTACTAAGG-3′. The expected size of amplified fragments is 447 bp.
    b. *Connexin 43 (Cx43)*, 5′-TGGGGGAAAGGCGTGAG-3′ and 5′-CTGCTGGCTCTG CTGGAAGGT-3′. The expected size of amplified fragments is 1.3 kbp.
    c. *Connexin 45 (Cx45)*, 5′-ATCATCCTGGTTGCAACTCC-3′ and 5′-CTCTTCATGG TCCTCTTCCG-3′. The expected size of amplified fragments is 168 bp.
17. Thermal cycler (e.g., GeneAmp PCR system 9600; PE Biosystems, Foster City, CA).
18. Electrophoresis-grade agarose (e.g., NuSieve 3:1 agarose; BioWhittaker Molecular Applications, Rockland, ME, cat. no. 50090).
19. 50X Stock solution of Tris-acetate-EDTA electrophoresis buffer: 2 *M* Tris-acetate, 100 m*M* EDTA, pH 8.5.
20. 20,000X Stock solution of ethidium bromide: 10 mg/mL in H$_2$O. Store in the dark at 4°C.
21. DNA molecular weight markers.
22. 6X Loading buffer: 30% glycerol, 0.25% bromphenol blue, 0.25% xylene cyanol.
23. Horizontal gel electrophoresis apparatus with DC power supply.

2.2. Immunofluorescent Labeling of Cxs in the ES Cells

1. Humidified staining box.
2. Phosphate-buffered saline (PBS).
3. 100% Ethanol, –10° to –20°C (solvent cooled in the freezer of a refrigerator).
4. Blocking solution: 5% (w/v) skim milk in PBS. Prepare just before use.
5. Antibody diluting solution: 1% bovine serum albumin (BSA), 0.1% Triton X-100, 0.1% NaN$_3$ in PBS. Store for up to 3 mo at 4°C.

6. Primary antibodies.
   a. Rabbit anti-Cx43 antibody (Zymed Laboratories, South San Francisco, CA, cat. no. 71-0700).
   b. Mouse monoclonal antibody against ventricular myosin heavy chain α/β (MHC-α/β) (Chemicon International, Temecula, CA, cat. no. MAB1552).
7. Secondary antibodies.
   a. Goat Alexa Fluor 488 anti-rabbit IgG conjugate (Molecular Probes, Eugene, OR, cat. no. A-11034).
   b. Goat Alexa Fluor 594 anti-mouse IgG conjugate (Molecular Probes, cat. no. A-11032).
8. Anti-quenching mountant (e.g., Vectorshield, Vector Laboratories, Burlingame, CA, cat. no. H-1000).
9. Nail polish for sealing coverslips onto dishes.

2.3. Gap Junctional Intercellular Communication Measured by Lucifer Yellow Dye Coupling

1. Inverted microscope (e.g., Olympus, IX70, Tokyo, Japan).
2. Micromanipulators (e.g., Narishige, WR-88, Tokyo, Japan).
3. Automatic injection system (e.g., Eppendorf, Transjector 5246, Hamburg, Germany).
4. Sterile femtotips (e.g., Eppendorf, 5242 952.008).
5. Microloader (e.g., Eppendorf, 5242 956.003).
6. 10% Solution (w/v) of Lucifer yellow CH (Sigma, St. Louis, MO, cat. no. L0259) dissolved in 0.33 M lithium chloride. Filter through a 0.22-μm filter (e.g., Ultrafree-MC; Millipore, Bedford, MA) by centrifugation and store for up to 6 mo in a screw-capped tube in the dark at room temperature.
7. Charge-coupled device (CCD) camera (e.g., Hamamatsu photonic System, Hamamatsu, Japan).
8. Video digitizer card installed on a personal computer.

2.4. Calcium Imaging Using Fluorescent Probes with Confocal Microscopy

1. Fluo-3 acetoxymethyl ester (fluo–3/AM) (Molecular Probes, cat. no. F-1242). Store in the dark at –20°C.
2. 6% (w/v) Stock solution of pluronic F-127 (Sigma, cat. no. P2443) in dimethyl sulfoxide (DMSO). Store in the dark for up to 6 mo at –20°C.
3. Tyrode's solution: 145 mM NaCl, 4 mM KCl, 1 mM MgCl$_2$, 1 mM CaCl$_2$, 10 mM D-glucose, and 10 mM HEPES (pH 7.3). Make fresh as required.
4. A confocal microscopy system equipped with an inverted microscope (e.g., Olympus, Fluoview).

3. Methods

The hanging drop culture method for cardiomyocytic differentiation of ES cells in vitro is described in detail in Chapter 13.

3.1. RT-PCR Analysis for Cx Expression in Differentiating ES Cells

3.1.1. RNA Extraction

1. Prepare 20–300 embryoid bodies for RNA extraction. If necessary, ES cell-derived contracting cardiac myocytes can be microdissected from surrounding nonmuscle cells using a micromanipulator with a fine needle under an inverted microscope.

2. Dissolve the cell pellet or embryoid bodies in solution D. Typically, apply about 100 μL solution D/well of 48-well tissue culture plates and 500 μL/3.5-cm dish. Shear through a 19-gauge needle.
3. For 1 volume of solution D, add 0.1 volume of 2 M sodium acetate (pH 4.0), 1 volume of water-saturated phenol, 0.2 volumes of chloroform–isoamyl alcohol (49.1). Vortex mix well after the addition of each solution.
4. Vortex mix for 15 s and then leave on ice for 15 min.
5. Centrifuge at 10,000g for 15 min at 4°C.
6. Transfer the upper aqueous phase into a new tube and add an equal volume of isopropyl alcohol. Mix well and leave the tube at –20°C for 45 min.
7. Centrifuge at 3000g for 10 min at 4°C.
8. Suspend the pellet in 2 mL of 4 M LiCl by vortex mixing. This is to solubilize contaminating polysaccharides.
9. Centrifuge at 3000g for 10 min.
10. Dissolve the resulting pellet in 0.3 mL of solution D and transfer into a 1.5-mL microcentrifuge tube.
11. Add 0.3 mL of isopropyl alcohol, mix, and keep the tube at –20°C for 45 min.
12. Centrifuge at 10,000g for 15 min at 4°C.
13. Suspend the pellet with 70% ethanol, centrifuge for 5 min at 10,000g, and then discard supernatant. Repeat this step once. Dry the pellet in the air for several minutes.
14. Dissolved the RNA pellet in 50 μL of DNase digestion buffer (40 mM Tris-HCl, pH 8.0, 10 mM NaCl, 6 mM MgCl$_2$) containing 0.04 U/μL of RNase-free DNase I and 0.04 U/μL of RNase inhibitor. Incubate the sample at 37°C for 15 min.
15. Extract the reaction with phenol–chloroform–isoamyl alcohol (25:24:1) once.
16. Recover the aqueous phase and add 5 μL of 2 M sodium acetate and 150 μL of ethanol, then chill at –70°C for 15 min.
17. Centrifuge at 10,000g for 10 min at 4°C.
18. Suspend the pellet with 70% ethanol, centrifuge for 5 min at 10,000g, and then discard supernatant. Repeat this step once. Dry the pellet in the air for several minutes.
19. Dissolve the pellet in an appropriate volume of DEPC-treated water.
20. Quantitate RNA by diluting in water and reading the A_{260} and A_{280}.

3.1.2. cDNA Synthesis

1. Reverse transcribe 0.5 μg of the total RNA using 2.5 μM of random 9-mer, 0.25 U/μL of avian myeloblastosis virus reverse transcriptase XL, 5 mM MgCl$_2$, 1 mM of dNTP mixture, and 1 U/μL of RNase inhibitor, in 20 μL of 1X RNA PCR buffer at 42°C for 30 min (*see* **Note 1**).

3.1.3. Polymerase Chain Reaction

1. Mix 5 μL of first-strand reaction, 200 nM of the appropriate primer, 2.5 mM MgCl$_2$ and 0.625 U of *Taq* DNA polymerase (Takara) in 25 μL of 1X RNA PCR buffer. Place the sample in the thermal cycler and carry out the amplification reaction. Our amplification protocol is 1 cycle at 94°C for 2 min, followed by 30 to 35 cycles of denaturation at 94°C for 30 s, annealing at 55°C for *Cx*45, at 56°C for *Cx*40, and at 57°C for *Cx*43 for 30 s, and extension at 72°C for 30 s (1 min for *Cx*43).
2. Electrophorese 10 μL of the amplified sample with 2 μL of 6X loading buffer on 4% agarose gel (NuSieve 3:1 agarose) containing ethidium bromide (final concentration, 0.5 μg/mL) (*see* **Fig. 1**).

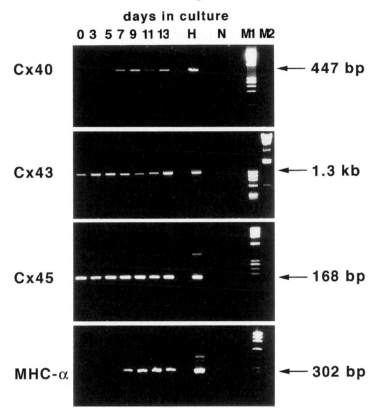

Fig. 1. RT-PCR analysis of *Cx*40, *Cx*43, *Cx*45, and MHC-α during cardiomyocytic differentiation of ES cells in vitro. H, sample form mouse heart as a positive control; N, negative control. Sample with no RNA was reverse-transcribed and PCR-amplified; M1, molecular weight marker. Product of φx 174 plasmid DNA cut by *Hae*III; M2, molecular weight marker. Product of λphage DNA cut by *Hind*III.

3.2. Immunofluorescent Labeling of Cxs in the ES Cells

The subject taken up here is visualization of the localization of *Cx* in ES cell-derived cardiomyocytes that express cardiac specific myosin heavy chain (e.g., ventricular MHC α/β). For this purpose, double immunofluorescent labeling of *Cx*43 and MHC α/β is performed using ES cells cultured in vitro.

1. Prepare embryoid bodies cultured on a 35-mm dish for d 7, 9, 11, and 13 (*see* **Note 2**).
2. Wash them twice with PBS and then fix in ethanol at –20°C for 10 min.
3. Block nonspecific binding using 5% skim milk in PBS for 20 min at room temperature.
4. Incubate cells with a mixture of a rabbit antibody against *Cx*43 (1/500 dilution) and a mouse monoclonal antibody against MHC α/β (1/5 dilution), overnight at 4°C.
5. Wash 3 times in PBS (5 min/wash).
6. Incubate cells with a mixture of goat Alexa 488 anti-rabbit IgG (1/1000 dilution) and goat Alexa 594 anti-mouse IgG (1/1000 dilution) for 1 h at room temperature.
7. Wash 3 times in PBS (5 min/wash).

8. Embed cells in Vectorshield and cover them with coverslips, which are sealed on dishes with nail polish.
9. Examine them under a fluorescence microscope and a confocal laser microscope.

3.3. Gap Junctional Intercellular Communication Measured by Lucifer Yellow Dye Coupling

1. Place the microscope with the phase-contrast equipment and micromanipulator on a vibration-free table.
2. Fill Femtotips with 1 μL of dye using microloader pipet tips matching 0.5- to 10-μL pipets.
3. Load the microinjection pipet into the instrument tube. Clamp the instrument tube on the instrument to the instrument tube holder of the micromanipulator. Connect the tube to the outputs of the microinjector.
4. Set the injection parameters, such as the compensation pressure (Pc) (pressures between 30 and 100 hPa are advisable when using Eppendorf's Fetotips), the injection pressure (Pi) (a pressure between 50 and 500 hPa is advisable when using Eppendorf's Femtotips), and the injection time (ti) (usually between 0.4 and 1.5 s).
5. Center the micropipet above the culture dish roughly by eye.
6. Lower the micropipet until it touches the culture medium.
7. Change to the lowest magnification available on the microscope (e.g., objective lens, ×4).
8. Focus on the cells.
9. Center the micropipet in the field of view in the microscope and lower it carefully, but not all the way to its focus (to avoid breaking the micropipet).
10. Change to the next higher magnification.
11. Repeat the steps 8–10 until you reach the working magnification (×40, objective lens). Now the cells should be in focus, and the micropipet should be centered but not in focus.
12. Lower the micropipet carefully by the manipulator until the tip comes just above the cell surface.
13. Fine position the capillary tip exactly above the cytoplasm that is to be injected.
14. Penetrate into the cell by further lowering the tip, and microinject the dye into the cell by pressing the "INJECT" key.
15. After injection, bring the tip just above the plane of the cell by turning the vertical knob of the manipulator and moving to the next cell.
16. Five minutes after the injection of dyes, monitor the extent of intercellular communication under fluorescence microscopy and record optical images on videotapes through a CCD camera. Digitize video images with a personal computer equipped with a video digitizer card.
17. Count the number of dye-transferring cells per injection of dye (*see* **Notes 3** and **4**).

3.4. Calcium Imaging Using Fluorescent Probes with Confocal Microscopy

1. Dissolve 50 μg of fluo-3/AM in 50 μL DMSO. Once prepared, this solution should preferably be used within 3 h.
2. Mix 25 μL of fluo-3/AM in DMSO with 12.5 μL of pluronic F-127 stock solution by vortex mixing.
3. Add 87.5 μL fetal calf serum (FCS) and mix well.
4. Add 250 μL Tyrode's solution and mix well (*see* **Note 5**).
5. Add 750 μL Tyrode's solution.
6. Wash the cells a few times with Tyrode's solution to remove excess culture medium.
7. Incubate the cells with the final loading solution for 30–60 min.

8. Wash the cells 3 times by incubation in Tyrode's solution for 5 min/wash.
9. View cells with epifluorescence optics. Confocal laser scanning microscopy (e.g., Fluoview, Olympus) can be used for imaging of fluo-3. The excitation wavelength of fluo-3 is 488 nm.

4. Notes

1. Both the 5′ PCR primer and the 3′ PCR primer for RT-PCR for *Cx*40 and *Cx*45 are designed to locate at the same exon (exon 2). To exclude the possibility that the bands obtained after RT-PCR are the result of contaminated genomic DNA, a negative experiment should be done for each sample without reverse transcriptase before PCR.
2. Video recording of the area, including beating cells before fixation for immunofluorescent staining, is helpful to identify whether positively stained areas of MHC-α/β correspond to areas of beating cells.
3. Microinjection of Lucifer yellow into a single ES cell is difficult, and Lucifer yellow is often injected into multiple cells. To confirm that Lucifer yellow is injected into a single ES cell, mix dextran-rhodamine (MW 10,000; Molecular Probes, cat. no. D-1824) with Lucifer yellow at the final concentration of 5% (w/v) and microinject the mixture into ES cells. Since dextran-rhodamine is too big to pass through gap junctions, if the fluorescent dyes are injected into a single cell, dextran-rhodamine remains in the single cell. For the exact measurement of gap junction function, count the number of dye-coupled cells only when microinjection into a single cell is successful.
4. A recent study on dye coupling in the early *Xenopus* embryo also showed that a rigorous and unambiguous demonstration of gap junctional intercellular communication demands the co-injection of permeant (Lucifer yellow or neurobiotin) and impermeant (dextran-rhodamine) tracers and the examination of paraffin or frozen sectioned specimens *(6)*. In case of dye coupling in ES cells, after co-injection of permeant and impermeant tracers, optical sectioning using confocal laser microscopy and subsequent three-dimensional reconstruction may enable one to get accurate results without sectioning.
5. Since fluo-3/AM is difficult to dissolve in water, dissolve fluo-3/AM in DMSO at the beginning. Fluo-3/AM in DMSO solution has to been mixed quickly with Tyrode's solution.

References

1. Kumar, N. M. and Gilula, N. B. (1996) The gap junction communication channel. *Cell* **84,** 381–388.
2. White, T. W. and Paul, D. L. (1999) Genetic diseases and gene knockouts reveal diverse connexin functions. *Annu. Rev. Physiol.* **61,** 283–310.
3. Oyamada, Y., Komatsu, K., Kimura, H., Mori, M., and Oyamada, M. (1996) Differential regulation of gap junction protein (connexin) genes during cardiomyocytic differentiation of mouse embryonic stem cells in vitro. *Exp. Cell Res.* **229,** 318–326.
4. Westfall, M. V., Pasyk, K. A., Yule, D. I., Samuelson, L. C., and Metzger, J. M. (1997) Ultrastructure and cell-cell coupling of cardiac myocytes differentiating in embryonic stem cell cultures. *Cell. Motil. Cytoskeleton* **36,** 43–54.
5. Oyamada, M., Oyamada, Y., Komatsu, K., Mori, M., and Takamatsu, T. (2000) In vitro cardiomyocytic differentiation of mouse embryonic stem cells deficient in gap junction protein connexin43. *Cardiac and Vascular Regeneration* **1,** 54–64.
6. Landesman, Y., Goodenough, D. A., and Paul, D. L. (2000) Gap junctional communication in the early *Xenopus* embryo. *J. Cell Biol.* **150,** 929–936.

8

Embryonic Stem Cell Differentiation as a Model to Study Hematopoietic and Endothelial Cell Development

Stuart T. Fraser, Minetaro Ogawa, Satomi Nishikawa, and Shin-Ichi Nishikawa

1. Introduction

1.1. Summary

With the attractive potential therapeutic uses of both endothelial and hematopoietic cells in medicine, much attention has been focused upon the differentiation of embryonic stem (ES) cells into mature cells of these lineages. In this chapter, we present a culture system that has successfully generated both complex endothelial structures and mature hematopoietic cells from undifferentiated ES cells in vitro. The simplicity of this system allows detailed analyses of factors and developmental steps essential in the generation of vascular structures and mature hematopoietic cell types.

1.2. Differentiation of Mesodermal, Endothelial, and Hematopoietic Cells in the Embryo

As the ES cell differentiation scheme presented in this chapter is based on developmental processes occurring within the embryo, an introduction to the generation of the mesoderm and its derivatives during embryogenesis is of use to the reader. The mesoderm is a sheet of cells located between the ectoderm and endoderm in the early developing embryo. In differentiating from an epithelial structure to a mesenchymal population, mesodermal cells lose expression of the cell adhesion molecule epithelial-cadherin (E-cadherin) (*1*). A subset of mesodermal cells up-regulate expression of vascular endothelial growth factor receptor 2, known in the mouse as fetal liver kinase 1 (Flk1). This Flk1$^+$ population includes endothelial cell precursors (*2*).

As the embryo develops, Flk1$^+$ cells in the yolk sac generate the blood islands that are composed of round clusters of hematopoietic cells surrounded by endothelial cells (*3*). Mice that lack Flk1 fail to generate blood islands and die early in development (*4*). Another mesodermal population, the paraxial mesoderm can be identified according to expression of platelet-derived growth factor receptor α (PDGFRα) (*5*). While this population can generate endothelial cells, it does not generate hematopoietic cells. The Flk1$^+$ (lateral) mesoderm, on the other hand, is a source of both endothelial

From: *Methods in Molecular Biology, vol. 185: Embryonic Stem Cells: Methods and Protocols*
Edited by: K. Turksen © Humana Press Inc., Totowa, NJ

and hematopoietic cells. During vasculogenesis, endothelial cells derived from both paraxial and lateral mesoderm form a vascular plexus throughout the yolk sac and embryo body *(6)*. This plexus is then modified into a highly complex branched vascular network during angiogenesis *(7)*.

The first hematopoietic cells to appear, the primitive erythrocytes, are thought to arise directly from the lateral mesoderm. Definitive erythroid cells, myeloid, and lymphoid cells are generated from endothelial cells expressing vascular endothelial–cadherin (VE-cadherin) *(8)*. Mice that lack the putative transcription factor, AML1/Runx1/Cbfa2, cannot generate definitive hematopoietic cells from endothelial cells and die early in embryogenesis *(9)*, clearly illustrating the importance of endothelial cells as a source of definitive hematopoietic cells. In early developmental stages, endothelial and hematopoietic cells co-express many markers, such as platelet endothelial cell adhesion molecule-1 (PECAM-1) *(10)*, CD34 *(11)*, AA4 *(12)*, and GSL I-B4 isolectin *(13)*, demonstrating the close relationship between endothelial and hematopoietic cell development, and thus making the separation of either cell type by surface phenotype difficult.

1.3. Differentiation of Mesoderm and Mesodermal Derivatives from ES Cells

1.3.1. Induction of Mesodermal Cells in the Absence of Embryoid Body Formation

It has been previously demonstrated that ES cells, when induced to differentiate in culture, recapitulate early embryonic events in vitro *(14–19)*. In these cultures, ES cells differentiate in complex three-dimensional cell aggregates termed embryoid bodies (EBs). To reduce the potential for interactions and, therefore, control more stringently the culture conditions, our group developed a culture system by which ES cells differentiated in the absence of EBs along a 2-dimensional plane. This system, therefore, lacked the structural complexity of the EB differentiation, and yet it effectively generated Flk1+ putative mesodermal cells *(20)*, indicating that the generation of mesodermal cells from ES cells did not require 3-dimensional cellular interactions or supportive feeder layers. Paraxial mesodermal cells (PDGFRα+ E-cadherin−) were also generated in this system, although the potential of these cells has not been investigated. As lateral mesodermal cells are a source of not just endothelial cells but also hematopoietic cells, we purified these cells using flow cytometry according to the expression of Flk1 in the absence of E-cadherin. **Fig. 1** illustrates the in vitro differentiation system used to generate mesodermal cells and, in turn, endothelial and hematopoietic cells. Such a system is far easier to monitor and manipulate than the EB system previously described.

1.3.2. Differentiation of Endothelial Cells from ES-Derived Flk1+ Mesodermal Cells

Endothelial cells can be generated from ES cells in EB *(15)*. However, endothelial cell development cannot easily be observed or manipulated in a complex 3-dimensional structure. To deal with this, our group extended the differentiation system presented above to the differentiation of endothelial cells from mesodermal cells. To enrich the endothelial population from nonmesodermal lineages, Flk1+ cells were sorted by flow cytometry and recultured *(20)*. During reculture of the Flk1+ cells, numerous mesodermally derived cell lineages appear. Therefore, we chose to further isolate

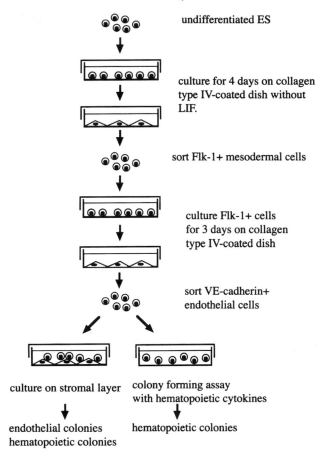

undifferentiated ES

culture for 4 days on collagen
type IV-coated dish without
LIF.

sort Flk-1+ mesodermal cells

culture Flk-1+ cells
for 3 days on collagen
type IV-coated dish

sort VE-cadherin+
endothelial cells

culture on stromal layer   colony forming assay
with hematopoietic cytokines

endothelial colonies   hematopoietic colonies
hematopoietic colonies

Fig. 1. In vitro differentiation of endothelial and hematopoietic cells from ES-derived mesodermal cells.

endothelial cells by purifying VE-cadherin⁺ cells by flow cytometry. VE-cadherin expression is highly restricted to endothelial cells *(21)* compared to other surface proteins, such as PECAM-1 *(10)*, making it a useful marker for isolating endothelial cells. An in vitro system that allows quantification of endothelial colony forming activity has also been developed *(22)*. Endothelial colony forming activity is strictly dependent on the stroma used, with the bone marrow-derived OP9 cell line *(23)* being the most supportive stroma of those examined so far *(22)*. While the supportive activity of these stroma has not been fully resolved, vascular endothelial growth factor (VEGF) is known to be expressed by OP9 and is essential in the growth of endothelial colonies.

1.3.3. Generation of Hematopoietic Cells from ES Cells

Hematopoietic cells can be generated from ES cells *(17)*. By culturing EBs, numerous hematopoietic cell types were generated. To avoid the complexity of the EB system and to improve the identification of developmental stage intermediate stages, ES cells were differentiated as a monolayer on OP9 stromal cells. This system resulted in the generation of all major hematopoietic cell types including lymphocytes *(24)*. As mesodermal cells are the source of the hematopoietic lineage, we assessed our two-dimensional feeder-layer free differentiation system for hematopoietic ability. As ectodermal cells are known to inhibit erythropoiesis *(25)*, Flk1$^+$ mesodermal cells were first isolated by cell sorting and then recultured. This resulted in the generation of hematopoietic cells in the absence of EB formation and feeder cells. As endothelial cells are know to be a source of definitive hematopoietic cells in the embryo, we also assessed the hematopoietic potential of VE-cadherin$^+$ cells from ES-derived Flk1$^+$ cells. VE-cadherin$^+$ cells, in contrast to their Flk1$^+$ precursors, required hematopoietic growth factors to generate hematopoietic cells.

2. Materials

2.1. ES Cell Lines

With numerous ES cell lines currently available, each line should first be examined for its ability to generate Flk1$^+$ mesodermal cells. The ES cell line most commonly used in our laboratory is CCE *(26)*. The methods presented below are optimized for CCE (*see* **Note 1**).

2.2. Reagents

2.2.1. Cell Culture Media

1. ES cell differentiation medium: (*see* **Note 2**).
   a. Alpha Modified Eagle's Medium (αMEM) (Gibco BRL, USA, cat. no. 11900-024) supplemented with 50 U/mL penicillin and 50 µg/mL streptomycin.
   b. Fetal calf serum (FCS) (Gibco BRL, cat. no. 26140-079), 10% final concentration (*see* **Note 3**).
   c. 2-Mercaptoethanol (Merck, Germany, cat. no. ZA11288), 5×10^{-5} *M* final concentration (*see* **Note 4**).
2. OP9 culture medium:
   a. αMEM supplemented with 50 U/mL penicillin and 50 µg/mL streptomycin.
   b. FCS (JRH Biosciences, KS, US, cat. no. 12003-78P), 20% final concentration (*see* **Note 5**).

2.3. Induction and Purification of Mesodermal Cells

1. Collagen type IV-coated 6-well plates, Biocoat (Becton Dickinson, MA, USA, cat. no. 354428) (*see* **Note 6**).
2. Cell dissociation buffer (Gibco BRL, cat. no. 13150-016).
3. Phosphate-buffered saline (PBS) lacking Ca^{2+} and Mg^{2+}: 2.89 g Na$_2$HPO$_4$ (12 H$_2$O), 0.2 g KH$_2$PO$_4$, 8.0 g NaCl, 0.2 g KCl, and distilled water up to 1000 mL.
4. Normal mouse serum (NMS). NMS can be prepared from mice in-house or can be purchased (OEM Co. Ltd., US, cat. no. 175-85340) (*see* **Note 7**).
5. Anti-Flk1 (AVAS12) *(27)* (Pharmingen, US, cat. no. 28185A) monoclonal antibody (mAb), fluorescently labeled.
6. Anti-E-cadherin (ECCD2) *(28)* MAb, conjugated to a second fluorochrome.
7. Hanks' buffered saline solution (HBSS) (Gibco BRL, cat. no. 14185-052) with 1% bovine serum albumin (BSA) (Sigma, US, cat. no. A-1253) (HBSS/BSA).
8. HBSS/BSA with 5 µg/mL propidium iodide (Sigma, U. S. A., cat. no. P-4170) (HBSS/BSA/PI).

2.4. Differentiation and Isolation of Endothelial Cells from ES-Derived Flk1⁺ Cells

1. Fluorescently labeled anti-VE-cadherin mAb (VECD1) *(29)*.
2. HBSS/BSA and HBSS/BSA/PI.
3. Falcon 6-well polystyrene culture dishes (Becton Dickinson, cat. no. 353046).

2.5. Generation of Hematopoietic Cells from ES-Derived Endothelial Cells

1. Mouse cytokines: Erythropoietin (Epo) (Roche Molecular Biochemicals, cat. no. 1276964), stem cell factor (SCF) (R & D Systems, US, cat. no. 455-MC), interleukin-3 (IL-3) (R & D Systems, cat. no. 403-ML), interleukin-7 (IL-7) (R & D Systems, cat. no. 407-ML), Flt3 ligand (cat. no. 427-FL).

3. Methods

3.1. Induction and Purification of Mesodermal Cells

1. Into each well of a 6-well collagen type IV-coated plate, add 1×10^4 undifferentiated ES cells.
2. Add 3 mL of differentiation medium to each well. Leave undisturbed for 4 d in a 37°C incubator with 5% CO_2 environment (*see* **Note 8**).
3. After 4 d, harvest the cells by first removing the medium, washing with PBS, and incubating at 37°C with 2 mL cell dissociation buffer for 20 min. Pipet cells from the surface of the plate.
4. Incubate single-cell suspensions with NMS for 20 min on ice. Normally, we incubate 1×10^7 cells in 100 µL NMS.
5. Add an appropriate concentration of fluorescently labeled anti-Flk1 and anti-E-cadherin mAbs to cell suspension in NMS and incubate for 20 min on ice. Wash the cells twice with HBSS/BSA. Resuspend the cells in HBSS/BSA/PI for dead cell exclusion.
6. Sort the Flk1⁺ E-cadherin⁻ cells. Mesodermal cells express Flk1 but not E-cadherin (*see* **Fig. 2A**).

3.2. Differentiation and Isolation of Endothelial Cells from ES-Derived Flk1⁺ Cells

1. Add 3×10^5 Flk1⁺ E-cadherin⁻ cells to each well of a collagen type IV- or gelatin-coated 6-well plate (*see* **Note 9**).
2. Add 3 mL ES cell differentiation medium.
3. Incubate at 37°C for 3 d under 5% CO_2.
4. Harvest cells using cell dissociation buffer as described in **Subheading 3.1, Step 1**.
5. Block with NMS for 20 min on ice.
6. Stain with an appropriate concentration of labeled anti-VE-cadherin mAb.
7. Analyze or sort by flow cytometry (*see* **Note 10** and **Fig. 2B**).

3.3. ES-Derived Endothelial Colony Assay

1. Three days prior to commencing Flk1⁺ cell culture, split one confluent 25-cm² flask of OP9 stromal cells into 3×6-well dishes.
2. Add 2 mL of OP9 medium to each well.
3. Three days later, when OP9 is confluent, add up to 5000 ES-derived Flk1⁺ cells per well.
4. After 3 d, sheets of cells can be seen growing on the OP9 stroma. These cultures can be maintained for several more days, but should not be allowed to overgrow (*see* **Note 11**).

A

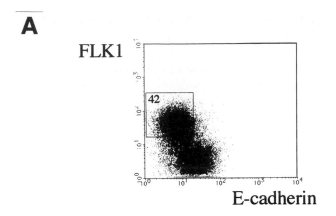

B

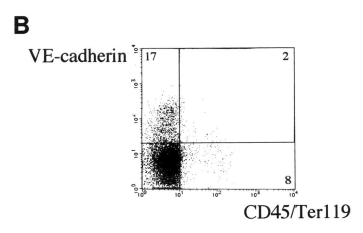

Fig. 2. Representative fluorescence-activated cell sorter (FACS) profiles of Flk1+ cell differentiation (**A**) after 4 d of culture on collagen type IV-coated dishes. VE-cadherin+ endothelial cell differentiation is compared to hematopoiesis (CD45/Ter119) after 3 d of culture of Flk1+ cells on collagen type IV-coated dishes (**B**). Boxed area in (**A**) represents the Flk1+ population sorted. Numbers in the sort region and quadrant indicate the frequency of the given population among the total live gated population.

3.4. Generation of Hematopoietic Cells from ES-Derived Endothelial Cells

3.4.1. Generating Primitive Erythroid Cells from ES Cells

1. Sort Flk1+ cells as described.
2. Add 1×10^4 cells to each well of a 6-well plate containing confluent OP9 stroma.
3. Add 3 mL ES cell differentiation medium/well supplemented with 2 U/mL Epo.
4. Analyze hematopoietic cells that have appeared after 4 d (*see* **Note 12**).

3.4.2. Generating Definitive Hematopoietic Lineages from ES-Derived VE-Cadherin+ Cells

1. Sort VE-cadherin+ cells as described.
2. Add 1×10^4 cells to a 25-cm^2 confluent flask of OP9 stromal cells.

3. Add 6 mL ES cell differentiation medium supplemented with cytokines. For erythroid–myeloid cultures, add 2 U/mL Epo, 100 U/mL SCF, and 200 U/mL IL-3. For B lymphoid cultures, we add 100 U/mL, 60 U/mL IL-7, and 50 U/mL Flt3 ligand (*see* **Note 13**).

4. Change the medium every 3–4 d. In general, definitive Ter-119$^+$ erythroid cells first appear in culture after 2–3 d. After 5 to 7 d, mature myeloid cells appear expressing the markers Gr-1 and Mac-1. CD19$^+$ B lymphoid cells appear after 10–14 d of culture (*see* **Note 14**).

4. Notes

4.1. Mesoderm Induction

1. CCE ES cells, when differentiated, generate Flk1$^+$ cells at a frequency of 30–40% of live gated cells.

2. We recommend using medium less than 4 wk old to obtain high yields of Flk1$^+$ cells.

3. FCS has a critical influence in obtaining high yields of Flk1$^+$ cells. Batch checks are therefore highly recommended for finding appropriate serum batches. Before commencing large-scale studies of ES cell differentiation into Flk1$^+$ cells, we checked 33 batches from 18 different companies. The frequency of Flk1$^+$ cells generated after 4 d in culture using different FCS batches ranged from 13 to 43%, highlighting the need for FCS batch checks.

4. 2-Mercaptoethanol is essential in obtaining both high yields of Flk1$^+$ cells and also later in hematopoietic cell differentiation, particularly during B lymphopoiesis.

5. It is highly recommended that FCS batch checks are conducted before large-scale use of OP9. The condition of OP9 is highly dependent upon FCS batch, and this stroma should be maintained carefully, as it easily loses its supportive activity. FCS can be checked by culturing OP9 for 10 passages in medium containing different FCS batches. The morphology of OP9 should be maintained over this period (i.e., large flat cells, not fibroblastic or transformed). Furthermore, the cell number should remain at 1–1.5×10^6 cells/25-cm^2 flask when confluent. Supportive activity is best analyzed by co-culture with Flk1$^+$ cells derived from ES cells, resulting in the generation of endothelial and hematopoietic cells as described in this chapter.

6. Several matrices have been analyzed for their ability to enhance mesoderm induction by ES cells including: gelatin, fibronectin, and types I and IV collagen. While all matrices generated mesodermal cells to some degree, type IV collagen resulted in the highest yield of Flk1$^+$ cells and was therefore routinely used.

7. NMS is used to prevent binding of the primary antibodies to the Fc receptors on the target cell surface.

8. Flk1$^+$ cells first appear after 3 d of differentiation. One day later, the frequency of Flk1$^+$ cells will have increased by 10-fold. By d 5 of culture, the frequency of Flk1$^+$ cells decreases, and endothelial markers begin to appear, indicating the differentiation of the mesodermal cells into endothelial cells. Therefore, d 4 was chosen as the optimal timepoint for isolating mesodermal cells. Within 4 d of differentiation, the cell number will increase 100-fold, from 1×10^4 cells/well to 1–3×10^6 cells/well.

9. Once sorted, Flk1$^+$ cells can differentiate into endothelial cells on either collagen type IV- or gelatin-coated dishes with similar efficiency. Gelatin has the advantage of being considerably cheaper than collagen type IV, as well as being easier for coating dishes. Gelatin solution (0.1%) can be prepared by dissolving 0.2 *g* of gelatin (Sigma, cat. no. G-2500) in 200 mL distilled water, followed by autoclaving. For coating dishes, apply enough solution to cover the bottom of the dish, leave for at least 2 h, then aspirate the gelatin, and use.

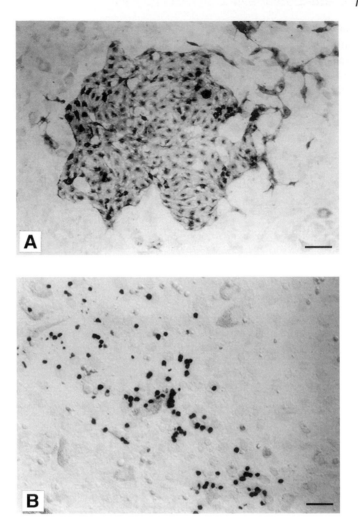

Fig. 3. Flk1$^+$ endothelial cells (**A**) and embryonic hemoglobin$^+$ primitive erythroid cells (**B**) can be generated from ES-derived Flk1$^+$ mesodermal precursors during co-culture on OP9 stromal cells.

10. In general, 3×10^5 Flk1$^+$ cells cultured on collagen type IV dishes for 3 d yield approximately $3–5 \times 10^5$ cells, of which 15–20% are usually VE-cadherin$^+$, although this can vary between ES cell lines examined.

11. Flk1$^+$ cells growing on stromal cells can generate numerous cell lineages. Immunocytochemistry is the easiest method for demonstrating the presence of endothelial cells in the cultures. Several markers such as VE-cadherin, PECAM-1, and Flk1 are suitable for immunostaining of endothelial colonies. Immunostaining of colonies in the culture dish is an approach commonly used by our group *(22,30)*. An example of an immunostained endothelial colony is presented in **Fig. 3A**.

12. Primitive erythropoiesis can be confirmed by retrieving the hematopoietic cells from the culture, preparing cytospots and immunostaining for embryonic hemoglobin. Embryonic hemoglobin$^+$ cells derived from Flk1$^+$ cells cultured on OP9 are shown in **Fig. 3B**. This

system is particularly useful in analyzing the potential of genetically modified (such as null mutant or transgenic) ES cells to generate primitive erythroid cells.

13. Flt3ligand has recently been shown to dramatically enhance the production of B lymphoid cells from ES cells cultured on OP9 stromal cells *(31)*.

14. OP9 is a useful stroma, as it is deficient in macrophage-colony stimulating factor (M-CSF) production, resulting in few macrophages in the culture *(23)*. If macrophages still grow excessively during ES cell differentiation into hematopoietic cells, hampering the development of other hematopoietic lineages, anti-M-CSF receptor (c-fms) monoclonal antibody, AFS98 *(32)* can be added to inhibit macrophage development.

Acknowledgments

This work is the accumulated knowledge of members, past and present, of the Department of Molecular Genetics, Kyoto University. Thanks must be given to members who have provided useful suggestions, including Drs. Jun Yamashita, Masanori Hirashima,Tetsuhiro Fujimoto, and Hisahiro Yoshida. The authors also wish to thank Dr. Kanako Yoshikawa for useful comments. This work was supported by grants from the Ministry of Education, Science, Sports and Culture (nos. 12219209, 10770139, and 12215071), the Japan Society for the Promotion of Science (no. 12670301), and the Ministry of Health and Welfare of Japan.

References

1. Burdsal, C. A., Damsky, C. H., and Pedersen, R. A. (1993) The role of E-cadherin and integrins in mesoderm differentiation and migration at the mammalian primitive streak. *Development* **118,** 829–844.

2. Yamaguchi, T. P., Dumont, D. J., Conlon, R. Λ., Breitman, M. L., and Rossant, J. (1993) flk-1, an flt-related receptor tyrosine kinase is an early marker for endothelial cell precursors. *Development* **118,** 489–498.

3. Moore, M. A. and Metcalf, D. (1970) Ontogeny of the haemopoietic system: yolk sac origin of in vivo and in vitro colony forming cells in the developing mouse embryo. *Br. J. Haematol.* **18,** 279–296.

4. Shalaby, F., Rossant, J., Yamaguchi, T. P., Gertsenstein, M., Wu, X. F., Breitman, M. L., and Schuh, A. C. (1995) Failure of blood-island formation and vasculogenesis in Flk1-deficient mice. *Nature* **376,** 62–66.

5. Mercola, M., Wang, C. Y., Kelly, J., Brownlee, C., Jackson-Grusby, L., Stiles, C., and Bowen-Pope, D. (1990) Selective expression of PDGF A and its receptor during early mouse embryogenesis. *Dev. Biol.* **138,** 114–122.

6. Risau, W. and Flamme, I. (1995) Vasculogenesis. *Annu. Rev. Cell Dev. Biol.* **11,** 73–91.

7. Coffin, D. J. and Poole, T. J. (1988) Embryonic vascular development: immunohistological identification of the origin and subsequent morphogenesis of the major vessel primordia in quail embryos. *Development* **102,** 735–748.

8. Nishikawa, S. I., Nishikawa, S., Kawamoto, H., Yoshida, H., Kizumoto, M., Kataoka, H., and Katsura, Y. (1998) In vitro generation of lymphohematopoietic cells from endothelial cells purified from murine embryos. *Immunity* **8,** 761–769.

9. North, T., Gu, T.-L., Stacy, T., Wang, Q., Howard, L., Binder, M., et al. (1999) Cbfa2 is required for the formation of intra-aortic hematopoietic clusters. *Development* **126,** 2563–2575.

10. Vecchi, A., Garlanda, C., Lampugnani, M. G., Resnati, M., Matteucci, C., Stoppacciaro, A., et al. Schnurch, H., Risau, W., Ruco, L., Mantovani A, et al (1994) Monoclonal antibodies

specific for endothelial cells of mouse blood vessels. Their application in the identification of adult and embryonic endothelium. *Eur. J. Cell Biol.* **63**, 247–254.

11. Young, P. E., Baumhueter, S., and Lasky, L. A. (1995) The sialomucin CD34 is expressed on hematopoietic cells and blood vessels during murine development. *Blood* **85**, 96–105.

12. Petrenko, O., Beavis, A., Klaine, M., Kittappa, R., Godin, I., and Lemischka, I. R. (1999) The molecular characterization of the fetal stem cell marker AA4. *Immunity* **10**, 691–700.

13. Nishikawa, S. I., Nishikawa, S., Fraser, S., Fujimoto, T., Yoshida, H., Hirashima, M., and Ogawa, M. A model for embryonic development of hemopoieitic stem cells through endothelial cells, in *Developmental biology of the hematopoietic system.* (Zon, L. ed.) Oxford Univ. Press, Oxford, UK, *in press.*

14. Risau, W., Sariola, H., Zerwes, H. G., Sasse, J., Ekblom, P., Kemler, R., and Doetschman, T. (1988) Vasculogenesis and angiogenesis in embryonic-stem-cell-derived embryoid bodies. *Development* **102**, 471–478.

15. Wiles, M. V. and Keller, G. (1991) Multiple hematopoietic lineages develop from embryonic stem (ES) cells in culture. *Development* **111**, 259–267.

16. Schmitt, R. M., Bruyns, E. and Snodgrass, H. R. (1991) Hematopoietic development of embryonic stem cells in vitro: cytokine and receptor gene expression. *Genes Dev.* **5**, 728–740.

17. Burkert, U., von Ruden, T., and Wagner, E. F. (1991) Early fetal hematopoietic development from in vitro differentiated embryonic stem cells. *New Biol.* **3**, 698–708.

18. Vittet, D., Prandini, M. H., Berthier, R., Schweitzer, A., Martin-Sisteron, H., Uzan, G., and Dejana, E. (1996) Embryonic stem cells differentiate in vitro to endothelial cells through successive maturation steps. *Blood* **88**, 3424–3231.

19. Kabrun, N., Buhring, H. J., Choi, K., Ullrich, A., Risau, W., and Keller, G. (1997) Flk1 expression defines a population of early embryonic hematopoietic precursors. *Development* **124**, 2039–2048.

20. Nishikawa, S.-I., Nishikawa, S., Hirashima, M., Matsuyoshi, N., and Kodama, H. (1998) Progressive lineage analysis by cell sorting and culture identifies FLK1⁺VE-cadherin⁺ cells at a diverging point of endothelial and hemopoietic lineages. *Development* **125**, 1747–1757.

21. Gory, S., Vernet, M., Laurent, M., Dejana, E., Dalmon, J., and Huber, P. (1999) The vascular endothelial-cadherin promoter directs endothelial-specific expression in transgenic mice. *Blood* **93**, 184–192.

22. Hirashima, M., Kataoka, H., Nishikawa, S., Matsuyoshi, N., and Nishikawa, S.-I. (1999) Maturation of embryonic stem cells into endothelial cells in an in vitro model of vasculo-genesis. *Blood* **93**, 1253–1263.

23. Kodama, H., Nose, M., Yamaguchi, Y., Tsunoda, J., Suda, T., Nishikawa, S., and Nishikawa, S.-I. (1992) In vitro proliferation of primitive hemopoietic stem cells supported by stromal cells: evidence for the presence of a mechanism(s) other than that involving c-kit receptor and its ligands. *J. Exp. Med.* **176**, 351–361.

24. Nakano, T., Kodama, H., and Honjo, T. (1994) Generation of lymphohematopoietic cells from embryonic stem cells in culture. *Science* **265**, 1098–1101.

25. Kessel, J. and Fabian, B. (1987) Inhibitory and stimulatory influences on mesodermal erythropoiesis in the early chick blastoderm. *Development* **101**, 45–49.

26. Robertson, E., Bradley, A., Kuehn, M., and Evans, M. (1986) Germ-line transmission of genes introduced into cultured pluripotential cells by retroviral vector. *Nature* **323**, 445–448.

27. Kataoka, H., Takakura, N., Nishikawa, S., Tsuchida, K., Kodama, H., Kunisada, T., et al. (1997) Expressions of PDGF receptor alpha, c-Kit and Flk1 genes clustering in mouse

chromosome 5 define distinct subsets of nascent mesodermal cells. *Dev. Growth Differ.* **39,** 729–740.

28. Shirayoshi, Y., Nose, A., Iwasaki, K., and Takeichi, M. (1986) N-linked oligosaccharides are not involved in the function of a cell-cell binding glycoprotein E-cadherin. *Cell Struct. Funct.* **11,** 245–252.

29. Matsuyoshi, N., Toda, K., Horiguchi, Y., Tanaka, T., Nakagawa, S., Takeichi, M., and Imamura, S. (1997) In vivo evidence of the critical role of cadherin-5 in murine vascular integrity. *Proc. Assoc. Am. Physicians* **109,** 362–371.

30. Ogawa, M., Kizumoto, M., Nishikawa, S., Fujimoto, T., Kodama, H., and Nishikawa, S.-I. (1999) Expression of alpha4-integrin defines the earliest precursor of hemopoietic cell lineage diverged from endothelial cells. *Blood* **93,** 1168–1177.

31. Cho, S. K., Webber, T. D., Carlyle, J. R., Nakano, T., Lewis, S. M., and Zuniga-Pflucker, J. C. (1999) Functional characterization of B lymphocytes generated in vitro from embryonic stem cells. *Proc. Natl. Acad. Sci. USA* **96,** 9797–9802.

32. Sudo, T., Nishikawa, S., Ogawa, M., Kataoka, H., Ohno, N., Izawa, A., et al. (1995) Functional hierarchy of c-kit and c-fms in intramarrow production of CFU-M. *Oncogene* **11,** 2469–2476.

9

Analysis of *Bcr-Abl* Function Using an In Vitro Embryonic Stem Cell Differentiation System

Takumi Era, Stephane Wong, and Owen N. Witte

1. Introduction

1.1. Chronic Myelogenous Leukemia: A Hematopoietic Stem Cell Disease

Numerous causative genes involved in human cancers, such as leukemias and lymphomas, have been identified *(1)*. Considerable evidence has accumulated showing that disruptions of these genes can affect hematopoietic cell differentiation and growth. However, it remains largely unknown how these molecules function directly in hematopoietic stem cells.

Chronic myelogenous leukemia (CML) is a hematopoietic stem cell disease associated with the specific chromosomal translocation, t(9;22) *(2)*. This abnormal chromosome generates a fusion gene product, *Bcr-Abl*, which encodes a constitutively active tyrosine kinase essential for the development of CML *(3–5)*.

Different in vitro experimental systems have been developed to demonstrate that *Bcr-Abl* expression can lead to immortalization, transformation, and anti-apoptosis. Specifically, enforced expression of *Bcr-Abl* results in immortalization and transformation of pro/pre-B lymphocytes and fibroblastic cells in vitro *(6,7)* and inhibits the apoptosis of leukemic cell lines induced by growth factor withdrawal *(8)*. In addition, this oncogene augments the formation of primary multilineage hematopoietic colonies in low growth factor conditions *(9)*. Retroviral gene transfer and transgenic expression in rodents have shown that *Bcr-Abl* can induce different types of leukemia with varying degrees of latency and penetrance *(10–13)*. These studies suggest that *Bcr-Abl* plays a role in the generation of human and murine leukemias. The level and timing of *Bcr-Abl* expression was not regulated in these types of studies, therefore, the analysis of these models has not determined the direct and immediate effects of *Bcr-Abl* on hematopoietic stem cells from those requiring secondary genetic or epigenetic changes selected during the pathogenic process.

1.2. The Combination of TET-Regulated Gene Expression with In Vitro ES Cell Differentiation: a Powerful System for the Analyses of Genetic Change on Hematopoietic Development

We needed an in vitro system that reproducibly produced stem cells capable of undergoing multilineage differentiation. Previous work defined an embryonic system

From: *Methods in Molecular Biology, vol. 185: Embryonic Stem Cells: Methods and Protocols*
Edited by: K. Turksen © Humana Press Inc., Totowa, NJ

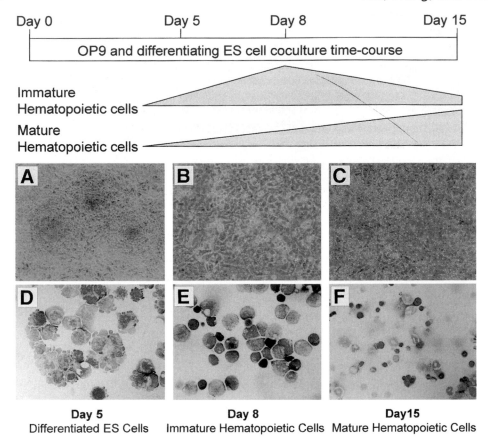

Fig. 1. Hematopoiesis of in vitro ES cell differentiation with M-CSF-deficient OP9 stromal cells. Day 5 ES differentiating cell colonies consist of immature mesoderm-like cells (**A** and **D**), while d 8 differentiating ES cells consist of immature hematopoietic cells (**B** and **E**). D 14 differentiating ES cells consist of mature erythrocytes and granulocytes (**C** and **F**). Magnification of panels A, B, and C at 100×. Magnification of panels D, E, and F at 500×. (*See* color plate 1, following p. 254).

(ES) cell co-cultivation strategy with the OP9 stromal cell line that is highly reproducible *(14–17)*. The OP9 stromal cell line is derived from the calvaria of osteopetrosis (op/op) mutant mice and cannot produce macrophage colony-stimulating factor (M-CSF). This is important because M-CSF directs the overproduction of macrophages, which inhibits hematopoietic development of other lineages during in vitro ES cell differentiation. The ES/OP9 co-culture system was chosen because a reasonable cell number (>1 × 10^6/culture) can be obtained to analyze hematopoietic development in modest amounts of fetal calf serum (FCS) without exogenous growth factors *(14–17)*. In this co-culture system, ES cells generate hemangioblasts (d 4 to 5), which produce immature hematopoietic stem and progenitor cells (d 7 to 8), and eventually mature blood cell elements (d 10 to 15) (**Fig. 1**).

   To directly investigate the effects of a gene such as *Bcr-Abl* in stem cell development, we needed a highly regulatable gene expression system. To control *Bcr-Abl* expression during hematopoietic development, we have combined an in vitro ES cell differentiation system with tetracycline (TET)-regulated expression *(18,19)*.

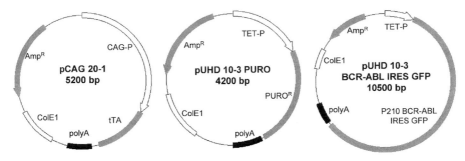

CAG-P: CMV-Enhancer + Chicken β-actin promoter
tTA: Tetracycline response element + 3 tandem repeats
of the VP-16 minimal transcriptional activation domain
(3x ^{435}PADALDDFDLDML^{447})
TET-P: Tetracycline controllable promoter
PURO^R: puromycin resistance gene
ColE1: replication origin of the plasmid
Amp^R: ampicillin resistance gene

Fig. 2. Constructs for TET-off system. pCAG20-1 plasmid carries a tTa gene driven by CAG promoter (left). pUHD 10-3 PURO has a puromycin resistance gene driven by TET responsive promoter (center). These two plasmids are used to establish parental ES cell lines (*see* text). pUHD 10-3 *Bcr-Abl* IRES GFP plasmid carries the p210 *Bcr-Abl* and GFP cDNAs driven by a TET responsive promoter and IRES, respectively (right).

In order to express the tetracycline transactivator (tTA), we needed to utilize a promoter system that was capable of expression in primitive ES cells, hematopoietic stem and progenitor cells, and throughout all the different intermediate and mature hematopoietic cell types produced in these cultures. Data from our laboratory and other groups have defined a composite control element comprised of the chicken actin promoter with a cytomegalovirus (CMV) enhancer (CAG; **Fig. 2**) that can drive gene expression in undifferentiated and differentiated ES cells *(19,21)*. We used a modified-tTA gene containing a tetracycline response element fused at its carboxy terminus with three minimal transcriptional domains of VP-16, FFF; 3× ^{435}PADALDDFDLDML^{447}. This form of the transactivator has less toxicity for cell growth than the previous forms *(20)*. In this system, gene expression is tightly regulated by TET concentration. In as short as 24 h, the TET system is able to induce gene expression by several logs (*see* Figs. 1 and 2 of **ref. *19***).

2. Materials

All culture medium supplemented with or without TET is made fresh and used within 1 wk of preparation. Tissue culture-treated dishes and multiwell plates are used for all cultures. We used the ES cell E14tg2a derived from the 129/Ola strain. This ES cell line does not require feeder cells for maintenance in an undifferentiated state *(21)*.

2.1. General Maintenance of ES Cells

1. Glasgow minimum essential medium (GMEM) (GIBCO/BRL, cat. no. 11710-035; or Bio Whittaker, cat. no. 12-739F), store at 4°C.

2. Fetal bovine serum (FBS) pretested for ES cells (*see* **Note 1**).
3. Phosphate-buffered saline (PBS) without calcium chloride and magnesium chloride.
4. L-glutamine (200 mM) 100X. Store at –20°C.
5. Nonessential amino acids solution (NEAA) (10 mM GIBCO/BRL, cat. no. 11140-050), 100×. Store at 4°C.
6. 2-Mercaptoethanol (2-ME) (Sigma, cat. no. M-7522): stock solution: 1000X (0.1 M). Add 0.1 mL 2-ME to 14.1 mL PBS and sterilize by 0.2-μm filter. Store up to 4 wk at 4°C. Final concentration in medium: 10^{-4} M.
7. 0.1% (w/v) Gelatin (Specialty Media, cat. no. ES-006-B).
8. 0.25% (w/v) Trypsin (GIBCO/BRL, cat. no. 25200-056) + 0.02% (w/v) ethlenediamine tetraacetic acid (EDTA) in PBS. Store at –20°C. Thaw just prior to use and store at 4°C.
9. Murine leukemia inhibitory factor (LIF) (10^6 U/mL; GIBCO/BRL, cat. no. 13275-011). 1000X. Store at 4°C for up to 6 mo.
10. Minimal essential medium (MEM) sodium pyruvate solution (100 mM GIBCO/BRL, cat. no. 11360-070), 100X. Store at 4°C.
11. TET (Sigma, cat. no. T-7660). Final concentration to be determined (*see* **Note 2**). Stock concentration at 1 mg/mL. Store at –20°C for 6 mo.
12. ES cell culture medium: GMEM + 10% FBS + 10^{-4} M 2-ME + 1 mM sodium pyruvate + 2 mM L-glutamine + 0.1 mM NEAA + 1000 U/mL LIF with or without TET. Store at 4°C.

2.2. Maintenance of OP9 Stromal Cell Line

1. Minimum essential medium α medium (αMEM) with ribonucleosides and deoxyribonucleosides (GIBCO/BRL, cat. no. 12571-063).
2. FBS pretested for OP9 culture (*see* **Note 1**).
3. PBS without calcium chloride and magnesium chloride.
4. L-glutamine (200 mM) 100X. Store at –20°C.
5. 0.05% (w/v) Trypsin + 0.004% (w/v) EDTA in PBS. Store at –20°C. Thaw just prior to use.
6. OP9 culture medium: αMEM + 20% FBS + 2 mM L-glutamine.

2.3. In vitro ES Cell Differentiation

1. αMEM.
2. FBS pretested for in vitro ES cell differentiation (*see* **Notes 1** and **3**).
3. PBS.
4. L-glutamine (200 mM).
5. 0.25% (w/v) Trypsin + 0.02% (w/v) EDTA in PBS. Store at –20°C. Thaw just prior to use.
6. TET. Final concentration to be determined (*see* **Note 2**). Stock concentration at 1 mg/mL. Store at –20°C.
7. Differentiation culture medium (αMEM + 20% FBS + 2 mM L-glutamine with or without TET).

2.4. TET System

2.4.1. Constructs for TET Regulatory Gene Expression System

The TET system requires two types of constructs, one carrying TET-regulated transactivators (tTa) and the other carrying a gene of interest with its expression driven by a TET response promoter *(18)*. The Tet-Off system was used here where the expression of *Bcr-Abl* was driven by a TET response promoter that is suppressed by the addition of TET.

The TET-regulated transactivator (tTA) expression vector, pCAG 20-1, was constructed by inserting a tTA cDNA downstream of the CAG promoter (**Fig. 2**) *(22)*. The TET-regulatable *Bcr-Abl* construct, pUHD 10-3 *Bcr-Abl* IRES GFP, was generated by inserting p210 *Bcr-Abl*, internal ribosomal entry site (IRES), and green fluorescence protein (GFP) genes downstream of the TET response promoter (**Fig. 2**) *(19)*. The pUHD10-3 puro vector was constructed by inserting the puromycin resistance gene downstream of TET response promoter (**Fig. 2**).

2.4.2. Drugs for TET Regulatory Gene Expression System

1. Puromycin (Sigma, cat. no. P8833) at 1 mg/mL in sterile distilled water (1000X). Final concentration in medium: 1 mg/mL. Store at –20°C for up to 1 yr.
2. Geneticin (G418) (GIBCO/BRL, cat. no. 11811-031) at 40 mg (activity)/mL PBS. Final concentration in medium: 200 mg/mL. Store at 4°C for up to 3 mo.
3. TET in sterile water at 1 mg/mL. Aliquot 0.5 mL/tube and store at –20°C for up to 1 yr.

3. Methods
3.1. Maintenance of Feeder-Independent ES Cells

The protocols on the maintenance of feeder-dependent ES cells were described previously *(23,24)*. We will describe the protocol for feeder-independent ES cell E14tg2a.

3.1.1. Gelatin Coating of Dishes

All dishes, flasks, and plates should be gelatinized before use.

1. Add enough 0.1% gelatin solution to cover the plate surface.
   6-well plate—1.5 mL/well.
   6-cm dish—3 mL.
   10-cm dish—8 mL.
2. Let the solution sit for at least 10 min at room temperature.
3. Aspirate the gelatin solution completely just before use.

3.1.2. Thawing of ES Cells

1. Thaw frozen vial at 37°C water bath with gentle agitation.
2. Transfer cell suspension into a 15-mL centrifuge tube, which contains 10 mL of 37°C prewarmed ES cell culture medium.
3. Spin down cells at 100*g* for 5 min at room temperature. Resuspend cells in 2 mL of ES cell culture medium.
4. Seed 1×10^6 ES cells/well of gelatin-coated 6-well plate with 2 mL ES cell culture medium.
5. Change entire medium daily until confluent.

3.1.3. Passage of ES Cells

1. Aspirate medium and wash cells twice with 37°C prewarmed PBS. Volume of PBS needed for:
   1 well of 6-well plate—2 mL/well.
   6-cm dish—3 mL.
   10-cm dish—6 mL.

2. To remove cells from dish, add 0.25% trypsin-EDTA. Incubate at 37°C for 5 min.
3. Add 5 mL of ES cell culture medium. Resuspend cells to a single-cell suspension by pipetting at least 5 times (*see* **Note 4**).
4. Spin down cells at 350*g* for 5 min at room temperature. Resuspend cells in ES cell culture medium. Seed ES cells at:

Seeding Number	Days Needed for Confluence	Confluent Cell Number
1×10^6/10-cm dish	3 d	$2–2.5 \times 10^7$/dish
3×10^6/10-cm dish	2 d	$2–2.5 \times 10^7$/dish
8×10^6/10-cm dish	1 d	$2–2.5 \times 10^7$/dish
1×10^5/well of 6-well plate	3 d	$2–2.5 \times 10^6$/well
5×10^5/well of 6-well plate	2 d	$2–2.5 \times 10^6$/well
8×10^5/well of 6-well plate	1 d	$2–2.5 \times 10^6$/well

5. Add TET into the culture when required. For TET concentration (*see* **Note 2**) as the expression level of *Bcr-Abl* is tightly regulated by TET concentration.
6. Daily complete medium change is required.
7. Cells should be passaged every 2 to 3 d as described above (*see* **Note 4**).

3.1.4. Cell Freezing

1. Prepare 2X freezing medium: 20% dimethyl sulfoxide (DMSO) + 80% ES cell culture medium. Keep on ice. Make fresh every time.
2. Remove cells from dish as in **Subheading 3.1.3.**
3. Resuspend 4×10^6 cells in 0.25 mL ice-cold 100% ES cell culture medium and keep on ice.
4. Add an equal amount of 2X freezing medium. Freeze the cells at –70°C overnight. The next day, transfer vials to a liquid nitrogen tank.

3.2. Generation of TET-Off Parental ES Cell Line

To establish a TET-off parental ES cell line, both constructs, pCAG 20-1 and pUHD10-3 Puro (*see* **Subheading 2.4.1.**), were transfected into ES cells by electroporation *(19,25)*. When the Tet-Off regulatory system is working, the expression of the puromycin resistance gene is suppressed by the presence of Tet, and cells will die in puromycin selection.

The procedure for the generation of tetracycline-off parental ES cell line is outlined in **Fig. 3**.

1. Prepare plasmids pCAG20-1 and pUHD10-3 PURO, in a molar ratio of 1:1 in sterile water as below:
   a. 50 µg pCAG20-1 and 40 µg pUHD10-3 PURO are linearized by *Sca*1 restriction enzyme.
   b. The DNA solution is extracted twice with phenol–chloroform and twice with chloroform.
   c. Precipitate the DNA with 100% ethanol and wash with 70% ethanol. Resuspend the dry DNA pellet in 25 µL sterile water.
2. Grow ES cells to 70–80% confluency in a 10-cm dish. Coat five 10-cm plates with gelatin before electroporation as described in **Subheading 3.1.1.**
3. Remove cells from the dish as in **Subheading 3.1.3.** and resuspend in 10 mL PBS/10-cm dish.
4. Take out 1.0×10^7 cells for each transfection, spin down, and resuspend in a final volume of 0.8 mL PBS.
5. Add 50 µg pCAG20-1 and 40 µg pUHD10-3 PURO DNA solution into ES cell suspension,

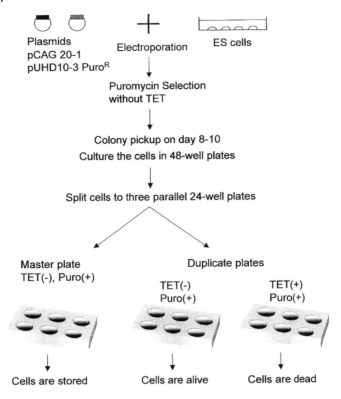

Fig. 3. Schematic protocol for the generation of TET-off parental ES cell line (*see* text).

mix gently, and place on ice for at least 1 min.

6. Transfer the cell suspension to a 0.4-cm cuvette for electroporation. Electroporate at a capacitance of 3.0 μFD and voltage of 0.8 kV (*see* **Note 5**).

7. Resuspend 1×10^7 cells in 50 mL 37°C prewarmed ES cell culture medium without TET. Seed 10 mL ES cell solution (2×10^6 ES cells) into one 10-cm gelatin-coated dish in the absence of TET (total five 10-cm dishes).

8. Start drug selection with 1 μg/mL Puromycin after 48 h. Change medium daily. Puromycin-resistance ES cell colonies can be visualized 8 to 10 d after transfection.

9. These puromycin-resistant colonies are picked by a p20 Pipetman into a 48-well plate containing puromycin medium without TET as previously described *(19,26)*. Change medium daily.

10. When the wells reach 80% confluency (usually 2 to 4 d later) split into three parallel 24-well plates containing puromycin medium. One plate will contain 1 μg/mL TET, and two plates will not contain TET.

11. The clones that survive in TET-free medium, but die in the presence of TET, are selected as TET-off parental ES cell line. Five clones are expanded and stored.

3.3. Establishment of TET-Regulated Bcr-Abl Expressing ES Cell Clone

1. The pUHD 10-3 *Bcr-Abl* IRES GFP vector and neomycin plasmid, pMC1NEO, are co-transfected at a 20:1 molar ratio into a Tet-Off parental ES cell line as described in **Subheading 3.2.**

2. The cells are selected in the presence of 1 μg/mL TET and 200 μg/mL G418.
3. On d 8 to 10, multiple G418-resistant colonies are picked as in **Subheading 3.2.**
4. When the wells reach 80% confluency, split into three parallel 24-well plates containing G418 medium. Two plates will contain TET, and one plate will not contain TET.
5. Clones tightly regulated by TET are initially analyzed for GFP expression by FACScan.
6. GFP-expressing clones are subsequently examined for *Bcr-Abl* expression by Western blotting in TET+ and TET– conditions.
7. To determine TET concentration required for different levels of *Bcr-Abl* expression (*see* **Note 2**).

3.4. Maintenance of OP9 Stromal Cell Line for In Vitro ES Cell Differentiation

3.4.1. Thawing of OP9 Cells

1. Thaw frozen vial in 37°C water bath with gentle agitation.
2. Transfer cell suspension into a 15-mL centrifuge tube, which contains 10 mL of 37°C prewarmed OP9 culture medium.
3. Spin down cells at 160g for 5 min at room temperature. Resuspend cells in 3 mL of OP9 culture medium.
4. Seed 5×10^5 OP9 cells/6-cm dish.

3.4.2. Passage of OP9

1. Aspirate medium and wash cells twice with PBS. Add 0.05% trypsin-EDTA and incubate at 37°C for 5 min.
2. Add 5 mL of OP9 culture medium into the dish.
3. Spin down cells at 160g for 5 min at room temperature. Resuspend cells in 2 mL OP9 culture medium.
4. We usually obtain: $7–8 \times 10^5$ OP9 cells/6-cm dish and $1.2–1.6 \times 10^6$ OP9 cells/10-cm dish.
5. Seed $3–4 \times 10^5$ OP9 cells/10-cm dish.
6. Change medium every 2 d. Split cells every 4 d.
7. OP9 cells should not be cultured for longer than 1 mo after thawing (*see* **Note 6**).

3.4.3. Cell Freezing

1. Prepare 2X freezing medium: 20% DMSO in 80% FBS, keep on ice.
2. Remove cells from dish as in **Subheading 3.4.2.**
3. Resuspend $6–8 \times 10^5$ cells in 0.25 mL ice-cold 100% FBS and keep on ice.
4. Add equal amount of 2X freezing medium to each sample.
5. Freeze the cells at –70°C overnight. The next day, transfer the vials to a liquid nitrogen tank.

3.5. In Vitro ES Cell Differentiation

The in vitro ES cell differntiation system used here requires continued culturing of ES cells without LIF on an OP9 cell layer. In this system, ES cells differentiate into primitive hemangioblasts by d 4 or 5. The primitive hemangioblasts produce immature hematopoietic stem cells and progenitors cells from d 5 to d 8. Mature blood cells of the myeloid and erythroid lineages are formed by d 14 or 15.

The direct effect of *Bcr-Abl* on immature hematopoietic stem cells and progenitors can be evaluated by inducing *Bcr-Abl* expression during d 5 to d 8 of ES cell differentiation. The direct effect of *Bcr-Abl* on mature myeloid and erythroid cells can be evaluated by inducing *Bcr-Abl* expression during d 8 to d 15 of ES cell differentiation.

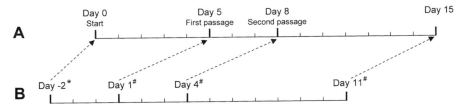

Fig. 4. Schedule of OP9 cells prepared for in vitro ES cell differentiation. (**A**) Time schedule of in vitro ES cell differentiation. (**B**) Splitting schedule of OP9 cells for differentiating ES cells. Each OP9 cell split provides stromal layers for the plating of ES cell differentiation cultures as indicated by arrows. *, each confluent 10-cm OP9 plate is split to one 6-well plate; #, each confluent 10-cm OP9 plate is split to four 10-cm plates or four 6-well plates.

3.5.1. Day 0 to Day 5 ES Cell Differentiation

1. Prepare 6-well plates with confluent OP9 cells as OP9 splitting schedule (**Fig. 4**).
2. Remove ES cells from dish as in **Subheading 3.1.3.**
3. Resuspend cells in 37°C prewarmed differentiation culture medium and count the cell number.
4. Aspirate OP9 culture medium from each well of a 6-well plate containing confluent OP9 cells.
5. Add 2 mL 37°C prewarmed differentiation culture medium to each well.
6. Ten thousand ES cells are transferred to each well of 6-well plate.
7. When TET is needed, add TET into the culture. For TET concentration, *see* **Note 2** as the expression level of *Bcr-Abl* is tightly regulated by the TET concentration.

3.5.2. Day 5 to Day 8 ES Cell Differentiation

Differentiated ES colonies are detectable 3 d after starting co-culture, and they continue to proliferate up to d 5 (**Fig. 1**). More than 90% of the ES colonies are differentiated on d 5. Differentiated ES cell colonies consist of large immature cells with a characteristic wide and basophilic cytoplasm (**Fig. 1**).

1. Aspirate the medium from each well. Rinse each well twice with 2 mL of PBS (*see* **Note 7**).
2. Add 0.5 mL of 0.25% trypsin-EDTA to each well. Incubate at 37°C for 5 min.
3. Add 5 mL of the differentiation culture medium into each well. Pipet 10 times to dissociate the cells.
4. Spin cells down at 350g for 5 min. Aspirate the supernatant and resuspend cell pellets in 10 mL of differentiation culture medium per three wells.
5. Transfer cell suspension into a new 10-cm dish and incubate in 37°C CO_2 incubator for 30 min to remove the adherent OP9 cells.
6. Collect floating cells by pipetting gently four times and spin down the cells as described in step 4. Resuspend cells in 4 mL of differentiation culture medium per three wells of harvested cells.
7. Seed ES cells as follows: 1×10^5 cells to each well of a 6-well plate containing OP9 cell layer, and 8×10^5 cells/10-cm dish containing OP9 cell layer.
8. Add TET to the cultures as required. For TET concentration, *see* **Note 2** as the expression level of *Bcr-Abl* is tightly regulated by the TET concentration.

3.5.3. ES Cell Differentiation of Day 8 to Day 15

Cell clusters representing immature hematopoietic blasts appear from d 5 to d 8 (**Fig. 1**). These immature cells are capable of forming multilineage hematopoietic colonies in methylcellulose *(19)*. Immature cells can also give rise to mature erythrocytes,

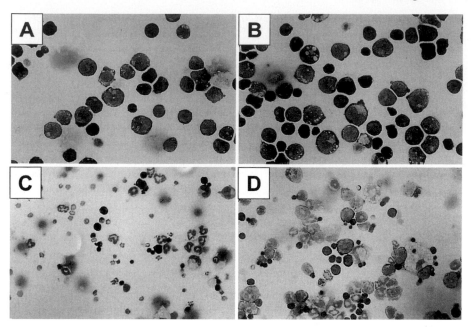

Fig. 5. Effect of *Bcr-Abl* expression on d 8 and d 15 hematopoietic cells. Giesma staining of d 8 and d 15 *Bcr-Abl* OFF-differentiated ES cells are shown in **(A)** and **(C)**, respectively. Giesma staining of d 8 and d 15 *Bcr-Abl* ON-differentiated ES cells are shown in **(B)** and **(D)**, respectively. Magnification at 500×. (*See* color plate 2, following p. 254).

granulocytes, megakaryocytes and B lymphocytes when cultured with OP9 and appropriate growth factors *(17,27)* by d 8.

1. Gently pipet the differentiating culture with the existing medium four times to collect hematopoietic cells.
2. Spin down cells at 350*g* for 10 min and resuspend the pellets in differentiation culture medium.
3. Cells $2–5 \times 10^5$ are recovered on d 8 ES cell differentiation when 8×10^5 cells are seeded in one 10-cm dish on d 5. Cell morphology and surface marker analysis show that immature hematopoietic cells are formed both under *Bcr-Abl* ON and OFF conditions (**Fig. 5A,B**) *(19)*. When *Bcr-Abl* is expressed from d 5, 2 to 3 times more cells are obtained as described previously *(19)*.
4. Seeding of d 8 differentiated ES cells onto an OP9 cell layer is more flexible than d 0 and d 5. A seeding range of $1–10 \times 10^4$ can be plated onto each well of a 6-well plate. A seeding range of $4–10 \times 10^5$ cells can be plated onto one 10-cm dish containing confluent OP9 cells.

3.5.4. Analysis of D 14 to 15

Hematopoietic cells continue to proliferate and differentiate to maturation from d 8 to d 14 to 15. On day 15, 10% of differentiated ES cells express CD11b, a myeloid marker, and 70% of the cells express TER119, an erythroid marker.

1. Generally, all hematopoietic cells can be collected from the culture by gently pipetting 4 times and harvesting the supernatant. Sometimes, adherent hematopoietic cells cannot be removed by gently pipetting, and the following protocol is necessary:

    a. After collecting nonadherent hematopoietic cells, rinse plates once with PBS.

    b. Add 0.5 mL of 37°C, 0.05% trypsin-EDTA onto each plate and incubate at 37°C for 1–2 min.

    c. Add differentiation culture medium (2 mL/well) and 5 mL/10-cm dish and pipet gently five times to dissociate the hematopoietic cells.

    d. Transfer cells into a new dish and incubate at 37°C in a CO_2 incubator for 30 min to remove adherent OP9 cells.

    e. Collect hematopoietic cells by gently pipetting five times.

2. Spin cells down at 350g for 10 min.
3. Aspirate the supernatant and resuspend the cell pellets in differentiation culture medium.
4. $1–2 \times 10^6$ cells are recovered on d 15 ES cell differentiation when $3–4 \times 10^5$ cells are seeded on a 10-cm dish on d 8. Mostly mature erythrocytes and some mature granulocytes are formed on d 15 (**Fig. 5C**). *Bcr-Abl* expression from d 5 to d 15 of differentiating cells leads to an accumulation of immature cells, myeloid cells, and a reduction of erythrocytes (**Fig. 5D**).

4. Notes

1. Test 10 different sera for both ES and OP9 cells. Select a serum lot that supports a good growth rate of ES and OP9 cells. Refer to Methods **Subheadings 3.1.3.** and **3.4.2.** for the growth rates of ES and OP9 cells, respectively.

   The serum lot should support good differentiation of ES cells. An important indicator for serum checking is the induction of ES cell differentiation. The percentage of differentiated ES colonies to undifferentiated ES colonies on d 5 is greater than 90%. Developmental markers during d 5 and d 15 are used to evaluate the differentiation state of ES cells. On d 5, 60% of differentiated ES cells express FLK-1, a receptor tyrosine kinase, which is a marker of hemaogioblasts. On d 15, 10% of differentiated ES cells express CD11b, a myeloid marker, and 70% of the cells express TER119 an erythroid marker.

2. The expression level of *Bcr-Abl* is controlled by the concentration of TET. A range of TET concentrations such as 0, 1, 3, 10, 30, 100, and 1000 ng/mL is tested to identify the dose response of *Bcr-Abl* expression. In our Bcr-Abl ES clones, 100 ng/mL TET is sufficient to completely suppress *Bcr-Abl* expression and does not inhibit the growth and differentiation of hematopoietic cells.

   A high concentration of 1 µg/mL TET inhibits hematopoietic development in in vitro ES cell differentiation.

3. FBS for in vitro ES cell differentiation requires heat inactivation at 56°C for 30 min. Neither refilteration nor centrifugation are needed after heat-inactivation.

4. ES cells have to be plated as single cells, otherwise, ES cells will differentiate even in the presence of LIF. A long term culture (>2 wk) of ES cells can accumulate genetic mutations and chromosome abnormality. Thaw new ES cells every 2 wk.

5. When you use ES cells other than E14tg2a, the conditions for electroporation may be different. For example, D3 and E14.1 ES cells are electroporated at a capacitance 500 µFD and a voltage 0.2–0.25 kV.

6. OP9 cells should not be cultured for longer than 1 mo after thawing. OP9 cells may lose the ability to support ES cell differentiation upon prolonged culture. Long-term culture of OP9 cells may stimulate their differentiation to fatty cells. OP9-differentiated fatty cells may inhibit in vitro ES cell differentiation. High passage of OP9 cells (>30 passages) are also easily transformed and may loose their ability to support ES cell differentiation. Use low passage OP9 (<30 passages) for experiments. Thaw new OP9 cells every mo.

7. In this system a very low amount of TET (3 ng/mL) greatly suppresses gene expression. To completely remove TET from culture, wash the culture at least twice.

References

1. Lowenberg, B., Downing, J. R., and Burnett, A. (1999) Acute myeloid leukemia. *N. Engl. J. Med.* **341,** 1051–1062.

2. Sawyers, C. L. (1999) Chronic myeloid leukemia. *N. Engl. J. Med.* **340,** 1330–1340.

3. Groffen, J., Stephenson, J. R., Heisterkamp, N., de Klein, A., Bartram, C. R., and Grosveld, G. (1984) Philadelphia chromosomal breakpoints are clustered within a limited region, bcr, on chromosome 22. *Cell* **36,** 93–99.

4. Konopka, J. B., Watanabe, S. M., and Witte, O. N. (1984) An alteration of the human c-abl protein in K562 leukemia cells unmasks associated tyrosine kinase activity. *Cell* **37,** 1035–1042.

5. Shtivelman, E., Lifshitz, B., Gale, R. P., Roe, B. A., and Canaani, E. (1985) Fused transcript of abl and bcr genes in chronic myelogenous leukaemia. *Nature* **315,** 550–554.

6. McLaughlin, J., Chianese, E., and Witte, O. N. (1987) In vitro transformation of immature hematopoietic cells by the P210 BCR/Abl oncogene product of the Philadelphia chromosome. *Proc. Natl. Acad. Sci. USA* **84,** 6558–6562.

7. Lugo, T. G., Pendergast, A., Muller, A. J., and Witte, O. N. (1990) Tyrosine kinase activity and transformation potency of bcr-abl oncogene products. *Science* **247,** 1079–1082.

8. Daley, G. Q. and Baltimore, D. (1988) Transformation of an interleukin 3-dependent hematopoietic cell line by the chronic myelogenous leukemia-specific P210{+bcr/abl} protein. *Proc. Natl. Acad. Sci. USA* **85,** 9312–9316.

9. Gishizky, M. L. and Witte, O. N. (1992) Initiation of deregulated growth of multipotent progenitor cells by bcr-abl in vitro. *Science* **256,** 836–839.

10. Daley, G. Q., Van Etten, R. A., and Baltimore, D. (1990) Induction of chronic myelogenous leukemia in mice by the P210$^{bcr/abl}$ gene of the Philadelphia chromosome. *Science* **247,** 824–830.

11. Kelliher, M. A., McLaughlin, J., Witte, O. N., and Rosenberg, N. (1990) Induction of a chronic myelogenous leukemia-like syndrome in mice with *v-abl* and *BCR/ABL*. *Proc. Natl. Acad. Sci. USA* **87,** 6649–6653.

12. Honda, H., Oda, H., Suzuki, T., Takahashi, T., Witte, O. N., Ozawa, K., et al. (1998) Development of acute lymphoblastic leukemia and myeloproliferative disorder in transgenic mice expressing p210bcr/abl: a novel transgenic model for human Ph1-positive leukemias. *Blood* **91,** 2067–2075.

13. Pear, W. S., Miller, J. P., Xu, L., Pui, J. C., Soffer, B., Quackenbush, R. C., et al. (1998) Efficient and rapid induction of a chronic myelogenous leukemia-like myeloproliferative disease in mice receiving P210 bcr/abl-transduced bone marrow. *Blood* **92,** 3780–3792.

14. Nakano, T., Kodama, H., and Honjo, T. (1994) Generation of lymphohematopoietic cells from embryonic stem cells in culture. *Science* **265,** 1098–1101.

15. Nakano, T. and Kodama, H. H. (1996) In vitro development of primitive and definitive erythrocytes from different precursors. *Science* **272,** 722–724.

16. Era, T., Takahashi, T., Sakai, K., Kawamura, K., and Nakano, T. (1997) Thrombopoietin enhances proliferation and differentiation of murine yolk sac erythroid progenitors. *Blood* **89,** 1207–1213.

17. Era, T., Takagi, T., Takahashi, T., Bories, J. C., and Nakano, T. (2000) Characterization of hematopoietic lineage-specific gene expression by ES cell in vitro differentiation induction system. *Blood* **95,** 870–878.

18. Gossen, M. and Bujard, H. (1992) Tight control of gene expression in mammalian cells by tetracycline-responsive promoters. *Proc. Natl. Acad. Sci. USA* **89,** 5547–5551.

19. Era, T. and Witte, O. N. (2000) Regulated expression of P210 Bcr-Abl during embryonic stem cell differentiation stimulates multipotential progenitor expansion and myeloid cell fate. *Proc. Natl. Acad. Sci. USA* **97,** 1737–1742.

20. Baron, U., Gossen, M., and Bujard, H. (1997) Tetracycline-controlled transcription in eukaryotes: novel transactivators with graded transactivation potential. *Nucleic Acids Res.* **25,** 2723–2729.
21. Niwa, H., Burdon, T., Chambers, I., and Smith, A. (1998) Self-renewal of pluripotent embryonic stem cells is mediated via activation of STAT3. *Genes Dev.* **12,** 2048–2060.
22. Niwa, H., Yamamura, K., and Miyazaki, J. (1991) Efficient selection for high-expression transfectants with a novel eukaryotic vector. *Gene* **108,** 193–199.
23. Robertson, E. J. (1987) Embryo-derived stem cells, in *Teratocacinomas and Embryonic Stem Cell: A Practical Approach* (Robertson, E. J., ed.), IRL Press, New York, pp. 71–112.
24. Wurst, W. and Joyner, A. L. (1993) *Gene Targeting: A Practical Approach*, vol. 126. IRL Press, New York.
25. Chambard, J. C. and Pognonec, P. (1998) A reliable way of obtaining stable inducible clones. *Nucleic Acids Res.* **26,** 3443–3444.
26. Ramirez-Solis, R., Davis, A. C., and Bradley, A. (1993) Gene targeting in embryonic stem cells. *Methods Enzymol.* **225,** 855–878.
27. Cho, S. K., Webber, T. D., Carlyle, J. R., Nakano, T., Lewis, S. M., and Zuniga-Pflucker, J. C. (1999) Functional characterization of B lymphocytes generated in vitro from embryonic stem cells. *Proc. Natl. Acad. Sci. USA* **96,** 9797–9802.

10

Embryonic Stem Cells as a Model for Studying Osteoclast Lineage Development

Toshiyuki Yamane, Takahiro Kunisada, and Shin-Ichi Hayashi

1. Introduction

Osteoclasts are cells specialized to resorb bone *(1,2)*. They originate from hematopoietic progenitors *(3,4)*, which are widely distributed throughout the body. Osteoclast progenitors differentiate into cells that are positive for calcitonin receptor and tartrate-resistant acid phosphatase (TRAP), which are specific markers for this lineage, and then they fuse with each other and form fully functional multinucleated large cells *(1,2)* (**Fig. 1**). TRAP-positive osteoclasts are found tightly attached to bone matrix *(1,2)*. Two cytokines, macrophage colony-stimulating factor (M-CSF) and receptor activator of NF-κB (RANK) ligand (RANKL; also known as OPGL, ODF, and TRANCE), are known to be responsible for this differentiation *(5–7)*. In this chapter, we describe the procedure for the induction of differentiation into the osteoclast lineage from murine embryonic stem (ES) cells *(8)*. We utilize co-culture systems with stromal cell lines, that is, bone marrow-derived ST2 cells *(9)* or newborn calvaria-derived M-CSF-deficient OP9 cells *(10)*. M-CSF is produced constitutively by ST2, and the expression of RANKL is induced on ST2 by 1,25-dihydroxyvitamin D$_3$ [1,25(OH)$_2$D$_3$] and dexamethasone (Dex) *(6)*.

First, we describe the protocol for inducing osteoclasts within ES cell-derived growing colonies (single-step culture) (**Fig. 2**) *(8)*. After culturing undifferentiated ES cells on the ST2 stromal cell line in the presence of 1,25(OH)$_2$D$_3$ and Dex for 11 d, multinucleated osteoclasts are generated at the periphery of the colonies (single-step culture) (**Fig. 3A,B**). Starting from single ES cells, the cells grow by piling up on each other thereby forming colonies. Hematopoietic cells are observed around the colonies after d 5. TRAP-positive mononuclear cells are first observed on d 8. Other cell lineages, for example, cardiac muscle cells and endothelial cells, are simultaneously observed with osteoclasts within the colonies. This is an appropriate system for examining the potential of ES cells to generate osteoclasts as these cultures are stable and enable us to assess the whole process of osteoclastogenesis without any manipulation except for regular medium changes.

Secondly, we describe a protocol to specifically induce osteoclasts using stepwise cultures (**Fig. 2**) *(8)*. As described by Nakano et al., ES cells generate colonies containing hematopoietic progenitors after 5 d of culturing on OP9 cells (**Fig. 3C**)

From: *Methods in Molecular Biology, vol. 185: Embryonic Stem Cells: Methods and Protocols*
Edited by: K. Turksen © Humana Press Inc., Totowa, NJ

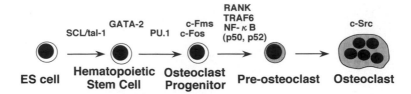

Fig. 1. Scheme describing the development of osteoclast lineage. The essential molecules expressed in this lineage are indicated at the stage where maturation is arrested by the loss of function. The cells positive for TRAP are shaded.

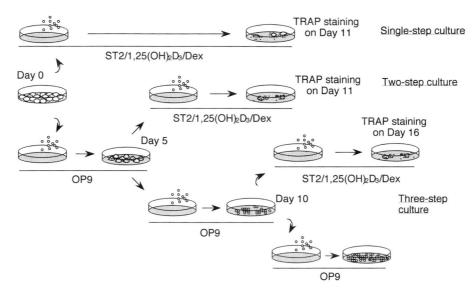

Fig. 2. Overview of the culture systems.

(11). When these colonies are dissociated to a single-cell suspension, transferred onto ST2, and further cultivated for 6 d in the presence of $1,25(OH)_2D_3$ and Dex, most of the cells that are observed on the secondary plates are osteoclasts (2-step culture) (**Fig. 2**), suggesting that the colonies on d 5 also contain osteoclast progenitors. This system allows the separation of the maturation step by M-CSF and RANKL on ST2 from the step that lead to the generation of osteoclast progenitors during the initial 5 d culture on OP9. Alternatively, when the dissociated cells prepared on d 5 from ES/OP9 co-cultures are transferred onto fresh OP9 cells and cultured for an additional 5 d, immature hematopoietic cells including osteoclast progenitors selectively propagate and form hematopoietic clusters on OP9 (**Fig. 3D**) *(11)*. Under this culture condition, very few cells have the tendency to pile up. When the hematopoietic cells are harvested by pipetting from the culture on d 10 and are cultured for an additional 6 d on ST2 in the presence of $1,25(OH)_2D_3$ and Dex, we could efficiently and selectively induce osteoclasts (3-step culture) (**Figs. 2** and **3E**). The number of osteoclast progenitors is increased 20-fold by the secondary step on OP9 *(12)*. Thus, the 3-step culture condition provides an excellent in vitro model system to study generation, expansion, and maturation of osteoclast progenitors.

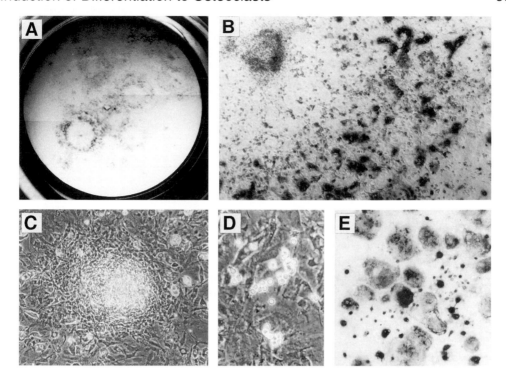

Fig. 3. Appearance of ES cell cultures. (**A** and **B**) Osteoclasts generated in the single-step culture. Whole view of a well of a 24-well plate (**A**) and high magnification view of a colony (**B**) are shown. Osteoclasts were stained based on their TRAP activity and are observed around the colonies. (**C**) A differentiated colony cultured for 5 d on OP9. (**D**) Hematopoietic cell cluster on d 10 on OP9. (**E**) Osteoclasts generated in the 3-step culture. Osteoclasts were stained based on their TRAP activity.

2. Materials

1. 0.25% trypsin + 0.5 mM EDTA in phosphate-buffered saline (PBS).
2. 0.1% trypsin + 0.5 mM EDTA in PBS.
3. 0.05% trypsin + 0.5 mM EDTA in PBS.

   For trypsin-EDTA solutions, we purchased 2.5% trypsin solution from a commercial supplier, for example from Gibco BRL (Grand Island, NY, cat. no. 15090-046). Dispense into aliquots and store at –20°C. Dilute to appropriate concentrations with PBS and supplement with EDTA. Store at 4°C. Use within 1 mo.
4. Dulbecco's Modified Eagle Medium (DMEM) (Gibco BRL, cat. no. 12800-017).
5. α-Minimum Essential Medium (α-MEM) (Gibco BRL, cat. no. 11900-024).
6. RPMI-1640 (Gibco BRL, cat. no. 31800-022).
7. Medium for the maintenance of ES cells: DMEM supplemented with 15% fetal calf serum (FCS) (*see* **Note 1**), 2 mM L-glutamine (Gibco BRL, cat. no. 25030-081), 1X nonessential amino acids (Gibco BRL, cat. no. 11140-050), 0.1 mM 2-mercaptoethanol (2-ME), 1000 U/mL recombinant leukemia inhibitory factor (LIF) or equivalent amounts of a culture supernatant from Chinese hmaster ovary (CHO) cells producing LIF, 50 μg/mL streptomycin, and 50 U/mL penicillin. Store at 4°C. Use within 1 mo.
8. Medium for embryonic fibroblasts: DMEM supplemented with 10% FCS, 50 μg/mL streptomycin, and 50 U/mL penicillin. Store at 4°C. Use within 2 mo.

Table 1
Volume of Medium for Culture Vessels

Culture vessels	Volume of medium (mL)
25-cm^2 flask	6
60-mm dish	5
100-mm dish	10
24-well plate	1
6-well plate	2

9. 0.1% Gelatin solution: Dissolve 0.5 g of gelatin (Sigma, St Louis, MO, cat. no. G2500) into 500 mL distilled water by autoclaving. Store at room temperature.
10. Mitomycin C (Sigma, cat. no. M0503): For 100X solution, dissolve the powder into sterile distilled water and adjust the concentration to 1 mg/mL just before use.
11. Medium for the maintenance of ST2: RPMI 1640 supplemented with 5% FCS, 50 μM 2-ME, 50 μg/mL streptomycin, and 50 U/mL penicillin. Store at 4°C. Use within 2 mo.
12. Medium for the maintenance of OP9: α-MEM supplemented with 20% FCS, 50 μg/mL streptomycin, and 50 U/mL penicillin. Store at 4°C. Use within 1 mo.
13. Medium for the differentiation of ES cells on ST2: α-MEM supplemented with 10% FCS, 50 μg/mL streptomycin, and 50 U/mL penicillin (*see* **Note 2**). Store at 4°C. Use within 2 mo.
14. 1,25(OH)$_2$D$_3$ (Biomol Research Laboratories, Plymouth Meeting, PA, cat. no. DM-200). Store at –70°C at 10^{-4} M in ethanol under light-tight conditions. Stable at least for six mo. Dilute from this stock for each use.
15. Dex (Sigma, cat. no. D4902). Store original stocks at 10^{-2} M in ethanol at –70°C. Prepare the stock at 10^{-3} M in ethanol at 4°C. Dilute from this stock for each use. Stable at least for 1 yr (stocks at –70°C) or 3 mo (stocks at 4°C).
16. Medium for differentiation of ES cells on OP9: α-MEM supplemented with 20% FCS, 50 μg/mL streptomycin, and 50 U/mL penicillin (*see* **Note 2**). Store at 4°C. Use within 2 mo.
17. TRAP staining solution: for 500 mL of solution, dissolve 5.75 g of sodium tartrate dihydrate, 6.8 g of sodium acetate trihydrate, and 250 mg of naphthol AS-MX phosphate (Sigma, cat. no. N-5000) in an appropriate vol of dH$_2$O. Adjust the pH to 5.0 with acetic acid and adjust the vol to 500 mL with distilled water. Store the stock in a dark glass bottle at 4°C. Just before use, dissolve fast red violet LB salt (Sigma, cat. no. F-3381) in the volume you need at the final concentration of 0.5 mg/mL.

3. Methods

For each manipulation, prewarm the medium to 37°C and do not leave cells out of the incubator for a long time. The cultures described here require a humidified incubator at 37°C in 5% CO$_2$. The volume of culture medium is summarized in **Table 1** for the culture vessels that appear in the protocol.

3.1. Maintenance of ES Cells

We ordinarily maintain ES cells on mitomycin C-treated embryonic fibroblasts (*see* **Note 3**). ES cells are cultured in the medium described in the material section.

1. Prepare mitomycin C-treated confluent embryonic fibroblasts on gelatin-coated dishes (*see* **Note 4**).
2. Thaw ES cells on mitomycin C-treated embryonic fibroblasts.
3. Grow them up to a subconfluent state.
4. Replace the medium of the ES cell cultures with fresh medium 2–4 h before passage.
5. Wash three times with PBS and trypsinize the cultures with 0.25% trypsin/0.5 m*M* EDTA for 5 min at 37°C. Add medium and dissociate the cell clump by pipetting up and down. Centrifuge at 120*g* for 5 min at 4°C, aspirate the supernatant, and suspend the cell pellet in the medium. Count the number of ES cells.
6. Replate them onto the freshly prepared mitomycin C-treated embryonic fibroblasts. For 60 or 100-mm dishes, inoculate 10^6 or 3×10^6 ES cells, respectively.
7. Change the ES cell culture media the day after passaging.
8. Two days after the passage, the cultures must reach a subconfluent state. Maintain ES cells by the above-mentioned manner (*see* **Note 5**).

3.2. Maintenance and Preparation of ST2 Stromal Cells

ST2 cells (*see* **Notes 6** and **7**) are maintained in the RPMI 1640 media supplemented with 5% FCS and 50 µ*M* 2-ME.

1. Trypsinize the confluent cells in a dish or flask with 0.05% trypsin/0.5 m*M* EDTA at 37°C for 1 to 2 min.
2. Add medium, dissociate the cells by pipetting, centrifuge them at 200*g* for 5 min, and split them 1:4.
3. Maintain by regularly passing them every 3 or 4 d.
4. To prepare feeder layers in 24-well plates, seed the cells of one confluent T25 flask or half of one confluent 100-mm dish into a 24-well plate. Two days later, the cells reach confluency and are ready to use for the differentiation of ES cells (*see* **Note 8**). Neither irradiation nor treatment with mitomycin C is needed.

3.3. Maintenance and Preparation of OP9 Stromal Cells

OP9 cells (*see* **Notes 7** and **9**) are maintained in α-MEM supplemented with 20% FCS.

1. Grow them to a subconfluent or confluent state.
2. Wash twice with PBS and trypsinize with 0.1% trypsin/0.5 m*M* EDTA at 37°C for 4 to 5 min.
3. Add medium, dissociate the cells well by pipetting up and down, centrifuge them at 200*g* for 5 min, and split them 1:4.
4. Change the medium 2 d later.
5. Passage them again the next day (subconfluent state) or 2 d after the medium change (confluent state).
6. To prepare a feeder layer in 6-well plates, seed half of the cells from a subconfluent or confluent 100-mm dish into a 6-well plate. Two days later, they reach confluency and are ready to use for the differentiation of ES cells (*see* **Note 8**). Neither irradiation nor treatment with mitomycin C is needed.

3.4. Single-Step Culture to Induce the Differentiation of Osteoclasts

1. Prepare ST2 feeder layer in 24-well plates as described in **Subheading 3.2.**
2. Grow ES cells to a subconfluent state as described in **Subheading 3.1.** (*see* **Note 10**).

3. Replace the medium of the ES cell cultures with fresh medium 2–4 h before inducing the differentiation.

4. Wash three times with PBS and trypsinize the culture with 0.25% trypsin/0.5 mM EDTA for 5 min at 37°C. Add α-MEM/10% FCS and dissociate the cell clump by pipetting up and down. Centrifuge at 120g for 5 min at 4°C, aspirate the supernatant, and suspend the cell pellet in α-MEM/10% FCS. Count the number of ES cells (*see* **Note 11**).

5. Dilute to the appropriate cell density in α-MEM/10% FCS (*see* **Note 12**). Aspirate the medium from the plates of ST2 feeder cells. Dispense 1 mL of the cell suspension into each well. Supplement with 10^{-8} M 1,25(OH)$_2$D$_3$ and 10^{-7} M Dex.

6. Cultivate in the incubator and change the culture medium every 2 or 3 d (*see* **Note 13**).

7. Stain the cultures for osteoclasts with a method detecting TRAP activity, as described in **Subheading 3.7.** on d 11, when mature osteoclasts have been generated (**Fig. 3A,B**).

3.5. Multistep Culture on OP9

1. Prepare an OP9 feeder layer in 6-well plates as described in **Subheading 3.3.**

2. Grow ES cells to a subconfluent state as described in **Subheading 3.1.** (*see* **Note 10**).

3. Replace the medium of ES cell cultures with fresh medium 2–4 h before inducing differentiation.

4. Wash three times with PBS and trypsinize with 0.25% trypsin/0.5 mM EDTA for 5 min at 37°C. Add α-MEM/20% FCS and dissociate the cell clump by pipetting up and down. Centrifuge at 120g for 5 min at 4°C, aspirate the supernatant, and suspend the cell pellet in α-MEM/20% FCS. Count the number of ES cells (*see* **Note 11**).

5. Dilute to 5×10^3 cells/mL with α-MEM/20% FCS. Aspirate the medium from the plates of OP9 feeder layers. Dispense 2 mL of cell suspension into each well (10^4 cells/well).

6. Maintain in the incubator at 37°C with 5% CO$_2$.

7. On d 2 or 3 of differentiation, replace half of the medium with fresh medium (*see* **Note 14**).

8. On d 5 of differentiation (*see* **Note 15**), colonies that have a differentiated appearance will be observed (**Fig. 3C**) (*see* **Note 16**). Wash the cultures 3 times with PBS, and trypsinize them with 0.5 mL of 0.25% trypsin/0.5 mM EDTA for 5 min at 37°C. Add α-MEM/20% FCS and dissociate the cell clump by pipetting up and down vigorously. Transfer the dissociated cells into 15-mL centrifuge tube and centrifuge at 350g for 5 min at 4°C. Cell aggregates may be generated after trypsinization. These aggregates consist mostly of OP9 cell debris. To precipitate these persistent cell aggregates, let stand for 4 to 5 min, then transfer the upper cell suspension into fresh tubes, and centrifuge. Aspirate the supernatant, suspend the cells in fresh medium, and count the cell number with a hemocytometer. Do not count OP9 cells; they are large and easily distinguished from ES cell-derived cells. About $1–2 \times 10^6$ cells are obtained per well. For 2-step cultures, refer to **Subheading 3.6.** For 3-step cultures, follow the steps below.

9. Dilute the cell suspension to 5×10^4 cells/mL with medium (α-MEM/20% FCS), and dispense 2 mL of the cell suspension into each well of a 6-well plate containing freshly prepared OP9 layers (1×10^5 cells/well).

10. Maintain in the incubator at 37°C with 5% CO$_2$.

11. On d 7 or 8 of differentiation, remove half of the medium gently from the culture plates and add 1 mL of fresh medium (*see* **Note 14**).

12. On d 10 of differentiation, hematopoietic clusters or colonies will have formed on the OP9 layers (**Fig. 3D**). Add 2 mL of fresh medium (α-MEM/20% FCS) to each well, and after pipetting vigorously, transfer the cells into centrifuge tubes, and let them stand for 4 to 5 min to precipitate debris of OP9 stromal cells. Transfer the supernatant into a fresh tube and centrifuge at 350g for 5 min at 4°C. Suspend the pellet in fresh medium and count the

cell number. About 10^5 hematopoietic cells will be obtained per well. For 3-step cultures, refer to **Subheading 3.6.** If you want to simultaneously analyze the other hematopoietic lineages, plate the cell suspension again onto fresh OP9 at 10^5 cells/well in 6-well plates (*see* **Note 17**).

3.6. Induction of Differentiation into Osteoclasts in Multistep Cultures

1. Prepare ST2 feeder layers in 24-well plates as described in **Subheading 3.2.**
2. For 2-step cultures (from step 8 in the previous section, **Subheading 3.5.**), dilute the cell suspension to $0.5–1 \times 10^4$ cells/mL with medium (α-MEM/10% FCS). For 3-step cultures (from step 12 in the previous section, **Subheading 3.5.**), dilute to 10^3 cells/mL.
3. Aspirate the medium from the plates of ST2 feeder layers. Dispense 1 mL of the cell suspension into each well. Supplement with 10^{-8} M 1,25(OH)$_2$D$_3$ and 10^{-7} M Dex.
4. Cultivate in the incubator, and change the culture medium every other day.
5. Stain the cultures for osteoclasts with a method detecting TRAP activity 6 d after seeding (on d 11 for 2-step cultures and on d 16 for 3-step cultures) when mature osteoclasts have been generated, as described in the next section (**Subheading 3.7.**) (**Fig. 3E**).

3.7. TRAP Staining

We ordinarily evaluate osteoclastogenesis by TRAP staining. It is also possible to check the functional activity of osteoclasts by the pit formation assay (*see* **Note 18**).

1. Aspirate the culture medium.
2. Add 1 mL of 10% formalin (3.7% formaldehyde) in PBS (v/v) to each well of the plates and fix them for 10 min at room temparature.
3. Wash once with PBS, aspirate, and cover with 0.5 mL of ethanol–acetone (50:50, v/v) for exactly 1 min at room temparature. After 1 min of treatment, immediately fill each well with PBS, aspirate the solution, and wash once more with PBS.
4. After aspiration of PBS, cover the fixed cells with 0.25 mL of TRAP staining solution and incubate for 10 min at room temparature. After the red color develops, wash the plates well with water. Insufficient washing will generate a high background staining.

4. Notes

1. Choose a lot of serum that well supports the growth of ES cells, does not generate differentiated cells, and has high plating efficiency. The culture at the clonal density without the addition of exogenous LIF would allow the selection of such lots. The batches of serum suitable for the maintenance of ES cells are also available from several commercial suppliers.
2. Batch-related variation of FCS affects the differentiation of ES cells. It is necessary to check the batches of the serum. In the single-step culture, most batches of serum support osteoclastogenesis. However, some lots are not good for the maintenance of ST2 feeder layer for 11 d, and the plating efficiency of ES cells or the number of osteoclasts generated on ST2 feeder layers varies according to the FCS lot. Carry out preliminary experiments to check the batches of the serum by seeding ES cells in the range of 10 to 1000 cells/well according to **Subheading 3.4.** (*see also* **Note 12**). Choose a lot of serum that is good for culturing ST2 and efficiently supports the formation of TRAP-positive osteoclasts. It should be noted that plating efficiency is not correlated with the ability to support osteoclastogenesis. For multistep cultures on OP9, choose lots of FCS that support the growth and differentiation of ES cells on OP9 (*see* **Subheading 3.5.** text and notes of

Methods section). Some serum does not support the selective propagation of hematopoietic cells after the replating due to taking nonhematopoietic cells into the secondary plates. To induce osteoclasts in multistep cultures (the secondary or tertiary step in the 2- or 3-step culture, respectively), we recommend using a batch of serum that has been used for the differentiation of osteoclasts from an in vivo source such as bone marrow cells, spleen cells, or peritoneal cells.

3. At least 3 ES cell lines D3 *(13)*, J1 *(14)*, and CCE *(15)*, gave similar results using our protocol. It is also possible to use ES cells that are adapted to feeder-independence.

4. Treat embryonic fibroblasts with 10 µg/mL mitomycin C for 2.5 h. After washing the cells, seed 10^6 or 3×10^6 cells into a 60- or 100-mm dish, respectively. It is also possible to freeze mitomycin C-treated cells. In this case, thaw the 6×10^6 cells from a stock vial into a 100-mm dish or three 60-mm dishes. Dishes should be coated with a 0.1% gelatin solution at 37°C for 30 min or at room temperature for 2 h. Remove the gelatin solution just before the plating of embryonic fibroblasts.

5. The passage numbers of ES cells should be kept as low as possible. Prepare sufficient amounts of young stocks and recover when needed.

6. ST2 cultures grown long periods sometimes change in appearance (i.e., changing to a more dendritic shape or senescent appearance). Discard such cultures and use freshly thawed young ST2 cell stocks. **Fig. 4A** shows the appearance of normal ST2.

7. ST2 (RCB0224) and OP9 (RCB1124) cells have been registered in the RIKEN Cell Bank (*http://www.rtc.riken.go.jp/CELL/HTML/RIKEN_Cell_Bank.html*). If the cells do not function well, please contact us.

8. For the purpose of preparing feeder layers, it is also possible to seed at higher or lower density, and culture for a shorter or longer times until the stromal cells are in a confluent state when ES cells are inoculated.

9. OP9 are highly sensitive to old media. Use fresh medium and change the medium 2 d after the passage. OP9 must not to be left in their confluent state for prolonged periods, because this makes it difficult to prepare the single-cell suspension. Pass them when they are subconfluent or have just reached confluency. Discard them if their growth is bad or if they have a senescent appearance. **Fig. 4B** shows the appearance of normal OP9.

10. Do not use ES cells that have just been thawed from frozen stocks. Use ES cells that have been cultured for at least 1 d after being thawed.

11. If you want to remove the embryonic fibroblasts, replate the ES cells on gelatin-coated dishes and incubate for 30 min in the culture medium, then collect the nonadherent cells by pipetting and count them. However, it does not cause any problems to directly seed the mixture of ES cells and embryonic fibroblasts onto ST2 or OP9 layers.

12. The efficiency with which ES cells form colonies on the ST2 feeder layer changes according to the lot of serum from only 1% to about 40%. Seed ES cells so that ≤20 colonies are generated per well in the 24-well plates. Seeding at higher density prevents osteoclastogenesis because spaces for osteoclasts on ST2 are occupied by the growing colonies. Plating efficiency also varies depending on the ES cell line used. Even the same cell line sometimes gives different plating efficiencies if it is obtained from different laboratories. Carry out preliminary experiments to check the plating efficiency using your lines and serum by seeding ES cells in the range from 10 to 1000 cells/well.

13. It is difficult to find ES cells on the first 1 or 2 d of differentiation. After this period, small ES cell colonies derived from single ES cells are discernible under the microscope. They will grow up to <3 mm (including the area of osteoclasts) in diameter. It has been observed that under these conditions, ST2 cells may generate adipocytes. This is not problematic and should not influence experimental outcomes.

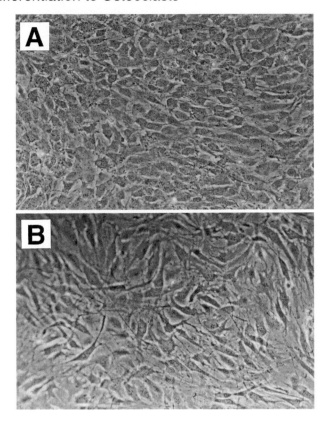

Fig. 4. Appearance of ST2 (**A**) and OP9 (**B**) cells.

14. Under these conditions, OP9 cells give rise to adipocyte cells, but this does not influence experimental outcomes.
15. When you replate the cultures in which the growth of ES cells on OP9 is not good on d 5, nonhematopoietic cells might be predominantly replated into the secondary plate. In such cases, delay the day of replating until the colonies grow up to sufficient level (*see* **Fig. 3C**).
16. Colonies that have an undifferentiated appearance are also observed on d 5 *(11)*. Such colonies may comprise, at most, one-third of the total colonies.
17. As described by Nakano et al. (1994), cultivation on OP9 for an additional 3 to 4 d generates various hematopoietic lineages, including the erythroid lineage, granulocytes, the mono/macrophage lineage, megakaryocytes, mast cells, and even lymphoid lineages (**Fig. 2**) *(11)*.
18. Pit formation assay on dentine slices should be carried out using cells from multistep cultures as described in **refs.** *8* and *12*. Although it is possible to check the osteoclast function by directly culturing undifferentiated ES cells and ST2 cells on the dentine slices in the presence of $1,25(OH)_2D_3$ and Dex, this method is not recommended as the results fluctuate greatly and are not suitable for quantitative analyses.

Acknowledgments

The culture system described here was developed in collaboration with Drs. Hidetoshi Yamazaki (Tottori University), Takumi Era (University of California, Los

Angeles), and Toru Nakano (Osaka University). Toshiyuki Yamane is a Research Fellow of the Japan Society for the Promotion of Science.

References

1. Mundy, G. R. and Roodman, G. D. (1987) Osteoclast ontogeny and function. *Bone Miner. Res.* **5,** 209–279.
2. Suda, T., Udagawa, N., and Takahashi, N. (1996) Cells of bone: osteoclast generation, in *Principles of bone biology* (Bilezikian, J. P., Raisz, L. G., and Roden, G. A., eds.), Academic Press, New York, NY, pp. 87–102.
3. Ash, P., Loutit, J. F., and Townsend, K. M. (1980) Osteoclasts derived from haematopoietic stem cells. *Nature* **283,** 669–670.
4. Hayashi, S.-I., Yamane, T., Miyamoto, A., Hemmi, H., Tagaya, H., Tanio, Y., et al. (1998) Commitment and differentiation of stem cells to the osteoclast lineage. *Biochem. Cell Biol.* **76,** 911–922.
5. Yoshida, H., Hayashi, S.-I., Kunisada, T., Ogawa, M., Nishikawa, S., Okamura, H., et al. (1990) The murine mutation osteopetrosis is in the coding region of the macrophage colony stimulating factor gene. *Nature* **345,** 442–444.
6. Yasuda, H., Shima, N., Nakagawa, N., Yamaguchi, K., Kinosaki, M., Mochizuki, S.-I., et al. (1998) Osteoclast differentiation factor is a ligand for osteoprotegerin/osteoclastogenesis-inhibitory factor and identical to TRANCE/RANKL. *Proc. Natl. Acad. Sci. USA* **95,** 3597–3602.
7. Lacey, D. L., Timms, E., Tan, H.-L., Kelley, M. J., Dunstan, C. R., Burgess, T., et al. (1998) Osteoprotegerin ligand is a cytokine that regulates osteoclast differentiation and activation. *Cell* **93,** 165–176.
8. Yamane, T., Kunisada, T., Yamazaki, H., Era, T., Nakano, T., and Hayashi, S. I. (1997) Development of osteoclasts from embryonic stem cells through a pathway that is *c-fms* but not *c-kit* dependent. *Blood* **90,** 3516–3523.
9. Ogawa, M., Nishikawa, S., Ikuta, K., Yamamura, F., Naito, M., Takahashi, K., and Nishikawa, S.-I. (1988) B cell ontogeny in murine embryo studied by a culture system with the monolayer of a stromal cell clone, ST2: B cell progenitor develops first in the embryonal body rather than in the yolk sac. *EMBO J.* **7,** 1337–1343.
10. Kodama, H., Nose, M., Niida, S., Nishikawa, S., and Nishikawa, S.-I. (1994) Involvement of the c-kit receptor in the adhesion of hematopoietic stem cells to stromal cells. *Exp. Hematol.* **22,** 979–984.
11. Nakano, T., Kodama, H., and Honjo, T. (1994) Generation of lymphohematopoietic cells from embryonic stem cells in culture. *Science* **265,** 1098–1101.
12. Yamane, T., Kunisada, T., Yamazaki, H., Nakano, T., Orkin, S. H., and Hayashi, S.-I. (2000) Sequential requirements for SCL/tal-1, GATA-2, M-CSF, and osteoclast differentiation factor/osteoprotegerin ligand in osteoclast development. *Exp. Hematol.* **28,** 833–840.
13. Doetschman, T. C., Eistetter, H., Katz, M., Schmidt, W., and Kemler, R. (1985) The in vitro development of blastocyst-derived embryonic stem cell lines: formation of visceral yolk sac, blood islands and myocardium. *J. Embryol. Exp. Morphol.* **87,** 27–45.
14. Li, E., Bestor, T. H., and Jaenisch, R. (1992) Targeted mutation of the DNA methyltransferase gene results in embryonic lethality. *Cell* **69,** 915–926.
15. Robertson, E., Bradley, A., Kuehn, M., and Evans, M. (1986) Germ-line transmission of genes introduced into cultured pluripotential cells by retroviral vector. *Nature* **323,** 445–448.

11

Differentiation of Embryonic Stem Cells as a Model to Study Gene Function During the Development of Adipose Cells

Christian Dani

1. Introduction

The white adipose tissue stores energy in the form of triglycerides in time of nutritional excess and releases free fatty acids during food deprivation. The adipose tissue mass is determined by the balance between energy intake and expenditure. Alterations of this steady state can lead to overweight and obesity, which is often accompanied by metabolic disorders associated with cardiovascular diseases such as hypertension and type II diabetes. The ongoing explosion in the incidence of obesity has focused attention on the development of adipose cells. Mainly, the in vitro system used to study adipogenesis is immortal preadipocyte cell lines *(1)*. However, these systems are limited for studies of early differentiation because they represent already determined cells. The commitment of embryonic stem (ES) cells into the adipocyte lincage offers the possibility to study the first steps of adipose cell development. In addition, the combination of genetic manipulations of undifferentiated ES cells and in vitro adipocyte differentiation facilitates elucidation of the role of genes expressed during adipose cell conversion. Terminal differentiation of preadipocytes into adipocytes is a multistep process. Several marker genes have been identified, and the hormonal regulation of these different genes has been studied in detail in recent years *(1)*. However, the requirement for adipogenesis of genes known to be expressed during the different stages of differentiation remains to be investigated (**Fig. 1**).

Peroxisome Proliferator-Activated Receptor γ (PPARγ) is member of the nuclear hormone receptor superfamily of ligand-activated transcriptional factors. PPARγ is expressed at high levels in adipose tissue, and several lines of evidence suggest that PPARγ plays a key role in the program of adipocyte differentiation *(2)*. Recently, we observed that activation of the membrane receptor for the Leukemia Inhibitory Factor (LIFR) promotes adipogenesis *(3)*. However, there has been no demonstration of the requirement of LIFR and PPARγ for differentiation. Both LIFR and PPARγ knock-out mice are not viable precluding the study of their role in the formation of fat *(3,4)*. The generation of LIFR$^{-/-}$ ES cells (**ref. 3**, and M. Li and A. G. Smith, unpublished results)

From: *Methods in Molecular Biology, vol. 185: Embryonic Stem Cells: Methods and Protocols*
Edited by: K. Turksen © Humana Press Inc., Totowa, NJ

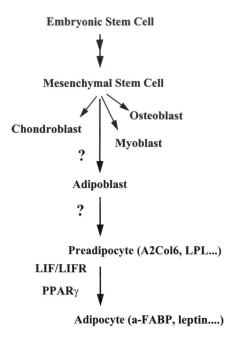

Fig. 1. Stages in the adipocyte development programm. A_2Col_6 and lipoprotein lipase (LPL) are markers of preadipose cells. a-FABP and leptin are markers of adipose cells. PPARγ and LIFR play a critical role in terminal differentiation. Question marks indicate that no regulatory genes involved in the commitment of stem cells towards the adipoblast lineage has been identify so far.

or PPARγ$^{-/-}$ ES cells *(4)* combined with conditions of culture to commit stem cells into the adipogenic pathway *(5)* allowed us *(3)* and Rosen and colleagues *(4)* to circumvent the lethality problem and to provide evidence that LIFR and PPARγ are important for the development of adipose cells.

The capacity of ES cells to undergo differentiation into several cell types in vitro enables the investigator to evaluate the specificity of the effect of the mutation. As the adipoblast and the skeletal myoblast come from the same mesenchymal stem cell precursor, it is interesting to compare the adipogenic and the myogenic capacity of mutant ES cells. Studies of the levels of expression of adipocyte-Fatty Acid Binding Protein (a-FABP), an adipocyte-specific gene, and of Myogenin, a skeletal myocyte gene, in outgrowths derived from LIFR-mutant ES cells and in those derived from wild-type ES cells, allowed us to evaluate in vitro the effect of LIFR mutation on the commitment of stem cells into the adipogenic and the skeletal myogenic lineages *(3,6)*.

The use of in vitro differentiation of ES cells to analyze the effects of mutations on adipogenesis is just beginning. However, the capacity of ES cells to undergo adipocyte differentiation in vitro provides a promising model for studying early differentiative events in adipogenesis and for identifying regulatory genes involved in the commitment of multipotent mesenchymal stem cell to the adipoblast lineage. In this chapter, an improved protocol to commit mouse ES cells into the adipogenic lineage at a high rate is detailed.

2. Materials

2.1. Maintenance of Mouse ES Cells

1. Complete growth medium (1X, stored at 4°C): to 300 mL autoclaved Milli-Q-plus water add:
   a. 40 mL of 10X Glasgow minimal essential medium (GMEM)/BHK21 (Gibco BRL, cat. no. 12541023), stored at 4°C.
   b. 13.2 mL of 7.5% Sodium bicarbonate (Gibco BRL, cat. no. 25080060), stored at 4°C.
   c. 4 mL of 100X Nonessential amino acids (Gibco BRL, cat. no. 11140035), stored at 4°C.
   d. 8 mL of 200 mM glutamine, 100 mM sodium pyruvate (Gibco BRL, cat. nos. 25030024 and 11360039), stored at –20°C.
   e. 0.4 mL of 0.1 M 2-Mercaptoethanol (stored at 4°C).
   f. 40 mL of Selected fetal calf serum (FCS) (S. A. Dutscher, France), stored at –20°C (*see* **Note 1**).
2. 0.1 M 2-Mercaptoethanol: add 100 µL 2-mercaptoethanol (Sigma, cat. no. M-7522) to 14 mL of sterile H$_2$O. Store up to 3 wk at 4°C.
3. LIF (*see* **Note 2**).
4. Phosphate-buffered saline (PBS) calcium and magnesium free: 0.17 M NaCl, 3.4 mM KCl, 4 mM Na$_2$H PO$_4$, and 2.4 mM KH$_2$PO$_4$, pH 7.4. Filter-sterilized.
5. Trypsin solution: Trypsin 1X is prepared by adding 1 mL of 2.5% trypsin (Gibco BRL, cat. no. 2509008), plus 1 mL of 100 mM EDTA and 1 mL of chicken serum to 100 mL PBS. Aliquot (10 mL) and store at –20°C. Thawed aliquots are stored at +4°C.
6. Gelatin 0.1%: purchase gelatin 2% (Sigma, cat. no. G-1393) and dilute to 0.1% with PBS. Store at 4°C.
7. Tissue culture 25-cm^2 flasks (Greiner, cat. no. 690175 or Corning, cat. no. 430168).

2.2. Differentiation into Adipocytes

1. Retinoic acid: All *trans* retinoic acid (RA) (Sigma, cat. no. R-2625) is diluted in the dark into dimethyl sulfoxide (DMSO) to prepare a 10 mM stock solution. Aliquot and store at –20°C. Subsequent dilutions of RA are performed in ethanol and used for one experiment only. After dilution into the culture media, the concentration of ethanol never exceeds 0.1%.
2. Differentiation medium: This medium consists of growth medium with selected serum and is supplemented with antibiotics, 0.5 µg/mL insulin and 2 nM triiodothyronine (*see* **Note 3**).
3. Insulin (Sigma, cat. no. I-5500) is prepared at 1 mg/mL in cold 0.01 N HCl. Mix gently and sterilize by filtration. Aliquot (1 mL) and store at –20°C. Thawed aliquots are stored at 4°C.
4. Triiodothyronine (Sigma, cat. no. T-2877) is prepared at 2 mM in ethanol (stock solution). Store at –20°C.
5. Antibiotics: 1000X penicillin–streptomycin (5000 IU/mL–5000 UG/mL; Gibco BRL). Aliquot (1 mL) and store at –20°C.
6. Bacteriological grade 100- and 60-mm Petri dishes (Greiner, cat. nos. 633185 and 628102).
7. Tissue culture 100-mm dishes (Corning, cat. no. 664160).

2.3. X-Gal and Oil-Red O Staining

1. Fix buffer: 0.25% glutaraldehyde (Sigma, cat. no. G-5882) in PBS supplemented with 2 mM MgCl$_2$ and 5 mM EGTA (pH 8.0). Store at 4°C.

2. Wash buffer: PBS supplemented with 2 mM MgCl$_2$. Store at 4°C.
3. 5-Bromo-4-chloro-3-indolyl-β-D-galactopyranoside (X–gal) staining solution:
   a. 1 mL of Potassium ferrocyanide (0.105 mg/mL; Sigma, cat. no. P-9387).
   b. 1 mL of Potassium ferricyanide (0.082 mg/mL; Sigma, cat. no. P-8131).
   c. 1 mL of X–gal (50 mg/mL; Biosolve Ltd., cat. no. 07102352).
   d. 50 mL of Wash buffer. Filter to remove crystals and store in the dark at 4°C.
4. Oil-Red O solution. Stock saturated solution: 0.5 % Oil-Red O (Sigma, cat. no. O-0625) in isopropanol. Working solution: mix 6 volumes of stock solution with 4 volumes H$_2$O. Mix and filter. Store at room temperature.
5. Store solution: 70% glycerol in H$_2$O (v/v).

2.4. RNA Preparation from Differentiating Embryoid Body Outgrowths

1. Guanidinium lysis buffer: 4 M guanidinium thiocyanate dissolved in 25 mM sodium citrate, pH 7.0, 0.5% sarcosyl. Just before to use, add 2-mercaptoethanol to a final concentration of 0.1 M.
2. Sterile 2.2-mL Eppendorf tubes.
3. Phenol saturated solution, pH 4.0 containing 0.1% 8-hydroxyquinoline (*see* **Note 4**).
4. Chloroform.
5. Isopropanol
6. 100% Ethanol.
7. 5 M NaCl. Sterilize.
8. TES buffer: 10 mM Tris-HCl, pH 7.4, 0.1 mM EDTA, 0.1% sodium dodecyl sulfate (SOS). Sterilize. Store at room temperature.

2.5. Reverse Transcription-Polymerase Chain Reaction

1. Reverse transcription polymerase chain reaction (RT-PCR) kit (avaible from several companies).
2. Pairs of primers used for detecting:
   a. a-FABP, an adipocyte-specific gene; 5′-GATGCCTTTGTGGGAACCTGG-3′ and 5′-TTCATCGAATTCCACGCCCAG-3′.
   b. Myogenin, a skeletal myocyte-specific gene: 5′-AGCTCCCTCAACCAGGAGGA-3′ and 5′-GGGCTCTCTGGACTCCATCT-3′.
   c. Chain α2 of collagen type VI (A2COL6), a gene preferentially expressed in mesenchymal cells: 5′-AACTTCGCCGTGGTCATCACTGACG-3′ and 5′-AGGAATCTCCAG-GCAGCTCACCTTG-3′.
   d. Hypoxanthine phosphoribosyltransferase (HPRT), as a standard to balance the amount of RNA and cDNA used, except for RNA prepared from E14TG2a ES cells, which are HPRT-deficient cells: 5′-GCTGGTGAAAAGGACCTCT-3′ and 5′-CACAG GACTAGAACACCTGC-3′.

3. Methods
3.1. Maintenance of ES Cells

Several previously published protocols describe procedures to maintain ES cells on fibroblast feeder layers. The conditions outlined below are applicable for the maintenance of feeder layer-independent ES cell lines, such as CGR8 *(7)*, E14TG2a *(8)*, or Zin40 *(9)*. These cells can be grown on gelatin-coated tissue culture flasks (*see* **Note 5**) and maintained in a multipotent undifferentiated state providing they are exposed to LIF (*see* **Note 2**).

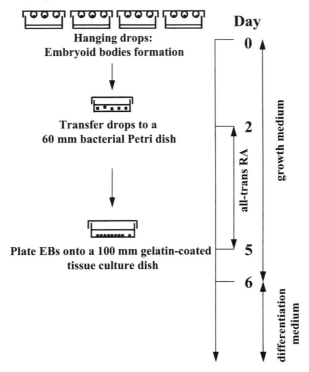

Fig. 2. Experimental protocol used for the commitment of ES cells into the adipocyte lineage.

1. For a 25-cm^2 flask, aspirate the medium off and wash twice with 5 mL of PBS. Aspirate off the PBS and add 1 mL of trypsin solution. Ensure the trypsin covers the cell monolayer and incubate at 37°C, 5% CO_2, for 2 to 3 min. Check under an inverted microscope that cells are correctly dissociated (*see* **Note 6**).
2. Add 5 mL of complete growth medium to stop trypsinization and suspend the cells by vigorous pipetting. Transfer the cells to a sterile tube and centrifuge at 250*g* for 5 min at room temperature.
3. Aspirate the medium off and resuspend the cell pellet with 5 mL of complete growth medium by pipetting up and down 2 to 3 times. Count the cells.
4. Add 10^6 ES cells into 10 mL of prewarmed complete growth medium containing LIF, then transfer to a freshly gelatinized 25-cm^2 flask.
5. Change the medium every day.
6. Trypsinize the cultures 2 d later as in step 1. Cultures should be subcultured before cells have reached confluence.

3.2. Differentiation of Embryoid Bodies into Adipocytes

Multilineage differentiation of ES cells is initiated by aggregation. The aggregates form structures known as embryoid bodies (EBs). To induce adipocyte lineage, the hanging drop method for the formation of EBs is routinely used (*see* **Note 7**). A schematic representation of the experimental protocol used for ES cell differentiation of ES cells into adipocytes is presented in **Fig. 2**.

1. Change media on ES cells with complete growth medium supplemented with LIF 2 h before subculture.
2. Aspirate medium off and wash twice with 5 mL of PBS. Aspirate off the PBS and add 1 mL of trypsin solution. Incubate at 37°C, 5% CO_2, for 2 to 3 min.
3. Add 5 mL of complete growth medium to stop trypsinization and suspend the cells by vigorous pipetting. Transfer the cells to a sterile tube and centrifuge at 250g for 5 min at room temperature.
4. Aspirate the medium off and resuspend the cell pellet into 10 mL of complete growth medium without LIF, but supplemented with antibiotics (*see* **Subheading 2.2.5.**). After cell counting, adjust the suspension to a concentration of 5×10^4 cells/mL.
5. Place aliquots of 20 μL of this suspension onto the lid of bacteriological grade dishes (*see* **Note 8**). This is defined as day 0 of EB formation.
6. Invert the lid and place it over the bottom of a bacteriological Petri dish filled with 8 mL PBS containing few drops of FCS. It is essential to cover the bottom of the dish with the liquid to prevent the evaporation of the hanging drops. When the lid is inverted, each drop hangs, and the cells fall to the bottom of the drop where they aggregate into a single clump (EB).
7. Two days later, remove the lid, invert it, and collect drops containing EBs in a conical sterile tube. Let it stand for 5 min at room temperature to allow the aggregates to sediment. Aspirate the supernatant and resuspend the pellet in 4 mL of complete growth medium supplemented with 10^{-7} M RA (*see* **Note 9**). Transfer the suspension into 60-mm bacteriological grade Petri dishes (*see* **Note 10**).
8. Incubate for 3 d in the presence of RA, changing the medium every day.
9. At d 5 after EB formation, change the medium without the addition of RA and plate 2–4 EBs/cm^2 in gelatinized-tissue culture dishes (*see* **Note 11**).
10. The day after plating, change the complete growth medium to differentiation medium. Change the medium every other day.
11. After 10–20 d in the differentiation medium, at least 50–70% of EB outgrowths should contain adipocyte colonies (*see* **Note 12**).

3.3. Identification of Adipose Cells in Outgrowth Cultures

3.3.1. X-Gal and Oil-Red O Staining

Among the different ES cell lines that we use, Zin40 displays ubiquitous nuclear expression of β-galactosidase (β-gal) in undifferentiated ES cells and derivatives in vitro *(5,9)*. Expression of the LacZ reporter gene under the control of a specific promoter enables the determination of cell specificity as well as the kinetics of expression of a gene of interest in differentiating outgrowths. Therefore, a procedure to detect β-gal activity in situ in ES cell-derived cultures is described.

Adipose cells with lipid droplets are easily visualized microscopically, especially under bright field illumination. Nonadipose cells containing structures resembling droplets are often detectable in untreated and RA-treated cultures. Therefore, it is essential to identify droplet-like structures as triglyceride droplets. Staining of cultures with Oil-Red O, a specific stain for triglycerides, gives a good indication of adipocyte differentiation.

A simple procedure to detect both β-gal activity and adipocytes containing triglycerides in situ in outgrowth cultures is described.

1. Aspirate medium and wash cells once with PBS.
2. Fix cells for 15 min at room temperature.

3. Wash twice for 10 min.
4. Stain with X-gal solution overnight at 37°C.
5. Aspirate X-gal solution and wash with H_2O.
6. Stain with Oil-Red O solution for 15 min.
7. Wash twice with H_2O. Cover cells with a film of storage solution (70% glycerol in H_2O).

β-gal and Oil-Red O staining can be seen either with phase contrast or bright field illumination.

3.3.2. RNA Preparation from EB Outgrowths

The procedure described is an adaptation of the single-step method previously published by Chomczynski and Sacchi *(10)*.

1. Wash cells with PBS.
2. Add 1.6 mL of guanidinium lysis buffer per 100-mm tissue culture dish containing EB outgrowths. Split cell lysate into 2 sterile 2.2-mL Eppendorf tubes. Vortex mix vigorously.
3. Add 0.8 mL of phenol-saturated solution, pH 4.0, per tube and mix by inversion. Then add 0.2 mL of chloroform. Shake vigorously for 10 sec and keep on ice for 15 min.
4. Centrifuge at 10,000g for 20 min at 4°C.
5. Carefully transfer the aqueous phase containing RNA (usually the upper phase) (*see* **Note 4**) to a fresh 2.2-mL tube. Take care to avoid any traces of interface material.
6. Repeat from step 3.
7. Transfer the aqueous phase to a fresh 2.2-mL tube. Add 0.8 mL of isopropanol, mix, and place at –20°C for at least 12 h.
8. Centrifuge at 10,000g for 20 min at 4°C.
9. Aspirate the supernatant off and collect the RNA pellets from both tubes in 0.3 mL of guanidinium lysis buffer.
10. Precipitate with 0.6 mL of ethanol at –20°C for 12 h.
11. Centrifuge at 10,000g for 20 min.
12. Carefully pour off the supernatant and dissolve the RNA pellet in 0.5 mL of TES (10 mM Tris-HCl, pH 7.4, 0.1 mM EDTA, 0.1% SDS).
13. Determine the RNA concentration (*see* **Note 13**). Store RNA at –20°C.

3.3.3. Analysis of Adipocyte Gene Expression

Expression of a-FABP and Myogenin in 20-d-old EB outgrowths can be detected by Northern blotting using 20 μg of total RNA. However, detection of the expression of these genes in early differentiating outgrowths requires a more sensitive method such as RT-PCR. **Table 1.** gives the temperatures of annealing for the PCR and the size of expected cDNAs, as well as the size of genomic DNA-derived contaminated bands. Gene expression can be detected either by a diagnostic ethidium bromide band or after blotting and hybridization with appropriate cDNA probes.

4. Notes

1. ES cell cultures may contain a proportion of "differentiated" cells that have lost their pluripotency. It is crucial to minimize this proportion of differentiated cells. This is achieved by the addition of LIF in a high quality culture medium, i.e., an adequate batch of serum. Identification of pluripotent stem cells is difficult unless one is familiar with the appropriate cellular morphology. Pluripotent stem cells: *(1)* are small, *(2)* have a

Table 1
Detection of Adipocyte- and Skeletal-Myocyte Specific Genes by PCR

Gene	Base Pairs		Annealing temperature (°C)
	cDNA	Genomic	
a-FABP	213	2400	56
Myogenin	500	1500	60
A2COL6	300	500	55
HPRT	249	1100	60

large nucleus containing prominent nucleoli structures, and *(3)* have minimal cytoplasm. Pluripotent stem cells, in contrast to differentiated cells, grow rapidly. Our experience has been to select a batch of serum, which is able to support the growth of stem cells, plate 10^6 cells/25-cm^2 flask in 10% of each set of FCS, supplemented with LIF, and sub-culture the cells every 2 d for 4 passages. For a high quality serum, a flask should yield $5–10 \times 10^6$ cells at each passage. Furthermore, no toxicity of the selected serum should be observed at a 30% concentration.

2. LIF is required to maintain pluripotent ES cells and is omitted to induce the commitment of ES cells towards the adipogenic lineage. To produce LIF, Cos7 cells are transiently transfected with a LIF-expressing construct (using standard techniques), and after 4 d of growth, the medium is collected. A titration of LIF activity in the medium conditioned by transfected Cos7 cells is then performed by testing several dilutions of the medium on ES cells plated at clonal density in 24-well plates. All these steps are described in detail in **ref. 12**. LIF is also commercially available (named ESGRO by Chemicon).

3. Serum added into the differentiation medium is preselected to support the terminal differentiation of preadipocyte clonal cell lines *(1)*. The addition of a PPARγ activator *(11)*, such as 0.5 μM thiazolidinedione BRL49653 (SmithKline Beecham Pharmaceuticals) in the differentiation medium dramatically stimulates the terminal differentiation of RA-treated EBs into adipocytes (Belmonte, N., Vernochet, C., and Dani, C., unpublished results). This compound is not commercially available. The regimen used to promote differentiation of 3T3 to L1 preadipose cells has recently been tested on RA-treated EBs *(4)*. With this regimen, 17-d-old EB outgrowths are treated with dexamethasone (DEX) (400 ng/mL) and methylisobutylxantine (MIX) (500 n*M*) for 2 d. Then, DEX and MIX are removed from the culture media, and outgrowths are maintained in differentiation medium. As expected, hormonal treatments, which have been previously proved to promote terminal differentiation of preadipocytes into adipocytes, are also efficient to induce terminal differentiation of RA-treated EBs. It is important to note that the treatment of EBs with RA is a prerequisite.

4. 8–hydroxyquinoline stains phenol yellow, which allows the unambigous identification of the phenol phase and the aqueous phase.

5. Attachment of ES cells to the substratum is susceptible to change according to the tissue culture material. The author uses Corning or Greiner tissue culture flasks and dishes. Coating is performed by covering the surface of a 25-cm^2 flask with 5 mL of 0.1% gelatin for 15 min at room temperature, followed by careful aspiration of the gelatin solution.

6. It is critical to produce a single cell suspension for subcultures. This is achieved by knocking the flask several times to ensure complete dissociation during the trypsin treatment.

7. The formation of EBs in mass culture (by maintaining pluripotent ES cells in suspension at a high density, i.e., 5×10^5 cells/mL in bacteriological grade Petri dish) leads subsequently to a low number of outgrowths containing adipocyte colonies.

8. We use a multipipettor with a sterile combitip dispensor (Eppendorf). Approximately 80 drops can be fitted on the lid of a 100-mm Petri dish.

9. Owing to the high instability of RA, the concentration of RA able to commit ES cells into the adipogenic lineage at a high rate should be determined for each new preparation of RA (try 10^{-8} to 10^{-6} M). RA is light-sensitive.

10. Bacterial grade Petri dishes are used to prevent cell attachment to the substrate. EBs have a tendency to attach to the bottom of the plastic dish. This phenomenon is reduced by changing the media daily. EBs that are firmly attached to the dish should be eliminated, as these kind of EBs seem to have no adipogenic capacity.

11. A higher density of EBs can lead to a decrease in the number of EB-containing adipocyte colonies. Two-day-old EBs from 4 lids of 100-mm Petri dishes are pooled into one 60-mm bacteriological grade Petri dish, then, after RA treatment, are plated into one 100-mm tissue culture dish (*see* **Note 13** and **Fig. 2**).

12. A wide variety of differentiated derivates, such as neurone-like cells, fibroblast-like cells, and unidentified cell types, appear over this period. Spontaneously beating cardiomyocytes should appear 1–5 d after plating the control culture (untreated with RA). At least 40% of EBs should contain beating cardiomyocytes. In contrast, few EBs should contain beating cells from RA-treated cultures. Large adipocyte colonies appear late in the RA-treated culture.

13. We routinely get 100–200 µg RNA from one 100-mm tissue culture plate containing 20-d-old EB outgrowths.

Acknowledgments

The work reported from the author's laboratory was supported by the Centre National de la Recherche Scientifique (grant no. UMR6543), the Institut National de la Santé et de la Recherche Médicale and the Association pour la Recherche Contre le Cancer (grant no. 9982).

References

1. Gregoire, F. M., Smas, C. M., and Sul, H. S. (1998) Understanding adipocyte differentiation. *Physiol. Rev.* **78**, 783–809.

2. Lowell, B. B. (1999) PPARγ: an essential regulator of adipogenesis and modulator of fat cell function. *Cell* **99**, 239–242.

3. Aubert, J., Dessolin, S., Belmonte, N., Li, M., McKenzie, F., Staccini, L., et al. (1999) Leukemia Inhibitory Factor and its receptor promote adipocyte differentiation via the mitogen-activated protein kinase cascade. *J. Biol. Chem.* **274**, 24,965–24,972.

4. Rosen, E. D., Sarraf, P., Troy, A. E., Bradwin, G., Moore, K., Milstone, D. S., et al. (1999) PPAR gamma is required for the differentiation of adipose tissue in vivo and in vitro. *Mol. Cell* **4**, 611–617.

5. Dani, C., Smith, A. G., Dessolin, S., Leroy, P., Staccini, L., Villageois. P., et al. (1997) Differentiation of embryonic stem cells into adipocytes in vitro. *J. Cell Sci.* **110**, 1279–1285.

6. Dani, C. (1999) Embryonic stem cell-derived adipogenesis. *Cells Tissues Organs* **165**, 173–180.

7. Mountford, P., Zevnik, B., Düwel, A., Nichols, J., Li, M., Dani, C., et al. (1994). Dicistronic targeting constructs: Reporters and modifiers of mammalian gene expression. *Proc. Natl. Acad. Sci. USA* **91**, 4303–4307.

8. Hooper, M. L, Hardy, K., Handyside, A., Hunter, S., and Monk, M. (1987) HPRT-deficient (Lesch-Nyhan) mouse embryos derived from germline colonisation by cultured cells. *Nature* **326,** 292–295.

9. Mountford, P. and Smith, A. G (1995) Internal ribosome entry site and dicistronic RNAs in mammalian transgenesis. *TIG* **11,** 179–184.

10. Chomczynski, P. and Sacchi, N. (1987) Single-step method of RNA isolation by acid guanidium thiocyanate-phenol-chloroform extraction. *Anal. Biochem.* **162,** 156–159.

11. Lehmann, J. M., Moore, L. B., Tracey, A., Smith-Oliver, T. A., Wilkison, W. O., Willson, T. M., and Kliewer, S. A. (1995) An antidiabetic thiazolidinedione is a high affinity ligand for peroxisome proliferator-activated receptor γ. *J. Biol. Chem.* **270,** 12,953–12,956.

12. Smith, A. G. (1991) Culture and differentiation of embryonic stem cells. *J. Tiss. Cult. Meth.* **13,** 89–94.

12

Embryonic Stem Cell Differentiation and the Vascular Lineage

Victoria L. Bautch

1. Introduction

The ability of mouse embryonic stem (ES) cells to undergo differentiation in vitro complements their ability to contribute to numerous tissues in vivo and provides a unique model system for aspects of early mammalian development. ES cells are differentiated in two major ways: *(i)* unmanipulated differentiation involves the removal of differentiation inhibitory factors, allowing the ES cells to undergo a programmed differentiation to form multiple cell types that provide developmental cues to each other; and *(ii)* manipulated differentiation begins with the removal of differentiation inhibitory factors, but at some point the cells are usually disaggregated and cultured with specific added factors to purify or to increase the proportion of cells that acquire a particular developmental fate. Examples of both kinds of differentiation are found in this volume. The protocols provided here are for unmanipulated differentiation, which reproducibly results in the development of a primitive vasculature. Endothclial cells typically comprise 15–20% of the differentiated ES cells.

Mouse ES cells were first differentiated in vitro using small clumps of ES cells in suspension culture in media containing human cord serum *(1–3)*. The ES cell clumps formed embryoid bodies (EBs) containing an outer and inner layer, and some EBs expanded to form a large lumen and were called cystic embryoid bodies (CEBs). These CEBs had hemoglobinized areas indicative of hematopoietic development. It was suggested that vascular development occured, analogous with the development of blood islands in the yolk sac *(1,2)*. Subsequently the presence of mouse endothelial cells and a primitive vasculature in CEBs was shown *(4)*, and the requirement for human cord serum was eliminated by using lot-tested fetal calf serum (FCS). A further technical modification involved reattaching the EBs to tissue culture plastic prior to overt differentiation of vascular tissue *(5)* (**Fig. 1**). This allowed for more sophisticated experimental manipulations of the early vasculature and for the introduction of quantitative assays of vascular development *(6,7)*.

A number of markers for the early mouse vasculature were identified and used to further characterize mouse vascular development in vivo and in ES cell cultures and CEBs. Platelet endothelial cell adhesion molecule-1 (PECAM-1) (also called CD31)

From: *Methods in Molecular Biology, vol. 185: Embryonic Stem Cells: Methods and Protocols*
Edited by: K. Turksen © Humana Press Inc., Totowa, NJ

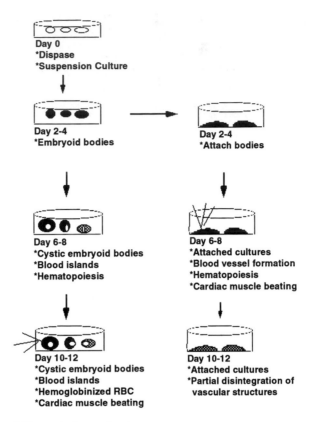

Fig. 1. In vitro differentiation of ES cells. Unmanipulated ES cell differentiation is accomplished by formation of CEBs in suspension culture (depicted in the left column) or by formation of attached cultures (depicted in the right column). Multiple cell types differentiate reproducibly using these protocols, somewhat analogous to differentiation of the mouse yolk sac. The lines in d 6–8 Attached Cultures and d 10–12 CEBs denote areas of beating due to formation of cardiac muscle.

has proven particularly useful, since it is a transmembrane protein expressed abundantly early in vascular development *(8–10)*. Other early markers are the vascular endothelial cell growth factor-A (VEGF) receptors flk-1 (VEGFR-2) and flt-1 (VEGFR-1), SCL/TAL, and vascular endothelial cadherin (VE-cadherin) *(11,12)*. Vascular markers that are expressed later, as angioblasts organize into blood vessels containing endothelial cells, include intercellular adhesion molecule-2 (ICAM-2), CD34, and Tie-2 *(7,11)*. Several groups have analyzed vascular marker expression as a function of time during ES cell differentiation using either fluorescence-activated cell sorter (FACS) analysis or in situ immunolocalization *(7,11)*, and while the details vary, the general trends are similar from study to study.

   An important criterion for a model system is that specific mutations have similar vascular phenotypes in vivo and in the model system. Vascular development during ES cell differentiation is perturbed by mutations with vascular phenotypes in vivo: both flk-1 and flt-1 homozygous mutant ES cells differentiate abnormally in the vascular compartment *(12–14)*, and VEGF-A mutant ES cells have reduced vascular development in both the hemizygous and homozygous mutations, consistent with in vivo data *(7)*. One exception is that flk-1 mutant ES cells do not contribute to

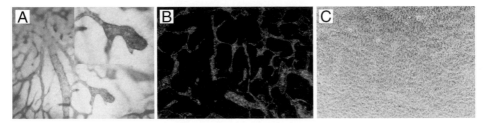

Fig. 2. Vascular development in vitro is similar to blood vessel formation in vivo. The vasculature of a mouse embryo (**A**) or the vasculature of an ES cell differentiated attached culture (**B,C**) was visualized by PECAM antibody staining. (**A**) The head vasculature of a d 10.5 mouse embryo, showing an area with a more bulbous vasculature (*see* inset). (**B,C**) ES cell attached culture differentiated to d 8, fixed and stained, showing a typical vascular pattern by immunofluorescence (**B**) and the corresponding phase photomicrograph (**C**), to emphasize that multiple cell types differentiate under these conditions.

hematopoietic development in vivo, but can produce primitive erythrocytes upon differentiation of ES cells *(12,15)*. It is thought that flk-1 is important for the initial migration of precursor cells to the yolk sac in vivo, and that in the ES differentiation model, the yolk sac cell types differentiate in close proximity, negating the migration requirement for hematopoiesis in vitro.

Thus, unmanipulated ES cell differentiation provides a reproducible developmental program that results in the differentiation of multiple cell types, many of which are found in the developing yolk sac. Specifically, vascular precursor cells are formed that mature into endothelial cells that organize into a primitive vasculature (**Fig. 2**). The shape of the primitive vasculature suggests that little or no remodeling occurs during ES cell differentiation, but clearly the initial stages of vasculogenesis to form blood vessels from precursor cells occur, and expansion of the primitive vasculature via angiogenic sprouting is also most likely a process that occurs during ES cell differentiation. The disadvantage of unmanipulated differentiation is that multiple cell types impose a requirement for careful in situ expression analysis, so vascular immunofluorescence staining and quantitation protocols are included in this chapter. However, I feel that by approximately recapitulating the in vivo environment in which vascular development occurs, with its attendant cell types producing important developmental signals, it is possible to analyze the roles of specific molecules in vascular development in an important but accessible context.

2. Materials

2.1. General Tissue Culture Materials (All Sterile)

1. 24-Well tissue culture dishes (Costar, cat. no. 3524).
2. 6- and 10-cm Tissue culture dishes (Corning, cat. no. 430166 [6 cm] and cat. no. 430167 [10 cm]).
3. Disposable pipets of sizes: 1 mL, 5 mL, 10 mL, and 25 mL.
4. Bacteriological Petri dishes (Fisher, cat. no. 543116, "made in Canada")—these dishes are the best we found to prevent sticking of EBs during the initial culture period.
5. Ca^{+2} and Mg^{+2}-Free phosphate-buffered saline (PBS).
6. 0.1% Gelatin (Difco, cat. no. 0143-17-9) Type A, porcine, Bloom factor 200, in PBS.

7. Monothioglycerol (Sigma, cat. no. M6145) cell culture tested. 100X stock is 32.5 μL to 50 mL PBS. Store at 4°C.
8. 1X Trypsin-EDTA: (Gibco, cat. no. 25300-054). Store at –20°C long term. Thawed aliquots can be kept at 4°C for several weeks.
9. Gentamicin (Gibco-BRL) (50 mg/mL gentamicin sulfate). 1000X stock, store at 4°C.

2.2. ES Cell Culture

All reagents and materials are sterile.

1. ES cell culture medium: 67% 5637 cell conditioned medium (5637 human bladder carcinoma cell line [ATCC, cat. no. HTB9], source of leukemia inhibitory fctor [LIF], *see* **Subheading 3.1., step 1** for collection protocol), Dulbecco's modified Eagle medium (DMEM)-H (Gibco-BRL), 17% FCS (lot tested from different manufacturers), 82 μ*M* monothioglycerol, 1X gentamicin (50 μg/mL). Store at 4°C and use within 1 mo.
2. Trypsin-EDTA solution: 0.25X trypsin-EDTA (diluted in PBS). Store stock at –20°C and working solution at 4°C for several wk.
3. Trypsin stop medium: 5637 cell conditioned medium: FCS (1 : 1).

2.3. Enzymatic Disruption and In Vitro Differentiation

All reagents and materials are sterile.

1. ES cell differentiation medium: DMEM-H, 20% lot-selected FCS, 150 μ*M* monothioglycerol (2 mL 100X stock/100 mL medium = 2X concentration), 1X gentamicin (50 μg/mL). Store at 4°C and use within 1 mo.
2. Dispase: (Boehringer Mannheim, cat. no. 295825) (Dispase Grade II, 2.4 U/mL). Dilute 1 : 1 in PBS just prior to use, so final concentration is 1.2 U/mL.
3. Medi-droppers: (Fisher, cat. no. 13-711 or Spectrum, cat. no. 180-875), autoclaved.

2.4. Materials for Fixation, Antibody Staining, and Imaging

2.4.1. Fixation

1. Methanol : acetone (1 : 1). Methanol and acetone are stored separately at –20°C and mixed just prior to use.
2. Fresh 4% paraformaldehyde (PFA): 2g PFA powder in 50 mL PBS (use a 50-mL conical tube). Heat solution to 60°C, shaking occasionally. When most of the solid is in solution, cool to room temperature, then filter through 0.45-μm filters fitted to a syringe. Extra 4% PFA can be stored at –20°C in aliquots and thawed on ice just prior to use.

2.4.2. Antibody Staining

1. Dilution buffer for all antibodies: 3% FCS, 0.1% NaN$_3$ in PBS.
2. Primary antibodies (partial list): rat anti-mouse PECAM-1 (Mec 13.3; Pharmingen) at 1 : 1000; rat anti-mouse ICAM-2 (3C4; Pharmingen) at 1 : 500; rat anti-mouse flk-1 (Ly-73; Pharmingen) at 1 : 100.
3. Secondary antibodies (partial list): donkey anti-rat B-phycoerythrin cross-absorbed (712-106-150; Jackson Immunoresearch) at 1 : 300 (for PECAM-1 and ICAM-2 detection); and donkey anti-rabbit IgG (H+L) tetramethyl rhodamine isothiocyanate (TRITC) cross-absorbed (711-025-152; Jackson Immunoresearch) at 1 : 100.

2.4.3. Imaging

1. Olympus IX-50 inverted microscope with epifluorescence and camera hook-up.
2. Photographic film (black and white).

3. Computer, Adobe Photoshop (version 4.0 or higher) and plug-ins (Image Processing Tool Kit, version 2.1, Reindeer Games, Asheville, NC).

3. Methods
3.1. ES Cell Culture

1. Collection of 5637 cell condition medium (CM): 5637 cells are grown to confluence in 15-cm tissue culture dishes in DMEM-H, 10% FCS, 1X monothioglycerol, and 1X gentamycin. At confluence, the medium is changed to the same formulation but with 5% FCS instead of 10% FCS. After 48–72 h, remove dishes from incubator to hood, where 50-mL conical tubes with loose caps are set up. Tilt each dish and remove the medium with a 25-mL pipet, and put into the 50-mL tubes. Add 30 mL fresh medium (DMEM-H, 5% FCS, 1X monothioglycerol, 1X gentamicin) to each dish and return to incubator. Balance the volumes in the 50-mL tubes and spin at 2000g for 10 min at 4°C. In the hood, remove the supernatant from each tube and filter using a 0.45-µm filter. When all the medium is filtered, place 3 mL in a 6-cm dish in the incubator and check after 24 h for any contamination. Do the collection every 2 to 3 d until you have 5 or 6 collections, then discard cells, pool collections, aliquot, and freeze at –20°C.
2. Prepare 0.1% gelatinized dishes by adding 2 mL/6-cm dish. Tilt the dish to cover the bottom. Incubate at 37°C for 1 h up to 1 wk. Check each dish under the microscope for bacterial or fungal contamination prior to use. If uncontaminated, aspirate the excess gelatin solution and add the appropriate amount of ES cell culture medium to each dish. Return to the incubator.
3. The ES cells should be passed every 3 to 4 d, when they are seen in shiny tight clumps (*see* **Note 1**). Remove the dishes from the incubator and aspirate off the medium. Wash 2 times with PBS and aspirate. Add 0.25X Trypsin-EDTA (diluted from 1X stock in PBS) to just cover the bottom of the dish (i.e., 1 mL for a 6-cm dish).
4. Place at 37°C until the majority of the ES cell clumps come up with gentle agitation (1–3 min). Stop the reaction by adding Trypsin stop solution (4 vol stop solution/vol Trypsin-EDTA). Gently draw the solution up and down a 10-mL pipet a few times to break up the clumps of cells.
5. Put 2 to 3 drops of the cell suspension into the 6-cm dish with medium in it. Observe the size of the clumps under a microscope. Try to have the cell clumps average about six cells each. If they are significantly larger, pipet further to break them up a bit. Move the dishes gently to disperse the ES cell clumps evenly throughout the dish and return to the 37°C incubator.

3.2. Enzymatic Disruption and In Vitro Differentiation
3.2.1. Enzymatic Disruption

All volumes are given assuming that one 6-cm dish of ES cell colonies is being processed. If using a larger dish or more dishes, adjust the volumes accordingly.

1. Choose the dish that has the best ES cell clumps to process for in vitro differentiation—the cell clumps should be flattened and differentiated on the very edge but round and shiny (aspects of undifferented ES cells) in the middle. Plates are incubated for 5 to 6 d without feeding after normal passage of ES cells.
2. Aspirate the medium from the dish. Rinse twice with PBS and aspirate. Add 1 mL cold Dispase (diluted just prior to use). Let sit at room temperature for 1 to 2 min. Check periodically by shaking the dish to see if the cell clumps have detached from the bottom.

3. When the majority of cell clumps have detached, use a 10-mL pipet to gently transfer the cell clumps to a 50-mL conical tube containing 35 mL PBS at room temperature. Invert the tube once gently to mix. Rinse the dish with 5 mL PBS and add to the contents of the 50-mL tube.

4. Let the tube sit until the cell clumps have settled to the bottom, then aspirate the liquid carefully to avoid disturbing the cell clumps. Add 30–40 mL PBS down the side of the tube, swirl gently to redistribute the cell clumps, and let settle again. Aspirate, repeat the PBS wash and aspiration, then add 5 mL differentiation medium down the side of the tube. With the residual PBS, the volume in the tube will be 6 to 7 mL.

5. Pipet 10 mL of differentiation medium into each of two bacteriological dishes (*see* **Note 2**). Pipet the cell clumps and medium from the 50-mL tube into a 25-mL pipet (*see* **Note 3**). Transfer them as equally as possible to the two bacteriological dishes, pulling in an air bubble to keep the clumps distributed in the medium.

6. Check the number of cell clumps in the dishes. The goal is to get very approximately 100 clumps per dish, since fewer is not an efficient use of medium, and more may promote aggregation of cell clumps. Incubate at 37°C in a humidified incubator with 5% CO_2.

3.2.2. In Vitro Differentiation

1. Change medium as required, at least every other day after Dispase treatment (*see* **Note 4**). To feed EBs, gently swirl the dish in a circular manner so that the EBs go to the center of the dish. Carefully aspirate most of the old medium from the dish. Gently add 10 mL fresh differentiation medium to each dish, being careful not to disrupt the EBs.

2. To form attached cultures (*see* **Note 5** for CEB production), use EBs that have been in suspension culture for 3 d (*see* **Note 6** for variations). Swirl the bacteriological dish so the EBs are fairly close together.

3. Dispense 1.5 mL of differentiation medium into each well of a 24-well tissue culture dish that is to be seeded (*see* **Note 7**). Use the sterile medidropper to move EBs from the dish to the wells of the 24-well dish. Generally, try to get 10–20 EBs into each well. The bodies can be estimated by covering the dish and holding it up to look through the bottom.

4. Spread the EBs evenly in the well. If they do not spread evenly by gentle rocking of the dish, it can be done by gently pipetting medium up and down in a well (use a Pipetman with a sterile tip). Keep the plate level as you return it to the incubator, as the EBs usually attach where they are left. Attachment generally occurs within a few hours. Incubate at 37°C in a humidified incubator with 5% CO_2 and feed with 2 mL fresh differentiation medium at least every other day until fixation.

3.3. Immunolocalization

1. Take dishes with attached cultures (*see* **Note 8** for timing) and aspirate the medium. Rinse twice with PBS and aspirate, then add 1 mL of the cold fixative/well of the 24-well dish (*see* **Note 9**). Incubate 5 min at room temperature.

2. Aspirate the fixative and rinse twice more in PBS. Either store at 4°C in PBS until ready to stain, or proceed with the staining protocol.

3. Add 1 mL of dilution buffer to wells in which the PBS has just been aspirated and incubate at 37°C for 30 min to 1 h.

4. Aspirate dilution buffer and add fresh dilution buffer, in which the primary antibody has been diluted. Incubate at 37°C for 1 to 2 h.

5. Rinse in three changes of dilution buffer, aspirating carefully each time (*see* **Note 10**).

6. Add fresh dilution buffer, in which the secondary antibody has been diluted, and incubate at 37°C for 1 h.

7. Rinse in three changes of PBS and store in PBS at 4°C.

8. To analyze the vasculature for quantitation, set up an inverted microscope outfitted with epifluorescence and a camera (*see* **Note 11**). Set up a protocol so that, using a 4× objective, you can take 6 or more frames of each well that are nonoverlapping (*see* **Note 12**).

9. To determine the percentage of vascular area, convert the data into digital images in Adobe Photoshop (*see* **Notes 13–16**).

4. Notes

1. ES cells maintained off feeder layers often look somewhat differentiated, especially around the edges of colonies. In our hands, different ES cell lines look more or less differentiated when kept under these conditions. We find that the passage of ES colonies in this state does not usually compromise the experiments—the differentiated cells presumably do not expand while the true ES cells remain pluripotent. While this protocol is not recommended for maintaining ES cells that will be reintroduced into mice, the in vitro differentiation process is not affected unless the majority of the cells are differentiated prior to enzymatic treatment.

2. It is imperative to use bacteriological dishes to prevent sticking of the cell clumps, so that they will form EBs in suspension culture. *See* **Subheading 2.1., item 4**, for a source of bacteriological dishes that have the least sticking of cell clumps in our hands. If this source is not available, check the dishes from 2 or 3 sources for sticking, since there are large variations in this parameter. Having said this, some of the cell clumps invariably stick to the bottom, and they are discarded after the EBs are moved to dishes for attachment.

3. It is important to use a 25-mL pipet here to prevent further mechanical disruption of the cell clumps.

4. We have found that timely feeding with fresh medium is the most important parameter for good differentiation and minimal cell death. We monitor the pH of the medium with phenol red, and if the medium on the cultures is light orange to yellow after 24 h, it is changed every day. We find 24-h feedings necessary for densely seeded wells, and sometimes for late days of a time course when there are a large number of cells in each well.

5. For CEB production, leave the EBs in the bacteriological dish and feed every other day (or every day, *see* **Note 4**) by swirling EBs/CEBs to the middle and aspirating off the old medium. If a dish has many cell clumps that have attached to the bottom, we sometimes transfer the EBs to a new bacteriological dish after 3 to 4 d of culture using a medidropper. After 3 to 4 d, there should be very few additional cells that stick to the bottom of the dish.

6. EBs can be plated into tissue culture dishes at any point, from right after the Dispase treatment (d 0) to d 4. By d 5, the EBs have usually started to become cystic, and, in any case, do not readily stick to the tissue culture dish and spread. In our hands, the best differentiation and development occurs when EBs are plated at d 3, but earlier times can be assayed as long as all cultures to be analyzed and compared are plated into tissue culure dishes on the same day.

7. Attached cultures can be set up in any size tissue culture dish from a 10-cm to a 48-well dish. The number of EBs is adjusted to fit the surface area. In general, we find that the area of a 24-well dish is suitable for antibody staining or *in situ* hybridization. The smaller wells sometimes have more lifting of the cell layer from the edges of the well after fixation and during the staining process, while larger wells waste expensive antibodies. However, for RNA analysis, we routinely plate in larger dishes, and for high-throughput screening *(16)*, the smaller wells suffice.

8. Using our protocols, we routinely visualize vascular development in cultures that have been differentiated for 8 d (using the day of Dispase treatment as d 0). We first see angioblasts at d 4–6, vessels forming d 6–8, and some expansion after d 8. A typical time course to analyze vascular development would cover d 5–8.

9. We use fresh cold methanol:acetone whenever possible, since this fixative produces good cell permeation, and it results in minimal lifting of the cell layers off the bottom. The PECAM, ICAM-2, and flk-1 antigens are all stable in this fixative. The fresh 4% PFA is less efficient at producing a permeable cell, and the cell layers are more prone to lift off the bottom of the dish (a problem that usually decreases with increasing age of the culture, as the attached areas get larger). However, certain antigens, such as CD34 and Mac-1 (to visualize macrophages), require this fixative.

10. As alluded to in **Note 9**, one common problem is lifting of the cell layers off the dish. This can occur at any time from the fixation step through the final wash of the antibody staining. This problem is most prevalent at early time points and when using the PFA fixative. We have found that the careful aspiration and addition of reagents, often using a pipet with a bulb rather than a vacuum trap, can minimize lifting. At early time points, we try to fix numerous wells so as to have staining choices. Finally, if the layers lift up, but can be kept relatively intact, the antibody staining will still work. When most of the PBS is carefully removed from the well, the layers will sit on the bottom and can be visualized microscopically.

11. You can either use black and white film (400 ASA) or a digital camera.

12. Make sure that each field has 100% cell coverage, since the total area will be considered the denominator to determine the percentage of vascular area. By using the protocol, a given area should be photographed even if there are no vessels or if they are not centered—this is for quantitative data, not aesthetics.

13. Using the appropriate tools in Adobe Photoshop, remove any light areas that are not vascular. It is important that all the light area be bona fide vascular staining. A good antibody stain with low background is essential, but even with that, we have noticed that the domes of cells that stay relatively thick, where the EBs first attach, often pick up nonspecific reactivity. This is hazy compared to the crisp vascular staining, so we have no trouble discerning the difference and removing the nonspecific areas. Of course, any antibody staining protocol should have control wells that are incubated with secondary antibody only for comparison.

14. Using the plug-ins provided by Reindeer Games, change the image to a binary mode. The computer can then determine the percent of the area that is white (corresponding to stained area) vs black (nonstained area).

15. The percentage stained area is averaged for all the fields of a given well, with no attempt to use error bars. It is given that there are variations in vascular coverage from field to field in a well, and the average is to prevent that bias. Once the averages for three or more wells of a given condition are calculated, the overall average and standard deviation are calculated using standard formulas.

16. The percentage area that is stained usually corresponds fairly well to independent assays of endothelial cell number such as FACS analysis. This is somewhat surprising given that most "areas" have two layers of endothelial cells, because most vascular structures have a lumen *(5)*. However, most areas are also likely to have multiple layers of nonendothelial cells as well. We never use the percentage area stained as an absolute indicator of the amount of vasculature, however, except in comparison to other wells analyzed the same way. We find significant changes among different mutant ES cell cultures compared to wild-type, and while the absolute numbers sometimes vary a bit from experiment to experiment, the trends do not change *(7,17)*.

References

1. Doetschman, T. C., Eistetter, H., Katz, M., Schmidt, W., and Kemler, R. (1985) The *in vitro* development of blastocyst-derived embryonic stem cell lines: formation of visceral yolk sac, blood islands and myocardium. *J. Embryol. Exp. Morph.* **87,** 27–45.

2. Risau, W., Sariola, H., Zerwes, H. G., Sasse, J., Ekblom, P., Kemler, R., and Doetschman, T. (1988) Vasculogenesis and angiogenesis in embryonic-stem-cell-derived embryoid bodies. *Development* **102**, 471–478.

3. Schmitt, R. M., Bruyns, E., and Snodgrass, H. R. (1991) Hematopoietic development of embryonic stem cells in vitro: cytokine and receptor gene expression. *Genes Dev.* **5**, 728–740.

4. Wang, R., Clark, R., and Bautch, V. L. (1992) Embryonic stem cell-derived cystic embryoid bodies form vascular channels: an *in vitro* model of blood vessel development. *Development* **114**, 303–316.

5. Bautch, V. L., Stanford, W. L., Rapoport, R., Russell, S., Byrum, R. S., and Futch, T. A. (1996) Blood island formation in attached cultures of murine embryonic stem cells. *Dev. Dyn.* **205**, 1–12.

6. Inamdar, M., Koch, T., Rapoport, R., Dixon, J. T., Probolus, J. A., Cram, E., and Bautch, V. L. (1997) Yolk sac-derived murine macrophage cell line has a counterpart during ES cell differentiation. *Dev. Dyn.* **210**, 487–497.

7. Bautch, V. L., Redick, S. D., Scalia, A., Harmaty, M., Carmeliet, P., and Rapoport, R. (2000) Characterization of the vasculogenic block in the absence of vascular endothelial growth factor-A. *Blood* **95**, 1979–1987.

8. Albelda, S. M., Muller, W. A., Buck, C. A., and Newman, P. J. (1991) Molecular and cellular properties of PECAM-1 (endoCAM/CD31): a novel vascular cell-cell adhesion molecule. *J. Cell Biol.* **114**, 1059–1068.

9. Piali, L., Albelda, S. M., Baldwin, H. S., Hammel, P., Ginsler, R. H., and Imhof, B. A. (1993) Murine platelet endothelial cell adhesion molecule (PECAM-1)/CD31 modulates β2 integrins on lymphocyte-activated killer cells. *Eur. J. Immunol.* **23**, 2464–2471.

10. Baldwin, H. S., Shen, H. M., Yan, H. C., DeLisser, H. M., Chung, A., Mickanin, C., et al. (1994) Platelet endothelial cell adhesion molecule-1 (PECAM/CD31): alternatively spliced, functionally distinct isoforms expressed during mammalian cardiovascular development. *Development* 120, 2539–2553.

11. Vittet, D., Prandini, M.-H., Berthier, R., Schweitzer, A., Martin-Sisteron, H., Uzan, G., and Dejana, E. (1996) Embryonic stem cells differentiate in vitro to endothelial cells through successive maturation steps. *Blood* **88**, 3424–3431.

12. Shalaby, F., Ho, J., Stanford, W. L., Fischer, K.-D., Schuh, A. C., Schwartz, L., et al. (1997) A requirement for Flk-1 in primitive and definitive hematopoiesis and vasculogenesis. *Cell* **89**, 981–990.

13. Schuh, A. C., Faloon, P., Hu, Q.-L., Bhimani, M., and Choi, K. (1999) In vitro hematopoietic and endothelial potential of *flk-1 –/–* embryonic stem cells and embryos. *Proc. Natl. Acad. Sci. USA* **96**, 2159–2164.

14. Fong, G.-H., Zhang, L., Bryce, D. M., and Peng, J. (1999) Increased hemangioblast commitment, not vascular disorganization, is the primary defect in *flt-1* knock-out mice. *Development* **126**, 3015–3025.

15. Hidaka, M., Stanford, W. L., and Bernstein, A. (1999) Conditional requirement for the Flk-1 receptor in the *in vitro* generation of early hematopoietic cells. *Proc. Natl. Acad. Sci. USA* **96**, 7370–7375.

16. Stanford, W. L., Caruana, G., Vallis, K. A., Inamdar, M., Hidaka, M., Bautch, V. L., and Bernstein, A. (1998) Expression trapping: identification of novel genes expressed in hematopoietic and endothelial lineages by gene trapping in ES cells. *Blood* **92**, 4622–4631.

17. Kearney, J.B., Ambler, C.A., Monaco, K., Johnson, N., Rapoport, R., and Bautch, V. L. (2000) The VEGF receptor flt-1 negatively regulates blood vessel formation by modulating endothelial cell division. (Submitted).

13

Embryonic Stem Cells as a Model to Study Cardiac, Skeletal Muscle, and Vascular Smooth Muscle Cell Differentiation

Anna M. Wobus, Kaomei Guan, Huang-Tian Yang, and Kenneth R. Boheler

1. Introduction

Embryonic stem (ES) cells, the undifferentiated cells of early embryos are established as permanent lines *(1,2)* and are characterized by their self-renewal capacity and the ability to retain their developmental capacity in vivo *(3)* and in vitro *(4–6)*. The pluripotent properties of ES cells are the basis of gene targeting technologies used to create mutant mouse strains with inactivated genes by homologous recombination *(7)*.

ES cells cultivated as embryo-like aggregates, called embryoid bodies (EBs), differentiate in vitro into cellular derivatives of all three primary germ layers of endodermal, ectodermal, and mesodermal origin. ES cell lines develop from an undifferentiated stage resembling cells of the early embryo into terminally differentiated stages of the cardiogenic *(6,8–11)*, myogenic *(11–13)*, neurogenic *(14–17)*, hematopoietic *(5,18,19)*, adipogenic *(20)*, or chondrogenic *(21)* lineage, as well as into epithelial *(22)*, endothelial *(23)*, and vascular smooth muscle (VSM) cells *(23–25)*. Terminally differentiated ES cells also show pharmacological and physiological properties of specialized cells: in vitro differentiated cardiomyocytes have characteristics typical of atrial-, ventricular-, Purkinje-, and pacemaker-like cells *(9,10,26,27)*, and neuronal cells are characterized by inhibitory and excitatory synapses *(14,17)*. Neuronal, cardiac, and VSM cells express functional receptors typical for each cell type *(8,14,25,26)*.

Differentiation protocols for the development of ES cells into cardiomyocytes, skeletal muscle, or VSM cells have been well established. The in vitro differentiation of ES cells allows investigators *(1)* to analyze developmental processes during the differentiation of stem cells into specialized cell types and early processes of commitment to specific lineages; *(2)* to study the effects of differentiation factors or xenobiotics on embryogenesis in vitro; *(3)* to investigate pharmacological effects on functionally active cardiac or VSM cells (which are otherwise not available from in vitro cultivated cells); and of potentially the greatest importance, *(4)* to establish strategies for cell and tissue therapy.

From: *Methods in Molecular Biology, vol. 185: Embryonic Stem Cells: Methods and Protocols*
Edited by: K. Turksen © Humana Press Inc., Totowa, NJ

The following parameters influence the developmental potency of ES cells in culture: *(1)* the number of cells differentiating in the EBs; *(2)* the media, quality of fetal calf serum (FCS), growth factors, and additives; *(3)* the ES cell lines used; and *(4)* the time of EB plating *(6)*. Genetic manipulation of ES cell lines through either "gain-" or "loss-of-function" strategies, when used in conjunction with established differentiation protocols, also permits the testing of specific hypotheses related to the development of these cell types *(26,28–30)*. The gain-of-function and loss-of-function analyses are excellent alternatives and substitutes to in vivo studies with transgenic animals to analyze the consequences of mutations on early embryogenesis and development. Loss-of-function analyses are especially helpful for the investigation of those mutations that result in early embryonal death of homozygous embryos *(22,24,28,29)*.

In principle, EB aggregates of ES cells develop into many differentiated cell types. To obtain maximal differentiation of a defined cell type, specific cell lines and cultivation conditions have to be used *(6)*.

In this chapter, we describe methods to differentiate ES cells into functionally active cardiac, skeletal muscle, and VSM cells and to characterize their phenotypes. Furthermore, strategies are presented for the genetic manipulation of ES cell differentiation.

2. Materials

2.1. Cells

The following cell lines have been used for in vitro differentiation into cardiac, skeletal muscle, and VSM cells:

1. Cardiac muscle differentiation: D3 *(4,8–11,26,29,31,32)*. R1 *(29)*, Bl17 *(8)*, AB1, AB2.1. *(24)*, CCE, and E14.1 *(33)*.
2. Skeletal muscle cell differentiation: D3 *(11,34)*, BLC6 *(12,13)*, AB1, and AB2.1 *(24)*.
3. VSM cell differentiation: D3 *(23,25)*, AB1, and AB 2.1 *(24)*.

In addition to ES cells, embryonic germ (EG) cells, i.e., EG-1 *(35,36)*, or embryonic carcinoma (EC) cells, i.e., P19 *(37)*, may be used for differentiation into cardiac and skeletal muscle cells.

2.2. Media, Reagents, and Stock Solutions

2.2.1. Solutions for Cell Culture (see **Notes 1** and **2**)

1. Dextran T500 (Amersham Pharmacia Biotech, cat. no. 17-0320-01).
2. 0.01 *M* Tris-HCl, pH 8.0.
3. Activated charcoal (Serva, cat. no. 30890.01).
4. Phosphate-buffered saline (PBS): containing 10 g NaCl, 0.25 g KCl, 1.44 g Na_2HPO_4, 0.25 g KH_2PO_4 x 2 H_2O/L, filter-sterilized through a 0.22-μm filter.
5. Trypsin solution: 0.2% trypsin 1:250 (Gibco BRL, cat. no. 27250-042) in PBS, filter-sterilized through a 0.22-μm filter.
6. Ethylenediaminetetra acetate (EDTA) solution: 0.02% EDTA (Sigma, cat. no. E6758) in PBS, filter-sterilized through a 0.22-μm filter.
7. Trypsin-EDTA: mix trypsin solution and EDTA solution at 1:1.
8. Gelatin solution: 1% gelatin (Fluka, cat. no. 48720) in double-distilled water, autoclaved, and diluted 1:10 with PBS. Coat tissue culture dishes with 0.1% gelatin solution for 1–24 h at 4°C before use.

9. Mitomycin C (MC) solution: dissolve 2 mg MC (Serva, cat. no. 29805.01) in 10 mL PBS, filter-sterilize. From this stock solution, dilute 300 µL into 6 mL of PBS (final concentration is 0.01 mg/mL). MC stock solution should be freshly prepared at weekly intervals and stored at 4°C.
   Caution: MC is carcinogenic.

10. β-mercaptoethanol (β-ME) (Serva, cat. no. 28625.01): prepare a stock solution from 7 µL of β-ME into 10 mL of PBS (stock concentration is 10 mM). Make fresh at weekly intervals and store at 4°C.

11. Cultivation medium I: Dulbecco's modified Eagle's medium (DMEM) (4.5 g/L glucose; Gibco BRL, cat. no. 52100-039) supplemented with 15% FCS for feeder layer cells (*see* **Note 1**).

12. Additives I: to 100 mL media, add 1 mL of 200 mM L-glutamine stock (100X) (Gibco BRL, cat. no. 25030-024), 1 mL of β-ME stock, 1 mL of nonessential amino acids (NEAA) stock (100×) (Gibco BRL, cat. no. 11140-035).

13. Additives II: to 400 mL medium, add 40 µL of a 3×10^{-4} M stock solution of Na-selenite (Sigma, cat. no. S5261), 10 mL of 7.5% stock solution of bovine serum albumin (BSA) (Gibco BRL, cat. no. 15260-037), and 1 mL of stock solution (4 mg/mL) of transferrin (Gibco BRL, cat. no. 13008-016).

14. Cultivation medium II: DMEM supplemented with 15% FCS (selected batches) and additives I for ES cell cultivation.

15. Differentiation medium I: DMEM (or Iscove's modification of DMEM [IMDM]) (Gibco BRL, cat. no. 42200-030) (*see* **Note 2**) supplemented with 20% FCS and additives I for EB differentiation into cardiac, skeletal muscle, and VSM cells.

16. Monothioglycerol (3-mercapto-1,2-propanediol]MTG]) (Sigma, cat. no. M6145): Prepare a stock solution from 13 µL of MTG into 1 mL of IMDM, filter-sterilized through a 0.22-µm filter. Make fresh before use. To 100 mL media, add 300 µL of stock solution (final concentration is 450 µM).

17. Differentiation medium II: DMEM supplemented with 15% dextran-coated charcoal-treated FCS (DCC-FCS) (*see* **Subheading 3.2.1.**) and additives I and additives II to analyze effects of growth factors on differentiation or for EB differentiation into neuronal or myogenic cells (line BLC6).

18. Retinoic acid (RA): prepare in the dark a 10^{-3} M stock solution of all-*trans* RA (Sigma, cat. no. R2625) in 96% ethanol or DMSO. Store aliquots at –20°C and use a fresh sample for each experiment.

19. Dibutyryl-cyclic adenosine monophosphate (db-cAMP) (Sigma, cat. no. D0627): dilute in double-distilled water to a 0.1 M stock solution. Store aliquots at –20°C for at least 1 mo.

20. Transforming growth factor β$_1$ (TGF β$_1$): prepare a 20 ng/µL stock solution of TGF β$_1$ (Strathmann Biotech, cat. no. TGFβ1-2) in 0.1 M acetic acid or 0.05 M HCl. Store aliquots in silanized glass tubes at –20 °C for at least 2 mo.

2.2.2. Solutions for Reverse Transcription Polymerase Chain Reaction (*see* **Notes 3** and **4**)

1. Diethyl pyrocarbonate-treated water (DEPC-H$_2$O): add 1 mL DEPC (Fluka, cat. no. 32490) to 1 L double-distilled or Milli-Q water (dH$_2$O) and stir overnight. DEPC is inactivated by heating to 100°C for 15 min or autoclaving for 15 min.

2. RNA lysis buffer: add 23.6 g of guanidinium thiocyanate to 5 mL of 250 mM Na-citrate, pH 7.0, 2.5 mL of 10% sarcosyl, and add DEPC-H$_2$O to a total volume of 49.5 mL, and mix carefully. Make fresh at monthly intervals. Add 1% β-ME before use.

3. 2 M Na-acetate, pH 4.0: dissolve 27.2 g of Na-acetate × 3 H_2O in 0.1% DEPC-H_2O, adjust the pH to 4.0 with glacial acetic acid, and adjust to 100 mL with DEPC-H_2O. Treat the buffer with 0.1% DEPC-H_2O at 37°C for at least 1 h and heat to 100°C or autoclave for 15 min.

4. Acidic phenol: phenol is saturated with DEPC-H_2O instead of Tris. The saturated acidic phenol contains 0.1% hydroxyquinoline (antioxidant, partial inhibitor of RNase, and a weak chelator of metal ions; its yellow color provides a convenient way to identify the organic phase). Store at 4°C for up to 2 mo.

5. Chloroform:Isoamylalcohol (24:1).

6. 75% Ethanol:prepare in DEPC-H_2O.

7. 25 mM $MgCl_2$ (Perkin Elmer, included in cat. no. N808-0161).

8. 10X Polymerase chain reaction (PCR) buffer II (Perkin Elmer, included in cat. no. N808-0161): 100 mM Tris-HCl, pH 8.3, 500 mM KCl.

9. RNase inhibitor (Perkin Elmer, cat. no. N808-0119): 20 U/µL.

10. Oligo d(T)$_{16}$ (Perkin Elmer, cat. no. N808-0128): 50 µM in 10 mM Tris-HCl, pH 8.3.

11. Random hexamers (Perkin Elmer, cat. no. N808-0127): 50 µM in 10 mM Tris-HCl, pH 8.3.

12. MuLV murine leukemia virus reverse transcriptase (Perkin Elmer, cat. no. N808-0018): 50 U/µL.

13. A*mpli*Taq® DNA polymerase (Perkin Elmer, included in cat. no. N808-0161): 5 U/µL.

14. 5 mM dNTP mixture: 100 mM of each dNTP (dGTP, dATP, dCTP, and dTTP) (Amersham Pharmacia Biotech, cat. no. 27-2035-01) dilute to 20 mM with DEPC-H_2O and freeze at –20°C. dNTP (5 mM) mixture is freshly made by mixing the equal volumes of 20 mM of each dNTP before use.

15. Select PCR primer pairs: dilute synthetic oligonucleotides to 10 mM with DEPC-H_2O and freeze at –20°C (a critical step in the PCR) *(38)*.

16. Glycogen: 20 mg/mL (Roche Diagnostics GmbH, cat. no. 901393).

17. 5 M NaCl: dissolve 29.2 g of NaCl in dH_2O, adjust to 100 mL with water and autoclave.

18. TE buffer: 10 mM Tris-HCl, 1 mM EDTA, pH 7.5, filter-sterilized through a 0.22-µm filter.

19. 6X Loading buffer: 0.25% bromophenol blue (Sigma, cat. no. B8026), 0.25% xylene cyanole FF (Sigma, cat. no. X4126), 30% glycerol in dH_2O.

20. 5X TBE: dissolve 54 g Tris-base and 27.5 g boric acid in dH_2O, add 20 mL of 0.5 M EDTA, pH 8.0, and adjust to 1 L with dH_2O.

21. Ethidium bromide aqueous solution (Serva, cat. no. 21251.01): 1% (w/v) = 10 mg/mL.

22. Agarose gels: melt electrophoresis grade agarose (Gibco BRL, cat. no. 15510-027) in 1X TBE by gentle boiling in a microwave oven. Cool to <60°C and pour into an agarose gel mold. Run small gels at around 80–100 V by using bromophenol blue and xylene cyanole FF in the stop mixture as an indicator of migration.

2.2.3. Solutions for Immunohistochemical Analysis

1. 3.7% Paraformaldehyde (PFA) (Serva, cat. no. 31628.01): dissolve 3.7 g PFA in PBS, adjust to 100 mL with PBS, heat the mixture to 95°C, stir until the solution becomes clear, and cool to room temperature.
   Caution: PFA is toxic. Work under the hood and use gloves.

2. Methanol: acetone (7:3) fixative.

3. 10% goat serum (Sigma, cat. no. G6767) or 1% BSA in PBS for blocking unspecific binding of antibodies.

4. Mounting medium: Vectashield (Vector, Burlingame, USA, cat. no. H-1000).

5. 0.02% Triton-X 100 in PBS.

6. 0.5% BSA in PBS for dilution of secondary antibodies.

2.2.4. Solutions for Single Cardiac Cell Isolation

1. Supplemented low-Ca^{2+} medium: 120 mM NaCl, 5.4 mM KCl, 5 mM $MgSO_4$, 1 mM EGTA, 5 mM Na pyruvate, 20 mM glucose, 20 mM taurine, 10 mM HEPES-NaOH, pH 6.9, at 24°C, supplemented with 1 mg/mL collagenase B (Roche Diagnostics GmbH, cat. no. 10888807) and 30 µM $CaCl_2$.
2. KB medium: 85 mM KCl, 30 mM K_2HPO_4, 5 mM $MgSO_4$, 1 mM EGTA, 5 mM Na-pyruvate, 5 mM creatine, 20 mM taurine, 20 mM glucose, freshly added 2 mM Na_2ATP, pH 7.2, at 24°C.

2.2.5. Solutions for Introduction of DNA into ES Cells

1. Plasmids ideally purified by cesium-chloride gradient ultracentrifugation or by ion-exchange chromatography. DNA should be suspended in H_2O or TE buffer, pH 8.0.
2. Restriction enzymes with appropriate digestion buffers as recommended by the manufacturer (e.g., Promega or Gibco BRL).
3. Phenol:chloroform:isoamyl alcohol (25:24:1, v/v/v). Phenol: equilibrated with 0.1 M Tris, pH 8.0, and containing 8-hydroxyquinoline (0.1%).
4. 3 M sodium-acetate, pH 5.2, sterilized by autoclaving.
5. 100 and 70% Ethanol for DNA precipitation and washes, respectively.
6. PBS: Dulbecco's PBS, containing 8 g NaCl, 0.2 g KCl, 0.2 g KH_2PO_4, 1.15 g Na_2HPO_4/L, sterilized by autoclaving (Gibco BRL, cat. no. 10010-023).
7. Leukemia inhibitory factor (LIF): 10^7 U/mL (Chemicon International, cat. no. ESG1107).
8. Cultivation medium II (*see* **Subheading 2.2.1.**) + 1000 U/mL LIF.
9. Geneticin sulfate (G418) stock solution (Gibco BRL, cat. no. 11811-023), puromycin (Clontech, cat. no. 8052-1), hygromycin B (Roche Products, cat. no. 843555), gancyclovir (Roche Products, this product is a controlled substance and is available only upon written request and justification). The concentration of each solution needed to kill ES cells should be determined for each batch of selection agent (for concentrations *see* **Table 1**).

2.2.6. Solutions for Analysis of Transfectants

2.2.6.1. RAPID DNA ISOLATION FROM ES CELLS

1. Cultivation medium I, PBS, 0.1% gelatin in PBS (*see* **Subheading 2.2.1.**).
2. Proteinase K (Sigma, cat. no. P2308): dissolve lyophilized proteinase K at 20 mg/mL in dH_2O. Store 125-µL aliquots at –20°C for up to 1 yr in sterile microcentrifuge tubes.
3. DNA lysis buffer (final concentrations): 10 mM Tris-HCl, pH 7.5, 10 mM EDTA, 10 mM NaCl, 0.5% sarcosyl, and 1 mg/mL proteinase K (added fresh).
4. NaCl/ethanol solution: 1.5 µL of 5 M NaCl to 100 µL of cold absolute ethanol. The salt will precipitate. Mix to obtain a uniform suspension.

2.2.6.2. PCR

1. 10X Mg^{2+}-free DNA polymerase buffer: 500 mM KCl, 100 mM Tris-HCl (pH 9.0 at 25°C), 1% Triton-X 100 (Promega, cat. no. M190G).
2. 20 mM dNTPs stock solution: dilute 100 mM stock solution (Amersham Pharmacia Biotech, cat. no. 27-2035-01), 1:5 with 10 mM Tris-HCl, pH 7.5.
3. 25 mM $MgCl_2$.
4. *Taq* DNA polymerase in buffer B (Promega, cat. no. M166B).
5. 10 µM PCR primer stock solutions: combine aliquots of primers for PCR and dilute to a concentration of 10 µM in TE. (Primer sets: β-globin as internal PCR control [sense primer: AGG TGA TAA CTG CCT TTA ACG A, antisense primer: CCC AGC ACA ATC ACG AT]; neomycin sequence primers [sense primer TAT TCG GCT ATG ACT GGG

Table 1
Selection Agents for Genetically Modified ES Cells

Marker gene	Selection	Drug concentration (μg/mL)
Neomycin phosphotransferase (neo)	Positive	G418 (300)
Puromycin N-acetyl transferase (pac)	Positive	Puromycin (0.7)
Hygromycin-B-phosphotransferase	Positive	Hygromycin-B (200)
Herpes simplex virus (HSV) thymidine kinase (HSV-tk)	Negative	Gancyclovir (2 μM)

The effective concentration of each selection agent should be tested on the ES cell line used for genetic manipulation. Typical test ranges are as follows: G418 (200–750 μg/mL); puromycin (0.4–1.0 μg/mL); hygromycin-B (100–400 μg/mL); and gancyclovir (1–3 μM).

CAC AA, antisense primer: AGC AAT ATC ACG GGT AGC CAA CG]; for puromycin [sense primer: CAG GAA GCT CCT CTG TGT CCT C, antisense primer: GCT TAT CCA GTG GAG TGC TGG GTT]). For design of other primers, use software such as Primer Express™ 1.5 from PE Applied Biosystems or MacVector 6.5 from Oxford Molecular Ltd.

6. Molecular weight markers: 100 bp DNA Ladder (Gibco BRL, cat. no. 15628-019).
7. Strain specific mouse genomic DNA (e.g., Jackson Laboratories, cat. no. 691).

2.2.6.3. SOUTHERN BLOT ANALYSIS

1. High concentration restriction enzymes and appropriate digestion buffers as recommended by the manufacturer (e.g., Gibco BRL or Promega).
2. RNase A (DNase-free) (Sigma, cat. no. R6513): dissolve lyophilized RNAse A to 10 mg/mL in sterile 5 mM Tris-HCl, pH 7.5. Place the preparation at 80°C for 15 min to inactivate any deoxyribonuclease activity. Prepare 100-μL aliquots and store at –20°C.
3. 20X Sodium chloride sodium phosphate EDTA (SSPE): 3.0 M NaCl, 0.2 M NaH$_2$PO$_4$, 0.02 M EDTA.
4. 20% Sodium dodecyl sulfate (SDS) (Oncor, cat. no. S4110).
5. Sheared salmon sperm DNA (5 Prime ↑ 3 Prime, cat. no. 5302-754688).
6. 50X Denhardt's solution: 1% Ficoll type 400 (Amersham Pharmacia, cat. no. 17-0400-01) 1% polyvinylpyrrolidone (PVP-40; Sigma), 1% BSA.
7. Molecular weight markers (λ DNA/*Hind*III fragments) (Gibco BRL, cat. no. 15612-013).
8. Ready-to-Go DNA labeling (-dCTP) kit (Amersham Pharmacia Biotech, cat. no. 27-9240-01).
9. ProbeQuant G-50 Micro Columns: (Amersham Pharmacia Biotech, cat. no. 27-5335-01).
10. Hybond-N$^+$ Nylon membrane (Amersham Pharmacia Biotech, cat. no. RPN203B).
11. Prehybridization solution: 50 mL containing 12.5 mL 20X SSPE, 5 mL 50X Denhardt's solution, 1.25 mL 20% SDS (add last), 31.1 mL dH$_2$O, and 0.15 mL of denatured 10 mg/mL sheared salmon sperm DNA.

2.3. Equipment

1. Tissue culture plates: 35 mm, 60 mm, 100 mm (Nunc or Falcon).
2. 6-well, 24-well (Falcon) and 96-well microwell plates (Nunc or Costar).
3. Pasteur pipets, 2-, 5-, 10-, 25-mL pipets.
4. Bacteriological Petri dishes (Greiner or Fisher Brand): 60 mm for EB mass culture, 100 mm for EB hanging drop culture.
5. Tissue culture plates (60 mm) with sterilized coverslips (n = 4) for immunofluorescence.
6. 2-mL Glass pipets for preparing single-cell suspensions.

7. For feeder layer culture: sterile dissecting instruments, screen or sieve (about 0.5 to 1 mm diameter), Erlenmeyer flasks with stir bars, centrifuge tubes.

8. Tissue culture incubator with 37°C and a 5% CO_2 atmosphere.

9. For pharmacological analysis, the inverted microscope Diaphot-TMD (Nikon) equipped with a 37°C heating plate and a CO_2 chamber is used; the computer-assisted imaging system (*see* **Subheading 3.3.3.**) is coupled via a one-chip charge-coupled device (CCD) camera (Sony, Japan) to a computer imaging station (Pentium CPU, 100 MHz) running the LUCIA Laboratory Imaging System including the "HEART" application (Nikon).

10. GeneQuant RNA/DNA calculator (Pharmacia LKB Biochrom Ltd.).

11. PCR apparatus: mastercycler gradient (Eppendorf) or Gene Amp PCR System 9700 (Perkin Elmer).

12. 0.5- and 1.5-mL Microtubes (Eppendorf) and 20-, 100-, and 1000-μL filtertips (Biozym).

13. Electrophoresis equipment (Bio-Rad).

14. A CN-TFX darkroom (MWG-Biotech GmbH) is coupled via GelPrint Workstation (MWG-Biotech) to a computer station (Peacock, 133 MHz) with digital graphic printer (Sony) running the PhotoFinish v3.0 program (MWG-Biotech GmbH). Alternatively, a Bio-Rad Gel Doc 1000 and Multi-Analyst Version 1.1 software can be used.

15. TINA2.08e software (raytest Isotopenmeßgeräte GmbH, Straubenhardt, FRG).

16. Inverted Confocal Laser Scanning Microscope (CLSM) LSM-410 (Carl Zeiss, Jena) equipped with an argon-ion laser.

17. Bio-Rad Gene Pulser II system with capacitance extender (Bio-Rad, cat. no. 165-2105) and Gene Pulser disposable cuvets, 0.4-cm electrode gap (Bio-Rad, cat. no. 165-2088).

18. UV Stratalinker (Stratagene, model 2400).

19. Multichannel pipets (Oxford Benchmate, multi 5–50 μL and 40–200 μL).

3. Methods

3.1. Cultivation of Undifferentiated ES Cells on Feeder Layer

3.1.1. Feeder Layer Culture

1. Remove embryos from a mouse pregnant for 15 to 17 d (i.e., NMRI or CD-1 outbred strains. For neomycin-resistant feeder cells, use MTK-neo^R or Zeta mice [NIA] or other available strains), rinse in PBS, and remove placenta and fetal membranes, head, liver, and heart. Rinse the carcasses in trypsin solution.

2. Mince the embryonic tissue in 5 mL of fresh trypsin solution and transfer to an Erlenmeyer flask containing a stir bar.

3. Stir on magnetic stirrer for 25 to 45 min (use longer incubation time if the embryos are older), filter the suspension through a sieve or a screen, add 10 mL of culture medium I and spin down.

4. Resuspend the pellet in about 3 mL of culture medium I and plate on 100-mm tissue culture plates (about 2×10^6 cells/100 mm-dish) containing 10 mL culture medium I, incubate at 37°C and 5% CO_2 for 24 h.

5. Change the medium to remove debris, erythrocytes, and unattached cellular aggregates, cultivate for an additional 1 to 2 d.

6. Passage the primary culture of mouse embryonic fibroblasts: split 1:2 to 1:3 on 100-mm tissue culture plates, grow in culture medium I for 1 to 3 d. The cells in passages 2–4 are most suitable as feeder layer for undifferentiated ES cells.

7. Incubate feeder layer cells with MC buffer for 2 to 3 h, aspirate the MC solution, wash three times with PBS, trypsinize feeder cells, and replate to new gelatin (0.1%)-treated microwell plates or to Petri dishes. Feeder layer cells prepared 1 day before ES cell subculture are optimal.

3.1.2. Culture of Undifferentiated ES Cells (see **Note 5**)

It is important to passage ES cells every 24 or 48 h. Do not cultivate longer than 48 h without passaging, or the cells may differentiate and be unsuitable for differentiation studies. Selected batches of FCS have to be used for ES cell culture (*see* **Note 6**).

1. Change the medium 1 to 2 h before passaging.
2. Aspirate the medium, add 2 mL of trypsin-EDTA, and incubate at room temperature for 30 to 60 s.
3. Remove carefully the trypsin-EDTA mixture and add 2 mL of fresh cultivation medium II.
4. Resuspend the cell population with a 2-mL glass pipet into a single-cell suspension and split 1:3 to 1:10 to freshly prepared (60 mm) feeder layer plates.

3.2. In Vitro Differentiation

3.2.1. Preparation of Growth Factor-Free FCS (DCC-FCS)

To analyze the influence of growth factors on ES cell differentiation, the medium should be free of high molecular weight proteins, like growth and differentiation factors. We use the DCC treatment of FCS. Prepare DCC-FCS as follows:

1. Dissolve 0.45 g dextran T500 in 1800 mL 0.01 M Tris-HCl, pH 8.0, add 4.50 g activated charcoal, and stir the mixture at 4°C in a tightly closed Erlenmeyer bottle overnight.
2. Inactivate FCS by incubation at 56°C for 30 min.
3. Fill 50 mL DCC solution in plastic centrifuge tubes and centrifuge at 2000g for 20 min, remove and discard the supernatant, repeat the procedure in the same tube without removing the pellet (= double pellet).
4. Add 50 mL of FCS to the tube with a double pellet, transfer the mixture to a clean glass bottle, and incubate for 45 min at 45°C in a water bath under shaking.
5. Centrifuge the mixture at 2000g for 20 min and transfer the supernatant to another centrifuge tube. Repeat steps 3–5.
6. Collect the FCS supernatant in a clean centrifuge tube and sterilize through a 0.22-µm filter (low protein binding) into sterile flasks. Add at required concentrations to differentiation medium II.

3.2.2. Differentiation Protocols (see **Note 7**)

For the development of ES cells into differentiated phenotypes, cells must be cultivated in 3-dimensional aggregates called embryoid bodies (EBs) by the hanging drop method *(8,39)* (**Fig. 1**), by mass culture *(4)* (*see* **Note 7**), or by differentiation in methylcellulose *(18,19)* (*see* **Note 7**). The differentiation of cardiac, skeletal, and VSM cells requires different conditions, and these may vary for the particular ES cell line used (*see* **Note 2**). In this chapter, differentiation protocols utilizing the hanging drop method are described.

1. Prepare a cell suspension containing a defined ES cell number of 400, 600, or 800 cells in 20 µL of differentiation medium (depending on the differentiation protocols, *see* **Subheadings 3.2.2.1.**, **3.2.2.2.**, and **3.2.2.3.**).
2. Place 20 µL drops ($n = 50–60$) of the ES cell suspension on the lids of 100-mm bacteriological Petri dishes containing 10 mL PBS.

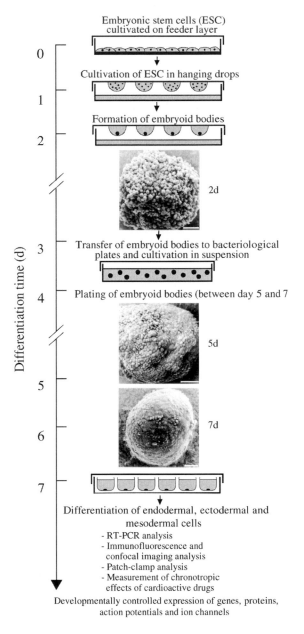

Fig. 1. Protocol for ES cell differentiation in vitro. Undifferentiated ES cells are cultivated as EBs in hanging drops for 2 days and in suspension for additional days followed by plating onto gelatin-coated tissue culture plates. The morphology of 2-, 5-, and 7-d-old EBs is shown by scanning electron microscopy (Bar = 50 μm). The cellular phenotypes derived from endodermal, ectodermal, and mesodermal lineages may be characterized by the following parameters: *(1)* gene expression patterns by semiquantitative RT-PCR; *(2)* protein formation by immunofluorescence and/or CLSM; *(3)* expression of action potentials and ion channels of excitable cells by patch-clamp analysis; and *(4)* chronotropic responses of cardiomyocytes.

3. Cultivate the ES cells in hanging drops for 2 d. The cells will aggregate and form one EB per drop.
4. Rinse the aggregates carefully from the lids with 2 mL of medium, transfer into a 60-mm bacteriological Petri dish with 5 mL of differentiation medium, and continue cultivation in suspension for 2 to 5 d until the time of plating.
5. Aspirate gelatin solution from pretreated 24-well microwell plates or 60-mm tissue culture dishes and add the appropriate differentiation medium (*see* step 1).
6. Transfer a single EB into each well of gelatin (0.1%)-coated microwell plates for morphological analysis, or transfer 20 to 40 EBs per dish onto 60-mm tissue culture dishes containing 4 coverslips (10 × 10 mm) for immunofluorescence, or 15 to 20 EBs onto 60-mm tissue culture dishes for reverse transcription PCR (RT-PCR) analysis of EB outgrowths.
7. Change the medium during EB differentiation every second or third day.
8. To characterize the EB outgrowths morphologically, calculate the percentage of EBs with the specific differentiated cell type (from EBs of at least 48 wells) or calculate the amount of the differentiated cell type as a percentage of the outgrowth area of each EB.

3.2.2.1. CARDIAC MUSCLE CELL DIFFERENTIATION

1. Use of 400–600 cells of ES cell lines D3, R1, or CCE for preparation of EBs is optimal for cardiac differentiation.
2. Culture with differentiation medium I (*see* **Subheading 2.2.1.**, **Note 2**).
3. Plate EBs onto gelatin-coated tissue culture plates at d 5 to 7. The first beating clusters in EBs can already be seen in 7-d-old EBs, but maximal cardiac differentiation is achieved after EB plating *(8,9,10)*.
4. For the investigation of early cardiac stages, plate EBs at d 5.

If EC cells are used (*see* **Subheading 2.1.**), they have to be induced to cardiac differentiation by treatment with 1% dimethyl sulfoxide (DMSO) between the first 2 d of EB development and plating at d 5 to 7 *(37)*. EC cells are cultivated without feeder cells and result after DMSO induction in a high number of cardiac cells.

3.2.2.2. SKELETAL MUSCLE CELL DIFFERENTIATION

1. Use of 600 cells of ES cell line BLC6 *(12)* or 800 cells of lines D3, R1, or EG-1 *(36)* per EB are optimal for myogenic differentiation.
2. Culture EBs with differentiation medium I or II (line BLC6).
3. Plate EBs at d 5. The first myoblasts appear 4 (line BLC6) or 5 to 6 d (D3 and R1, respectively) after EB plating. Skeletal muscle cells begin to fuse into myotubes in the EB outgrowths 1 to 2 d later.
4. A specific differentiation induction of skeletal muscle cells from ES *(34)* or EC *(40)* cells is achieved by 10^{-8} *M* RA or 1% DMSO.

3.2.2.3. VASCULAR SMOOTH MUSCLE CELL DIFFERENTIATION

1. ES cells of line D3 (*n* = 800) are differentiated as EBs in hanging drops in differentiation medium I.
2. Plate EBs at d 7 and induce differentiation of VSM cells by treatment with 10^{-8} *M* RA and 0.5×10^{-3} *M* db-cAMP between d 7 and 11 after plating (duration and treatment time has to be optimized for each cell line).
3. The first spontaneously contracting VSM cells, which express the vascular-specific splice variant of the VSM myosin heavy chain (MHC) gene, appear in the EBs around 1 wk after plating *(25)*.

4. Change the medium during the differentiation period every day or every second day *(25)*. A similar VSM cell induction is achieved by cultivating D3 cells ($n = 600$) as EBs in hanging drops in differentiation medium II containing 2 ng/mL TGF β_1 from d 0 to 5, and plating of EBs at d 5. The first spontaneously contracting VSM cells appear in EBs 10 d after plating, and maximal VSM cell differentiation (60%) is achieved at d 5 + 24 to 5 + 28.

Alternatively, ES (AB1 or AB2.1) cells ($n = 400$) were differentiated in hanging drops in M15 (DMEM plus 15% FCS, 0.1 mM β-ME, 2 mM L-glutamine, 0.05 mg/mL streptomycin, and 0.03 mg/mL penicillin). After plating at d 4.5, the medium is partially exchanged every third day. Maximal VSM cell differentiation (30%) is achieved at days 4.5 + 17 to 4.5 + 19 *(24)*.

3.3. Characterization of Differentiated Phenotypes

3.3.1. Semiquantitative RT-PCR Analysis (see **Note 8**)

3.3.1.1. PREPARATION OF CELL SAMPLES (*SEE* **NOTE 9**)

The transcripts of genes, which are specifically expressed during ES cell differentiation, are analyzed by RT-PCR with primers of tissue-specific genes (**Table 2**). The following steps are used to harvest ES cells or EB outgrowths:

1. Discard the medium and wash twice with PBS.
2. Add 400 µL of RNA lysis buffer per 60-mm culture dish. Allow the lysis buffer to spread across the surface of the dish and transfer the lysate into a 1.5-mL microtube.

The following steps are used to harvest EBs from suspension:

1. Collect EBs by centrifugation at 2000g for 3 min.
2. Wash the EBs twice by resuspension in PBS.
3. Add 100 µL of lysis buffer per 10 EBs and lyse the cells completely.
4. Store samples at –20° or –80°C.

3.3.1.2. ISOLATION OF TOTAL RNA (*SEE* **NOTE 10**)

The method described here is based on the use of a chaotropic agent (guanidine salt) for disruption of cells and inactivation of ribonucleases *(41)*.

1. Thaw lysate (400 µL) and vortex mix for 15 s.
2. Add 40 µL (1/10 vol) of 2 M Na-acetate, pH 4.0. Mix carefully.
3. Add 400 µL of acidic phenol and vortex mix vigorously.
4. Add 80 µL of chloroform : isoamylalcohol (24 : 1) and vortex mix again.
5. Store for 15 min on ice.
6. Separate the organic and aqueous phases by centrifugation at 16,000g for 10 min at room temperature.
7. Transfer the upper aqueous phase carefully to a fresh tube, add an equal volume of isopropanol and mix well. Store for 1 h at –20°C (*see* **Note 11**).
8. Centrifuge at 16,000g for 10 min at room temperature. Carefully discard the supernatant.
9. Dissolve the pellet in 300 µL of lysis buffer. If the pellet is difficult to dissolve, heat to 65°C for several minutes. Add an equal vol (300 µL) of isopropanol and mix well. Store at –20°C for 1 h.
10. Centrifuge at 16,000g for 10 min at room temperature. Carefully discard the supernatant.
11. Wash the pellet with 500 µL of 75% ice-cold ethanol (made with DEPC-H_2O), vortex mix briefly, recentrifuge at 16,000g for 10 min, discard supernatant, and allow the pellet of nucleic acid to dry in the air.

12. Dissolve RNA pellet in 30 μL of DEPC-H_2O and freeze at –80°C.
13. Dilute 1 μL of RNA with 100 μL of DEPC-H_2O, measure OD_{260} and the concentration of RNA using GeneQuant RNA/DNA calculator or a suitable spectrophotometer, adjust all samples to the same RNA concentration (i.e., 0.2 μg/μL) with DEPC-H_2O, and measure again to ensure the same RNA concentration of all samples. The yield of RNA from EBs ($n = 20$) is in the range of 20 to 100 μg.

3.3.1.3. REVERSE TRANSCRIPTION REACTIONS (*SEE* NOTE 12)

All RT and PCR solutions are available from commercial suppliers in ready-to-use form. RT reactions are performed in 20 μL of reaction volumes using 0.5-mL microcentrifuge tubes.

1. Label one PCR tube for each sample and appropriate controls. Add the same amount of RNA (0.5–1.0 μg in 3 μL) to each tube.
2. Prepare the following RT-mastermixture for 25 reactions (or a smaller quantity as required) containing: 100 μL of DEPC-H_2O, 50 μL of 10X PCR buffer II, 100 μL of 25 mM $MgCl_2$, 100 μL of 5 mM dNTPs mixture, 25 μL of RNase inhibitor, 25 μL of specific antisense primers or random hexamers or oligo d(T)$_{16}$ and 25 μL of MuLV reverse transcriptase to a total vol of 425 μL.
3. Add 17 μL of RT-mastermixture to each tube, mix carefully, and centrifuge briefly.
4. Transfer the tubes to a thermal cycler and perform RT reactions for 1 h at 42°C and then heat to 99°C for 5 min.
5. Cool the samples to 4°C or store at –20°C until use.

3.3.1.4. POLYMERASE CHAIN REACTIONS (*SEE* NOTE 13)

1. Prepare a PCR-mastermixture for 25 reactions (or a smaller quantity as required) containing: 825 μL of ddH_2O, 120 μL of 10X PCR buffer II, 90 μL of 25 mM $MgCl_2$, 40 μL of dNTPs mixture, 50 μL of 10 μM 5′ sense primer of target gene, 50 μL of 10 μM 3′ antisense primer of target gene, 12.5 μL of AmpliTaq® DNA polymerase, 12.5 μL of DMSO to a total vol of 1200 μL.
2. Label new PCR tubes and add 2.0 μL of RT reaction product to each tube as template DNA.
3. Add 48 μL of PCR-mastermixture to each tube, vortex mix and centrifuge briefly.
4. Transfer the tubes to a thermal cycler. Amplify the cDNA through 25–40 thermal cycles. Standard conditions are denaturation at 95°C for 40 s, annealing at 55°–68°C for 40 s and extension at 72°C for 40 s. The conditions depend on the primers and thermal cycler used.
5. Run a parallel reaction containing 2.0 μL of RT reaction product and 48 μL of PCR-mastermixture with primers of the internal standard gene (i.e., β-tubulin or hypoxanthine-guanine phosphoribosyltransfcrase [HPRT]) instead of the target gene.
6. Cool the samples to 4°C and store at –20°C.

3.3.1.5. POST-PCR TREATMENT OF SAMPLES

1. Transfer the PCR products to 1.5-mL microtubes.
2. Add 2.5 μL of a 1:4 mixture of glycogen:5 M NaCl and 150 μL of ice-cold ethanol to each tube.
3. Incubate at –20°C for at least 1 h and centrifuge at 16,000g for 15 min.
4. Dissolve the pellet in 25 μL of TE buffer, add 5 μL of 6X loading buffer, and store at 4°C.

3.3.1.6. ELECTROPHORESIS AND QUANTITATIVE ANALYSIS OF GENE EXPRESSION

1. Separate one third of each PCR (10 µL) by electrophoresis on a 2% agarose gel in 1X TBE containing 0.35 µg/mL of ethidium bromide at 5–10 V/cm for 70 to 100 min.
2. Illuminate the gel by UV light and obtain a digital image.
3. Quantitate the ethidium bromide fluorescence signals of gels. We use the TINA2.08e software (raytest Isotopenmeßgeräte GmbH) or Multi-Analyst Version 1.1 software (Bio-Rad) to evaluate the relative mRNA levels of the target gene in relation to the internal standard gene.

3.3.2. Immunofluorescence Analysis (see **Note 14**)

The formation of tissue-restricted proteins in the EB outgrowths is analyzed by immunofluorescence with a normally equipped fluorescence microscope or with a CLSM (*see* **Note 14**).

3.3.2.1. SINGLE-CELL ISOLATION OF ES CELL-DERIVED CARDIAC CELLS

For a better demonstration of the structural organization of intracellular, especially sarcomeric proteins, ES cell-derived cardiomyocytes are isolated as single cells by the following procedure (a modification of the method described in **ref. *42***):

1. Isolate the beating areas of EBs (*n* = 10) mechanically using a microscalpel under an inverted microscope and collect the tissue in PBS Dulbecco in a centrifuge tube at room temperature. Centrifuge at 1000*g* for 1 min and aspirate the supernatant.
2. Incubate the pellet in collagenase B-supplemented low-Ca^{2+} medium at 37°C for 25–45 min dependent on the collagenase activity. For the isolation of cardiac clusters, shorten incubation time to 10 to 20 min.
3. Aspirate the enzyme solution, resuspend the cell pellet in about 200 µL of KB medium and incubate at 37°C for 60–90 min.
4. Transfer the cell suspension into tissue culture plates containing gelatin-coated slides and incubate in differentiation medium I at 37°C overnight. The KB medium is diluted at least 1:10 with differentiation medium I.
5. Change the medium to differentiation medium I; cardiomyocytes begin rhythmical contractions and are ready for immunostaining after a recovery time of 24 h.

3.3.2.2. IMMUNOFLUORESCENCE FOR THE DETECTION OF TISSUE-RESTRICTED PROTEINS

For characterization of ES cell-differentiated phenotypes, monoclonal antibodies (mAbs) against tissue-specific intermediate filament proteins or sarcomeric proteins are suitable (*see* **Table 3**).

1. Rinse coverslips containing EB outgrowths twice with PBS.
2. Fix cells onto coverslips with methanol: acetone (7:3) at –20°C for 10 min, or alternatively, with 3.7% paraformaldehyde in PBS at room temperature for 10 min (depending on the antibody used).
3. Rinse coverslips twice with PBS at room temperature for 5 min.
4. Incubate the cells with 10% goat serum in PBS in a humidified chamber at room temperature for 30 to 60 min to prevent unspecific immunostaining.
5. Incubate with the primary antibody at 37°C for 30 to 60 min, or at 4°C overnight (final concentration according to manufacturers' instructions).

Table 2
Primers Used for RT-PCR

Cell types	Genes	Primer sequences (5'→3')	Size	Annealing temperature	References
Cardiac cells	Cardiac α-MHC	CTGCTGGAGAGGTTATTCCTCG GGAAGAGTGAGCGGCGCATCAAGG	301 bp	64°C	(26,27,29,36)
	Cardiac β-MHC	TGCAAAGGCTCCAGGTCTGAGGGC GCCAACACCAACCTGTCCAAGTTC	205 bp	64°C	(27,29)
	Myasin light chain isoform 2V (MLC-2V)	TGTGGGTCACCTGAGGCTGTGGTTCAG GAAGGCTGACTATGTCCGGGAGATGC	189 bp	60°C	(26,27,29,36)
	Atrial natriuretic factor (ANF)	TGATAGATGAAGGCAGGAAGCCGC AGGATTGGAGCCCAGAGTGGACTAGG	203 bp	64°C	(27,29,60)
	Nkx 2.5	CGACGGAAGCCACGCGTGCT CCGCTGTCGCTTGCACTTG	181 bp	60°C	(61)
Skeletal muscle cells	Myf5	TGCTGTTCTTTCGGGACCAGACAGG GGAGATCCTCAGGAATGCCATCCGC	132 bp	65°C	(12,28,30)
	Myogenin	CAACCAGGAGGAGCGCGATCTCCG AGGCGCTGTGGGAGTTGCATTCACT	85 bp	60°C	(12,28,36,62)
	MyoD	ATGCTGGACAGGCAGTCGAGGC GCTCTGATGGCATGATGGATTACAGCG	144 bp	65°C	(12,28,30)
	Myf6	GAGGGTGCGGATTTCCTGCGCACC GGAGGCTGAGGCATCCACGTTTGC	117 bp	60°C	(12,28,62)
	M-cadherin	AACTGGAGCGTCAGCCAGATTAACG GCGCGGCAAACAGGATGAGAAC	386 bp	56°C	(13,28)
Smooth muscle cells	smooth muscle myosin heavy chain (SM-MHC)	GGATGCCACCACAGCCAAGTA TGGTGTGGGTCCCTTCAGAGA	497 bp	60°C	(25)
Internal standards	β-tubulin	GGAACATAGCCGTAAACTGC TCACTGTGCCTGAACTTACC	317 bp	60°C	(27–29,36,63)
	HPRT M1/P1	CGCTCATCTTAGGCTTTGTATTTGGC AGTTCTTTGCTGACCTGCTGGATTAC	447 bp	60°C	(26,28,64)
	HPRT M2/P19	GCCTGTATCCAACACTTCG AGCGTCGTGATTAGCGATG	502 bp	64°C	(64)

6. Rinse coverslips with PBS 3 times at room temperature for 5 min.
7. Incubate with the secondary antibody (i.e., dilute Dichlorotriazinyl/Amino Fluorescein [DTAF]-labeled goat anti-mouse IgG 1:100 in PBS with 0.5% BSA; or prepare DTAF-labeled goat anti-rat IgG, at 12 µg protein/mL final concentration, depending on the primary antibody) in a humidified chamber at 37°C for 45 to 60 min.
8. Rinse coverslips twice with PBS at room temperature for 5 min.
9. Rinse coverslips quickly with distilled water at room temperature.
10. Embed coverslips in mounting medium and analyze immunolabeled cells with a conventional fluorescence or CLSM.

3.3.3. Pharmacological Analysis of ES Cell-Derived Cardiac Cells

Cardiomyocytes differentiated from ES or EC cells develop cardiac-specific physiological properties, as well as cardiac-specific receptors and signal transduction mechanisms *(8–10,37)*. Therefore, ES cell-derived cardiomyocytes are suitable to measure chronotropic effects of cardioactive substances (*see* **Fig. 2**).

1. Plate EBs separately onto 24-well microwell plates at d 5 or 7, cultivate the EB outgrowths for further 5 to 7 d until 85 to 100% of the EBs contain clusters of rhythmically contracting cardiomyocytes. Beating cardiomyocytes should comprise 5 to 30% of the EB outgrowths area. Change medium 1 d before measurements and add exactly 1 mL of medium per well.
2. Place the 24-well microwell plate on the inverted microscope equipped with a 37°C heating plate and a CO_2 incubation chamber, localize independently beating areas ($n = 20$), and measure the spontaneous beating frequency.
   Alternatively, select different areas ($n = 20$) of pulsating cardiomyocytes by visual control under the inverted microscope coupled via a one-chip CCD camera to a computer imaging station (Pentium CPU, 100MHz) running the LUCIA Imaging System including the "HEART" application (Nikon). The coordinates (x, y, z) of selected areas are collected by the LUCIA HEART System, and the spontaneous beating activity is automatically determined for each area.
3. Add different concentrations of the test substance and incubate for 3 min before measurement. Determine dose-dependent effects of the beating frequency after cumulative application of increasing concentrations of the test substance by adding positive chronotropic drugs (i.e., BayK 8644 may be used as a positive control) or negative chronotropic drugs (i.e., diltiazem may be used as a positive control) at final concentrations in the range of about 10^{-9} to 10^{-5} M (depending on the test substances).
4. Calculate the mean values (≥ standard error of the mean) of beats per minute for each data point from the pulsation rates of ES cell-derived cardiomyocytes with and without addition of drugs. Test for significance by the Mann-Whitney U-test and calculate dose-response curves. Alternatively, the final processing of the data is done by the LUCIA HEART Imaging System resulting in dose-response curves of chronotropic activity.

3.4. Genetic Modulation of Differentiation (see Note 15)

The existence of ES cells has allowed the creation of almost any kind of mutation in any mouse gene. The study of genetically modified ES cells and their differentiation in vitro has thus provided a novel approach to study the development or study the effects of mutations in the genome. ES cells can be used to study promoter elements instead of using transgenic animals *(43)*. Additionally, random insertion of exogenous DNA into single sites in the mammalian genome (gene trapping) provides a genome-wide

1. Selection of beating clusters in EB outgrowths

↓

2. Evaluation of basal level of beating frequency by the
LUCIA 'Heart' imaging system

↓

3. Pharmacological analysis of cardioactive drugs by automatic measurement
of beating activity after cumulative drug exposure

BayK 8644   Diltiazem

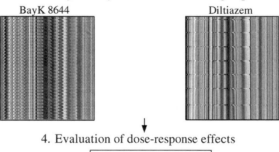

↓

4. Evaluation of dose-response effects

Fig. 2. A computer-assisted imaging system for analysis of pharmacological responses of ES cell-differentiated cardiomyocytes. ES cells are cultivated via EBs into the cardiogenic lineage. EBs are then plated onto tissue culture plates, areas of beating cardiomyocytes are selected (**1**), and the coordinates are digitized and stored by the computer-assisted imaging system LUCIA HEART (**2**). The levels of beating frequencies of cardiomyocytes are automatically measured before (**2**) (basal level) and after cumulative addition of positively or negatively acting cardiotropic agents (**3**), and dose-response curves are evaluated (**4**).

strategy for functional genomics. When performed in tandem with the developmental potential of ES cells to differentiate into distinct cell lineage (expression trapping), such techniques are useful for the identification of novel genes expressed in developing systems. Finally, the Cre recombinase/loxP system *(44,45)* has permitted loss-of-function analyses in vitro and subsequent rescue of targeted alleles for examination of individual gene products and their function during development (*see* **Note 16**).

3.4.1. Introduction of DNA into ES Cells (see **Note 17**)

3.4.1.1. PREPARATION OF DNA

3.4.1.1.1. DNA for Stable Transfections

1. Super-coiled plasmid DNA, prepared by cesium-chloride centrifugation or ion-exchange chromotography, needs to be appropriately linearized to permit integration into the genome. DNA linearization and preparation should be performed under sterile conditions.
2. Linearize a minimum of 20 µg selection vector and 40–200 µg targeting vector containing a selection cassette in the construct (*see* **Note 15** and **Table 1** and **Subheading 3.4.1.2., step 6**) with the appropriate digestion buffer and restriction enzymes (ensure linearization is complete). Bring the final vol to 100 µL with TE buffer.
3. Extract with an equal volume of phenol : chloroform : isoamyl alcohol. Mix until an emulsion appears, separate aqueous and phenolic phases by centrifugation in a microcentrifuge at 4200g for 2–5 min at room temperature.
4. Transfer the aqueous (top) phase to a fresh tube.
5. To the tube containing the phenolic phase, add 100 µL of TE buffer. Mix and centrifuge as before. Remove the aqueous phase and combine with what is already in the fresh Eppendorf tube and increase the volume to 0.5 mL (*see* **Note 18**).
6. To concentrate the samples, add 1/10 vol 3 *M* Na-acetate and 2.2 vol of ice-cold ethanol. Precipitate overnight at –20°C or for 1 h at –70°C. Centrifuge Eppendorf tubes in a microfuge at 10000g at 4°C for 15 min.
7. Remove the supernatant and add 200 µL of 70% ethanol. Mix briefly and centrifuge for 5 min at 10000g at 4°C.
8. Remove supernatant and let air-dry in the laminar flow hood. Dissolve the pellets completely in PBS at a concentration of 1–5 mg/mL.

3.4.1.1.2. DNA for Transient Transfections

1. Prepare supercoiled plasmid DNA by standard methods of either cesium-chloride centrifugation or ion-exchange chromotography. Use sterile techniques. (*see* **Notes 16, 17, and 18**).
2. To concentrate the solution, precipitate DNA with 1/10 vol 3 *M* Na acetate and 2.2 volumes of ethanol at –20°C overnight.
3. Centrifuge the DNA precipitate in an Eppendorf microfuge at 12500g for 15 min at 4°C.
4. Discard the supernatant and add 200 µL of 70% ethanol. Centrifuge as above but for 3–5 min.
5. Discard the supernatant and let air-dry in the laminar flow hood. Dissolve the pellets completely in sterile PBS at a concentration of 1–5 mg/mL.

3.4.1.2. ELECTROPORATION OF CELLS *(46,47)*

1. Change the media of the actively dividing ES cells at least 1 to 2 h before harvesting.
2. Under sterile conditions, harvest the ES cells with trypsin-EDTA and centrifuge the suspension at 2000g.

Table 3
Antibodies for the Analysis of Tissue-Restricted Proteins
on ES Cell-Derived EB Outgrowths

Cell types	Cell-specific antigens	Antibody	References
Cardiac and	**Sarcomeric proteins:**		
skeletal	Titin (Z-disk)	T11, T12	*(28,36,65,66)*
muscle	Titin (M-band)	T51	*(65,67)*
cells	α-Actinin	653	*(29,65,68)*
	Myomesin	MyBB78	*(65,69)*
	Sarcomeric MHC	MF20	*(28,29,65,70)*
	α-Sarcomeric actin	5C5	*(29,65,71)*
	M-protein	MpAA241	*(65,69)*
	Cardiac-specific proteins:		
	α-Cardiac MHC	BA-G5	*(10,34,72)*
	Cardiac troponin T	M7	*(65,73)*
	Skeletal muscle-specific proteins:		
	Nebulin	Nb2	*(28,66)*
	Slow MHC	S58	*(74)*
	Fast MHC	F59	*(74)*
	Slow myosin binding protein C	α-sMyBP-C	*(28,75)*
	M-cadherin		*(13)*
Smooth muscle	Smooth muscle α-actin	1A4	*(25,76)*
cells	SM-MHC (fast)	MY-32	*(25,77)*

3. Wash the cell pellet thoroughly with PBS (three washes with 10 mL PBS each time), centrifuging after each wash.

4. Before the final centrifugation, take an aliquot of cells and count them using a hemocytometer. (From a 60-mm plate, between 3 and 5×10^6 ES cells should be harvested.) Resuspend the cell pellet to 4.0–5.0×10^6 cells/mL in PBS. Cell viability should be 90–95%, and the feeder cells should comprise no more than 5% of the total cell number.

5. Transfer 0.8 mL of the cell suspension to a cuvette.

6. For stable transfections, add 10–20 µg of linearized plasmid DNA to the cuvette. For homologous recombination, the amount of DNA needed is greater because of the larger size plasmid (for every 1000 bp of plasmid, add 2 to 4 µg of linearized plasmid). Alternatively, for transient transfections, add supercoiled plasmid to the cells at a concentration of 3–5 nM (for every 1000 bp of plasmid, add 0.8 to 1.5 µg of linearized plasmid). For transient transfections in which loss-of-function or gain-of-function is to be achieved by either deletion of loxP flanked sequences or insertion of DNA sequences into loxP sites, respectively. We recommend cotransfection of the Cre recombinase containing plasmid with either the loxP flanking positive selection cassettes or loxP flanked gain-of-function constructs. Add the DNA to the cuvettes at a molar ratio of 1:1 (for every 1000 bp of plasmid, add 2 to 4 µg of plasmid), for inducible knock-out or knock-in of a loxP flanked sequence (*see* **Note 16**).

7. Once the DNA has been added to the cells, mix the solution thoroughly by gentle trituration. Let the cells and DNA incubate for 10 min on ice.

8. Mix DNA and cells by gentle pipetting and recap the cuvette.

9. Electroporate the cells by delivering a pulse of 250 V at 500 μF. Record the exact voltages, conductance, and pulse duration. After electroporation, the solution will contain viscous material.

10. One minute after delivering the pulse, transfer half of the contents of the cuvette to each of the two culture dishes containing feeder cells and 15 mL of cultivation medium plus 1000 U LIF/mL. Be very gentle during the transfer process, trying not to damage the fragile cells. Let incubate for 40–48 h (37°C, 5% CO_2).

3.4.1.3. Cultivation and Selection of Clones

1. Replace medium with fresh medium containing selection drug. Change the medium every 2–4 d during the selection period (*see* **Note 15**).

2. After a further 8–12 d, resistant colonies of ES cells should become visible (to the naked eye). Harvest individual colonies and transfer to 96-well plates containing 20 μL trypsin-EDTA at room temperature. Once 20 to 50 individual colonies have been transferred, incubate the plate at 37°C for 5 min. Triturate thoroughly and transfer cells to 96-well plates containing feeder cells, fresh media, and the selection agent.

3. Individual wells should be monitored daily, ensuring that the media does not change color. Expand the cell population by passaging to 24-well plates. Let grow for 24–48 h. Freeze half the cells in the well and use the other half for either continued expansion or preparation of DNA.

3.4.2. Analysis of Transfectants

3.4.2.1. Rapid Isolation of DNA from ES Cells (48)

1. Plate an aliquot of trypsinized ES cell clones (*see* **Subheading 3.4.1.3.**) into 4 wells each of a 96-well plate pretreated with 0.1% gelatin.

2. Grow in cultivation media without LIF, changing media frequently to prevent pH changes, to confluence.

3. When the cells are confluent, remove the media by aspiration. Wash the wells twice with PBS.

4. Add 50 μL of DNA lysis buffer to each well. Seal the 96-well plate with parafilm and place in a tupperware box where the bottom is covered in water. Incubate the plate overnight at 55°C in a humid atmosphere.

5. The next day, prepare a mixture of NaCl and ethanol and add 100 μL of the mixture to every well using a multichannel pipet.

6. Leave the 96-well plate on the bench at room temperature for 20–30 min or until the solution becomes transparent.

7. Invert the plate gently onto paper towels to discard the solution. Gently add to the side of the wells 150 μL ice-cold 70% ethanol with the multichannel pipet. (Do not add ethanol directly on the precipitated DNA.) Invert the plate, as before, to discard the 70% ethanol.

8. Repeat the 70% ethanol washes, being particular gentle. After the second wash, most of the wells become transparent, and it is possible to see the DNA on the bottom of the well. Some of the wells may still contain salt (white as opposed to translucent). If so, just wash these wells with 70% ethanol again.

9. After the final washing, invert the plate, and let it partially dry for a few minutes.

10. The DNA is now ready to use or can be stored until needed. For storage, seal the plates with parafilm and place at –20°C.

11. For immediate use, add 30 μL sterile H_2O into each well and dissolve the DNA very well.

12. For PCR, 0.5–1 µL of this DNA is usually sufficient for genotyping. For Southern blotting, a vol of 25 µL is needed. Approximately 5 µg DNA can be obtained from a single fully confluent well of a 96-well plate. This quantity is sufficient for digestion with one restriction enzyme and preparation of a Southern blot.

3.4.2.2. PCR ANALYSIS *(49)*

1. The PCR conditions should be established with a positive control and with a plasmid DNA template with characteristics similar to that predicted from the chromosomal structure after random DNA insertion events or targeted events (homologous recombination or Cre recombinase-mediated insertion into pre-existing loxP sites). Dilute the DNA template to 50 pg/µL in TE.
2. Random insertion events: design a primer set from the gene product of interest. Software such as Primer Express 1.5 is useful for this.
3. Primers for targeted events: in addition to the primer set designed in step 2, design a primer set that has one primer located either upstream or downstream of the targeting DNA sequence and one primer located in the targeting sequence. (This amplification is however dependent on the distance between the primers. If the primer located out of the targeting construct is more than 3 kb away from the internal primer, we recommend Southern blot analysis for genotyping.)
4. Set up parallel PCRs with different Mg^{2+} concentrations and containing primers for β-globin. In a duplicate set of reactions, prepare the same mixes, but omit primers for β-globin. On occasion, β-globin primers interfer with the amplification and a separate primer set is chosen.

10X Mg^{2+} free DNA polymerase buffer	5.0 µL
dNTPs, 20 m*M* dilution	1.0 µL
± β-Globin primer set	1.0 µL
Primer set for gene product of interest	1.0 µL
Template DNA (10–50 pg/µL)	1.0 µL
Mouse genomic DNA (0.5 µg/µL)	1.0 µL
$MgCl_2$	1.8 µL
Taq DNA polymerase (1 U/µL)	0.7 µL
dH$_2$O	to 50.0 µL

5. Amplify the DNA according to the following cycling parameters (*see* **Note 19**).

Initial denaturation:	94°C for 5 min
Primary cycles:	94°C for 45 s
	55°C for 45 s
	72°C for 45 s
	Repeat cycle 24 times
Final step:	72°C for 5 min
	4°C for infinity.

The PCR can be stored at room temperature. Elongation and denaturing times should be increased for longer DNA products (>1000 bp).

6. Volume (1/10) of the PCR should be transferred to a fresh Eppendorf tube containing 17.5 µL of TE buffer and supplemented with 2.5 µL loading buffer.
7. To analyze the PCR products, load the products on a 1.0% agarose gel. Include molecular weight markers to determine the size of the amplified products. Amplification results (*see* **Note 19**) should have two DNA products: one for β-globin (control product) at 800 bp and one for neomycin–puromycin–or another selection cassette (**Fig. 3B**).

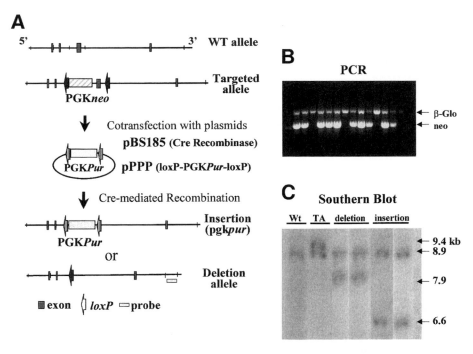

Fig. 3. Site-specific ES cell insertion and excision by Cre recombinase—*lox*P recombination. **(A)** A gene locus in ES cells was targeted by homologous recombination to insert a sequence consisting of a PGK-*neo*R cassette flanked by two *lox*P sites. The resulting targeting event resulted in a clonal ES cell line containing a wild-type (WT) allele and a targeted allele. This ES clonal cell line was then transiently transfected with two supercoiled plasmids: pBS185 (containing a cytomegalovirus [CMV] promoter-driven Cre recombinase) and pPPP (containing a PGK-*Pur*R cassette flanked by two *lox*P sites) (*see* **Subheading 3.4.1.1.2.** and **Subheading 3.4.1.2.**). The cells were allowed to recover for 48 h, followed by selection with appropriate antibiotics. Resistant colonies were isolated and expanded (*see* **Subheading 3.4.1.3.**). The potential Cre recombinase-mediated insertion or deletion events are indicated in the diagram. **(B)** Genotyping by PCR (*see* **Subheading 3.4.2.2.**) was performed on the clonal lines to select for cells that had lost the neomycin resistance cassette. An internal control (β-globin [β-Glo]) amplification was included for each DNA preparation to ensure against false negatives (*see* **Subheading 2.2.6.2., step 6**). **(C)** These lines were then tested by Southern analysis (*see* **Subheading 3.4.2.3.**) with *Eco*RI to identify those clones that had undergone deletion or insertion events. The probe used for Southern analysis was generated from a DNA fragment outside of the original targeting (homologous recombination) vector (see panel **A**). Four distinct bands can be identified: *(1)* an 8.9-kb band corresponding to the WT allele *(2)* a 9.4-kb band of the original targeted allele containing the neomycin resistance cassette; *(3)* a 7.9-kb band where the neomycin resistance cassette has been lost and the flanking *lox*P sites have recombined (deletion); and *(4)* a 6.6-kb band generated by digestion of the newly inserted Cre recombinase-targeted allele. The PGK-*Pur* cassette contained an internal *Eco*RI site leading to the smaller fragment. These targeted clonal ES cell lines were subsequently differentiated, and the phenotype was examined.

3.4.2.3. SOUTHERN BLOT ANALYSIS *(50)*

1. Digestions of DNA (*see* **Subheading 3.4.2.1.**) can be performed either in the 96-well plates (if doing all wells) or in microfuge tubes if the contents of the well are transferred before digestions.
2. To the 25 μL of genomic DNA, add 3 μL of 10× restriction enzyme digestion buffer, 1 μL RNase A, and 1 μL of the appropriate high concentration restriction enzyme (40 U/reaction). Mix contents gently. Do not sheer the DNA.
3. Incubate the digestion at 37°C overnight in a humidified atmosphere.
4. The following day, add another 20 U restriction enzyme and incubate at 37°C for another 2–3 h.
5. Prepare a 0.7% agarose gel containing 1X TBE and ethidium bromide.
6. Stop the reaction with the addition of loading buffer. Load the samples and molecular weight markers on the agarose gel and fractionate the DNA products by electrophoresis. For the best resolution, run the gel at low voltage (0.7 V/cm) in 0.5X TBE.
7. Place the gel on an UV transilluminator and photograph. Record the migration distances for the molecular weight markers.
8. Transfer the gel to a glass baking dish and depurinate the DNA by soaking the gel for 5–10 min in 0.2 N HCl. This is important particularly for improved transfer of large DNA fragments (>8 kb in length).
9. Rinse the gel with distilled water. Soak the gel in strong base (1.5 M NaCl, 0.5 M NaOH) for 45 min to denature the DNA.
10. Rinse the gel briefly in dH$_2$O for 5 min. Neutralize the solution by soaking it in 1.0 M Tris, 1.5 M NaCl (pH 7.4) for 30 min.
11. Transfer the genomic DNA to Hybond-N$^+$ Nylon membrane by capillary transfer (*see* **Note 20**).
12. Air-dry the membrane, then fix the DNA to the membrane by using UV Stratalinker (120,000 μJ/cm^2).
13. Put the membrane into a prehybridization solution and hybridize at 65°C for at least 2 h with agitation.
14. Prepare the probe by using 25–50 ng denatured cDNA and Ready-to-go DNA beads (-dCTP) to make a probe with specific activity approx 10^9 cpm/μg DNA, and add the denatured probe (heated to 95°C for 5 min and quenched on ice) to the prehybridization buffer at an activity of 0.5–2 × 10^6 counts per minute (cpm)/10 mL (*see* **Note 21**).
15. Hybridize the probe to the membrane DNA overnight at 65°C.
16. Wash membrane as follows:
    a. Twice with 2X SSPE buffer; 0.1% SDS at room temperature for 20 min.
    b. Once with 1X SSPE buffer; 0.1% SDS at room temperature for 15 min.
    c. Once with 1X SSPE buffer; 0.1% SDS at 50°C for 15 min.
    d. Once with 0.5X SSPE buffer; 0.1% SDS at 60°C for 15–30 min (monitor the radioactivity, if it is still very radioactive, go to next wash).
    e. Once with 0.2X SSPE buffer; 0.1% SDS at 65°C for 15–30 min.
17. Expose to X-ray film for 1–10 d at –80°C or use a phosphoimager.
18. Once the cells have been expanded, frozen stocks prepared, and genotyping complete (stable clonal lines), analysis of gain- or loss-of-function tests can begin (*see* **Subheadings 3.2.** and **3.3.**).

3.4.3. Gain-of-Function and Loss-of-Function Analysis In Vitro

The developmental pattern of EBs differentiated in vitro may be modulated by exogenous factors, i.e., RA *(26,34)*, growth factors *(33)*, or by genetic means: the targeted inactivation of genes (equals loss-of-function) *(24,29)* or the overexpression of genes in ES cells (equals gain-of-function) *(30)*. (For differentiation and characterization of transgenic ES cells, *see* **Subheadings 3.2.** and **3.3.**).

The in vitro differentiation of mutant pluripotent ES cells has been used as an alternative and supplement to in vivo studies to analyze the phenotypes of mutant cells during early embryonic development. The strategy is especially useful in those cases where the mutation results in early embryonic death in vivo. This has been successfully used in the analysis of cellular differentiation of i.e., desmin- and β_1 integrin-deficient ES cells *(24,29)*.

4. Notes

1. ES cell lines should be cultivated without antibiotics. In some cases (i.e., for selection procedures), the addition of a penicillin–streptomycin mixture or gentamycin (Gibco BRL, 1 mL of stock solution to 100 mL medium) may be helpful.
2. Both DMEM and IMDM can be used for efficient cardiac and myogenic differentiation of ES cells as D3 and R1. If IMDM is used, in additives I MTG (final concentration 450 μM) instead of β-ME is used for differentiation.
3. DEPC is a suspected carcinogen and should be handled with care.
4. If possible, the solutions should be treated with 0.1% DEPC at 37°C for 1 h, and then heated to 100°C for 15 min or autoclaved for 15 min. DEPC reacts rapidly with amines and cannot be used to treat solutions containing buffers, such as Tris.
5. Whereas EC cell lines are cultivated without feeder cells, ES and EG cell lines need feeder cells for growth in the undifferentiated state. Some ES or EG cell lines (e.g., EG-1 cells) *(35)* need both feeder cells and LIF to keep growth in the undifferentiated state. LIF is commercially available (Chemicon International) or may be prepared from LIF expression vectors *(51–53)*.
6. Good quality FCS is critical for long-term culture of ES cells, and failure to acquire good quality serum may be one reason why ES cells fail to differentiate appropriately. Extensive serum testing is necessary, therefore, to achieve good results. The most sensitive tests for sera include: *(1)* comparative plating efficiencies at 10, 15, and 30% serum concentrations; *(2)* alkaline phosphatase activity in undifferentiated ES cells (Vector Blue Alkaline phosphatase substrate kit III, SK-5300; Vector Laboratories); and *(3)* test of in vitro differentiation capacity after 3 to 5 passages in selected serum *(54)*.
7. For preparation of EBs, three different protocols may be used: the hanging drop method *(8,39)*, the mass culture *(4)*, or the methylcellulose technique *(18,19)*. The hanging drop method generates EBs of a defined cell number (and size). Therefore, this technique is used for developmental studies, because the differentiation pattern is dependent on the number of ES cells that differentiate within the EBs. For mass culture, plate 5×10^5 to 2×10^6 cells (depending on ES cell lines used) into 60-mm bacteriological Petri dishes containing 5 mL differentiation medium. After 2 d, let the aggregates settle in a centrifuge tube, remove medium, and carefully transfer the aggregates with 5 mL fresh differentiation medium into a new bacteriological dish. Change the medium every second day. Mass cultures of EBs may be used for differentiation of a large number of cells. For hematopoietic differentiation, the methylcellulose method is used *(18,19)*. Methylcellulose (e.g., MethoCult H4100; Stem Cell Technologies Inc., Vancouver, B.C.) is added to the differentiation medium at a final concentration of 0.9%.
8. Use gloves and filtertips throughout the whole procedure.
9. Do not leave RNA lysis buffer in culture dishes longer than 5 min, as polystyrene is not resistant to lysis buffer.

10. mRNA isolation from small samples of cells or tissues (i.e., 1 EB) can be performed using the Dynabeads® mRNA DIRECT™ Micro kit (Dynal, cat. no. 610.21).

11. Never mix and disturb the organic and the aqueous phases.

12. For RT-PCR, rTth DNA polymerase can also be used as both reverse transcriptase and DNA polymerase *(55)*. In this case, the components of both RT- and PCR-mastermixture are different from using MuLV reverse transcriptase and *Taq* DNA polymerase.

13. Semiquantitative RT-PCR is used to detect the relative levels of mRNA expression and includes at least two sets of primer pairs in separate or co-amplification reactions. A "housekeeping" gene, i.e., β-tubulin or HPRT, is used as an internal standard to control variations in product abundance caused by differences in individual RT reaction and PCR efficiencies. The co-amplification reactions of two primer sets can be performed by using the primer-dropping method *(56)*. In this method, the final yield of more efficiently amplified templates is reduced below saturating levels by using fewer PCR amplification cycles than for less efficiently amplified templates. This difference in amplification, between the templates with variable starting amounts, is achieved by dropping primer sets into the reactions at distinct cycle numbers. The first primer sets for less efficiently amplified templates are added at the beginning of reactions, the equal aliquots (4 μL) of second primer sets for more efficiently amplified templates are added at the appropriate cycle number (preliminary titration experiments for optimal cycle numbers have to be performed before). The relative signal strengths of PCR products of the target gene and the internal standard gene can be controlled by using varying numbers of cycles in multiplex reactions *(56)*.

14. CLSM analysis can be used to study EBs. For immunofluorescence analysis of EBs cultivated in suspension, it is necessary to use a CLSM, because EBs are three-dimensional aggregates, which require an extended depth of focus. Some EBs are up to 400 μm in diameter. Therefore, EBs are scanned in thin sections (0.5–10 μm) using the appropriate filter combinations depending on the fluorescent dyes used.

15. The use of random insertion events or gene targeting technologies to ES cells in cell culture enables studies of gain- or loss-of-function. Targeting events are extremely powerful, because individual genes can be altered or modified, and the consequences on an individual gene product can be studied both in vitro and/or, if desired, in vivo. The difficulty for a traditional "knock-out" ES cell line is the need to target alleles on both chromosomes of a diploid cell. Increased concentrations of selective agent are required for appropriate selection or the use of widespread DNA screening. For selection of a double-targeting event (i.e., targeting of both alleles), use approximately 2.5 μg/mL puromycin and 1 mg/mL G418 *(57)*. In cases in which gene targeting results in embryonic lethality in mice, such a technique in ES cells affords the opportunity of examining loss-of-function on individual cell types.

16. LoxP sites are 34-bp sequences containing an 8-bp core region and flanking 13-bp inverted sequences. The bacteriophage P1 Cre-recombinase recognizes specific loxP recognition sequences, and in an ATP independent manner, specifically catalyzes recombination events (deletion, insertion, inversion, and translocation). Insertion events require a single loxP site; whereas the other recombinantion events require 2 sites. The orientation–direction of the central 8-bp cores determine the type of recombination event between 2 sites. Authentic loxP sites are not found in mammalian genomes, but when inserted randomly or targeted into mammalian systems are recognized by Cre-recombinase. All four recombination events can be catalyzed in either dividing or nondividing mammalian cells, and, when used appropriately, can be used to make gene chimeras between the endogenous mouse gene and cDNAs targeted to the loxP sites.

17. When DNA is introduced into ES cells by electroporation, transfection of a single DNA molecule containing selection-cassettes such as neo[R] or pur[R] (*see* **Table 1**) or cotransfection

with one plasmid containing the gain-of-function construct and a second with a selection cassette can be performed. Information on production of targeting cassettes can be found in **ref.** *58*.

18. In cases in which the sterility of the DNA is of concern, filter-sterilize the DNA by passing a solution containing the DNA through a pre-wet 0.22-μm filter. In the absence of filter-sterilization, it is recommended that the plasmid be ethanol-precipitated 2 times to minimize potential contamination.

19. To obtain the optimal PCR conditions, not only optimal Mg^{2+} concentration, but also optimal annealing temperatures, cycle numbers, and elongation temperatures have to be tested. To test for optimal temperatures, dilute the template to 1 pg/μL in TE buffer. Set up the PCRs as given in the text, but with the Mg^{2+} concentration that gave the best results. Amplify with different annealing temperatures: 58°, 61°, 63°C, or, if no fragment was seen in the initial amplifications, reduce the temperatures to 48°, 51°, and 53°C. Determine which conditions give the maximum amount of amplified product on an agarose gel. To obtain optimal cycle number, different numbers of cycles of 24, 29, 34, and 39 have to be tested. Afterwards, using the annealing temperature and cycle number that give the best amplification (higher temperature and lower cycle number, if several reactions gave similar results), repeat the amplifications with different elongation temperatures: 65°, 68°, 71°, and 75°C. Test the products on a gel to determine what temperature gives the best amplification result.

20. Capillary transfers can be set up as follows: place a support inside a large baking dish filled with transfer buffer (10X SSPE). Put a piece of Whatman 3MM paper on the support, and let the two ends of the paper fall into the transfer buffer. Invert the gel, and place it on the Whatman paper. Place a piece of Hybond-N+ Nylon membrane cut to the same size on the gel. Then, place several pieces (5–7) of 3MM paper and a stack of paper towels (5–8 cm high), just smaller than the membrane, on top. Put a glass plate on top of the entire stack and weigh it down with a 500–g weight. Transfer overnight *(59)*.

21. The DNA used in the probe preparation must be taken from sequences outside (up- or down-stream) of the targeting construct. Otherwise, it cannot differentiate between a homologous recombination event and a random integration event.

Acknowledgments

We are grateful to Mrs. S. Sommerfeld, K. Meier, O. Weiβ and L. O'Neill for expert technical assistance in the establishment of the differentiation protocols and to D. Riordon and S. Brugh for genetic analyses. We wish to thank Drs. J. Rohwedel, now at the University of Lübeck; J. Hescheler, University of Köln; M. Drab, Max-Delbrück-Center Berlin-Buch; Katja Prelle, Gene Center, Munich, Germany; V. Maltsev, now at the Henry Ford Heart and Vascular Institute, Detroit, MI, USA, and C. Strübing, now at the Div. of Cardiovascular Research, Children's Hospital, Boston, MA, USA, for collaboration. We also thank Dr. Christine Mummery, Hubrecht Laboratory, University of Utrecht, The Netherlands, for the protocol of DCC-FCS preparation. The work was supported by grants of the Deutsche Forschungsgemeinschaft (SFB 366, Wo 1/1-3) and Fonds der Chemischen Industrie.

References

1. Evans, M. J. and Kaufman, M. H. (1981) Establishment in culture of pluripotential stem cells from mouse embryos. *Nature* **291,** 154–156.

2. Martin, G. (1981) Isolation of a pluripotent cell line from early mouse embryos cultured in medium conditioned by teratocarcinoma cells. *Proc. Natl. Acad. Sci. USA* **78,** 7634–7638.

3. Bradley, A., Evans, M., Kaufman, M. H., and Robertson, E. (1984) Formation of germ-line chimaeras from embryo-derived teratocarcinoma cell lines. *Nature* **309,** 255–256.

4. Doetschman, T. C., Eistetter, H. R., Katz, M., Schmidt, W., and Kemler, R. (1985) The in vitro development of blastocyst-derived embryonic stem cell lines: formation of visceral yolk sac, blood islands and myocardium. *J. Embryol. Exp. Morphol.* **87,** 27–45.

5. Keller, G. (1995) In vitro differentiation of embryonic stem cells. *Curr. Opin. Cell Biol.* **7,** 862–869.

6. Wobus, A., Rohwedel, J., Strübing, C., Jin S., Adler, K., Maltsev, V., and Hescheler J. (1997) In vitro differentiation of embryonic stem cells, in *Methods in Developmental Toxicology and Biology* (Klug, E. and Thiel, R., eds.), Blackwell Science, Berlin, pp. 1–17.

7. Thomas, K. R. and Capecchi, M. R. (1987) Site-directed mutagenesis by gene targeting in mouse embryo-derived stem cells. *Cell* **51,** 503–512.

8. Wobus, A. M., Wallukat, G., and Hescheler, J. (1991) Pluripotent mouse embryonic stem cells are able to differentiate into cardiomyocytes expressing chronotropic responses to adrenergic and cholinergic agents and Ca^{2+} channel blockers. *Differentiation* **48,** 173–182.

9. Maltsev, V. A., Rohwedel, J., Hescheler, J., and Wobus, A. M. (1993) Embryonic stem cells differentiate in vitro into cardiomyocytes representing sinusnodal, atrial and ventricular cell types. *Mech. Dev.* **44,** 41–50.

10. Maltsev, V. A., Wobus, A. M., Rohwedel, J., Bader, M., and Hescheler, J. (1994) Cardio-myocytes differentiated in vitro from embryonic stem cells developmentally express cardiac-specific genes and ionic currents. *Circ. Res.* **75,** 233–244.

11. Miller-Hance, W. C., LaCorbiere, M., Fuller, S. J., Evans, S. M., Lyons, G., Schmidt, C., et al. (1993) In vitro chamber specification during embryonic stem cell cardiogenesis. *J. Biol. Chem.* **268,** 25244–25252.

12. Rohwedel, J., Maltsev, V., Bober, E., Arnold, H.-H., Hescheler, J., and Wobus, A. M. (1994) Muscle cell differentiation of embryonic stem cells reflects myogenesis in vivo: developmentally regulated expression of myogenic determination genes and functional expression of ionic currents. *Dev. Biol.* **164,** 87–101.

13. Rose, O., Rohwedel, J., Reinhardt, S., Bachmann, M., Cramer, M., Rotter, M., et al. (1994) Expression of M-cadherin protein in myogenic cells during prenatal mouse development and differentiation of embryonic stem cells in culture. *Dev. Dyn.* **201,** 245–259.

14. Strübing, C., Ahnert-Hilger, G., Jin, S., Wiedenmann, B., Hescheler, J., and Wobus, A. M. (1995) Differentiation of pluripotent embryonic stem cells into the neuronal lineage in vitro gives rise to mature inhibitory and excitatory neurons. *Mech. Dev.* **53,** 275–287.

15. Bain, G., Kitchens, D., Yao, M., Huettner, J. E., and Gottlieb, D. I. (1995) Embryonic stem cells express neuronal properties in vitro. *Dev. Biol.* **168,** 342–357.

16. Fraichard, A., Chassande, O., Bilbaut, G., Dehay, C., Savatier, P., and Samarut, J. (1995) In vitro differentiation of embryonic stem cells into glial cells and functional neurons. *J. Cell Sci.* **108,** 3181–3188.

17. Okabe, S., Forsberg-Nilsson, K., Spiro, A. C., Segal, M., and McKay, R. D. G. (1996) Development of neuronal precursor cells and functional postmitotic neurons from embry-onic stem cells in vitro. *Mech. Dev.* **59,** 89–102.

18. Wiles, M. V. and Keller, G. (1991) Multiple hematopoietic lineages develop from embryonic stem (ES) cells in culture. *Development* **111,** 259–267.

19. Hole, N. and Smith, A. G. (1994) Embryonic stem cells and hematopoiesis, in *Culture of Hematopoietic Cells*. (Freshney, R. I., Pragnell, I. B., and Freshney, M. G. eds.), Wiley-Liss, Inc. New York, pp. 235–249.

20. Dani, C., Smith, A. G., Dessolin, S., Leroy, P., Staccini, L., Villageois, P., et al. (1997) Dif-ferentiation of embryonic stem cells into adipocytes in vitro. *J. Cell Sci.* **110,** 1279–1285.

21. Kramer, J., Hegert, C., Guan, K., Wobus, A. M., Müller, P. K., and Rohwedel, J. (2000) Embryonic stem cell-derived chondrogenic differentiation in vitro: activation by BMP-2 and BMP-4. *Mech. Dev.* **92,** 193–205.

22. Bagutti, C., Wobus, A. M., Fässler, R., and Watt, F. (1996) Differentiation of embryonal stem cells into keratinocytes: comparison of wild-type and β_1 integrin-deficient cells. *Dev. Biol.* **179,** 184–196.

23. Risau, W., Sariola, H., Zerwes, H.-G., Sasse, J., Ekblom, P., Kemler, R., and Doetschman, T. (1988) Vasculogenesis and angiogenesis in embryonic stem cell-derived embryoid bodies. *Development* **102,** 471–478.

24. Weitzer, G., Milner, D. J., Kim, J. U., Bradley, A., and Capetanaki, Y. (1995) Cytoskeletal control of myogenesis: a desmin null mutation blocks the myogenic pathway during embryonic stem cell differentiation. *Dev. Biol.* **172,** 422–439.

25. Drab, M., Haller, H., Bychkow, R., Erdmann, B., Lindschau, C., Haase, H., et al. (1997) From totipotent embryonic stem cells to spontaneously contracting vascular smooth muscle cells: a retinoic acid and db-cAMP in vitro differentiation model. *FASEB J.* **11,** 905–915.

26. Wobus, A. M., Guan, K., Jin, S., Wellner, M.-C., Rohwedel, J., Ji, G., et al. (1997) Retinoic acid accelerates embryonic stem cell-derived cardiac differentiation and enhances development of ventricular cardiomyocytes. *J. Mol. Cell. Cardiol.* **29,** 1525–1539.

27. Hescheler, J., Fleischmann, B. K., Lentini, S., Maltsev, V. A., Rohwedel, J., Wobus, A. M., and Addicks, K. (1997) Embryonic stem cells: a model to study structural and functional properties in cardiomyogenesis. *Cardiovasc. Res.* **36,** 149–162.

28. Rohwedel, J., Guan, K., Zuschratter, W., Jin, S., Ahnert-Hilger, G., Fürst, D. O., et al. (1998) Loss of β_1 integrin function results in a retardation of myogenic, but an acceleration of neuronal differentiation of embryonic stem (ES) cells in vitro. *Dev. Biol.* **201,** 167–184.

29. Fässler, R., Rohwedel, J., Maltsev, V., Bloch, W., Lentini, S., Guan, K., et al. (1996) Differentiation and integrity of cardiac muscle cells are impaired in the absence of β_1 integrin. *J. Cell Sci.* **109,** 2989–2999.

30. Rohwedel, J., Horak, V., Hebrok, M., Füchtbauer, E.-M., and Wobus, A. M. (1995) M-twist expression inhibits embryonic stem cell-derived myogenic differentiation in vitro. *Exp. Cell Res.* **220,** 92–100.

31. Robbins, J., Gulick, J., Sanchez, A., Howles, P., and Doetschman, T. (1990) Mouse embryonic stem cells express the cardiac myosin heavy chain genes during development in vitro. *J. Biol. Chem.* **265,** 11905–11909.

32. Sanchez, A., Jones, W. K., Gulick, J., Doetschman, T., and Robbins, J. (1991) Myosin heavy chain gene expression in mouse embryoid bodies. *J. Biol. Chem.* **266,** 22419–22426.

33. Johansson, B. M. and Wiles, M. W. (1995) Evidence for involvement of activin A and bone morphogenetic protein 4 in mammalian mesoderm and hematopoietic development. *Mol. Cell. Biol.* **15,** 141–151.

34. Wobus, A. M., Rohwedel, J., Maltsev, V., and Hescheler, J. (1994) In vitro differentiation of embryonic stem cells into cardiomyocytes or skeletal muscle cells is specifically modulated by retinoic acid. *Roux's Arch. Dev. Biol.* **204,** 36–45.

35. Stewart, C. L., Gadi, I., and Bhatt, H. (1994) Stem cells from primordial germ cells can reenter the germ line. *Dev. Biol.* **161,** 626–628.

36. Rohwedel, J., Sehlmeyer, U., Shan, J., Meister, A., and Wobus, A. M. (1996) Primordial germ cell-derived mouse embryonic germ (EG) cells in vitro resemble undifferentiated stem cells with respect to differentiation capacity and cell cycle distribution. *Cell Biol. Intern.* **20,** 579–587.

37. Wobus, A. M., Kleppisch, T., Maltsev, V., and Hescheler, J. (1994) Cardiomyocyte-like cells differentiated in vitro from embryonic carcinoma cells P19 are characterized by

functional expression of adrenoceptors and Ca^{2+} channels. *In Vitro Cell. Dev. Biol.* **30A,** 425–434.

38. Trower, M. K. and Elgar, G. S. (1994) PCR cloning using T-vectors, in *Protocols for Gene Analysis. Methods in Molecular Biology*, vol. 31. Humana Press, Totowa, N.J., pp. 19–33.

39. Rudnicki, M. A. and McBurney M. W. (1987) Cell culture methods and induction of differentiation of embryonal carcinoma cell lines, in *Teratocarcinomas and Embryonic Stem Cells—a Practical Approach* (Robertson, E. J., ed.), IRL Press, Oxford, pp. 19–49.

40. Edwards, M. K. S., Harris, J. F., and McBurney, M. W. (1983) Induced muscle differentiation in an embryonal carcinoma cell line. *Mol. Cell. Biol.* **3,** 2280–2286.

41. Chomczynski, P. and Sacchi, N. (1987) Single-step method of RNA isolation by acid guanidinium thiocyanate-phenol-chloroform extraction. *Anal. Biochem.* **162,** 156–159.

42. Isenberg, G. and Klöckner, U. (1982) Calcium-tolerant ventricular myocytes prepared by preincubation in a "KB medium". *Pflügers Arch.* **395,** 6–18.

43. Pari, G., Jardine, K., and McBurney, M. W. (1991) Multiple CArG boxes in the human cardiac actin gene promoter required for expression in embryonic cardiac muscle cells developing in vitro from embryonal cardcinoma cells. *Mol. Cell. Biol.* **11,** 4796–4803.

44. Sauer, B. and Henderson, N. (1989) Cre-stimulated recombination at loxP-containing DNA sequences placed into the mammalian genome. *Nucleic Acids Res.* **17,** 147–161.

45. Sauer, B. and Henderson, N. (1990) Targeted insertion of exogenous DNA into the eukaryotic genome by the Cre recombinase. *New Biol.* **2,** 441–449.

46. Chu, G., Hayakawa, H., and Berg, P. (1987) Electroporation for the efficient transfection of mammalian cells with DNA. *Nucleic Acids Res.* **15,** 1311–1326.

47. Potter, H. (1988) Electroporation in biology: methods, application, and instrumentation. *Anal. Biochem.* **174,** 361–373.

48. Ramirez-Salis, R., Davis, A. C., and Bradley, A. (1993) Gene targeting in ES cells, in *Guide to Techniques in Mouse Development*, vol. 225. (Wassarman, P. M. and DePamphelis, M. L., eds.), Academic Press, New York, pp. 855–878.

49. Saiki, R. K., Gelfand, D. H., Stoffel, S., Scharf, S. J., Higuchi, R., Horn, G. T., et al. (1988) Primer-directed enzymatic amplification of DNA with a thermostable DNA polymerase. *Science* **239,** 487–491.

50. Southern, E. M. (1975) Detection of specific sequences among DNA fragments separated by gel electrophoresis. *J. Mol. Biol.* **98,** 503–517.

51. Smith, A. G., Heath, J. K., Donaldson, D. D., Wong, G. G., Moreau, J., Stahl, M., and Rogers, D. (1988) Inhibition of pluripotential embryonic stem cell differentiation by purified polypeptides. *Nature* **336,** 688–690.

52. Smith, D. B. and Johnson, K. S. (1988) Single-step purification of polypeptides expressed in *Escherichia coli* as fusions with glutathione S-transferase. *Gene* **67,** 31–40.

53. Gearing, D. P., Nicola, N. A., Metcalf, D., Foote, S., Wilson, T. A., Gough, N. M., and Williams, R. L. (1989) Production of leukemia inhibitory factor in *Escherichia coli* by a novel procedure and its use in maintaining embryonic stem cells in culture. *BioTechnology* **7,** 1157–1161.

54. Robertson, E. J. (1987) Embryo-derived stem cell lines, in *Teratocarcinoma and Embryonic Stem Cells—a Practical Approach* (Robertson, E. J., ed.), IRL Press, Oxford, pp. 71–112.

55. Myers, T. W. and Gelfand, D. H. (1991) Reverse transcription and DNA amplification by a *Thermus thermophilus* DNA polymerase. *Biochemistry* **30,** 7661–7666.

56. Wong, H., Anderson, W. D., Cheng, T., and Riabowol, K. T. (1994): Monitoring mRNA expression by polymerase chain reaction: the "primer-dropping" method. *Anal. Biochem.* **223,** 251–258.

57. Mortensen, R. M., Conner, D. A., Chao, S., Geisterfer-Lowrance, A. A., and Seidman, J. G. (1992) Production of homozygous mutant ES cells with a single targeting construct. *Mol. Cell. Biol.* **12,** 2391–2395.

58. Hasty, P. and Bradley, A. (1994) Gene targeting vectors for mammalian cells, in *Gene Targeting—A Practical Approach* (Joyner, A. L., ed.), (The Practical Approach Series). IRL Press, Oxford, pp. 1–32.

59. Sambrook, J., Fritsch, E. F., and Maniatis, T. (1989) Analysis and cloning of eukaryotic genomic DNA, in *Molecular Cloning—A Laboratory Manual*, 2nd ed., CSH Laboratory Press, Cold Spring Harbor, N.Y., pp. 9.34–9.37.

60. Seidman, C. E., Bloch, K. D., Klein, K. A., Smith, J. A., and Seidman, J. G. (1984) Nucleotide sequences of the human and mouse atrial natriuretic factor genes. *Science* **226,** 1206–1209.

61. Lints, T. J., Parsons, L. M., Hartley, L., Lyons, I., and Harvey, R. P. (1993) *Nkx-2.5*: a novel murine homeobox gene expressed in early heart progenitor cells and their myogenic descendants. *Development* **119,** 419–431.

62. Montarras, D., Chelly, J., Bober, E., Arnold, H., Ott, M.-O., Gros, F., and Pinset, C. (1991) Developmental patterns in the expression of *Myf5*, MyoD, *myogenin* and *MRF4* during myogenesis. *New Biol.* **3,** 592–600.

63. Wang, D., Villasante, A., Lewis, S. A., and Cowan, N. J. (1986) The mammalian β-tubulin repertoire: hematopoietic expression of a novel, heterologous β-tubulin isotype. *J. Cell Biol.* **103,** 1903–1910.

64. Konecki, D. S., Brennand, J., Fuscoe, J. C., Caskey, C. T., and Chinault, A. C. (1982) Hypoxanthine-guanine phosphoribosyltransferase genes of mouse and Chinese hamster: construction and sequence analysis of cDNA recombinants. *Nucleic Acids Res.* **10,** 6763–6775.

65. Guan, K., Fürst, D. O., and Wobus, A. M. (1999) Modulation of sarcomere organization during embryonic stem cell-derived cardiomyocyte differentiation. *Eur. J. Cell Biol.* **87,** 813–823.

66. Fürst, D. O., Osborn, M., Nave, R., and Weber, K. (1988) The organization of titin filaments in the half-sarcomere revealed by monoclonal antibodies in immunelectron microscopy: a map of the nonrepetitive epitopes starting at the Z-line extends close to the M-line. *J. Cell Biol.* **106,** 1563–1572.

67. Obermann, W. M., Gautel, M., Steiner, F., van der Ven, P. F., Weber, K., and Fürst, D. O. (1996) The structure of the sarcomeric M band: localization of defined domains of myomesin, M-protein, and the 250-kD carboxy-terminal region of titin by immunoelectron microscopy. *J. Cell Biol.* **134,** 1441–1453.

68. Van der Ven, P. F., Obermann, W. M., Lemka, B., Gautel, M., Weber, K., and Fürst, D. O. (2000) The characterization of muscle filamin isoforms suggests a possible role of γ-filamin/ABP-L in sarcomeric Z-disc formation. *Cell Motil. Cytoskeleton* **45,** 149–162.

69. Vinkemeier, U., Obermann, W., Weber, K., and Fürst, D. O. (1993) The globular head domain of titin extends into the center of the sarcomeric M band. cDNA cloning, epitope mapping and immunoelectron microscopy of two titin-associated proteins. *J. Cell Sci.* **106,** 319–330.

70. Bader, D., Masaki, T., and Fischman, D. A. (1982) Immunochemical analysis of myosin heavy chain during avian myogenesis in vivo and in vitro. *J. Cell Biol.* **95,** 763–770.

71. Skalli, O., Gabbiani, G., Babai, F., Seemayer, T. A., Pizzolato, G., and Schurch, W. (1988) Intermediate filament proteins and actin isoforms as markers for soft tissue tumor differentiation and origin. II. Rhabdomyosarcomas. *Am. J. Pathol.* **130,** 515–531.

72. Rudnicki, M. A., Jackowski, G., Saggin, L., and McBurney, M. W. (1990) Actin and myosin expression during development of cardiac muscle from cultured embryonal carcinoma cells. *Dev. Biol.* **138,** 348–358.

73. Müller-Bardorf, M., Freitag, H., Scheffold, T., Remppis, A., Kubler, W., and Katus, H. A. (1995) Development and characterization of a rapid assay for bedside determinations of cardiac troponin T. *Circulation* **92,** 2869–2875.

74. Crow, M. T. and Stockdale, F. E. (1986) The developmental program of fast myosin heavy chain expression in avian skeletal muscles. *Dev. Biol.* **118,** 333–342.

75. Gautel, M., Fürst, D. O., Cocco, A., and Schiaffino, S. (1998) Isoform transitions of the myosin binding protein C family in developing human and mouse muscles: lack of isoform transcomplementation in cardiac muscle. *Circ. Res.* **82,** 124–129.

76. Skalli, O., Ropraz, P., Trzeciak, A., Benzonana, G., Gillessen, D., and Gabbiani, G. (1986) A monoclonal antibody against α-smooth muscle actin: a new probe for smooth muscle differentiation. *J. Cell Biol.* **103,** 2787–2796.

77. Naumann, K. and Pette, D. (1994) Effects of chronic stimulation with different impulse patterns on the expression of myosin isoforms in rat myotube cultures. *Differentiation* **55,** 203–211.

14

Cardiomyocyte Enrichment in Differentiating ES Cell Cultures: *Strategies and Applications*

Kishore B.S. Pasumarthi and Loren J. Field

1. Introduction

Advances in our understanding of cardiomyocyte cell biology have been dependent largely upon the ability to generate primary cultures from enzymatically dispersed fetal, neonatal, or adult hearts. Primary cardiomyocyte cultures recapitulate many of the physiologic and molecular attributes found in intact hearts at the corresponding developmental stage. Moreover, these cultures are readily amenable to a wide variety of physical, physiologic, and molecular analyses. Gene transfer approaches including traditional calcium phosphate and lipofection techniques, as well as viral transduction with recombinant retro-, adeno-, or adeno-associated viruses are also readily accomplished. In light of these attributes, primary cardiomyocyte cultures constitute an extremely versatile experimental system.

Despite their tremendous experimental utility, primary cardiomyocytes cultures are not without weaknesses. For example, it is very difficult to generate large-scale cultures of adult cardiomyocytes which retain a differentiated phenotype. Consequently, studies designed to address issues pertaining to adult cardiac biology frequently utilize primary cultures of fetal or neonatal cardiomyocytes. Given the marked physiologic differences observed in vivo between fetal–neonatal cardiomyocytes as compared to terminally differentiated adult cells, interpretation of such experiments can be problematic. Perhaps the greatest weaknesses of primary cardiomyocyte cultures is that they are subject to strict temporal constraints. This is a result of the relatively rapid cell cycle withdrawal that cardiomyocytes undergo in culture, in contrast to the sustained proliferation observed for noncardiomyocytes (i.e., fibroblasts, vascular smooth muscle cells, etc.). Although enriched cardiomyocyte preparations can be generated, the resulting cultures typically will be contaminated with noncardiomyocytes. Treatments designed to block proliferation of the contaminating cells (e.g., exposure to Ara-C or UV irradiation) can frequently have a deleterious effect on the resident cardiomyocytes as well. Efforts to circumvent this intrinsic limitation have typically relied on enhancing cardiomyocyte proliferative capacity through the targeted expression of oncogenes or proto-oncogenes. Although this approach has yielded a number of cardiomyocyte cell lines which retained varying degrees of differentiation (reviewed in **ref. *1***),

From: *Methods in Molecular Biology, vol. 185: Embryonic Stem Cells: Methods and Protocols*
Edited by: K. Turksen © Humana Press Inc., Totowa, NJ

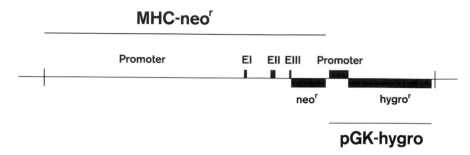

Fig. 1. Structure of the MHC-neo^r/pGK-hygro^r transgene. The transgene comprises an α-MHC-aminoglycoside phosphotransferase (MHC-neo^r) transcriptional unit and pGK-hygro^r transcriptional unit on a common pBM20 vector backbone. The α-MHC promoter consisted of 4.5 kb of 5′ flanking sequence and 1 kb of the gene encompassing exons 1 through 3 up to, but not including, the initiation codon *(19)*. The aminoglycoside phosphotransferase (neo^r) cDNA was subcloned from pMC1-neo poly(A)(Stratagene, La Jolla, CA). The pGK-hygromycin sequences were described previously *(20)*.

sustained oncogene expression prohibits studies on many aspects related to the terminal differentiation program.

In vitro differentiation of embryonic stem (ES) cells provides an alternative source of cardiomyocytes for study in tissue culture. ES cells are derived from the inner cell mass of preimplantation embryos and, when grown under appropriate conditions, can be propagated in an undifferentiated state indefinitely. Culturing ES cell in suspension and in the absence of differentiation inhibitors, such as leukemia inhibitory factor (LIF), results in the formation of multicellular structures called embryoid bodies (EBs), which reproducibly contain most if not all ectoderm-, endoderm-, and mesoderm-derived cell lineages *(2,3*; see also other chapters in this series). In many instances, regions of cardiomyogenesis are readily apparent, as evidenced by the presence of spontaneous contractile activity. Numerous studies have shown that cardiogenic induction in ES cells faithfully recapitulates the physical and molecular properties of developing myocardium in vivo *(4–11)*. In addition, cell cycle withdrawal and terminal differentiation in ES-derived cardiomyocytes appears to parallel closely that which occurs during normal development *(12)*.

In this report, we describe a relatively simple genetic enrichment approach that facilitates the generation of highly enriched cultures of cardiomyocytes from differentiating ES cells *(13)*. The approach relies on the use of two transcriptional units introduced into undifferentiated ES cells on a common vector backbone (*see* **Fig. 1**). The first transcriptional unit comprises a promoter expressed in undifferentiated ES cells linked to a marker gene suitable for enrichment of cells carrying the DNA (in our example, the phosphoglycerate kinase [PGK] promoter and a cDNA encoding resistance to hygromycin [hygro^r] were used; the transcriptional unit is designated pGK-hygro^r). The second transcriptional unit comprises a cell lineage restricted promoter linked to a marker gene suitable for enrichment of the desired cells (in our example, the cardiomyocyte-restricted α-cardiac myosin heavy chain [α-MHC] promoter and a cDNA encoding aminoglycoside phosphotransferase were used; the transcriptional unit is designated MHC-neo^r).

Using this approach, the generation of highly enriched cardiomyocyte cultures is experimentally quite simple (**Fig. 2**). Undifferentiated ES cells are transfected with the MHC-neo^r/pGK-hygro^r construct. Cells incorporating the DNA are enriched based on their resistance to hygromycin. Differentiation is then induced, and once evidence of cardiomyogenesis is observed (i.e., spontaneous contractile activity), the cultures are treated with geneticin (G418). Because the α-MHC promoter is only active in cardiomyocytes, only these cells express aminoglycoside phosphotransferase and survive G418 treatment (**Fig. 2**). Although the example presented utilizes antibiotic resistance as the basis of the enrichment, a wide variety of analogous marker genes–enrichment protocols can readily be used (e.g., green fluorescent protein [GFP] targeted expression of cell surface markers, which could be used in conjunction with fluorescence-activated cell sorting [FACS] protocols, etc.).

The genetic enrichment approach has the advantage that very long-term cultures of terminally differentiated cardiomyocytes can be generated, since noncardiomyocytes are eliminated from the culture. Moreover, the approach is easily amenable to gene transfer, either prior to differentiation or, alternatively, after the generation of terminally differentiated cells. In addition, the genetic enrichment approach is applicable to all cell lineages derived from ES cells, as well as to all multipotent stem cell systems. Here, we present a detailed description of the genetic enrichment protocols used to produce essentially pure populations of cardiomyocytes from differentiating ES cells. The Methods section is divided into subsections describing: *(1)* the transfection and selection of undifferentiated ES cells; *(2)* the "en mass" differentiation of the selected cells; *(3)* the generation of highly enriched cardiomyocyte cultures; and *(4)* the use of periodic acid Schiff's (PAS) staining to visualize colonies of ES-derived cardiomyocytes. In addition, we provide a practical example of the use of the system to generate cardiomyocytes of sufficient purity for intracardiac engraftment (an emerging protocol aimed at restoring systolic function in diseased hearts). We also provide an example of how the system can be utilized to perform an ES-derived cardiomyocyte colony growth assay, which provides a relatively rapid throughput system to identify genes that impact on cardiomyocyte cell cycle regulation.

2. Materials

1. ES cell line ES-D3 (obtained from the American Type Culture Collection, Rockville, MD).
2. pMHC-neo^r/PGK-hygro^r plasmid.
3. Restriction enzymes *Xho*I and *Hind*III.
4. Geneclean kit (Bio 101, cat. no. 1001-400).
5. Cell culture dishes (Corning, cat. no. 430293).
6. ES cell grade fetal bovine serum (FBS) (Gibco, cat. no. 10439-024).
7. Dulbecco's modified Eagle's medium (DMEM) (Sigma, cat. no. D-6546).
8. ESGRO/mLIF (murine leukemia inhibitory factors) (Chemicon International, cat. no. ESG-1107).
9. Nonessential amino acids (Gibco, cat. no. 11140-050), L-Glutamine (Gibco, cat. no. 25030-081), and Penicillin–Streptomycin solutions (Gibco, cat. no. 15070-063).
10. 2-Mercaptoethanol (Sigma, cat. no. M-7522).
11. Phosphate-buffered saline (PBS) (Sigma, cat. no. D-8537).
12. Trypsin solution: 0.025% trypsin (1:250) (Gibco, cat. no. 27250-042), 1 mM EDTA and 1% chicken serum (Sigma, cat. no. 16110-082) in sterile PBS.
13. Electroporator (Gibco, cat. no. 71600-19).

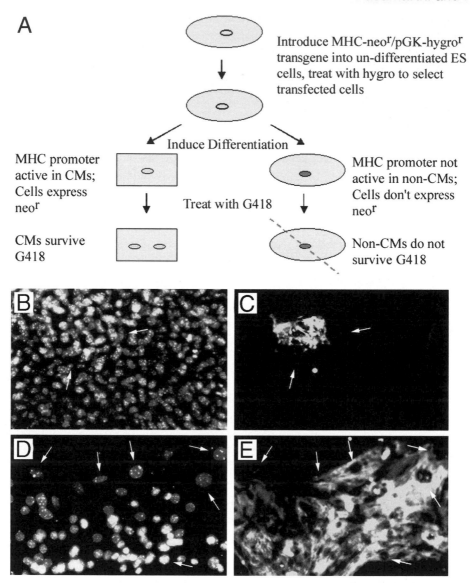

Fig. 2. **(A)** Schematic diagram of the genetic enrichment approach. **(B)** Hoechst epifluorescence of a nonselected culture of ES cells carrying the MHC-neo[r]/pGK-hygromycin transgene 16 d after differentiation was induced. Note the high density of cells present in the field. **(C)** Antisarcomeric MHC immunofluorescence (green signal) of the same field depicted in panel **B**. Note that only a small percentage of the cells are cardiomyocytes (arrows demarcate the same group of cells in panels **B** and **C**). **(D)** Hoechst epifluorescence of a G418-selected culture of ES cells carrying the MHC-neo[r]/pGK-hygromycin transgene 16 d postcardiogenic induction; note the reduction in total cell number as compared to the un-selected cultures depicted in panel **B**. **(E)** Antisarcomeric myosin immunofluorescence (green signal) of the same field depicted in panel **D**. Note that all of the cells present in the G418-selected culture express sarcomeric myosin (arrows demarcate the same group of cells in panels **D** and **E**). (*See* color plate 3, following p. 254).

14. Hygromycin B (Calbiochem, cat. no. 400051).
15. G418 (Roche, cat. no. 1464990).
16. Falcon bacterial Petri dishes (Becton Dickinson, cat. no. 1029).
17. ES growth medium: DMEM containing 15% heat-inactivated FBS, 0.1 mM nonessential amino acids, 2 mM glutamine, 50 U/mL penicillin, 50 μg/mL streptomycin, 0.1 mM 2-mercaptoethanol, and 10^3 U/mL LIF.
18. Differentiation Medium A: DMEM containing 15% heat-inactivated FBS, 0.1 mM nonessential amino acids, 2 mM glutamine, 50 U/mL penicillin, 50 μg/mL streptomycin, and 0.1 mM 2-mercaptoethanol.
19. Differentiation Medium B: DMEM containing 20% heat-inactivated FBS, 0.1 mM nonessential amino acids, 2 mM glutamine, 50 U/mL penicillin, 50 μg/mL streptomycin, and 0.1 mM 2-mercaptoethanol.
20. Differentiation Medium C: DMEM containing 20% heat-inactivated FBS, -.1 mM nonessential amino acids, 2 mM glutamine, 50 U/mL penicillin, 50 μg/mL streptomycin, 0.1 mM 2-mercaptoethanol, and 200 μg/mL G418.
21. Formaldehyde (37%) (Sigma cat. no. F-1268).
22. Ethanol (95%).
23. Periodic acid (Sigma, cat. no. P-7875).
24. Schiff's reagent (Sigma, cat. no. 395-2-016) stored at 4°C until expiration date.
25. Sodium metabisulfite solution: 1% sodium metabisulfite (Sigma, cat. no. S-1516), 0.05 N HCl in distilled water; made fresh.

3. Methods

3.1. Transfection and Selection of Undifferentiated ES Cells

1. Prior to transfection of ES cells, digest the selection cassette (pMHC-neo^r/PGK-hygro^r) with *XhoI*/*Hind*III and isolate the 8.8-kb fragment containing the entire MHC-neo^r/PGK-hygro^r sequence using a Gene cleankit.
2. ES cells are routinely maintained in an undifferentiated state by culturing them in the ES Growth Medium.
3. Dissociate cells using trypsin, count, and resuspend 4×10^6 cells in 0.8 mL of ES Growth Medium. Transfer the cells into an electroporation chamber and leave on ice.
4. Mix MHC-neo^r/PGK-hygro^r DNA (1 μg) and 25 μg sonicated salmon testes DNA in a total volume of 70 μL, add this mixture to the cells, and leave the electroporation chamber on ice for 15 min.
5. Electroporate the cells (180 V, 800 μF) and leave on ice for 15 min.
6. Plate the cells in 100-mm Corning dishes (6×10^5 cells/dish) in ES Growth Medium for 24 h.
7. Aspirate the medium the next day and switch the cells to ES Growth Medium supplemented with 200 μg/mL hygromycin B.
8. Change the medium daily and select transfected cells over a period of 7 d. Cells may be trypsinized and replated into new dishes if a plate becomes confluent.

3.2. "En Mass" Differentiation of Transfected ES Cells

1. After 7 d of hygromycin selection, dissociate the cells using trypsin and plate 4×10^6 cells in a 100-mm bacterial Petri dish in 10 mL of Differentiation Medium A. Cells will grow in suspension under these conditions (*see* **Note 1**).
2. Supplement the cells with 5 mL of Differentiation Medium A on the next day, to facilitate EB formation.

3. On the third day, transfer the medium containing EBs using a 10-mL pipet into a sterile 50-mL cell culture tube and allow the EBs to settle by gravity. Aspirate the medium, resuspend the EBs in 10 mL of fresh Differentiation Medium A, and plate in a new bacterial Petri dish.
4. Supplement cells with 5 mL of Differentiation Medium A on the next day (*see* **Note 2**).
5. Collect EBs on the fifth day by gravity and resuspend in 10 mL of Differentiation Medium B. Plate EBs in 100-mm Corning cell culture dishes at different dilutions (1:2, 1:5, 1:10 etc).
6. Change the medium daily; regions of cardiogenesis can be readily identified by the presence of spontaneous contractile activity within 4–6 d of EB attachment.

3.3. Selection of Cardiomyocyte Restricted Lineages

1. For enrichment of cardiomyocyte restricted lineages, cultures exhibiting spontaneous contractile activity are grown in Differentiation Medium C.
2. Cultures can be grown in Differentiation Medium C for as long as required to eliminate noncardiomyocytes (*see* **Note 3**).

3.4. Use of PAS Staining to Visualize Colonies of ES-Derived Cardiomyocytes

1. For a rapid visualization of areas of cardiogenic induction, the PAS reaction can be used. Cardiomyocytes are rich in glycogen, and the PAS reaction is based on the oxidative action of periodic acid on glycol groups present in glucose residues giving rise to aldehyde groups. These aldehyde groups react with Schiff's reagent producing a new complex compound with a purple or magenta color.
2. Aspirate the medium, rinse cells in PBS, and fix with formyl alcohol (9:1 mixture of 95% ethanol and formaldehyde) for 15 min at room temperature.
3. Aspirate the fixative, rinse once with 95% ethanol, air-dry the plate, and rinse twice with tap water.
4. Add 1% periodic acid to the plate and incubate for 10 min at room temperature.
5. Rinse twice with tap water.
6. Add Schiff's reagent and incubate for 10 min in the hood (*see* **Note 4**).
7. Rinse the plate with sodium metabisulfite solution for 2 min in the hood; repeat three times.
8. Rinse twice with tap water.
9. Areas with cardiogenic induction can be readily visualized by intense purple color staining (*see* **Fig. 3**).

3.5. Practial Examples Illustrating the Utility of the Technique

3.5.1. Example 1: Generation of Intracardiac Grafts with ES-Derived Cardiomyocytes

Many forms of cardiovascular disease are characterized by progressive loss of cardiomyocytes, which result in decreased systolic function. The ability to increase the number of functional cardiomyocytes in diseased hearts might have a positive impact of systolic function (*14*). One approach to accomplish this relies on the physical delivery of donor cardiomyocytes into the diseased heart, with the expectation that such cells can form stable functional grafts. Proof of concept studies with donor cells from mouse, dog, and rat demonstrated that fetal cardiomyocytes can form stable intracardiac grafts, and at least in the case of the mouse model (*15,16*), ultrastructural analyses revealed the presence of physical attributes necessary for force and action potential propagation

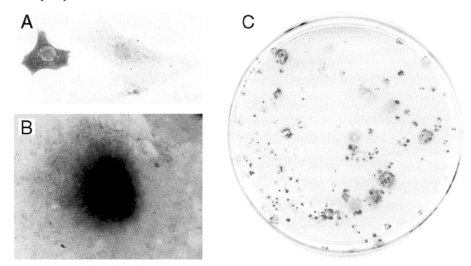

Fig. 3. PAS staining provides rapid assessment of cardiomyocyte yield in differentiating ES cells. (**A**) PAS staining of enzymatically dispersed cells prepared from a differentiating ES cell culture. Image shows a cardiomyocyte that stained dark purple due to its high glycogen content. In contrast, the nonmyocyte present in the same field was not stained. (**B**) Image of a PAS-stained cardiomyocyte colony following the genetic enrichment protocol. (**C**) Image of a PAS-stained tissue culture dish of ES-derived cardiomyocytes following the genetic enrichment protocol. Note the presence of many independent colonies. (*See* color plate 4, following p. 254).

between donor and host cardiomyocytes (i.e., fascia adherens, desmosomes, and gap junctions).

While intracardiac engraftment of fetal cardiomyocytes holds great promise for functional augmentation in diseased hearts, identification of a suitable source of donor cells is quite problematic for clinical application in humans. Cardiomyocytes derived from differentiating ES cells could constitute a suitable surrogate source of donor cells for therapeutic intracardiac engraftment. Accordingly, a proof of concept experiment was initiated in mice (*13*). Cardiomyocytes derived from ES cells using the genetic selection approach described in this Subheading were digested with trypsin, and approx 1 $\times 10^4$ cells were delivered into the left ventricular free wall of dystrophic adult muscular dystrophy (mdx) recipient mice. The mdx mice harbor a mutation in the dystrophin gene and show no immune reactivity to antidystrophin antibodies. Consequently, the fate of the engrafted cardiomyocytes, which expressed a wild-type dystrophin gene, could easily be monitored by antidystrophin immune histology. Phase contrast microscopic examination of cryosections from the recipient hearts revealed that engrafted regions frequently exhibited normal myocardial topography (**Fig. 4A**). Immune cytologic assays with antidystrophin antibody revealed the presence of dystrophin-positive G418-selected cardiomyocytes (**Fig. 4B**). Comparison of phase contrast and antidystrophin images revealed the presence of myofibers in the engrafted ES-derived cardiomyocytes.

Additional analyses with other antidystrophin antibodies, as well as transgene-specific polymerase chain reaction (PCR) analysis of material harvested from the graft-bearing regions of the cryosections, confirmed that the dystrophic immune reactivity

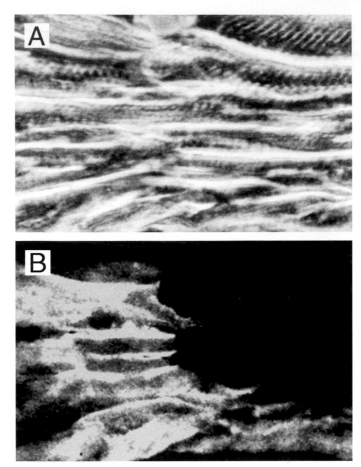

Fig. 4. Genetically enriched cardiomyocytes form stable intracardiac grafts. Phase contrast image **(A)** and antidystrophin immune fluorescence **(B)** of the same field from an mdx heart engrafted with G418-selected cardiomyocytes. Dystrophin immune reactivity appears as a green signal in panel **B**. (*See* color plate 5, following p. 254).

corresponded to engrafted ES-derived cardiomyocytes. Grafts were observed for as long as 7 wk post-implantation, which was the latest time point analyzed. Thus, the use of the genetic enrichment approach provides a suitable source of donor cardiomyocytes for intracardiac engraftment. Current research efforts are focused on enhancing graft size in experimental models of heart disease. If this approach is successful at restoring partial or complete systolic function in appropriate animal models, the recent development of human ES *(17)* raises the possibility that the approach can be tested therapeutically.

3.5.2. Example 2: Development of an ES-Derived Cardiomyocyte Colony Growth Assay

Cell cycle reactivation (i.e., induced proliferation) constitutes an alternative strategy through which to increase the number of functional cardiomyocytes in a diseased heart. The premise of the approach is that increased myocardial mass resulting from the proliferation of differentiated functional cardiomyocytes will augment systolic function. The successful application of this approach requires the identification of gene products that are capable of reactivating the cell cycle in genetically naive terminally differentiated cardiomyocytes. It has proven to be somewhat difficult to identify exploitable cardiomyocyte cell cycle regulatory genes. Although adenoviral-mediated gene transfer in primary cultures have identified some genes that promote cardiomyocyte growth, long-term analyses of the transfected cardiomyocytes in these cultures was limited because of the intrinsic proliferative capacity of the resident noncardiomyocytes. The use of transgenic mice with targeted expression in the heart or, alternatively, the use of retroviral transfection in developing avian hearts has identified several genes that are capable of inducing cardiomyocyte proliferation *(1,18)*. However, these latter approaches are both time-consuming and can incur significant expenses related to animal husbandry charges.

By incorporating a slight modification of the genetic enrichment protocol, coupled with the use of a muscle-specific cytologic stain suitable for tissue culture dishes, it should be possible to generate an ES-derived cardiomyocyte colony growth assay. Such an assay would provide a comparatively high-throughput, as compared to in vivo gene transfer approaches, and also would not be subject to the temporal constraints encountered with gene transfer approaches in primary cardiomyocyte cultures. As a proof of concept, we compared the activities of simian virus 40 (SV40) Large T-Antigen (T-Ag) vs the adenoviral E1A oncoproteins directly in the ES-derived cardiomyocyte colony growth assay. Previous studies, performed largely in transgenic mice with either wild-type or a conditionally active mutant T-Ag, have shown that expression of T-Ag gene (TAG) in the atrial or ventricular myocardium during fetal life is sufficient to induce sustained cardiomyocyte proliferation, while expression after terminal differentiation does not induce cell cycle activation. In contrast, studies utilizing adenoviral transduction of primary cardiomyocytes indicate that expression of E1A at any point of cardiac development can induce cell cycle reentry (as evidenced by the initiation of DNA synthesis), followed by a rapid apoptotic response.

To compare the activities of T-Ag and E1A directly in the same system, undifferentiated ES cells were cotransfected with the MHC-neo^r/pGK-hygro^r and an MHC-T-Ag expression construct or, alternatively, with the MHC-neo^r/pGK-hygro^r and an MHC-E1A expression construct. Control cultures were transfected with the MHC-neo^r/pGK-hygro^r transgene only. After hygromycin selection, differentiation was induced, and cardiomyocytes were enriched as described above. Following 52 d of culture in G418 (a total of 60 d following the induction of differentiation), the dishes were rinsed, fixed, and stained with PAS as described above. The effect of oncogene expression was easily scored by simple visualization (**Fig. 5**). The density of cardiomyocyte colonies in the control dishes (MHC-neo^r/pGK-hygro^r transgene only) is used for reference. In good agreement with the data obtained via adenoviral transfection of primary cardiomyocytes, targeted expression of E1A in the ES-derived cardiomyocyte growth assay induced apoptosis, as evidenced by the marked reduction of cardiomyocyte colonies, as well as the presence of pronounced DNA fragmentation (**Fig. 5**). In contrast, targeted expression of T-Ag in the ES-derived cardiomyocyte growth assay resulted in fulminate proliferation, with the dish appearing as a nearly confluent

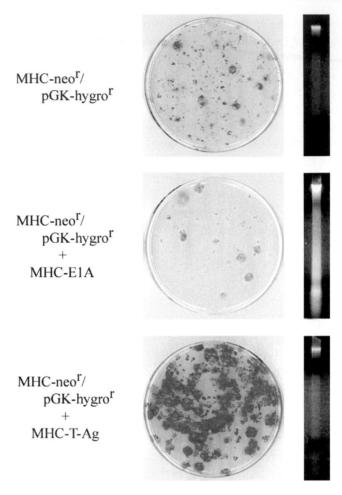

MHC-neo^r/
pGK-hygro^r

MHC-neo^r/
pGK-hygro^r
+
MHC-E1A

MHC-neo^r/
pGK-hygro^r
+
MHC-T-Ag

Fig. 5. Use of the ES-derived cardiomyocyte colony growth assay to monitor the effects of gene transfer on cardiomyocyte proliferation. Undifferentiated ES cells were transfected with the MHC-neo^r/pGK-hygro^r alone or in combination with MHC-E1A or MHC-T-Ag expression constructs. Transfected cells were processed via the genetic enrichment protocol as described in the text, and at 60 d after differentiation, the cultures were fixed and stained with PAS. In addition, DNA fragmentation was monitored as an indirect indication for apoptosis. DNA prepared from parallel cultures was analyzed by agarose gel electrophoresis and visualized by ethidium bromide staining and UV illumination. Note the marked decrease in cardiomyocyte colonies in cells transfected with the MHC-E1A construct and the marked increase in cardio-myocyte colonies in cells transfected with the MHC-T-Ag construct as compared to the control dish. Note also the presence of extensive DNA fragmentation (as evidenced by the small molecular weight species) in DNA prepared from MHC-E1A transfected cells, but not in the control nor the MHC-T-Ag transfected cultures. (*See* color plate 6, following p. 254).

synchronously beating mass (**Fig. 5**). In agreement with results obtained in transgenic animals, T-Ag-induced proliferation was not accompanied by pronounced cardiomyo-cyte dedifferentiation, nor was an apoptotic response apparent (**Fig. 5**). Thus, the ES-derived cardiomyocyte growth assay described here faithfully recapitulates the results of gene transfer observed in other experimental systems and appears to be suitable for relatively high-throughput screens of the activity of cardiomyocyte cell cycle regulatory proteins.

3.6. Conclusions

The genetic enrichment approach described in this chapter facilitates the generation of essentially pure cardiomyocyte cultures. Importantly, several groups have already used the approach in cardiomyocytes *(13,19)*, attesting to both its reproducibility and relative ease of utilization. Moreover, it is clear that the enrichment protocol works in other cell lineages, as evidenced by the recent generation of relatively pure cultures of neurons *(20)* as well as insulin-secreting pancreatic β-cells *(21)*. This approach should thus be useful for generation of a wide variety ES-derived cells suitable for both in vitro and in vivo applications.

4. Notes

1. Test several types/batches of bacterial Petri dishes to screen for lots that exhibit minimal EB attachment.
2. Change medium at all stages with care to avoid detachment of differentiating ES cell clusters.
3. The genetic enrichment approach described above is highly reproducible, relatively straightforward in nature, and can yield cardiomyocyte cultures in excess of 99% purity. As indicated in the Introduction, it is well established that ES-derived cardiomyocytes are highly differentiated. Similarly, molecular, immune cytologic, and ultrastructural analyses all indicate that genetically enriched cardiomyocyte cultures share these attributes *(13)*. Moreover, cardiomyocytes from traditional murine ES cultures, as well as those from the genetically enriched cultures, follow cell cycle withdrawal and terminal differentiation programs, which are temporally similar to the programs observed during mouse embryonic development *(12)*. Finally, the cultures can be maintained in vitro for as long as 11 mo while still retaining spontaneous contractile activity, thereby eliminating most of the temporal restraints encountered with traditional primary cardiomyocyte cultures.
4. Filter Schiff's reagent prior to use, bring it to room temperature, and take precautions not to inhale vapors. Incubations with Schiff's reagent and subsequent washes during PAS staining should be performed in the hood.

References

1. Pasumarthi, K. B. S. and Field, L. J. (2001) Strategies to identify cardiomyocyte cell cycle regulatory genes, in *Molecular Approaches to Heart Failure Therapy* (Hasenfuss, G. and Marban, E., eds.), Steinkopff, Verlag, Dormstadt. 333–351.
2. Doetschman, T. C., Eistetter, H., Katz, M., Schmidt, W., and Kemler, R. (1985) The in vitro development of blastocyst-derived embryonic stem cell lines: formation of visceral yolk sac, blood islands and myocardium. *J. Embryol. Exp. Morphol.* **87**, 27–45.
3. Risau, W., Sariola, H., Zerwes, H. G., Sasse, J., Ekblom, P., Kemler, R., and Doetschman, T. (1988) Vasculogenesis and angiogenesis in embryonic-stem-cell-derived embryoid bodies. *Development* **102**, 471–478.
4. Sanchez, A., Jones, W. K., Gulick, J., Doetschman, T., and Robbins, J. (1991) Myosin heavy chain gene expression in mouse embryoid bodies. An in vitro developmental study. *J. Biol. Chem.* **266**, 22,419–22,426.
5. Muthuchamy, M., Pajak, L., Howles, P. L., Doetschman, T., and Wieczorek, D. F. (1993) Developmental analysis of tropomyosin gene expression in embryonic stem cells and mouse embryos. *Mol. Cell. Biol.* **13**, 3311–3323.

6. Miller-Hance, W. C., LaCorbiere, M., Fuller, S. J., Evans, S. M., Lyons, G., Schmidt, C., et al. (1993) In vitro chamber specification during embryonic stem cell cardiogenesis. Expression of the ventricular myosin light chain-2 gene is independent of heart tube formation. *J. Biol. Chem.* **268,** 25,244–25,252.

7. Ganim, J. R., Luo, W., Ponniah, S., Grupp, I., Kim, H. W., Ferguson, D. G., et al. (1992) Mouse phospholamban gene expression during development in vivo and in vitro. *Circ. Res.* **71,** 1021–1030.

8. Boer, P. H. (1994) Activation of the gene for type-b natriuretic factor in mouse stem cell cultures induced for cardiac myogenesis. *Biochem. Biophys. Res. Commun.* **199,** 954–961.

9. Metzger, J. M., Lin, W.-I., and Samuelson, L. C. (1994) Transition in cardiac contractile sensitivity to calcium during the in vitro differentiation of mouse embryonic stem cells. *J. Cell Biol.* **126,** 701–711.

10. Maltsev, V. A., Rohwedel, J., Hescheler, J., and Wobus, A. M. (1993) Embryonic stem cells differentiate in vitro into cardiomyocytes representing sinusnodal, atrial and ventricular cell types. *Mech. Dev.* **44,** 41–50.

11. Wobus, A. M., Wallukat, G., and Hescheler, J. (1991) Pluripotent mouse embryonic stem cells are able to differentiate into cardiomyocytes expressing chronotropic responses to adrenergic and cholinergic agents and Ca^{++} channel blockers. *Differentiation* **48,** 173–182.

12. Klug, M. G., Soonpaa, M. H., and Field, L. J. (1995) DNA synthesis and multinucleation in embryonic stem cell-derived cardiomyocytes. *Am. J. Physiol.* **269,** H1913–H1921.

13. Klug, M. G., Soonpaa, M. H., Koh, G. Y., and Field, L. J. (1996) Genetically selected cardiomyocytes from differentiating embryonic stem cells form stable intracardiac grafts. *J. Clin. Invest.* **98,** 216–224.

14. Reinlib, L. and Field, L. J. (2000) Transplantation: Future Therapy for Cardiovascular Disease? An NHLBI Workshop. *Circulation* **101,** e182–e187.

15. Soonpaa, M. H., Koh, G. Y., Klug, M. G., and Field, L. J. (1994) Formation of nascent intercalated discs between grafted fetal cardiomyocytes and host myocardium. *Science* **264,** 98–101.

16. Koh, G. Y., Soonpaa, M. H. Klug, M. G., Pride, H. P., Zipes, D. P. Cooper, B. J., and Field, L. J. (1995) Stable fetal cardiomyocyte grafts in the hearts of dystrophic mice and dogs. *J. Clin. Invset.* **96,** 2034–2042.

17. Thomson, J. A., Itskovitz-Eldor, J., Shapiro, S. S., Waknitz, M. A., Swiergiel, J. J., Marshall, V. S., and Jones, J. M. (1998) Embryonic stem cell lines derived from human blastocysts. *Science* **282,** 1145–1147.

18. Franklin, M. and Field, L. J. (1997) Use of Transgenic Animals in Cardiovascular Toxicology Testing, in *Comprehensive Toxicology, Vol. 6. Cardiovascular Toxicology* (Bishop, S. P., McQueen, C. A., and Gandolfi, A. J., eds.), Cambridge University Press, Cambridge, U. K., pp. 201–212.

19. Minamino, T., Yujiri, T., Papst, P. J., Chan, E. D., Johnson, G. L., and Terada, N. (1999) MEKK1 suppresses oxidative stress-induced apoptosis of embryonic stem cell-derived cardiac myocytes. *Proc. Natl. Acad. Sci. USA* **96,** 15,127–15,132.

20. Li, M., Pevny, L., Lovell-Badge, R., and Smith, A. (1998) Generation of purified neural precursors from embryonic stem cells by lineage selection. *Curr. Biol.* **8,** 971–974.

21. Soria, B., Roche, E., Berna, G., Leon-Quinto, T., Reig, J. A., and Martin, F. (2000) Insulin-secreting cells derived from embryonic stem cells normalize glycemia in strptozotocin-induced diabetic mice. *Diabetes* **40,** 157–162.

15

Embryonic Stem Cells as a Model for the Physiological Analysis of the Cardiovascular System

Jüyrgen Hescheler, Maria Wartenberg, Bernd K. Fleischmann, Kathrin Banach, Helmut Acker, and Heinrich Sauer

1. Introduction

Embryonic stem (ES) cells have the potential to proliferate infinitely in vitro in an undifferentiated and pluripotent state, thereby maintaining a relatively normal and stable karyotype even with continual passaging. Remarkably, in vivo, ES cells can be reincorporated into normal embryonic development by transfer into a host blastocyst or aggregation with blastomere stage embryos. They can contribute to all tissues in the resulting chimeras including gametes. When cultivated in vitro, ES cells differentiate under appropriate cell culture conditions, i.e., in the absence of leukemia inhibitory factor (LIF) into cell types of all three germ layers: endoderm, ectoderm and mesoderm *(1)*. However, these differentiation processes occur only when ES cells are cultivated in suspension culture in which they grow to multicellular spheroidal tissues, termed embryoid bodies (EBs).

In recent years, a large number of different cell types have been described to differentiate within EBs. These include hematopoietic *(2,3)* and endothelial cells *(4,5)*, cartilage *(6)*, neurons *(7–9)* as well as smooth *(10)*, skeletal *(11)* and cardiac *(12)* muscle cells. The capacity of ES cells to differentiate into cell types of the mesodermal cell lineage has been extensively used by us and others to investigate the molecular and physiological events occurring during the process of differentiation into cells of the cardiovascular system, i.e., cardiac and endothelial cells *(4,5,12–17)*. We have observed cardiac development within EBs as early as 7 d after formation of the aggregates, which correlates well with the murine embryo, where the first beating is seen on day E8.5 to E9.5 (i.e., 8.5 to 9.5 d postcoitum) *(18)*. It was found that the earliest detectable cardiomyocytes (stage 0) were not beating, but expressed already voltage-dependent L-type Ca^{2+} channels at low density *(15)*. During the further developmental stages (stage 1–4), spontaneous contracting activity occurs, and the increasing number of different ion channels causes a diversification of cardiac phenotypes finally leading to the known specialized cardiomyocytes that are found in the neonatal heart, i.e., ventricular-like, atrial-like, and sinus-nodal-like, as well as Purkinje-like cardiomyocytes *(19,20)*. Cardiomyogenesis in EBs is paralleled by the development of vascular structures

From: *Methods in Molecular Biology, vol. 185: Embryonic Stem Cells: Methods and Protocols*
Edited by: K. Turksen © Humana Press Inc., Totowa, NJ

starting on d 5 of differentiation and resulting in the formation of hollow capillary-like tubules within 3 to 4 d, which improves the supply of nutrients and oxygen to EBs, as well as its export of catabolic end products *(4)*.

The development of a primitive cardiovascular system giving the EB an embryo-like organization has raised the idea of using the EB to screen for embryotoxic and teratogenic agents *(21,22)*. Furthermore, EBs have been recently introduced as a novel in vitro assay system to study the effectiveness of anti-angiogenic agents, which have been proven to efficiently inhibit tumor growth in in vitro as well as in vivo anticancer trials *(4)*. With the recent isolation of ES cells of human origin by Thomson and coworkers *(23)*, a renewable tissue culture source of human cells capable of differentiating into a variety of different cell types is available. Their capacity to differentiate into cardiac and endothelial cells may be potentially useful for tissue transplantation and cell replacement strategies and may be exploited for the treatment of infarcted hearts and other cardiac disorders.

Research by use of the ES cell technology requires sophisticated techniques for tissue cultivation and physiological as well as biophysical cell analysis. In this chapter, we describe recent advances of our laboratory in cell culture protocols and techniques for the analysis of the function of the cardiovascular system developed in EBs. We report on the recently developed spinner flask technique, that allows mass culturing of EBs with highly synchronized and efficient differentiation of cardiac and endothelial cells. For the investigation of diffusion processes via the vascular structures differentiated in EBs, the optical probe technique was elaborated. This technique permits the quantification of fluorescent tracers in the depth of the three-dimensional tissue by an optical sectioning routine based on confocal laser scanning microscopy and a mathematical algorithm to correct for the attenuation of fluorescence light in the depth of the tissue. Furthermore, we report on electrophysiological techniques for the characterization of the differentiation of cardiac cell lineages, i.e., the patch-clamp technique and the micro-electrode array (MEA) technique. The patch-clamp technique has provided deep insights into the formation of electrical activity and the underlying ion channels and signaling cascades during cardiomyogenesis. Whole mount EBs can be plated onto MEAs, and the electrical signals of the field potentials are recorded over several days from a multitude of electrodes beneath the spontaneously contracting tissue. The MEA technique has been proven fruitful for the analysis of action potential propagation, the characterization of pacemaker activity, the development of intercellular communication, and the analysis of arrythmia within cardiac cell areas of spontaneously beating EBs *(24,25)*.

2. Materials

2.1. Maintenance of Mouse ES Cells

1. Complete growth medium (1X stored at 4°C). To 760 mL Iscove's modified Dulbecco's medium (Gibco BRL, Karlsruhe, Germany, cat. no. 42200-048) add: 10 mL Glutamax I (100X) (Gibco, BRL, cat. no. 35050-038) 10 mL penicillin–streptomycin solution (Gibco BRL, cat. no. 15070-022) 10 mL minimum essential medium (MEM) nonessential amino acid solution (100X) (Gibco BRL, cat. no. 11140-035), 100 μM 2-mercaptoethanol (Sigma, Deisenhofen, Germany, cat. no. M6250), and 200 mL fetal calf serum (FCS) (Gibco BRL, cat. no. 10270106).

2. Phosphate-buffered saline (PBS) Ca^{2+} and Mg^{2+}-free: 137 mM NaCl, 2.7 mM KCl, 10.1 mM Na_2HPO_4, 2 mM KH_2PO_4, pH 7.4, filter-sterilized.
3. Trypsin-EDTA in PBS (Gibco, BRL, cat. no. 45300-019).

2.2. Spinner Flasks

1. Cellspin stirrer system (Integra Biosciences Fernwald, Germany) equipped with 250-mL spinner flasks.
2. Sigmacote® solution (Sigma, cat. no. SL-2), stored at 4°C.
3. 5 N NaOH solution.

2.3. Investigation of Fluorochrome Distributions and Diffusion Coefficients

1. Cell culture medium: to 860 mL Nutrient Mixture Ham's F-10 medium (Gibco BRL, cat. no. 81200) add: 10 mL Glutamax I (100X) (Gibco BRL, cat. no. 35050-038), 10 mL penicillin–streptomycin solution (Gibco BRL, cat. no. 15070-022), 10 mL MEM nonessential amino acid solution (100X) (Gibco BRL, cat. no.11140-035), 100 μM 2-mercaptoethanol (Sigma, cat. no. M6250), and 100 mL FCS (Gibco BRL, cat. no. 10270106).
2. PBS, Ca^{2+}- and Mg^{2+}-free: 137 mM NaCl, 2.7 mM KCl, 10.1 mM Na_2HPO_4, 2 mM KH_2PO_4, pH 7.4, filter sterilized.
3. E1-buffer: 135 mM NaCl, 5.4 mM KCl, 1.8 mM $CaCl_2$, 1 mM $MgCl_2$, 10 mM glucose, 10 mM HEPES, pH 7.4, filter-sterilized.
4. Doxorubicin (Sigma, cat. no. D1515).

2.4. Isolation of Cardiomyocytes from EBs

1. Dissociation buffer: 120 mM NaCl, 5.4 mM KCl, 5 mM $MgSO_4$, 0.03 mM $CaCl_2$, 5 mM Na pyruvate, 20 mM glucose, 20 mM taurine, 10 mM HEPES, pH 6.9, with NaOH.
2. KB buffer: 85 mM KCl, 30 mM K_2HPO_4, 5 mM $MgSO_4$, 1 mM EDTA, 2 mM Na_2ATP, 5 mM pyruvate, 5 mM creatine, 20 mM taurine, 20 mM glucose, pH 7.2, with KOH.
3. Collagenase B (Boehringer Mannheim, Mannheim, Germany, cat. no. 1088807).

2.5. Patch-Clamp Recording

1. Patch-clamp amplifier, model Axopatch 200A (Axon Instruments, Foster City, CA).
2. Data aquisition software, Iso2 (MFK, Niedernhausen, Germany).
3. Extracellular solution for the recording of I_{Ca} : 135 mM NaCl, 5 mM KCl, 10 mM $CaCl_2$, 2 mM $MgCl_2$, 5 mM HEPES, 10 mM glucose, pH 7.4, with NaOH.
4. Pipet solution for the recording of I_{Ca}: 55 mM CsCl, 80 mM Cs_2SO_4, 2 mM $MgCl_2$, 10 mM HEPES, 10 mM EGTA, 1 mM $CaCl_2$, 5 mM ATP (Mg), pH 7.4, with CsOH.
5. Amphotericin B (Sigma, cat. no. A4888) solution. Dissolve amphotericin B (7.5 mg) as stock solution in 125 μL dimethyl sulfoxide (DMSO).
6. Borosilicate glass capillaries with filament (Clark Electromedical Instruments, Reading, UK, cat. no. GC150TF-10).

2.6. Ca²⁺ Imaging

1. Computer-controlled monochromator (TIL Photonics, Planegg, Germany).
2. Inverted microscope (Axiovert 135M; Carl Zeiss, Jena, Germany) equipped with a 40X oil-immersion objective (Zeiss), a fura-2 filterblock (TIL Photonics), a 470/525-nm interference filter, and an intensified charge-coupled device (CCD) camera (Thetha, München, Germany).

3. Software for data recording and analysis, Fucal fluorescence software package (TIL Photonics).
4. Fura-2,AM, fura-2-free acid (Molecular Probes, Eugene, OR, cat. no. F-1201).

3. Methods

3.1. Mass Culture Techniques for EBs: the Spinner Flask Technique

Routine screening of embryotoxic and/or anti-angiogenic agents, as well as transplantation of organ-specific cells, requires large numbers of EBs differentiated from ES cells. In most laboratories, the hanging drop method is applied, which was previously developed by Wobus and coworkers *(26)*, and has been proven useful for the characterization of the developmental aspects of cardiomyogenesis *(27)*, skeletal *(11)*, and smooth muscle *(10)* development, as well as neurogenesis *(8)*. However, this technique is limited by the low yield of EBs, since a single hanging drop has to be prepared manually for each EB. To achieve mass cultures of EBs with a high efficiency of differentiation, a spinner flask culture technique was recently developed in our laboratory. It was observed that, by using this technique, large amounts of EBs (about 1000/spinner flask) could be cultivated. EBs cultivated in spinner flask culture grew to a significantly larger size, as compared with cultures of EBs in methylcellulose-supplemented cell culture medium or cultivation using the liquid overlay technique, thereby indicating an improved supply with oxygen and nutrients (**Fig. 1**). The spinner flask technique proved efficient for the differentiation of cells of the endothelial cell and the cardiac cell lineage. For the evaluation of the differentiation of the endothelial cell lineage, EBs were fixed after different times of cultivation and immunostained using an antibody directed against platelet endothelial cell adhesion molecule (PECAM-1). It was demonstrated that 100% of EBs differentiated vascular structures with a more synchronized time course of differentiation as compared to conventional cell culture protocols (**Fig. 2A**). The spinner flask technique proved similarly efficient for the differentiation of the cardiac cell lineage. When EBs were cultivated using the spinner flask technique and were plated to cell culture dishes after 7 d, approx 90% of the EBs developed spontaneous beating areas of cardiomyocytes, which were positive for cardiac-specific contractile proteins (**Fig. 2B**).

3.1.1. Procedure for Spinner Flask Preparation Prior to the Addition of ES Cells

1. Wash clean spinner flasks with excessive Milli-Q-plus water and dry for 1 h at 60°C.
2. Siliconize spinner flasks by moistening the interior as well as the mallets with Sigmacote. Excessive Sigmacote is removed from the flasks using a 10-mL glass pipet.
3. Dry the silicon coat in an oven for 1 h at 120°C.
4. Rinse spinner flasks three times with 250 mL Milli-Q-plus water and autoclave subsequently.
5. Moisten the interior of the flasks with 20 mL of complete Iscove's medium prior to the addition of ES cells; exchange the medium for 125 mL of complete Iscove's medium.

3.1.2. Procedure for the Cleaning of the Spinner Flask at the End of the Experiment

After the end of the experiment, the spinner flasks have to be cleaned prior to the inoculation with fresh ES cells.

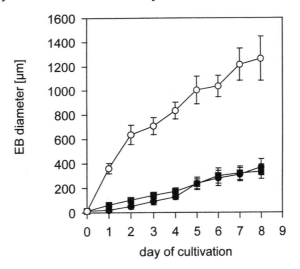

Fig. 1. Growth kinetics of EBs cultivated in either spinner flask culture (○) or Petri dish suspension culture in normal cell culture medium (■) or methylcellulose-supplemented medium (●). Reproduced with permission from *(4)*.

1. Remove the old medium with residual EBs and cells.
2. Wash the flasks with 70% ethanol and subsequently with 1 L water.
3. Remove the silicon coat by adding 250 mL 5 N NaOH to the spinner flasks for a maximum of 12 h.
4. Remove NaOH and wash the flasks with at least 5 L water. The interior of the flasks is thoroughly cleaned with a brush. Subsequently, rinse the flasks with 1 L of Milli-Q-plus water.

3.1.3. Inoculation of Spinner Flasks with ES Cells

1. Wash ES cells grown in 6-cm cell culture Petri dishes once with 0.2% trypsin and 0.05% EDTA in PBS.
2. Remove the trypsin solution and incubate ES cells for 5 min with 2 mL 0.2% trypsin and 0.05% EDTA.
3. Triturate the cells with a 2-mL glass pipet until the ES cell clusters are dissociated and a single-cell suspension is achieved (control under a microscope).
4. Prepare spinner flasks (as described under **Subheading 3.1.1.**) and seed ES cells at a density of 1×10^7 cells/mL in 125 mL complete Iscove's cell culture medium. Stir at a speed of 20 rotations/minute. The stirring direction is reversed every 1440°.
5. Add 125 mL Iscove's complete cell culture medium after 24 h to yield a final volume of 250 mL.
6. Exchange 125 mL of the cell culture medium every day.
7. For the differentiation of cardiomyocytes, EBs are carefully removed from the spinner flask by the use of 10-mL plastic pipets to avoid any cell injury. They are subsequently plated to 10-cm cell culture dishes filled with 20 mL complete Iscove's cell culture medium. After 24–48 h, spontaneous beating activity of the EBs indicates the differentiation of cardiac cells.

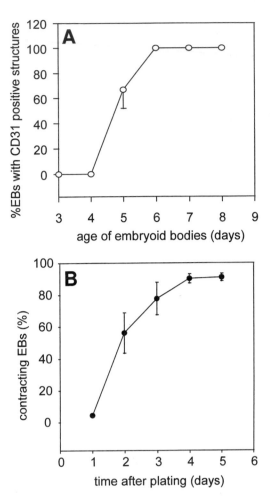

Fig. 2. Time-course of endothelial (**A**) and cardiac (**B**) cell differentiation in EBs. Endothelial cell differentiation was evaluated by anti-PECAM-1 immunohistochemistry and confocal laser scanning microscopy. For cardiac cell differentiation, EBs were cultivated for 7 d in spinner flasks. They were subsequently plated onto glass coverslips, and the EBs exerting spontaneous contracting activity were counted.

3.2. Investigation of the Distribution of Fluorescent Molecules in the 3-Dimensional Tissue of EBs: the Optical Probe Technique

Fluorescent tracers can be used in multicellular tissues to study diffusion processes. Moreover, fluorescent indicators are generally applied for the evaluation of physiological parameters in cell cultures and tissue slices, e.g., intracellular pH, Ca^{2+}, Mg^{2+}, Na^+, and Cl^- ion concentrations, nitric oxide, and reactive oxygen species, as well as the membrane potential. The optical probe technique was recently developed in our laboratory for the study of the diffusion properties of fluorochromes in thick biological specimens, the distribution of vital–lethal fluorescence dyes, and fluorescent anthracyclines, i.e., doxorubicin in the depth of the tissue as well as the fluorescence distribution of fluorescent ion indicators *(28,29)*. The optical probe technique (*see* **Fig. 3**) is based on confocal laser scanning microscopy and an optical sectioning routine

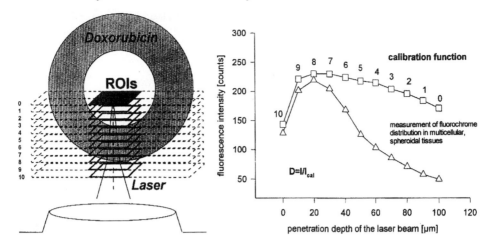

Fig. 3. Principle of the optical probe technique. The left side of the diagram shows a scheme of the objective and confocal laser beam scanning 11 consecutive optical sections (dashed lines) starting from the center of a doxorubicin-stained spherical multicellular tissue and moving towards the periphery. The mean field fluorescence intensity is determined in ROIs of about 600 μm^2 in the center of each optical section (traced squares). The right side of the diagram shows tracings of fluorescence values plotted vs the penetration depth of the laser beam in cither densely packed cell pellets (open squares) or the spherical multicellular tissue (open triangles). The numbers following the tracings of fluorescence in the depth of the tissue denote the sequence of optical sections. The fluorescence distribution coefficient $D = I/I_{Cal}$ was calculated as the ratio of fluorescence intensity I in a discrete depth from the periphery of the spheroid and the fluorescence intensity I_{Cal} obtained equidistantly in the densely packed cell pellet. Note that optical sectioning starts in the center of the multicellular object. Reproduced with permission from *(28)*.

and allows the determination of fluorescence distributions in the tissue down to a depth of approximately 300 µm. A mathematical algorithm was elaborated, which corrects for light absorbtion and scattering, and allows the quantification of fluorescence in the depth of a 3-dimensional tissue. By use of the optical probe technique, we recently determined the diffusion coefficients of the fluorescent anthracycline doxorubicin in the vital tissue of vascularized EBs, in EBs treated with anti-angiogenic agents, and in avascular tumor spheroids *(4,30)*.

3.2.1. Dissociation of Single Cells from EBs and Multicellular Tumor Spheroids and Production of Densely Packed Cell Pellets

1. Wash either EBs or multicellular tumor spheroids in PBS and place the tissues in a Petri dish containing 1 mL 0.2% trypsin and 0.05% EDTA in PBS.
2. Incubate the objects for 10 min at 37°C and agitate every 30 s.
3. Triturate the tissues with a 1-mL Eppendorf pipet and stop the enzymatic reaction by addition of 4 mL fresh cell culture medium.
4. Centrifugate at 80g for 5 min and resuspend the cells in cell culture medium.
5. For the preparation of densely packed cell pellets, incubate the dissociated cells for 4 h in bicarbonate-buffered F10 medium supplemented with 10% FCS at 37°C.
6. Triturate the cells and incubate for 2 h with different concentrations of the respective fluorescence dye ranging from 2–60 µM.

7. After dye loading, wash the cells 3 times in serum-free HEPES-buffered F10 medium.
8. Centrifugate for 10 min at 80g and store the cell pellets at room temperature prior to examination.

3.2.2. Determination of Fluorescence Distribution Coefficients

1. Stain either EBs or tumor spheroids for 1–2 h with the respective fluorescence dye. Wash the culture objects in E1 buffer and transfer them in a liquid vol of 300 µL on 24 × 60 mm coverslips to the stage of an inverted confocal laser scanning microscope (LSM 410; Carl Zeiss, Jena, Germany).
2. Adjust the motor commands of the stepper motor of the confocal setup to steps of 10 µm in z-direction. Adjust the pinhole settings of the confocal setup to achieve a full-width half-maximum of 8 µm.
3. Select a 25X oil-immersion corrected objective (Plan-Neofluar; Carl Zeiss) with a numerical aperture of 0.8.
4. Perform single x-y-scans in selected regions of interest (ROIs) in the center of the spherical objects and move in 10 µm steps to the periphery of the object. The area of the ROIs is selected to 600 µm^2 (40 × 40 pixel). Each ROI is scanned once in 0.064 seconds. A whole z-series of 16 ROIs is achieved in approx 7 s, the convolution time of the stepper motor thereby being the rate limiting step.
5. Determine the mean field intensity of the fluorescence signal in each ROI and plot the fluorescence intensity data as a function of the penetration depth of the laser beam in the tissue (see open triangles in **Fig. 3**).
6. In order to correct the measured fluorescence intensity traces for fluorochrome and depth-dependent light absorption and scattering, determine attenuation coefficients C using homogenously stained densely packed cell pellets (see open squares in **Fig. 3**). This results in a fluorescence distribution coefficient $D = I/I_{cal}$, which is calculated as the ratio of fluorescence intensity I, at a discrete depth from the periphery of the spheroid, and the fluorescence intensity I_{cal} obtained equidistantly in the densely packed cell pellet. The distribution coefficients become $D = 1$, when the tissue is homogenously stained, comparable to the situation in the densely packed cell pellet. When the tissue is less stained or unstained, the distribution coefficient shows values lower than 1. In heterogenously stained samples the coefficient D differs in every optical section z.

3.2.3. Correction Procedure for Light Attenuation in the Depth of the Tissue

The correction for light attenuation within multicellular tissues loaded with doxorubicin is performed using the following algorithm:

$$\ln (I_{corr}) = \ln (I) + C \cdot D(z) \cdot z$$

where I_{corr} is the fluorescence intensity after correction for light attenuation in a discrete depth z within the fluorochrome containing specimen, and I is the fluorescence intensity before correction. The fluorescence distribution coefficient $D(z) = I/I_{cal}$ is obtained by dividing the measured fluorescence intensity I at a discrete depth z of the laser beam in the tissue by the respective intensity value I_{cal} in the calibration function, i.e., the light attenuation function of the densely packed cell suspension. D ranges between 0 and 1. It approximates 1 in homogenously stained cell layers ($I = I_{cal}$) and approximates 0 where consecutive optical sections reach unstained cell layers.

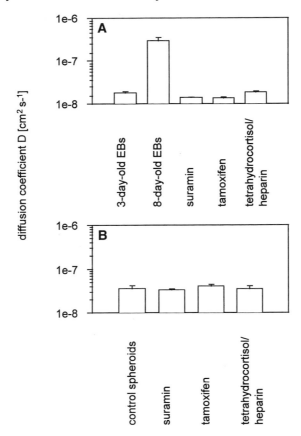

Fig. 4. Diffusion coefficients D for doxorubicin as evaluated by the optical probe technique. (**A**) Diffusion coefficients for doxorubicin were determined in 3-d-old avascular EBs, 8-d-old vascularized EBs, and 8-d-old EBs treated with antiangiogenic agents. (**B**) Diffusion coefficients for doxorubicin were determined in 8-d-old avascular multicellular prostate tumor spheroids (control) and 8-d-old spheroids treated with antiangiogenic agents. Reproduced with permission from *(4)*.

3.2.4. Evaluation of Dynamic Processes of Fluorochrome Distribution: Use of the Optical Probe Technique for the Study of Diffusion

By use of the optical probe technique, diffusion processes of fluorescent molecules in 3-dimensional tissues can be determined, and diffusion coefficients can be calculated from these data. We have recently used this approach to determine the diffusion coefficients for doxorubicin in the vascularized tissue of EBs and EBs treated with anti-angiogenic agents, and for comparison in the avascular tissue of multicellular tumor spheroids *(4,30)* (**Fig. 4**). For diffusion studies, EBs and multicellular tumor spheroids are incubated at room temperature for 10, 30, and 60 min with 10 µM doxorubicin. They are then washed, and doxorubicin fluorescence is determined by the optical probe technique as described in **Subheading 3.2.4.** The time-dependent fluorescence increase is recorded in a depth of 80 µm from the spheroid periphery.

For the determination of diffusion coefficients, the doxorubicin distribution is evaluated by use of the optical probe technique after a 60-min incubation with doxorubicin. From the maximal diffusion distance of doxorubicin fluorescence from the tissue periphery, diffusion coefficients are calculated according to the Einstein-Smoluchovski equation: $D = x^2/2t$, where D is the diffusion coefficient, x is the maximal diffusion distance of doxorubicin from the spheroid periphery, and t is the diffusion time of 60 min.

3.3. The Patch-Clamp Technique and Ca^{2+} Imaging for the Single-Cell Analysis of Cardiomyocytes

One distinct advantage of the investigation of ES cell-derived cardiomyocytes is that, unlike in mammal-derived embryonic cardiomyocytes, their functional phenotype can be investigated starting from very early stages of cardiomyogenesis *(14,20)*. Using patch-clamp as well as single-cell Ca^{2+} imaging techniques allows, for the first time, the ability to characterize the correlation between ion channel expression and the electrical activity of early embryonic cardiomyocytes. Thus, by using these techniques not only the detailed development-dependent expression of ion channels could be described, but also their possible involvement in the determination of action potential shape and electrical activity *(19)*. This also enables the functional characterization of cardiomyocytes deficient in genes (i.e., β1 integrin) *(31)*, thus resulting in early embryonic death.

Because ion channels can be used as sensitive tools for the analysis of signaling cascades *(32,33)*, we have used in our studies the voltage-dependent L-type Ca^{2+} current (I_{Ca}) as well as the hyperpolarization-activated nonselective cation current (I_f) to investigate the establishment of various signaling cascades, in particular muscarinic and β-adrenergic regulation (*see* **Fig. 5**). These studies demonstrate that the inhibitory muscarinic regulation is present starting from early stages (stage 0 to 1), whereas the β-adrenergic regulation is becoming only fully established at later stages (stage 3 to 4) of development *(16,17,34)*, which is similar to findings in mammals *(35)*. Furthermore, we could also identify a development-dependent switch in the muscarinic signaling cascade *(16)*. These studies are particularly interesting because pathologically transformed cells can recapitulate their embryonic phenotype.

To successfully perform these physiological experiments on isolated cells, however, cell cultivation, as well as isolation procedures, need to be well established to provide appropriate cell quality. In addition, because within the EB all different cell types are generated, it is mandatory to have clear criteria to identify the cell type of interest or to alter cell culture conditions for the preferential generation of one tissue type, as shown for neuronal differentiation *(36)*. In the past, spontaneously beating cells were recognized as cardiomyocytes *(19)* and used for patch-clamp experiments. However, this approach excluded from the analysis cardiomyocytes prior to the initiation of beating and also differentiated ventricular cardiomyocytes known to lack spontaneous electrical activity *(19)*. We have, therefore, established stably transfected ES cell lines, in which the expression of live reporter genes, such as the enhanced green fluorescent protein (EGFP), is under control of a tissue-specific promoter, allowing recognition of the cells of interest based on their fluorescence under excitation light *(15)*.

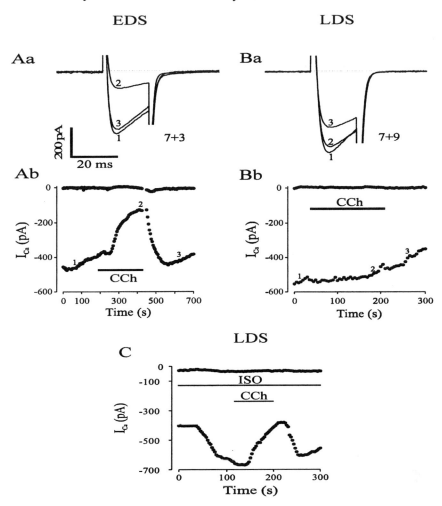

Fig. 5. Basal I_{Ca} was strongly depressed upon application of the muscarinic agonist carbachol (CCh, 1 μM) in early development stage cardiomyocytes (**Aa, Ab**), whereas no effect of CCh on basal I_{Ca} was detected in late stage cardiomyocytes (**Ba, Bb**). In contrast, the time course of I_{Ca} amplitude in a late stage cell shows that isoprenaline (0.1 μM) strongly stimulates I_{Ca}, and CCh (1 μM) depresses I_{Ca} (**C**). I_{Ca} was evoked by applying 20 ms lasting depolarizing voltage steps to 0 mV from a holding potential of –50 mV. Reproduced with permission from *(16)*.

While electrophysiological experiments have been performed on isolated ES cell-derived skeletal muscle cells *(11)*, smooth muscle cells *(10)*, and neurons within the whole EB *(9)*, we will focus on cardiomyocytes, because most data have been accumulated on this topic.

3.3.1. Isolation of Cardiomyocytes from EBs

Cardiomyocytes are derived from pluripotent ES cell lines (i.e., D3;R1). EBs are generated by cultivating the ES cells first for 2 d in hanging drops (400 cells/20 μL), then for 5 d in suspension, and finally plated for 1–15 d on gelatin-coated glass coverslips. About 12–24 h after plating, spontaneously contracting cell clusters appear.

1. For the preparation of isolated cardiomyocytes, dissect beating areas of 20–30 EBs under the microscope using a microscalpel.
2. Disperse the tissue by enzymatic digestion using collagenase B (0.5–1 mg/mL dissolved in dissociation buffer) (*see also* **refs. *19,37***).
3. Sediment the digested tissue by gravity and remove the supernatant.
4. Resuspend cells in 200 µL KB buffer and incubate for 1 h at 37°C.
5. Triturate the tissue until a homogenous cell suspension is achieved.
6. Plate the cell suspension onto gelatine-coated glass coverslips in 24-multiwell dishes filled with cell culture medium and store in an incubator. Within the first 12 h, cells attach to the glass surface. Ideally, the isolated cells are functionally characterized 24–48 h after dissociation (*see* **Note 1**).

3.3.2. Patch-Clamp Recording

ES cell-derived cardiomyocytes are well suited for electrophysiological experiments since they are small (cellular capacitance ranging between 10–25 pF) and lack the T-tubular system at least during early stages *(39)*. This leads to an improvement of the quality of voltage clamp conditions and facilitates intracellular dialysis. Moreover, upon appropriate enzymatic dissociation, gigaohm-seals are relatively easily obtained even without fire-polishing the pipet tip. In order to preserve the biophysical characteristics and the function of the ion channels of interest, it is recommended to perform the experiments at a temperature of 35–37°C. The pipets are made on a DMZ Universal Puller (DMZ, München, Germany) from 1.5-mm borosilicate glass capillaries with filaments yielding a resistance between 2–4 MΩ. The composition of the different recording solutions varies in dependence of the ion channel of interest. For the recording of I_{Ca}, the solutions described in **Subheading 2.5.** are used. The series resistance reaches a steady-state level of 10–20 MΩ within 5–15 min after obtaining the gigaohm seal. For the recording of I_{Ca}, voltage-clamped cells are held at –50 mV, and trains of depolarizing pulses lasting 20 ms are applied to a test potential of 0 mV at a frequency of 0.2 Hz (*see* **Note 2**). Current–voltage (I/V) relationships are determined by applying 150 ms lasting depolarizing voltage steps from test potentials of –40 mV to +40 mV in 10-mV steps.

3.3.3. Ca²⁺ Imaging

For the recording of changes in the intracellular Ca^{2+} concentration of ES cell-derived cardiomyocytes, different approaches can be taken. Our group has had good experience with single-cell Ca^{2+} imaging techniques using a monochromator as the fluorescence excitation source. For the exclusion of movement artifacts, we have preferentially used the ratiometric dye fura-2 *(41)*.

1. Load cardiomyocytes on coverslips either with fura-2-acid (50 µM, equilibration after 3–5 min dialysis) via the patch-pipet or with the cell permeate form fura-2,AM (1 µM) for 12–15 min at 37°C in cell culture medium. In case organelle loading is observed, fura-2,AM loading at room temperature for about 30 min is recommended.
2. Transfer the coverslips to a temperature-controlled recording chamber and superfuse for 10 min by gravity at a rate of 1 mL/min with extracellular solution (*see* **Subheading 2.5.**). A 90% volume exchange should be achieved within approx 10 s.

3. For monitoring changes of the intracellular Ca^{2+} concentration ($[Ca^{2+}]_i$) use oil immersion objectives, preferentially 40X. Monochromic excitation light (340, 380 nm) is generated by the monochromator at a frequency of 50 Hz. The emitted fluorescence is imaged through a 470-nm interference filter using an intensified CCD camera connected to the TV port of the microscope. Fluorescence images (50–100 ms exposure time) are acquired at different rates.

4. For conversion of the ratiometric data into $(Ca^{2+})_i$, *in situ* calibration is performed according to Grynkiewicz *(41)*. R_{max} is obtained in the presence of ionomycin and 10 mM Ca^{2+}, and R_{min} is in the presence of excess EGTA. Representative values are determined in a series of experiments and these average values are used to calculate $[Ca^{2+}]_i$. Average values used for R_{max}, R_{min}, and $F380_{max}/F380_{min}$ are 4.3, 0.44, and 3.94, respectively, for embryonic stem cells, and 2.18, 0.33, 3.57, respectively, for EBs and the dissociation constant is assumed to be 224. When changes of $[Ca^{2+}]_i$ during spontaneous contractions are monitored, fast acquisition rates (50–100 Hz) are required for temporal resolution. To avoid dye bleaching at these high acquisition rates, we have used neutrodensity filters or defocusing of the excitation light source.

3.3.4. Experiments Using Intracellular Dialysis of Various Substances

For the investigation of the modulation of ion channels and/or the functional expression of inositoltriphosophate (IP$_3$)-sensitive Ca^{2+} stores, substances such as cAMP, cGMP, protein kinase inhibitor, catalytic subunit of protein kinase A, GTP-γ-S, as well as IP$_3$ are added to the patch pipet. After obtaining a gigaohm seal, the modulation, for example, of I_{Ca} or I_f can be monitored separately or in combined voltage protocols using appropriate solutions *(16,17)*. In order to observe changes occurring during intracellular dialysis with the above mentioned substances, it is mandatory to initiate current recordings immediately upon obtaining the whole cell conformation.

When changes of $[Ca^{2+}]_i$ are monitored while dialyzing the cell with fura-2-acid, excitation wavelengths of 360/390 nm are used. The ensuing changes of the dye concentration are taken into account by monitoring the fluorescence emission intensity at the isobestic wavelength (360 nm), which is independent of $[Ca^{2+}]_i$ *(41)*.

3.4. MEAs to Monitor the Development of Excitability in Clusters of Cardiomyocytes Differentiating from ES Cells

MEAs find broad application in the study of excitable multicellular preparations. In rat hippocampus slice preparations, local field potentials and spike measurements allow the analysis of network activity with high spatiotemporal resolution *(42)*. It enables a tissue-dependent description of long-term potentiation or the long-term effect of drug application.

Different techniques were applied up to now to describe the excitation spread in preparations of cardiac muscle cells. Rohr and coworkers *(43)*, for example, examined the excitation spread in monolayer cultures of neonatal cardiomyocytes on a cellular level by the help of voltage-sensitive dyes and multisite optical mapping. With this approach, they analyzed the dependence of excitation spread on tissue geometry and excitability *(44)*. In whole organ preparations *(45)*, excitation spread can be monitored by optical mapping, thereby resolving the involvement of the cardiac structure in the occurrence of arrhythmia or reentrant phenomena.

In the approach described in this chapter, an in vitro system is used to differentiate ES cells into beating clusters of cardiomyocytes and cultivation on substrate-integrated microelectrodes allows for the long-term recordings on a single preparation. Electrophysiological studies on single cells isolated from these clusters of cardiomyocytes *(19,24)* could demonstrate that the cells express ion channels characteristic for cardiac muscle cells. The action potentials recorded with the current clamp technique correlate with those described for sinus node, atrium, or ventricle, underlining the ability of ES cells to differentiate into various tissue forms *(27)*. Cultivation of the ES cells on MEAs will enable us to combine long-term culture of differentiating cardiomyocytes with the characterization of their electrical activity.

3.4.1. Tissue Culture

In culture, ES cells are maintained on feeder layers with the addition of LIF. For the preparation of EBs, a cell suspension is prepared and then cultivated in hanging drops as described previously *(26)*. After 2 d, the drops are washed off the lid of the culture dish, and the 3-dimensional cell aggregates are maintained in suspension for another 5 d. A schematic drawing of the cultivation steps is shown in **Fig. 6**. After 7 d, the EBs are transferred into the culture dishes containing the MEA recording system. The culture dishes are filled with Iscove's medium, and, by careful manipulation with a pipet, the EBs are placed in the middle of the electrode array. After 1 d of incubation, the EBs attach to the bottom of the dish and start to grow out, they form a 3-dimensional tissue layer on top of the electrodes. At d 9, spontaneously beating areas of cardiomyocytes can be observed, and field potentials can be recorded.

3.4.2. Electrophysiological Recording by Using MEAs

We use the recording system of Multi Channel Systems (Reutlingen, Germany), which consists of electrode arrays integrated in a culture dish, a heated recording system, a pre- and filter-amplifier, an A/D board (MC_Card), and a computer program (MCRack) for the acquisition and analysis of data.

The MEA consists of 60 substrate-integrated Ti/TiN electrodes with a diameter of 30 μm, arranged in an 8×8 array with an interelectrode spacing of 200 μm (*see* **Fig. 6**). The array is integrated in a reusable culture dish with an insulating cover of silicon nitrite. The system can be used for recordings from acute tissue preparations as well as for long-term culture. The data are sampled (up to 50 kHz) and can be analyzed on-line from all 60 channels with a PC-based acquisition system (MC_Rack).

3.4.3. What Do We Learn from MEA Recordings?

The MEA electrodes record a field potential that reflects the electrical activity of the cluster of cells growing on top of the electrode. Action potentials can be detected as negative spikes. In **Fig. 7** on the left, original voltage traces, representing a 2-s recording from cardiomyocytes, differentiated from ES cells can be seen. Each voltage trace represents the recording from one microelectrode. A large area of spontaneously active cells covered the electrode field. Corresponding to the location of this area, electrical activity could be detected on almost 3/4 of the electrodes. Two areas with distinct beating frequencies can be distinguished. By the analysis of the delay in the occurrence of the spike at the different electrodes, the direction of excitation spread can

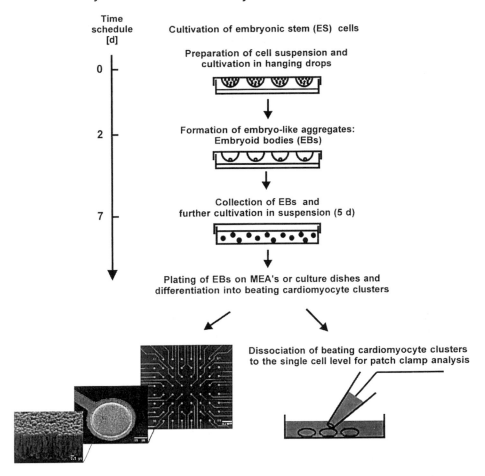

Fig. 6. Cultivation of ES cell in the hanging drop technique. After 7 d of differentiation in the 3-dimensional embryo-like structure of the EB, the EBs are plated for further differentiation. On the lower left, the MEA is shown with additional magnification of one single Ti/TiN electrode. Besides using the whole 3-dimensional tissue for MEA analysis, differentiated cardiomyocytes can be dissociated and used for patch-clamp analysis.

be determined. The origin of excitation, as well as the direction of excitation spread, is stable in the differentiated clusters of cardiomyocytes. Moreover, Igelmund and coworkers *(25)* could show that failures of propagation occur at narrow tissue pathways and by the Ca^{2+} channel blocker nimodipine.

On the left side of **Fig. 7**, the original voltage trace of one electrode is magnified. The different components of the spike, representing the excitation or action potential of the cells covering the electrode can be seen. By the application of specific ion channel blockers, the different components of the spike can be characterized. Whereas the negative spike amplitude depends on the voltage-dependent Na^+ current, the duration of the spike plateau can be modulated by Ca^{2+} and K^+ channel blockers. The careful characterization of the signal will allow us to determine the effect of pharmacological agents on multicellular preparations of cardiomyocytes at the ion channel level. Besides the analysis of wild-type cells, the ES cell technique also allows us to examine cells

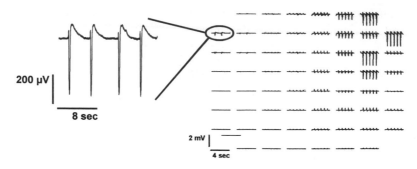

Fig. 7. Electrical activity recorded from cardiomyocytes differentiated from ES cells. On the right, original voltage traces recorded with the MEA system are shown. Each trace represents the recording from one microelectrode. On the left, a voltage trace is shown in higher magnification. The different components of the spike, like the maximum negative and the following maximum positive deflection, can be identified. These two components determine the plateau length of the spike.

that are genetically manipulated. In this way, we could also follow up the development of beating clusters of cardiomyocytes that are homozygous negative for the expression of Connexin 43, which is the major gap junction protein of ventricular muscle cells. These experiments are of special interest, since homozygous mice of this type are not viable *(46)*.

Since the MEA system is noninvasive, the cells in culture are not damaged by the experimental procedure. This enables us to follow up a single culture for several days and weeks and to monitor the excitability and excitation spread under developmental aspects or under the long-term influence of pharmacologic agents. Therefore, the ES cells in combination with the MEA provide an ideal system for the description of excitability and excitation spread in 3-dimensional preparations of cardiomyocytes.

3.5. Conclusion

There is an increasing number of techniques to evaluate the fundamental properties of ES cells differentiating within EBs. It should be mentioned that the functional analysis of differentiating ES cells is of increasing importance, since functionally intact cells are needed in approaches that use organotypic cells derived from ES cells as a source for cell or tissue transplantation. Scientific emphasis should, therefore, be drawn, not only on cell culture protocols that yield specific cell lineages with reproducible efficiency, but also on the basic physiological characterization of the ES cell-derived organotypic cells.

4. Notes

1. Since the functional characteristics of the isolated cardiomyocytes critically depend on their differentiation stage, the time after plating defines the stage of the development of the cells (for detail see **ref. 38**). For the preparation of very early stage (stage 0–1) cardiomyocytes, cells are plated for 2 d, for early stages 3 to 4 d and for the differentiated stage in which cells display action potentials typical of nodal-, atrial-, or ventricular-like cardiomyocytes, a plating time of 9–15 d is required. Although after longer plating times, cardiomyocytes can be still detected, enzymatic dispersion is increasingly complicated by the accumulation

of interstitial tissue requiring longer digestion periods with collagenase. This results in a significant deterioration of cell quality and poorer quality of electrophysiological data.

2. Under the above described conditions, pronounced run-down of I_{Ca} is observed. We have, therefore, used the perforated patch-clamp technique *(40)* using amphotericin B. The pipet is backfilled with amphotericin B containing intracellular solution, and the pipet tip is shortly put into amphotericin-free intracellular solution. The final concentration of amphotericin B in the pipet solution is 1 mg/mL.

References

1. Evans, M. J. and Kaufman, M. H. (1981) Establishment in culture of pluripotential cells from mouse embryos. *Nature* **292,** 154–156.
2. Wiles, M. V. and Keller, G. (1991) Multiple hematopoietic lineages develop from embryonic stem (ES) cells in culture. *Development* **111,** 259–267.
3. Choi, K., Kennedy, M., Kazarov, A., Papadimitriou, J. C., and Keller, G. (1998) A common precursor for hematopoietic and endothelial cells. *Development* **125,** 725–732.
4. Wartenberg, M., Gunther, J., Hescheler, J., and Sauer, H. (1998) The embryoid body as a novel in vitro assay system for antiangiogenic agents. *Lab. Invest.* **78,** 1301–1314.
5. Vittet, D., Prandini, M. H., Berthier, R., Schweitzer, A., Martin-Sisteron, H., Uzan, G., and Dejana, E. (1996) Embryonic stem cells differentiate in vitro to endothelial cells through successive maturation steps. *Blood* **88,** 3424–3431.
6. Kramer, J., Hegert, C., Guan, K., Wobus, A. M., Muller, P. K., and Rohwedel, J. (2000) Embryonic stem cell-derived chondrogenic differentiation in vitro: activation by BMP 2 and BMP-4. *Mech. Dev.* **92,** 193–205.
7. Bain, G., Kitchens, D., Yao, M., Huettner, J. E., and Gottlieb, D. I. (1995) Embryonic stem cells express neuronal properties in vitro. *Dev. Biol.* **168,** 342–357.
8. Strubing, C., Ahnert-Hilger, G., Shan, J., Wiedenmann, B., Hescheler, J., and Wobus, A. M. (1995) Differentiation of pluripotent embryonic stem cells into the neuronal lineage in vitro gives rise to mature inhibitory and excitatory neurons. *Mech. Dev.* **53,** 275–287.
9. Strubing, C., Rohwedel, J., Ahnert-Hilger, G., Wiedenmann, B., Hescheler, J., and Wobus, A. M. (1997) Development of G protein-mediated Ca^{2+} channel regulation in mouse embryonic stem cell-derived neurons. *Eur. J. Neurosci.* **9,** 824–832.
10. Drab, M., Haller, H., Bychkov, R., Erdmann, B., Lindschau, C., Haase, H., et al. (1997) From totipotent embryonic stem cells to spontaneously contracting smooth muscle cells: a retinoic acid and db-cAMP in vitro differentiation model. *FASEB J.* **11,** 905–915.
11. Rohwedel, J., Maltsev, V., Bober, E., Arnold, H. H., Hescheler, J., and Wobus, A. M. (1994) Muscle cell differentiation of embryonic stem cells reflects myogenesis in vivo: developmentally regulated expression of myogenic determination genes and functional expression of ionic currents. *Dev. Biol.* **164,** 87–101.
12. Westfall, M. V., Samuelson, L. C., and Metzger, J. M. (1996) Troponin I isoform expression is developmentally regulated in differentiating embryonic stem cell-derived cardiac myocytes. *Dev. Dyn.* **206,** 24–38.
13. Metzger, J. M., Lin, W. I., and Samuelson, L. C. (1996) Vital staining of cardiac myocytes during embryonic stem cell cardiogenesis in vitro. *Circ. Res.* **78,** 547–552.
14. Wobus, A. M., Wallukat, G., and Hescheler, J. (1991) Pluripotent mouse embryonic stem cells are able to differentiate into cardiomyocytes expressing chronotropic responses to adrenergic and cholinergic agents and Ca^{2+} channel blockers. *Differentiation* **48,** 173–182.
15. Kolossov, E., Fleischmann, B. K., Liu, Q., Bloch, W., Viatchenko-Karpinski, S., Manzke, O., et al. (1998) Functional characteristics of ES cell-derived cardiac precursor cells identified by tissue-specific expression of the green fluorescent protein. *J. Cell Biol.* **143,** 2045–2056.

16. Ji, G. J., Fleischmann, B. K., Bloch, W., Feelisch, M., Andressen, C., Addicks, K., and Hescheler, J. (1999) Regulation of the L-type Ca^{2+} channel during cardiomyogenesis: switch from NO to adenylyl cyclase-mediated inhibition. *FASEB J.* **13,** 313–324.

17. Abi-Gerges, N., Ji, G. J., Lu, Z. J., Fischmeister, R., Hescheler, J., and Fleischmann, B. K. (2000) Functional expression and regulation of the hyperpolarization activated non-selective cation current in embryonic stem cell-derived cardiomyocytes. *J. Physiol. (Lond)* **523,** 377–389.

18. Fishman, M. C. and Chien, K. R. (1997) Fashioning the vertebrate heart: earliest embryonic decisions. *Development* **124,** 2099–2117.

19. Maltsev, V. A., Wobus, A. M., Rohwedel, J., Bader, M., and Hescheler, J. (1994) Cardiomyocytes differentiated in vitro from embryonic stem cells developmentally express cardiac-specific genes and ionic currents. *Circ. Res.* **75,** 233–244.

20. Hescheler, J., Fleischmann, B. K., Lentini, S., Maltsev, V. A., Rohwedel, J., Wobus, A. M., and Addicks, K. (1997) Embryonic stem cells: a model to study structural and functional properties in cardiomyogenesis. *Cardiovasc. Res.* **36,** 149–162.

21. Laschinski, G., Vogel, R., and Spielmann, H. (1991) Cytotoxicity test using blastocyst-derived euploid embryonal stem cells: a new approach to in vitro teratogenesis screening. *Reprod. Toxicol.* **5,** 57–64.

22. Scholz, G., Pohl, I., Genschow, E., Klemm, M., and Spielmann, H. (1999) Embryotoxicity screening using embryonic stem cells in vitro: correlation to in vivo teratogenicity. *Cells Tissues Organs* **165,** 203–211.

23. Thomson, J. A., Itskovitz-Eldor, J., Shapiro, S. S., Waknitz, M. A., Swiergiel, J. J., Marshall, V. S., and Jones, J. M. (1998) Embryonic stem cell lines derived from human blastocysts. *Science* **282,** 1145–1147.

24. Gryshchenko, O., Fischer, I. R., Dittrich, M., Viatchenko-Karpinski, S., Soest, J., Bohm-Pinger, M. M., et al. (1999) Role of ATP-dependent K(+) channels in the electrical excitability of early embryonic stem cell-derived cardiomyocytes. *J. Cell Sci.* **112,** 2903–2912.

25. Igelmund, P., Fleischmann, B. K., Fischer, I. R., Soest, J., Gryshchenko, O., Bohm-Pinger, M. M., et al. (1999) Action potential propagation failures in long-term recordings from embryonic stem cell-derived cardiomyocytes in tissue culture. *Pflugers Arch.* **437,** 669–679.

26. Wobus, A. M., Holzhausen, H., Jakel, P., and Schoneich, J. (1984) Characterization of a pluripotent stem cell line derived from a mouse embryo. *Exp. Cell Res.* **152,** 212–219.

27. Maltsev, V. A., Rohwedel, J., Hescheler, J., and Wobus, A. M. (1993) Embryonic stem cells differentiate in vitro into cardiomyocytes representing sinusnodal, atrial and ventricular cell types. *Mech. Dev.* **44,** 41–50.

28. Wartenberg, M., Hescheler, J., Acker, H., Diedershagen, H., and Sauer, H. (1998) Doxorubicin distribution in multicellular prostate cancer spheroids evaluated by confocal laser scanning microscopy and the "optical probe technique." *Cytometry* **31,** 137–145.

29. Wartenberg, M. and Acker, H. (1995) Quantitative recording of vitality patterns in living multicellular spheroids by confocal microscopy. *Micron* **26,** 395–404.

30. Sauer, H., Gunther, J., Hescheler, J., and Wartenberg, M. (2000) Thalidomide inhibits angiogenesis in embryoid bodies by the generation of hydroxyl radicals. *Am. J. Pathol.* **156,** 151–158.

31. Fassler, R., Rohwedel, J., Maltsev, V., Bloch, W., Lentini, S., Guan, K., et al. (1996) Differentiation and integrity of cardiac muscle cells are impaired in the absence of beta 1 integrin. *J. Cell Sci.* **109,** 2989–2999.

32. Reuter, H. (1983) Calcium channel modulation by neurotransmitters, enzymes and drugs. *Nature* **301,** 569–574.

33. Trautwein, W. and Hescheler, J. (1990) Regulation of cardiac L-type calcium current by phosphorylation and G proteins. *Annu. Rev. Physiol.* **52,** 257–274.
34. Maltsev, V. A., Ji, G. J., Wobus, A. M., Fleischmann, B. K., and Hescheler, J. (1999) Establishment of beta-adrenergic modulation of L-type Ca^{2+} current in the early stages of cardiomyocyte development. *Circ. Res.* **84,** 136–145.
35. Slotkin, T. A., Lau, C., and Seidler, F. J. (1994) Beta-adrenergic receptor overexpression in the fetal rat: distribution, receptor subtypes, and coupling to adenylate cyclase activity via G-proteins. *Toxicol. Appl. Pharmacol.* **129,** 223–234.
36. Okabe, S., Forsberg-Nilsson, K., Spiro, A. C., Segal, M., and McKay, R. D. (1996) Development of neuronal precursor cells and functional postmitotic neurons from embryonic stem cells in vitro. *Mech. Dev.* **59,** 89–102.
37. Isenberg, G. and Klockner, U. (1982) Calcium tolerant ventricular myocytes prepared by preincubation in a "KB medium." *Pflugers Arch.* **395,** 6–18.
38. Hescheler, J., Fleischmann, B. K., Wartenberg, M., Bloch, W., Kolossov, E., Ji, G., et al. (1999) Establishment of ionic channels and signalling cascades in the embryonic stem cell-derived primitive endoderm and cardiovascular system. *Cells Tissues Organs* **165,** 153–164.
39. Hamill, O. P., Marty, A., Neher, E., Sakmann, B., and Sigworth, F. J. (1981) Improved patch-clamp techniques for high-resolution current recording from cells and cell-free membrane patches. *Pflugers Arch.* **391,** 85–100.
40. Korn, S. J. and Horn, R. (1989) Influence of sodium-calcium exchange on calcium current rundown and the duration of calcium-dependent chloride currents in pituitary cells, studied with whole cell and perforated patch recording. *J. Gen. Physiol.* **94,** 789–812.
41. Grynkiewicz, G., Poenie, M., and Tsien, R. Y. (1985) A new generation of Ca^{2+} indicators with greatly improved fluorescence properties. *J. Biol. Chem.* **260,** 3440–3450.
42. Egert, U., Schlosshauer, B., Fennrich, S., Nisch, W., Fejtl, M., Knott, T., et al. (1998) A novel organotypic long-term culture of the rat hippocampus on substrate-integrated multielectrode arrays. *Brain Res. Brain Res. Protoc.* **2,** 229–242.
43. Rohr, S., Kucera, J. P., and Kleber, A. G. (1998) Slow conduction in cardiac tissue, I: effects of a reduction of excitability versus a reduction of electrical coupling on microconduction. *Circ. Res.* **83,** 781–794.
44. Kucera, J. P., Kleber, A. G., and Rohr, S. (1998) Slow conduction in cardiac tissue, II: effects of branching tissue geometry. *Circ. Res.* **83,** 795–805.
45. Morley, G. E., Vaidya, D., Samie, F. H., Lo, C., Delmar, M., and Jalife, J. (1999) Characterization of conduction in the ventricles of normal and heterozygous Cx43 knockout mice using optical mapping (see comments). *J. Cardiovasc. Electrophysiol.* **10,** 1361–1375.
46. Banach, K., Egert, U., and Hescheler, J. (2000) Excitation spread between heart cells derived from embryonic stem (ES) cells. *Pflugers Arch.* **439(Suppl. 6),** P13–P19.

16

Isolation of Lineage-Restricted Neural Precursors from Cultured ES Cells

Tahmina Mujtaba and Mahendra S. Rao

1. Introduction

Embryonic stem (ES) cells are derived from undifferentiated cells present at the inner cell mass at the blastula stage of development *(1,2)*. In the normal course of development, these cells give rise to the primitive ectoderm, which in turn gives rise to the ectoderm, mesoderm, and endoderm through the process of gastrulation. Under the influence of the mesoderm, the midline ectoderm differentiates into the nervous system *(3)*. Undifferentiated epiblast cells can be obtained from dissociated blastulas and cultured on mitotically arrested mouse embryonic fibroblast feeder layers and/or in the presence of the cytokine leukemia inhibitory factor (LIF). The epiblast cells expand and give rise to colonies of stem cells that are undifferentiated, nontransformed, have stable diploid karyotype and can proliferate indefinitely. They can integrate within blastocysts or with morulas to generate chimeric animals *(4)*. These cell lines, called ES cells, are pluripotent and generate all embryonic derivatives including germ cells (cxcept the trophoblast–placenta). Similar cells can also be generated from primordial germ cells (PGC) present in the fetus. Though the initial culture conditions are slightly different, the resulting stem cells, termed embryonic germ (EG) cells are indistinguishable from ES cells in most of their properties *(5,6)*.

ES cells grown in the absence of LIF on nonadhesive substrates form embryoid bodies (EBs), i.e., ball-like multicellular aggregates. If the aggregates, after a few days in culture, are plated onto adhesive substrates, they give rise to multiple cell types, which include muscle-like, neuron-like, and hematopoietic-like cells *(2,7–9)*. This suggests that the pathways instituted during normal development can be mimicked to some extent in vitro and allows access to early precursor populations normally found during development. The idea that ES cells can be used as a reservoir of cells from which differentiated cells can be harvested as required has been validated in rodent models. Several groups have shown that cell surface markers *(10)*, manipulation of culture conditions *(11)*, or utilizing tissue-specific promoters *(12)* can be used to isolate neural stem cells or more restricted precursors. Recently, oligodendrocyte precursor cells, isolated from ES cells, have been shown to be functionally useful in a rat model

From: *Methods in Molecular Biology, vol. 185: Embryonic Stem Cells: Methods and Protocols*
Edited by: K. Turksen © Humana Press Inc., Totowa, NJ

of demyelination *(13)*. McKay and colleagues *(11)* have manipulated culture conditions to isolate multipotent stem cells from ES cell cultures, and Li and colleagues *(12)* have used an elegant selection strategy to isolate neural precursor cells.

In this chapter, we describe methods for isolating neural precursor cells from ES cell cultures.

2. Materials

All reagents and materials in ES cell culture must be sterile. All culture medium supplemented with growth factors should be used within 2 wk of preparation or as stated otherwise.

2.1. Feeder Layers Culture

1. Mouse fibroblast line STO (ATTC, accession no. CRL-1503).
2. STO fibroblast media (SFM) store at 4°C. The following obtained as stock solutions are added to Dulbecco's modified Eagle media (DMEM) (Gibco BRL, cat. no. 10564) to give final indicated concentrations: 10% fetal calf serum (FCS), 1% pen–strep.
3. Feeder layer media (FM) store at 4°C. The following are added to DMEM to give a final concentration of the following: 10% FCS, 1% pen–strep (Gibco BRL, cat. no. 15070-063), 1 µg/mL mitomycin C (Sigma, cat. no. M-4287) (*see* **Note 1**).
4. Passaging cells: phosphate-buffered saline calcium- and magnesium- free (PBS-CMF), pH 7.4, trypsin (0.05% trypsin and 0.53 m*M* EDTA).
5. Freezing media: 90% SFM media, 10% dimethyl sulfoxide (DMSO) (Sigma, cat. no. D2650).

2.2. Undifferentiated ES Cells

1. ES-D3 cell line (ATCC, accession no. CRL-1934). J1, CJ7, and R1 cell lines have also been used with similar results (*see* **Note 2**).
2. ES cell media (ES media) store at 4°C. The following are added to DMEM to give the final indicated concentrations: 20% fetal bovine serum (FBS), 0.1% 2-mercaptoethanol (Sigma, cat. no. M-3148), 1000 U/mL LIF (Gibco BRL, cat. no. 13310-016), 1% pen–strep.
3. Gelatin-coated tissue culture dishes: prepare 0.1% (w/v) solution of gelatin (Type A; Sigma cat. no. G-1890) in distilled water. Incubate the dishes for 1 h at room temperature, aspirate the gelatin, and rinse with medium once before use.
4. Freezing media: 90% ES media plus 10% DMSO.

2.3. Formation of EBs

1. ES media: as described in **Subheading 2.2.**
2. Neuroepithelial stem cell (NEP) basal media (store at 4°C): add the following to DMEM/F12 (Gibco BRL, cat. no. 11320-033) to the final indicated concentrations: B27 (50X stock) (Gibco BRL, cat. no. 17504-010), N2 (100× stock) (Gibco BRL, cat. no. 17502-014), 1 mg/mL BSA (Sigma, cat. no. A-2153), 20 ng/mL fibroblast growth factor-2 (FGF-2) (PeproTech, cat. no. 100-B). The FGF-2 solution is made without the addition of heparin, 1% pen–strep.
3. Fibronectin-coated tissue culture dishes: make 20 µg/mL Fibronectin (Sigma, cat. no. F-1141) solution in distilled water. Coat the dishes with fibronectin and leave overnight at 4°C. Rinse the dishes with medium once before use.

2.3.1. For Isolation of ES-Derived Neuron-Restricted Precursors and Glial-Restricted Precursors by Immunopanning

1. NEP basal media: as described in **Subheading 2.3.** with the following additional supplements for neuronal-restricted precursors (NRPs): neurotrophic factor-3 (NT-3) (20 ng/mL) and FGF-2 (10 ng/mL, in addition to the 20 ng/mL that is present in the basal medium) and FGF-2 (10 ng/mL) alone for glial-restricted precursors (GRPs).
2. 100-mm Bacteriological Petri dishes.
3. Coating antibody: anti-mouse IgM unlabeled (Southern Biotechnology, cat. no. 1020-01): make 10 μg/mL solution of the antibody in 50 mM Tris-HCl, pH 9.5.
4. PBS-CMF.
5. Cell scrapers.
6. Embryonic neural cell adhesion molecule (E-NCAM) hybridoma supernatant (Developmental Studies Hybridoma Bank), A2B5 hybridoma supernatant (Developmental Studies Hybridoma Bank).
7. Poly-L-lysine–laminin-coated tissue culture dishes: coat the culture dishes with 15 μg/mL solution of poly-L-lysine (Sigma, cat. no. P-1274) (make up the solution in distilled water and store at 4°C) for 30 min to 1 h at room temperature. Remove the poly-L-lysine, rinse the dish once with distilled water, and incubate with laminin (20 μg/mL in distilled water; Gibco BRL, cat. no. 23017-015) overnight at 4°C. Rinse the dishes with medium once just before plating the cells (*see* **Note 3**).
8. Fibronectin-coated dishes: as described in **Subheading 2.3.**

2.3.2. Magnetic Bead Sorting

1. Magnetic cell separators, such as Mini/MidiMACS (Miltenyi Biotec, 425-01).
2. Positive selection column MS$^+$/LS$^+$ column (Miltenyi Biotec, cat. no. 422-01/424-01).
3. Buffer 1 (elution buffer): PBS-CMF, pH 7.2, supplemented with 0.5% bovine serum albumin (BSA) and 2 mM EDTA.
4. Buffer 2 (blocking buffer): PBS-CMF, supplemented with 1% BSA.
5. Primary antibodies: E-NCAM and A2B5.
6. Rat anti-mouse IgM microbeads (Miltenyi Biotec, cat. no. 473-02)
7. 0.02 mM EDTA-Hank's balanced salt solution (HBSS): to dissociate the cells.
8. Poly-L-lysine–laminin double-coated dishes for NRPs and GRPs.
9. Fibronectin-coated dishes: as described in **Subheading 2.3., step 3**.

2.3.3. Maturation of ES-Derived NRPs

1. NEP basal media: as mentioned in **Subheading 2.3.**, except FGF-2 is omitted from the media.
2. All-*trans* retinoic acid (RA) (Sigma, cat. no. R-2625) light sensitive, store at –20°C): Prepare a 1000X stock solution in 95% ethanol. Add the stock solution directly to the culture medium to obtain a final concentration of 1 μM. The shelf life of the stock solution is 10 d.
3. Poly-L-lysine–laminin double-coated dishes: prepare the dishes as described in **Subheading 2.3.1.**

2.3.4. Maturation of ES-Derived GRPs

1. NEP basal media: As described in **Subheading 2.3.**
2. Platelet-derived growth factor BB (PDGF) (Upstate Biotechnology, cat. no. 01-305): make 5 μg/mL of stock solution in 10 mM acetic acid containing 1 mg/mL BSA for maximum

recovery. Store the stock solution as single use working aliquots at –20°C. Add the stock solution directly to the culture medium to get a desired concentration of 10 ng/mL. The shelf life of the stock solution is 1 yr at –20°C.

3. 3,3,5-Triiodo thyronine (T3) (Sigma, cat. no. T-6397): Make 20 μg/mL stock solution in DMEM. Store the stock solution at –20°C. Use at a final concentration of 30 ng/mL.

4. FBS (Gibco BRL, cat. no. 26140-079): use 10% FBS to induce astrocytic differentiation. Serum quality is critical for ES cells and varies from lot to lot. Test different serum lots by running several experiments in parallel and ordering in bulk the most suitable sera lot. Serum can be stored as working aliquots for a period of up to 2 yr at –20°C.

5. Ciliary neurotrophic factor (CNTF) (PeproTech, cat. no. 450-50): make 5 μg/mL stock solution and store as single use working aliquots at –20°C. Add the stock solution directly to the culture dish to get the desired concentration of 10 ng/mL. The shelf life of the stock solution is 3 mo at –20°C.

6. Poly-L-lysine–laminin double-coated dishes: prepare the dishes as described in **Subheading 2.3.1.**

2.4. For Immunocytochemistry

1. Primary antibodies
   a. E-NCAM, A2B5, O4 (Chemicon, cat. no. MAB345), galactocerebroside (Gal-C) (Boehringer Mannheim, cat. no. 1351 621).
   b. β-III Tubulin (Sigma, cat. no. T-8660), microtubule-associated protein-2 (MAP-2) (Sigma, cat. no. M-4403), neurofilament-150 (Chemicon, cat. no. AB1981), polyclonal glial fibrillary acid protein (GFAP) (DAKO, cat. no. Z-0334) (*see* **Note 4**).
   c. Gamma amino butyric acid (GABA) (Signature Immunologics, cat. no. YY 100), GAD (Chemicon, cat. no. AB 108), Glycine (Signature Immunologics, cat. no. G 100).

2. Secondary Antibodies:
   a. Goat anti-mouse IgM ((fluorescein isothiocyanate [FITC], cat. no. 1020-02, or tetra-methylrhodamine isothiocyanate [TRITC]-conjugated, cat. no. 1020-03, from Southern Biotechnology).
   b. Goat anti-mouse IgG (FITC, cat. no. 115-096-062 or TRITC-conjugated, cat. no. 115-026-062, from Jackson ImmunoResearch). IgG1 (FITC, cat. no. 1070-02, or TRITC-conjugated, cat. no. 1070-03, from Southern Biotechnology) and IgG2b (FITC, cat. no. 1090-02, or TRITC-conjugated, cat. no. 1090-03, from Southern Biotechnology).
   c. Goat anti-rabbit IgG (FITC, cat. no. 4030-02, or TRITC-conjugated, cat. no. 4030-03, from Southern Biotechnology).
   d. Horseradish peroxidase (HRP)-conjugated goat anti-mouse IgG (cat. no. 115-036-062, from Jackson Immunochemicals).
   e. Peroxidase substrate kit diaminobenzidine tetrahydrochloride [DAB] (cat. no. SK-4100, from Vector Laboratories) (*see* **Note 5**).

3. Blocking Buffer To PBS-CMF, add the following to give the final indicated concentrations: 0.1% Triton X-100, 1 mg/mL BSA, 5% goat serum (*see* **Note 6**).

4. Fixatives
   a. 4% Paraformaldehyde in phosphate buffer (*see* **Note 7**).
   b. Glutaraldehyde fix: make 2.5% glutaraldehyde (G-5882, grade 1; Sigma) in 1% paraformaldehyde solution. The solution can be kept for 15 d at 4°C. This fixative is used for small molecules, for example neurotransmitters (*see* **Note 8**).

3. Methods

The stem cells are grown exclusively on feeder layers of mitotically inactive fibroblasts. The most commonly used are mouse embryonic and STO fibroblasts. The cells can be rendered mitotically inactive either by exposure to gamma irradiation or mitomycin C treatment (for more details on feeder layers *see* **ref. *14***).

3.1. Routine Culture of STO Fibroblasts

STO fibroblasts have simple growth requirements and are easy to grow.

1. Thaw out the ATCC vial and plate the cells on 10-cm tissue culture dishes (Falcon) in 10 mL of SFM. Within 2 d, the dishes should be confluent. Ideally, one 10-cm dish yields around $6–8 \times 10^6$ cells.
2. Aspirate the medium and wash the cells once with 5 mL PBS-CMF.
3. Add 3 mL of trypsin-EDTA and incubate the dish at 37°C for 5 min. Neutralize the trypsin with medium containing 5% FCS. Vigorously triturate the cells by pipetting up and down to get single-cell suspension.
4. Resuspend the cells in 50 mL of SFM medium and dispense the cell suspension into five 10-cm dishes.
5. Subculture the cells again upon confluence by splitting them 1:5 (approximately 1.5×10^6 cells/dish). Keep track of passage number (*see* **Note 9**).

3.1.1. Preparation of Feeder Layers from STO Cells

1. Aspirate the medium from confluent STO cells and replace the medium with the freshly prepared FM for a minimum of 2 h.
2. Meanwhile, coat 10-cm tissue culture dishes with 10 mL of 0.1% gelatin solution for 1 h at room temperature.
3. Aspirate the SFM from STO cells. Wash the dishes 3 times with 5 mL of PBS-CMF. Trypsinize the cells with 3 mL trypsin-EDTA for 5 min and collect the cells in DMEM containing 10% FCS.
4. Collect the cells into 15-mL plastic centrifuge tubes and spin them at 183*g* for 5 min. Resuspend the cells in FM and count using a hemocytometer.
5. Dispense the cell suspension onto the gelatin-coated dishes at a plating density of 5×10^5 cells/cm². Incubate the dishes at 37°C.
6. The cells should attach within 20 min and spread out to give a monolayer 12 h later. If the cultures look too sparse, then additional mitomycin-treated cells can be added. The feeder layers can be maintained for 10 days after preparation.

3.2. Maintenance of Undifferentiated ES Cells

The stem cells grow rapidly, dividing every 18–24 h. It is advisable to keep the cells at relatively high densities to ensure a high rate of cell division, which minimizes spontaneous differentiation.

3.2.1. Subculturing of Undifferentiated ES Cells

1. Passage the cells when the dishes are confluent. Passage of a 60-mm dish is given below.
2. Aspirate the medium and wash the dish with 3 mL PBS-CMF. Add 2 mL trypsin-EDTA to the dish and incubate it at 37°C for 3–5 min. Check the dish to see if the cells are coming off.
3. Neutralize the trypsin with ES medium and remove the cells from the dish by gently triturating with a 1-mL pipet tip. Observe the cells under the microscope to ensure you have single cells. If you see aggregates, then triturate the cells with a fire-polished glass pasteur pipet.
4. Collect cells in a sterile plastic centrifuge tube and spin at 183*g* for 5 min.

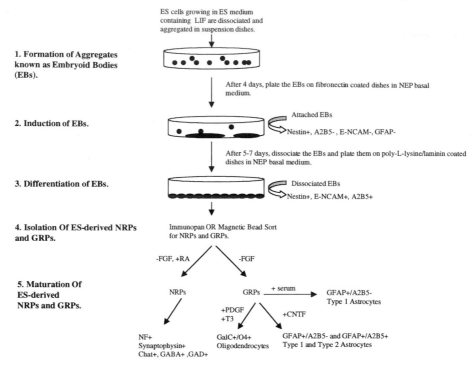

Fig. 1. Differentiation of ES cells: The process of generating purified populations of cells from undifferentiated cell cultures is summarized.

5. Discard the supernatant and resuspend the pellet in 1 mL ES medium and plate the cells on feeder layers containing 2 mL ES medium. Add approx 1×10^6 cells/dish.

6. Change the medium every other day.

7. The cells will be confluent within 3 to 4 days. At this point, you can either freeze them, repassage them, or allow them to differentiate into EBs.

3.3. Formation of EBs

1. Detach the undifferentiated cells from the feeder layers by using 0.05% trypsin-EDTA solution.

2. Trypsinize the cells for 3 min; neutralize the trypsin with medium containing serum (*see* **Note 10**). Spin the cells at 183*g* for 5 min.

3. Discard the supernatant and resuspend the pellet in a small amount of medium to which no LIF has been added.

4. Seed approximately 5×10^3 cells/mL in suspension dishes or in nonadhesive dishes in ES medium. Under these conditions, cells do not attach to the dish surface and readily form floating aggregates known as EBs (*see* **Fig. 1**).

5. Observe these aggregates under a light microscope. During the first 2 d, the aggregates are small with irregular outlines. By day 4–6, the aggregates become larger and spherical with smoother outlines (*see* **Note 11**).

6. After the aggregates become larger and spherical, spin the EBs at 183*g* for 3 min, aspirate the medium gently and resuspend the aggregates in 500 µL of ES medium. The aggregates are now ready to be plated on fibronectin-coated dishes.

7. Within 24 h, the EBs attach to the substratum (*see* **Note 12**). At this point, the aggregates look like tightly packed epithelial cells.

8. The next day, replace the ES medium with NEP basal medium and change 70% of the medium every 2 d.
9. After 5–7 days in culture, a large proportion of cells change their morphology to small elongated cells, which resemble the native neuroepithelial precursor cells.

3.4. Differentiation of EBs into NRPs and GRPs

Stain the 5- to 7-d-old EBs for nestin, a marker for undifferentiated stem cells, and for E-NCAM and A2B5, the lineage markers for NRPs and GRPs, respectively. The small elongated cells should be positive for nestin and negative for the other markers tested. The cells can now be induced to differentiate into NRPs and GRPs by the following protocol.

1. Add 1 mL trypsin to the 5- to 7-d-old EBs for 2 min at 37°C.
2. Neutralize the trypsin using excess medium and gently triturate to a single-cell suspension.
3. Plate the cells on poly-L-lysine and laminin-coated dishes in 1 mL NEP basal medium. The density of the cells should be approx 2000–5000 cells/35-mm dish.
4. Change the medium every 2 d.
5. 5–7 d later, stain for E-NCAM and A2B5. Around 30–35% of the cells will express E-NCAM, and approx 40% of the cells will express A2B5.
6. At this point, the ES cultures can be enriched for neuronal (E-NCAM) and glial (A2B5) populations. The 2 enrichment procedures; immunopanning and magnetic bead sorting are described below in **Subheadings 3.4.1.** and **3.4.2.**, respectively. In addition to these 2 isolation protocols, the cells can also be isolated by fluorescence-activated cell sorter (FACS) sorting and retroviral labeling (*see* **Note 13**).

3.4.1. Isolation of ES Derived NRPs by Immunopanning (**Fig. 2**)

3.4.1.1. Preparation of Panning Dishes

Onc night before use (*see* **Note 14**):

1. Preincubate panning dishes (100-mm polystyrene bacteriological Petri dishes) with coating antibody. For E-NCAM, goat anti-mouse IgM (10 µg/mL in 50 mM Tris-HCl, pH 9.5) is used.
2. Add 12 mL of the coating solution to the panning dish. This volume is sufficient to cover the 100-mm dish.
3. Incubate the dishes overnight at 4°C.
   2 h before use:
4. Rinse dishes 3 times with 4 mL PBS-CMF.
5. Incubate the washed dishes with 1:1 dilution of E-NCAM hybridoma supernatant in NEP basal medium.
6. Incubate for 1 h at room temperature. Just before use, rinse 3 times with PBS-CMF. Do not allow the dish to dry between washes.
7. Add 8 mL NEP basal medium to the 100-mm panning dish. The dish is now ready for the cells to be immunopanned.

3.4.1.2. Panning Procedure

1. Triturate the cells into a single-cell suspension and add the cell suspension to the culture medium. Swirl the panning dish to distribute the cells evenly.
2. Allow to stand for 1 h at room temperature. Observe on a phase contrast microscope. When the dish is lightly tapped, unbound moving cells should be observed.

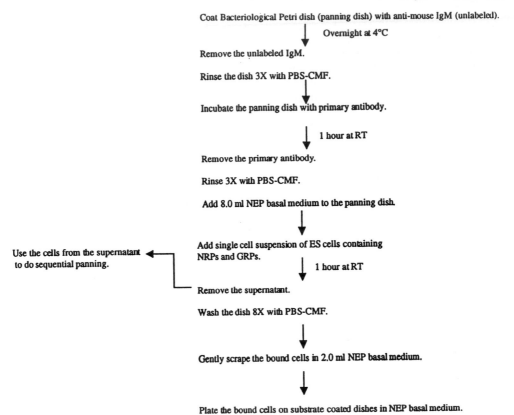

Coat Bacteriological Petri dish (panning dish) with anti-mouse IgM (unlabeled).

Overnight at 4°C

Remove the unlabeled IgM.

Rinse the dish 3X with PBS-CMF.

Incubate the panning dish with primary antibody.

1 hour at RT

Remove the primary antibody.

Rinse 3X with PBS-CMF.

Add 8.0 ml NEP basal medium to the panning dish.

Use the cells from the supernatant to do sequential panning.

Add single cell suspension of ES cells containing NRPs and GRPs.

1 hour at RT

Remove the supernatant.

Wash the dish 8X with PBS-CMF.

Gently scrape the bound cells in 2.0 ml NEP basal medium.

Plate the bound cells on substrate coated dishes in NEP basal medium.

Fig. 2. Immunopanning of cells: A flowchart summarizing the process of panning is shown. Cells are harvested, dissociated into a single cell suspension and selected by allowing them to adhere to a plate coated with an antibody that will recognize a epitope specific to the cell type of interest.

3. Wash the dish gently 8 times with PBS-CMF (*see* **Note 15**). The desired cells are bound to the dish. Gently scrape the cells off with a cell scraper in 2 mL NEP basal medium.
4. Plate out the cells at desired density on poly-L-lysine–laminin-coated 35-mm dishes in NEP basal medium. The medium is further supplemented with NT-3 (10 ng/mL) for the survival of neurons.
5. Check the efficiency of the panning by staining with E-NCAM antibody. At this stage, cells have the morphology of immature neurons, i.e., cells with small cell bodies and short processes. They divide rapidly in response to the mitogen (FGF-2) in the medium and form tightly packed clusters.

3.4.2. Isolation of ES-Derived NRPs by Magnetic Bead Sorting

1. Incubate the cells with 1:1 dilution E-NCAM antibody with NEP basal medium for 1 h at room temperature (*see* **Note 16**).
2. Rinse the cells with buffer 1.
3. Add 1 mL of 0.02 mM EDTA-HBSS solution to the cells. Incubate for 5–7 min at 37°C.
4. Spin the cells at 183g for 5 min.
5. Resuspend approximately 10^7 cells in 80 μL of buffer 1.

6. Add 20 µL of magnetic cell sorting (MACS) rat anti-mouse IgM microbeads per 10^7 cells. Mix well by gentle trituration 4 to 5 times.
7. Incubate the cells for 15 min at 4°C (*see* **Note 17**).
8. Mix the cells well by gentle trituration, bring the sample volume to 500 µL with buffer 1, and separate on MiniMacs separation system (*see* **Note 18**).
9. Place the MS⁺/LS⁺ column in magnetic field.
10. Incubate the column for 1 h with ice-cold buffer 2 to block nonspecific binding sites.
11. Wash the column 3 times with 3 mL of buffer 1. Discard the effluent.
12. Load magnetically labeled cell suspension (10^8 cells/500 µL) to the column (*see* **Fig. 3**), allow the negative cells to pass through at a rate of 1 vol/3 min (*see* **Note 19**).
13. Wash the column 3 times with 1 mL of the buffer each time at a faster elution rate: 5 vol/min. Always allow the entire amount of the buffer to flow through the column before adding the new buffer. Collect the total effluent as negative fraction.
14. Remove the column from the separation unit and place the column on a 15-mL centrifuge tube and firmly flush out the positive fraction with 5 mL ice-cold buffer1, using the plunger supplied with the column (*see* **Note 20**).
15. Plate the negative and positive fractions on poly-L-lysine–laminin double-coated dishes. Let them grow for 3 to 4 h and then stain them with FITC or TRITC-conjugated anti-mouse IgM to ascertain the efficiency of sorting.

3.5. Maturation of ES-Derived NRPs

To enhance neuronal differentiation, the mitogen FGF-2 is withdrawn and all-*trans* RA is added to the cultures. The cultures can be maintained without significant cell death in NEP basal medium for more than 2 wk.

1. Let the panned NRPs grow to subconfluency (40–50% confluency is sufficient) on poly-L-lysine–laminin-coated dishes in NEP basal medium.
2. Withdraw the FGF-2 from the cultures and add RA (1 µ*M*).
3. Add RA and NT-3 every day and change 70% of the medium every 2 d.
4. By d 3, the cells have small cell bodies with comparatively longer neurites emerging from them.
5. By d 6, the cell bodies become bigger and form an extensive neuritic network.
6. Stain these neuronal-looking cells with antibodies specific to neurons, i.e., β-III tubulin, MAP-2, NF-150. To check the specificity of neuronal staining, it is recommended to stain a dish of ES cell-derived GRPs with these neuron-specific antibodies. In the NRP dish, the anti-MAP-2 will stain short thick dendritic processes and cell bodies, while the NF-150 will stain the longer and thinner axons. This will confirm that both classes of neurites are present. In the GRP dish, there will be no staining, as the culture conditions do not support the survival and maturation of neurons.
7. To see if these neuronal cells can synthesize neurotransmitters culture the cells for an additional 4 d, fix, and stain for glutamate, GABA, and glycine. There will be a large subset of glutamate-positive neurons with smaller subsets of glycine- and GABA-positive neurons present in these cultures.

3.6. Isolation of ES-Derived GRPs

The protocol described below is for a single 35-mm tissue culture dish.

1. Trypsinize the 5-d-old EBs growing in NEP basal medium by using 500 µL of trypsin-EDTA.
2. Incubate for 3 min at 37°C.

Mixed population
of cells

Magnetic stand to
retain labeled cells

Flow through-
unlabeled cells

Flow through-
labeled cells

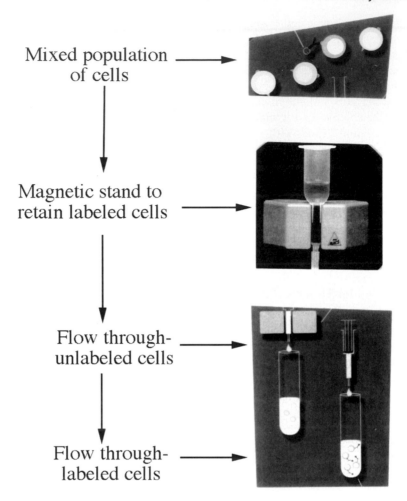

Fig. 3. Magnetic bead sorting: A flowchart summarizing the process of magnetic bead sorting is shown. Note: Cell clumps need to be avoided and the flow rate needs to be calibrated to ensure purity. (*See* color plate 7, following p. 254).

3. Neutralize the trypsin with excess medium.
4. Spin the cells at low speed 183g for 5 min.
5. Resuspend the cells in 500 µL of NEP basal media (*see* **Note 21**). Triturate the cells gently to get a single-cell suspension and plate the cells on poly-L-lysine–laminin-coated dishes.
6. Add 1 mL NEP basal medium containing PDGF (10 ng/mL) and additional FGF-2 (10 ng/mL).
7. Feed the cells daily with FGF-2 and PDGF and change the medium every 2 d.
8. The cells can be subpassaged and expanded in culture by splitting 1:4 (approx 2000–5000 cells/35-mm dish).
9. After 7–14 d in culture, an isomorphous population of round to bipolar cells are seen. Stain these cells with the monoclonal A2B5 antibody (*see* **Subheading 3.8.1.** for staining). This particular antibody recognizes a membrane epitope typically expressed by glial cells. These conditions yield approx 30–40% A2B5-positive cells. At this point, the cultures can be enriched for the A2B5-positive cells by immunopanning or magnetic bead sorting.

3.6.1. Isolation of ES-Derived GRPs by Immunopanning

Follow the panning protocol described for E-NCAM panning with the following changes.

1. Pan for E-NCAM (*see* **Subheading 3.4.1.**).
2. Coat a bacteriological Petri dish with 10 µg/mL of unlabeled IgM antibody in 12 mL of Tris buffer overnight.
3. On the next morning, rinse the dish 3 times with PBS-CMF and incubate with monoclonal A2B5 diluted 1:1 with NEP basal media.
4. After 1 h, remove the A2B5 antibody (*see* **Note 22**) and rinse the dish 3 times with PBS-CMF.
5. Add 8 mL of NEP basal medium to the new A2B5 panning dish and then add the supernatant containing unbound cells from the E-NCAM panned dish. The supernatant from E-NCAM panning is used because it is depleted of neuronal precursors and enriched for glial precursors. It is also essential to perform sequential panning because a small subset of A2B5 cells is also β-III tubulin positive (a neuronal marker). Thus, sequential panning is recommended to get rid of A2B5-positive cells that might also be neuronal.
6. Incubate at room temperature for 1 h (*see* **Note 23**).
7. Wash the dish 8 times with 3 mL PBS-CMF.
8. Scrape the bound cells gently with a cell scraper in 3 mL of media.
9. Collect the cells in plastic centrifuge tube and spin at 183g for 5 min.
10. Resuspend the bound cells in small volume of media.
11. Plate approx 2000–5000 cells/35-mm fibronectin-coated dishes containing 1 mL NEP basal media supplemented with FGF and PDGF.
12. Check the panning efficiency by staining both the bound and unbound fractions with A2B5 antibody.
13. Cells at this stage are bipotential and can differentiate into oligodendrocytes and astrocytes with growth factor withdrawal. They can be frozen and thawed without losing their potential for differentiation.

3.6.2. Isolation of ES-Derived GRPs by Magnetic Bead Sorting

Follow the protocol for E-NCAM magnetic bead sorting (*see* **Subheading 3.4.2.**) with the following changes.

1. Use 1:1 dilution of monoclonal A2B5 with NEP basal media instead of E-NCAM.
2. Plate the positive and negative fractions on fibronectin-coated dishes.

3.7. Maturation of ES-Derived GRPs

The procedure described below is for a 35-mm tissue culture dish.
1. Add 500 µL yrypsin-EDTA to the dish containing A2B5-positive cells.
2. Incubate the dish for 3 min at 37°C.
3. Neutralize the trypsin by adding excess medium and spin at 183g for 5 min.
4. Resuspend the cells by gentle trituration in 100 µL of the medium.
5. Plate the cells on poly-L-lysine–laminin-coated dishes in 1 mL NEP basal media.
6. For differentiation into mature oligodendrocytes, FGF-2 is withdrawn and T3 (30 ng/mL) is added to the medium.
7. 7–14 d after growth factor withdrawal and addition of T3, around 40% of the cells are O4 and Gal-C positive (antibodies recognizing oligodendrocyte-specific glycolipids). These cells look multipolar with large cell bodies and short processes.

8. For astrocytic differentiation, withdraw growth factors from two 35-mm dishes and add 10% serum to one dish and CNTF (10 ng/mL) to the other. After 7–14 d, the cells in the serum-treated dish will differentiate almost entirely into fibroblast-looking flat cells. Stain these cells with A2B5 and GFAP. Almost all cells will be A2B5$^-$ and GFAP$^+$ astrocytes, a characteristic of Type1 astrocytes. Stain the CNTF dish for A2B5 and GFAP. Fifty percent of the cells will be positive for both A2B5 and GFAP (Type 2 astrocytes). These double-positive cells bear processes and appear different from the other 50% of the cells, which are Type1 astrocytes.

3.8. Immunocytochemical Stainings

The protocol described below is for a 35-mm tissue culture dish.

3.8.1. Cell Surface Stainings

1. E-NCAM, A2B5, Gal-C, and O4 are cell surface markers.
2. Dilute the antibodies in NEP basal medium (dilute the hybridoma supernatants 1:1 or 1:2, and the other antibodies as per the manufacturer's recommendation) (*see* **Note 24**).
3. Aspirate the medium from the dishes and incubate the cells with 1 mL of primary antibodies diluted in NEP basal media for 1 h at room temperature.
4. After 1 h, remove the primary antibody and wash the dish gently three times with DMEM.
5. Dilute the appropriate secondary antibody (1:200–1:500) in NEP basal media and incubate the cells for 20 min at room temperature.
6. Rinse the dish three times and observe the staining under fluorescent microscope with the appropriate filter set.

3.8.2. Cytoplasmic Stainings

The following procedure is for one 35-mm dish.

3.8.2.1. FLUORESCENCE STAININGS

1. Rinse the dish once with DMEM.
2. Fix the cells with either 2% paraformaldehyde or glutaraldehyde depending upon the type of antigens used. The glutaraldehyde fix is used for the neurotransmitter stainings (*see* **Note 25**).
3. Rinse the dish three times with blocking buffer.
4. Incubate the dish with the primary antibody diluted in blocking buffer for 4 h at room temperature or overnight at 4°C.
5. Rinse the dish three times with blocking buffer.
6. Incubate the dish with appropriate secondary antibody (1:100–1:200) diluted in blocking buffer for 20 min at room temperature.
7. Remove the secondary and rinse the dish three times with the blocking buffer and observe the staining under a fluorescent microscope with the appropriate filter set.

3.8.2.2. HRP STAININGS

1. Follow steps 1–5 as detailed in **Subheading 3.8.2.1.**
2. Incubate the dish with HRP-conjugated secondary for 1 h at room temperature.
3. Remove the secondary, rinse the dish thoroughly with blocking buffer first and then with Tris-imidazole buffer, pH 7.2.
4. Develop the reaction using the DAB substrate kit as per the manufacturer's directions. The actual duration of this step is quite variable, and is determined by direct evaluation of cells under a microscope for optimal coloration.

5. Rinse the dishes 3 times with Tris-imidazole buffer and observe the chromogen reaction under a light microscope (*see* **Note 26**).

3.8.2.3. DOUBLE AND TRIPLE LABELINGS

In many cases, localizing a single antigen is not sufficient. Double or triple labeling is done to further confirm the phenotype of an antigen. Double labeling can be performed with either cell surface markers or with a combination of cell surface and cytoplasmic markers. An example of each is given below.

1. Double labeling with cell surface markers:
   a. Remove the media from the dish and incubate the dish with a 1:1 dilution of Gal-C for 1 h at room temperature.
   b. Rinse the cells 3 times with DMEM.
   c. Add FITC-labeled, anti-mouse IgG secondary (1:100 dilution) for 30 min at room temperature.
   d. Remove the secondary and rinse the dish 3 times with DMEM.
   e. Observe the dish under the fluorescent microscope to see the FITC-labeled, Gal-C positive cells.
   f. Incubate the dish with an appropriate dilution of the second primary antibody, e.g., O4 for 1 h at room temperature.
   g. Rinse the dish three times with DMEM.
   h. Add TRITC-labeled, anti-mouse IgM secondary for 30 min at room temperature (*see* **Note 27**).
   i. Rinse the dish three times with DMEM.
   j. Fix the cells with fixative of choice.
   k. Take double exposures, using the appropriate filter set to record the results of the stainings.
2. Triple labeling with cell surface and cytoplasmic markers.
   a. Perform the live cell labeling with a mouse monoclonal as primary antibody and with TRITC as the secondary (to the monoclonal).
   b. Fix the dish and incubate with a second monoclonal antibody with subtype specificity with respect to the secondary antibody. For example, if your first secondary is IgG, use a FITC-labeled IgG2b.
   c. Incubate the cells with the third primary, such as a rabbit polyclonal antibody and use a secondary, which excites at a different wavelength from FITC and TRITC, such as Alexa 350.
   d. Take double or triple exposures using the appropriate filter set to record the results of double and triple stainings.

4. Notes

1. Dissolve 2 mg of mitomycin C in 1 mL of PBS, and add 1 mg of mitomycin C to 100 mL of medium. The mitomycin C solution should be made fresh each time and added to medium just prior to use.
2. All experiments in this chapter have been performed with ES-D3 cell line. If contamination of feeder cells is a problem in running some assays, then a feeder-independent subline of CCE line of ES cells is available. This line is capable of differentiating into cells of hematopoietic lineage and therefore has extensive developmental potential.
3. The substrate solutions can be reused for at least 15–20 d, if stored immediately after use at 4°C.

4. Freeze the primary antibodies in working aliquots to avoid excessive freeze-thaw cycles, which deteriorates the antibody and results in increased background. Avoid frost-free freezers.

5. DAB is a potential carcinogen. Care must be taken in handling of this material. Neutralize the used DAB solution with concentrated bleach and discard the solution in labeled containers for later disposal.

6. For blocking, use normal serum from host species providing the secondary antibodies.

7. Dissolve 4 g of paraformaldehyde in 100 mL 0.2 *M* phosphate buffer. Stir the solution at 60°C, clear the solution by adding 200 μL of 5 *M* NaOH. Filter through Whatman no. 1 paper and store in aliquots at –20°C. Paraformaldehyde will dissolve in approx 20 min with heating and stirring. Make the solution in a fume hood.

8. Depending upon the amount of native antigen in the cells, lower the levels of glutaraldehyde with concomitant increase in paraformaldehyde. The lower levels of glutaraldehyde often yield excellent signals. Exercise caution when preparing.

9. Overconfluence will select for cells that have lost contact-mediated growth inhibition. These cells rapidly overgrow and will not serve as good feeder layers.

10. The neutralization ratio is normally one volume of trypsin to two volumes of neutralization media.

11. Successful inductions can be done by culturing the aggregates at this stage, as the cells are still totipotent.

12. It should be noted that some of the aggregates will not attach to the substrate, and the proportion of unattached aggregates varies.

13. Induced cells can be selected by either using live cell labeling as described or directing expression of a reporter gene (cell surface marker, fluorescent label, or antibiotic resistance) and subsequently isolating by FACS sorting, immunopanning, magnetic bead isolation, immunodepletion, or killing uninfected cells. We have chosen to use immunopanning or magnetic bead sorting, as these methods work well with small numbers of cells. For larger numbers of cells, FACS sorting is more efficient. ES cells tend to be fragile, and we have noted a 50% or greater loss after FACS sorting. Retrovirally driven expression of GFP or an antibiotic under a cell type-specific promoter works well for cell isolations but involves several additional steps.

14. Panning dishes can also be prepared a week in advance. Seal the dishes with parafilm and store at 4°C until further use.

15. The panning dish is washed a number of times to get rid of loosely bound cells. The number of washes is important as each wash increases the efficiency of purification. We consistently get approx 90% purification after eight washes.

16. The primary antibody should be titrated carefully. If the dilution is too high, magnetic labeling will not be sufficient to retain the wanted cells on the column. On the other hand, if too concentrated a dilution is used, nonspecific binding will occur.

17. To avoid capping of antibodies on the cell surface during labeling, it is recommended to work fast, keep the cells cold, and use only cold solutions. Higher temperatures and longer incubation times at this step lead to nonspecific cell labeling.

18. Use a maximum of $10^8/500$ μL of the buffer. The volume of the buffer should be adjusted if a higher number of cells are used.

19. The rate of flow can be adjusted by varying the amount of pressure applied on the plunger, which is supplied with the column.

20. To increase the purity of the magnetic separation, cells can be passed over a second freshly prepared column.

21. Resuspension in lower volumes is recommended to get better single cell dissociation.

22. Store the used antibody at 4°C. It can be reused 3 to 4 times.

23. Always pan at room temperature. Never pan at 37°C, as this promotes activation of cell adhesion mechanisms and allows unwanted contaminating cell types to adhere to the panning dish.
24. If background is a problem, check the dilution of your hybridoma supernatant; a higher dilution may help reduce the background while still retaining the specific staining.
25. Never use gluteraldehyde with FITC label, as it gives high background.
26. Sometimes it is useful to add tap water to the final colored product, because ions present in the water can help intensify the color.
27. To avoid cross reaction of the secondaries, it is recommended to do sequential stainings with IgG first and then IgM. Fixation with fixatives such as 2% paraformaldehyde between steps also helps reduce the cross reactivity of the immunoglobulins. Finish the IgG staining, incubate the cells with second primary antibody, fix, and then perform the secondary IgM staining.

Acknowledgments

This work was supported by a National Institute of Health (NIH) FIRST award and a Muscular Dystrophy Association (MDA) research grant to M.S.R. We thank all members of our laboratories for constant stimulating discussions. Several of the antibodies used were obtained from the developmental studies hybridoma bank (DHSB). DHSB is supported by a grant from National Institute of Neurological Disorders and Stroke (NINDS). M.S.R. thanks Dr. Seema Rao for her constant support through all phases of this project.

References

1. Evans, M. J. and Kaufman, M. H. (1981) Establishment in culture of pluripotential cells from mouse embryos. *Nature* **292,** 154–156.
2. Martin, G. R. (1981) Isolation of a pluripotent cell line from early mouse embryos cultured in medium conditioned by teratocarcinoma stem cells. *Proc. Natl. Acad. Sci. USA* **78,** 7634–7638.
3. Gilbert. S. F. (1988) *Developmental biology.* Sinauer, Sunderland, MA.
4. Bradley, A., Evans, M., Kaufman, M. H., and Robertson, E. (1984) Formation of germ-line chimaeras from embryo-derived teratocarcinoma cell lines. *Nature* **309,** 255–256.
5. Stewart, C. L., Gadi, I., and Bhatt, H. (1994) Stem cells from primordial germ cells can reenter the germ line. *Dev. Biol.* **161,** 626–628.
6. Resnick, J. L., Bixler, L. S., Cheng, L., and Donovan, P. J. (1992) Long-term proliferation of mouse primordial germ cells in culture. *Nature* **359,** 550–551.
7. Doetschman, T. C., Eistetter, H., Katz, M., Schmidt, W., and Kemler, R. (1985) The in vitro development of blastocyst-derived embryonic stem cell lines: formation of visceral yolk sac, blood islands and myocardium. *J. Embryol. Exp. Morphol.* **87,** 27–45.
8. Bain, G., Kitchens, D., Yao, M., Huettner, J. E., and Gottlieb, D. I. (1995) Embryonic stem cells express neuronal properties in vitro. *Dev. Biol.* **168,** 342–357.
9. Burkert, U., von Ruden, T., and Wagner, E. F. (1991) Early fetal hematopoietic development from in vitro differentiated embryonic stem cells. *New Biol.* **3,** 698–708.
10. Mujtaba, T., Piper, D. R., Kalyani, A., Groves, A. K., Lucero, M. T., and Rao, M. S. (1999) Lineage-restricted neural precursors can be isolated from both the mouse neural tube and cultured ES cells. *Dev. Biol.* **214,** 113–127.
11. Okabe, S., Forsberg-Nilsson, K., Spiro, A. C., Segal, M., and McKay, R. D. (1996) Development of neuronal precursor cells and functional postmitotic neurons from embryonic stem cells in vitro. *Mech. Dev.* **59,** 89–102.

12. Li, M., Pevny, L., Lovell-Badge, R., and Smith, A. (1998) Generation of purified neural precursors from embryonic stem cells by lineage selection. *Curr. Biol.* **8,** 971–974.

13. Brustle, O., Jones, K. N., Learish, R. D., Karram, K., Choudhary, K., Wiestler, O. D., et al. (1999) Embryonic stem cell-derived glial precursors: a source of myelinating transplants. *Science* **285,** 754–756.

14. Robertson, E. J. (1987) *Embryo-derived stem cell lines* (Robertson, E. J., ed.), IRL Press, Washington, D.C., pp. 71–112.

17

Lineage Selection for Generation and Amplification of Neural Precursor Cells

Meng Li

1. Introduction

Embryonic stem (ES) cells are derived from the epiblast of mouse blastocyst. They can repopulate all cell lineages in vivo and can differentiate into a wide variety of cell types in vitro during embryoid body (EB) formation *(1)*. ES cells have been shown to generate both neurones and glial cells *(2,3)*. During the course of ES cell differentiation, neural precursors that express nestin and/or sox1 and sox2 appear first, these are followed by βtubulin 3 and neurofilament-expressing neurons and, subsequently, glial fibrillary-acid protein (GFAP) or O4 positive glial cells *(4–8)*. These results suggest that ES cell-derived neural system can be used for experimental dissection of various aspects of mammalian neural development. If extended to humans, in vitro-generated neural cells could also be used as source for transplantation-based cell therapy.

However, at the present time, the process of ES cell differentiation can not be directed into a single lineage. EB differentiation is disorganized and heterogeneous in nature, and the cultures contain a large number of non-neural cell types. This presents significant problems. First, inductive or trophic signals may be masked or suppressed in such a complex cellular environment and, therefore, limit the use of this system to define the controls of neural cell commitment. Second, the presence of inappropriate cells may compromise the differentiation of desired neurochemical phenotypes in grafts following transplantation. We have, therefore, developed a method by which highly purified neural precursors can be isolated *(5)*. Unlike many cells of the hematopoietic system, which can be isolated by immmunopurification, no surface markers have been identified to date to be specifically expressed by neural precursors. Therefore, we adopted a genetic approach. Although it is described in this chapter as a specific application for neural cells, lineage selection is in principle applicable as a general approach for isolation of other cell types.

The strategy is based on targeted integration of a selection marker, β*geo*, into the *sox2* gene (**Fig. 1**) (Avilion, A., Nicolis, S., Perez, L. P., Vivian, N. and Lovell-Badge, R., unpublished), the expression of which is largely restricted to the developing

From: *Methods in Molecular Biology, vol. 185: Embryonic Stem Cells: Methods and Protocols*
Edited by: K. Turksen © Humana Press Inc., Totowa, NJ

I. Targeting marker into neural precursor-specific gene in ES cells

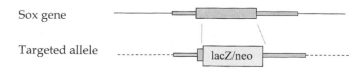

II. Elimination of non-neural cells from differentiating ES cell cultures

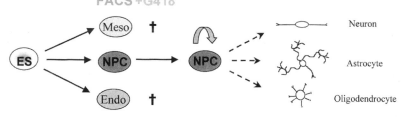

Fig. 1. Neural stem cell selection strategy. (**I**) To target a promoter-less reporter–selection marker into a neuroepithelial-specific gene by homologous recombination. In this application, β*geo*, a *lacZ-neo* fusion is integrated into the *sox2* locus, the expression of which is largely restricted to the developing neuroepithelium. (**II**) To apply either drug selection or fluorescence-activated cell sorter (FACS) sorting as positive selection of cells of interest. (*See* color plate 8, following p. 254).

neuroepithelium. In targeted cells, β*geo* is expressed in neural precursors but not other differentiated ES cell progeny. By applying G418 to the EBs, we isolated a population of morphologically immature cells that express neural precursor markers; sox1, sox2, and nestin. Subpopulations of these cells express region specific marker genes such as *mash-1*, *math* genes, and *pax* genes, suggesting that they may include distinct classes of neural progenitors corresponding to their normal counterpart during embryo development (*5*). *Sox*-selected cells differentiate into neurons and glia efficiently in the absence of mitogen. If, on the other hand, mitogen such as FGF2 is provided in the culture medium, these cells proliferate in culture while retaining their capacity for neuronal and glial differentiation (**Fig. 2**).

The ability to generate and culture pure populations of neural precursors via in vitro differentiation, combined with genetic manipulation of ES cells, provide the foundation for novel assays for analyzing inductive molecules and gene function in neural stem cell fate determination and differentiation. This model also offers a new route for the isolation of novel genes governing neural commitment. In a human setting, ES cell differentiation and lineage selection strategy may ultimately offer means to generate potentially limitless numbers of well-defined neural cells for clinical transplantation.

In this chapter, the protocols we use to isolate and manipulate ES cell-derived neural precursors are detailed. Procedures to identify those precursor cells and their differentiated progeny are also described. The basic differentiation protocol is derived from that of Bain et al. with some modifications (*2*). In our hands this protocol can yield more than 50% neuronal differentiation even without lineage selection. We have used several *sox2*-targeted ES cell lines for lineage selection, either derived from E14TG2a or CCE parental ES cells. These lines gave comparable results in terms of

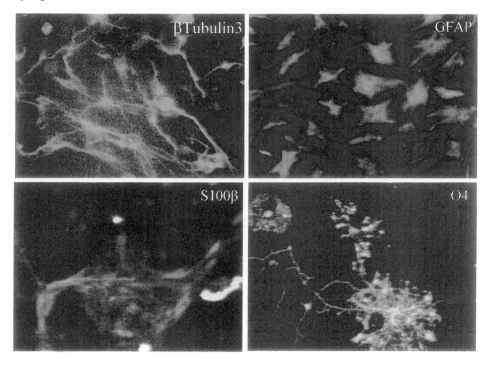

Fig. 2. ES cell-derived neurons and glia following *Sox2* selection. ES cells with a targeted insertion of β*geo* into the *Sox2* gene were induced to differentiate in EBs and selected by exposure to G418 for 48 h. EBs were dissociated and plated in DMEM/F12 plus N2 supplement. The βtubulin3 staining shown was a 4-d culture. GFAP, O4, and S100β staining were performed in 10-d cultures, which contain 5% FCS in the culture medium. The same magnification was used for all staining. (*See* color plate 9, following p. 254).

the β-galactosidase (β-gal) reporter, neuronal, and glial marker expression pattern following EB differentiation.

2. Materials

2.1 ES Cell Maintenance

1. ES cell complete growth medium. To make 440 mL of 1X complete medium, add the following components to 335 mL of sterile ultrapure water (Elgastat prima 4 system, Elga):
   a. 40 mL of 10X Glasgow modified Eagle's medium (GMEM (Gibco BRL, cat. no. 12541-025), store at 4°C).
   b. 13.2 mL of 7.5% Sodium bicarbonate (Sigma, cat. no. S-5761), store at 4°C).
   c. 4 mL of 100 m*M* glutamine (Gibco BRL, cat. no. 25030-123), store at –20°C).
   d. 4 mL of 50 m*M* sodium pyruvate mixture (Gibco BRL, cat. no. 11360-039), store at –20°C).
   e. 4 mL of 100X nonessential amino acids (Gibco BRL, cat. no. 11140-035), store at 4°C).
   f. 400 μL of 0.1 *M* 2-Mercaptoethanol (Sigma, cat. no. M-7522), store up to 4 wk at 4°C).
   g. 40 mL of fetal calf serum (FCS), store at –20°C (*see* **Note 1**).

2. Leukaemia inhibitory factor (LIF) (*see* **Note 2**). Store at –20°C.
3. Phosphate-buffered saline (PBS), calcium- and magnesium-free (Oxoid, cat. no. BR014G). 0.17 M NaCl, 3.4 mM KCl, 4 mM Na$_2$H PO$_4$, and 2.4 mM KH$_2$PO$_4$, pH 7.4. Dissolve 10 PBS tablets in 1 L ultrapure pure water, filter-sterilize, and store at room temperature.
4. 1X Trypsin solution: 0.25% (w/v) trypsin (Gibco BRL, cat. no. 25090-028). 1 mM ethylenediamine tetracetic acid (EDTA) (Sigma, cat. no. E-6758), 1% chicken serum in PBS (Sigma, cat. no. C-5405). Filter-sterilize, aliquot into 20 mL, and store at –20°C. Once thawed, keep up to 4 wk at 4°C.
5. Gelatin (Sigma, cat. no. G-1890): make 1% sterilized stock solution in ultrapure water and dilute to 0.1% with PBS. Store at 4°C.
6. 25-cm^2 plastic culture flasks (Nunc); 2–, 5–, 10–, and 25–mL pipets and centrifugation tubes.

2.2. In Vitro Differentiation and Culture of Neural Cells

1. All-*trans* retinoic acid (RA) (Sigma, cat. no. R-2625). Make stock solution in dimethyl sulfoxide (DMSO) at 10 mM. Aliquot into 50 µL and store at –20°C. Thaw and use only for a single experiment. Protect from light at all times.
2. ES cell complete medium as described in **Subheading 2.1., step 1**.
3. Geneticin (G418) (Roche, cat. no. 1 464 973). Make 200 mg/mL stock in PBS. Aliquot into 1 mL and store at –20°C.
4. 90-mm Diameter bacteriological grade Petri dishes.
5. Dulbecco's modified Eagle medium (DMEM)/F12 (1:1) with L-glutamine (Gibco BRL, cat. no. 32500-019). Prepare from powder according to manufacturer's instructions using ultrapure H$_2$O. Filter-sterilize and store at 4°C up to 1 mo.
6. Modified N2 supplement. 1X solution contains: 25 µg/mL bovine insulin (Sigma, cat. no. I-1882), 100 µg/mL human apo-transferrin (Sigma, cat. no. T-1147), 6 ng/mL progesterone (Sigma, cat. no. P-8783), 16 µg/mL putrescine (Sigma, cat. no. P-5780), 30 nM selenium chloride (Sigma, cat. no. S-5261), and 50 µg/mL bovine serum albumin (BSA) (Gibco BRL, cat. no. 15260-011). Make 100X N2 stock and store in 1-mL aliquots at –20°C. When required, dilute the stock to 1X into DMEM/F12 basal medium. This medium should be stored at 4°C and used within 2 wk.
7. Neurobasal medium (Gibco BRL, cat. no. 21103-049).
8. B27 supplement (Gibco BRL, cat. no. 17504-044).
9. Trypsin 4X: 1% trypsin, 4 mM EDTA, 4% chicken serum in PBS. Aliquot into 20 mL and store at –20°C. Once thawed, keep at 4°C.
10. Fibroblast growth factor 2 (FGF2) (PeproTech, cat. no. 100-18B). Reconstitute lyophilized protein in complete medium at 1 mg/mL. Store in 10–20 µL aliquots at –20°C.
11. Chick embryo extract (CEE) (Gibco BRL, cat. no. 15115-017). Rehydrate with 10 mL sterile deionized distilled water. Store in 1-mL aliquots at –20°C.
12. Poly-D-lysine (PDL) (30–70 kDa; Sigma, cat. no. P-7280). Make stock solution of 1 mg/mL with PBS. Store in 200-µL aliquots at –20°C.
13. Laminin (Sigma, cat. no. L-2020). Store in 20-µL aliquots at –20°C. Use once.
14. Tissue culture plates (Nunc).

2.3. Histochemical Staining for β-Gal

1. Wash buffer: 2 mM MgCl$_2$ in 0.1 M phosphate buffer, pH 7.2. Stable at 4°C for many months.
2. Fix buffer: 0.25% glutaraldehyde (Sigma, cat. no. G-5882), 5 mM EGTA (pH 8.0), and 2 mM MgCl$_2$ in 0.1 M phosphate buffer, pH 7.2. Stable at 4°C for many months.

3. X-Gal stain solution: X-gal stock: 5-bromo-4-chloro-3-indolyl-β-D-galactopyranoside (Sigma, cat. no. B-4252) dissolved in N,N-dimethylformamide (Sigma, cat. no. D-4254) at 50 mg/mL. Store at –20°C. Stain base solution: 1.64 mg/mL of $K_3Fe(CN)_6$ (Sigma, cat. no. P-8131), 2.1 mg/mL of $K_4Fe(CN)_6$ (Sigma, cat. no. P-9387), in wash buffer. Store at 4°C. The stain buffer is made up by diluting X-gal stock to 1 mg/mL into the stain base solution.

2.4. Immunocytochemistry

1. Tris-buffered saline (TBS). Make up 1× solution from 10× stock. One liter of 10X stock contains: 14.8 g $Na_2HPO_4 \cdot 2\ H_2O$, 4.3 g KH_2PO_4, 70 g NaCl, 50 g Tris-base, and 2 g NaN_3. Store at room temperature.
2. 4% Paraformaldehyde (PFA) in PBS. Store at –20°C.
3. Acidified ethanol: 95% ethanol, 5% glacial acetic acid. Store at –20°C.
4. Normal serum (e.g., goat serum). Store at –20°C.
5. Triton X-100 (Fisher, cat. no. T/3751).

2.5. Intra-Uterine Injection

1. Tribromoethanol anaesthetic (Avertin). Ingredients for 200 mL: 2.5 g 2,2,2-tribromoethanol (Aldrich, cat. no. T4, 840-2), 5 mL 2-methyl-2-butanol (Aldrich, cat. no. 24, 048-6) and distilled water *(11)*. Stable at –20°C for many months.
2. 70% Alcohol and cotton swabs.
3. Pulled glass capillary pipet (Harvard apparatus) attached to mouth pipet.
4. Dissecting instruments: small scissors, small blunt forceps, and scalpel.
5. Surgical suture (5.0 silk), needle (1/2 inch curved very fine), needle holder.
6. Saline, muslin, and 15-cm Petri dish as container.
7. Fiber optic light.
8. 37°C Warming pad.

3. Methods
3.1. Culture of Undifferentiated ES Cells

We routinely maintain ES cells in gelatin-coated tissue culture flasks without feeder cells (**ref. 9** and *see* **Note 3**). To passage ES cells cultured in 25-cm^2 flasks:

1. Draw off medium using aspirator.
2. Wash cells twice with 5 mL PBS.
3. Add 1 mL of 1X trypsin. Ensure the trypsin solution covers the cell monolayer and place in the incubator for 2 to 3 min.
4. Knock the flask several times to dissociate the cells and check under inverted microscope to ensure that cells have dissociated.
5. Add 5 mL complete medium and suspend the cells by pipetting up and down 3 times.
6. Spin down cells in bench centrifuge for 3 min at 259g force.
7. Resuspend cells in 5 mL complete medium and count the number of cells.
8. Add 10^6 ES cells in 10 mL complete medium with 100 U/mL of LIF to a fresh gelatin-coated 25-cm^2 flasks.

3.2. Preparation of Substrate-Coated Plastics

1. Dilute PDL to 10 µg/mL with PBS.
2. Apply just enough volume of PDL solution to cover tissue plates and leave at room temperature for 20 min.

3. Withdraw excess PDL using an aspirator.
4. Rinse plates twice with PBS. PDL-coated plastics can be stored dry at room temperature for several months.
5. Coat with laminin (minimum volume to cover the dish) at a concentration of 2–10 µg/mL in PBS for at least 20 min at room temperature. Alternatively, leave plates overnight at 4°C. Aspirate immediately before plating cells.

3.3. In Vitro Differentiation into Neural Cells

1. Dissociate ES cells by trypsinization as described in **Subheading 3.1., steps 1–6**.
2. Wash cells with complete medium in the absence of LIF (*see* **Note 4**).
3. Resuspend the cell pellet in complete medium at a concentration of 5–7 × 10^5 cells/mL.
4. Using a plastic pipet, place 10 mL ES cells in each 90-mm bacterial grade dish (*see* **Note 5**). This is defined as d 0 of EB differentiation.
5. On d 2, collect EBs into a conical tube. Allow 5 min for EBs to sediment at the bottom of the tube, and then remove the supernatant with an aspirator. Add fresh medium and transfer EBs to a new bacterial dish.
6. On d 4, repeat the above step and add all-trans RA to a final concentration of 10^{-6} *M*.
7. On d 6, repeat step 6. For lineage selection using Sox2βgeo targeted ES cells, in addition to all trans RA, apply 200 µg/mL of G418 to EBs (*see* **Note 6**).
8. Maintain EBs for 2 more days.

3.4. Plating and Culture of ES Cell-Derived Neural Precursors

1. On d 8 of the differentiation program, gently collect EBs into a conical tube, remove the medium with an aspirator. Wash EBs twice with PBS and dissociate them by incubating with 500 µL of 4× trypsin for 5 min in a 37°C water bath.
2. Add 5 mL complete medium to the cells to stop trypsinization. Pellet the cells by centrifugation (259*g* for 5 min).
3. Resuspend the pellet with fresh complete medium and disassociate EBs by gentle pipetting with a glass pipet. Allow 5 min for undigested EBs to sediment at the bottom of the tube, then carefully remove the upper portion that contains mainly single cells.
4. Count the cells and calculate the amount required for plating (*see* **Note 7**). Spin cells down.
5. Resuspend cells in DMEM/F12 with 1X N2. Add 20 ng/mL of FGF2, 1 µg/mL of heparin, 2% CEE, and 200 µg/mL of G418 to the cells and plate onto tissue culture plastics pretreated with PDL and laminin (*see* **Note 7**).
6. Supplement FGF2 and CEE every day at a concentration of 5 ng/mL and 0.5%, respectively. Change 2/3 volume of the medium every other day (*see* **Note 8**).
7. After 3 to 4 d, carefully remove medium using aspirator. Wash cells twice with PBS and incubate cells with minimal volume of PBS at 37°C for around 10 min. Add DMEM/F12 with 1X N2 and disassociate cells by brief pipetting, then count the cells (*see* **Note 9**).
8. Spin cells down and replate the cells under the same conditions as **step 5**.

3.5. Differentiation into Neuronal and Glial Cells

3.5.1. Differentiation Directly from Plated EBs

1. Plate EBs at 1 to 2 × 10^5 cells/cm^2 in DMEM/F12 with N2 supplement.
2. 1 to 2 d later, replace 1/2 of the medium with Neurobasal medium supplemented with B27 (*see* **Note 10**).
3. Every 3 d, change half of the medium with fresh neurobasal/B27 medium.
4. Fix cells for immunostaining 1 to 2 wk after plating (*see* **Note 11**).

3.5.2. Differentiation Following In Vitro Expansion

1. Remove growth factors and G418, then rinse cells with fresh DMEM/F12/N2 medium.
2. Culture cells in DMEM/F12 with N2 in the presence of 10^{-6} M of all-trans RA, 1 ng/mL of FGF2, and 0.5% FCS (*see* **Note 12**).
3. Cells can be analyzed 6 to 10 d later.

3.6. Characterization of ES Cell-Derived Neural Precursors

3.6.1. Histochemical Staining for β-Gal Reporter

We have shown that the βgeo selection marker–reporter is down-regulated when neural precursors proceed for terminal differentiation *(5)*. β-Gal staining can, therefore, be used as a simple method to detect the presence of neural precursor cells. This is useful for monitoring the efficiency of precursor cell generation or maintenance under different conditions.

1. Rinse cells twice with wash buffer or PBS.
2. Fix cells for 5 min at room tempeature.
3. Wash cells 3 times with wash buffer or PBS for 5 min each.
4. Incubate cells with X-gal stain solution overnight at 37°C.
5. Wash cells with PBS and observe under a microscope.

Stained cells can be stored dry indefinitely at room temperature.

3.6.2. Immunocytochemical Staining for Neural Markers

We have used a variety of immunocytochemical markers to identify neural precursors and differentiated neurons and glia derived from ES cells (**Table 1**). Most of the antibodies are available commercially. Below is a basic protocol we follow.

1. Rinse cells twice with TBS.
2. Fix with 4% PFA at room temperature for 10 min.
3. Wash cells twice with TBS, 5 min each. For intracellular antigens, permeabilize cells with 0.1% Triton X-100 in TBS for 20 min (*see* **Note 13**).
4. Block with 1% BSA and 1% normal serum (from the same species of the second antibody) in TBS for 10–20 min.
5. Incubate cells with primary antibody in TBS plus 1% normal serum for at least 1 h at room temperature (or overnight at 4°C).
6. Wash with TBS 3 times for 10 min.
7. Incubate cells with fluorescence-conjugated secondary antibody for 1 h in TBS at room temperature.
8. Wash with TBS 3 times for 10 min.
9. Mount cells with Vectashield and observe under microscope.

3.6.3. Intra-Uterine Injection into Fetal Brain

Following transplantation into embryonic brain, neuroepithelial precursors can be incorporated and undergo differentiation into cell types appropriate to the graft site. This technique can, therefore, be used to address whether ES cell-derived neural precursors can migrate and differentiate appropriately in response to distinct brain environments. The surgery is traditionally performed in rat, but we have adapted the technique for mice (**refs.** *11,12*, and Ying, Q. L. and Smith, A., unpublished). We

Table 1
Immunocytochemical Markers Used to Identify Neural Cells

Cell types	Antigens/Antibodies	Sources
ES	SSEA-1[a]	DSHB[b]
Stem/Precursor	Nestin	Chemicon
Stem/Precursor	Sox1, sox2[c]	Lovell-Badge R[d]
Stem/radial glia	RC2	DSHB
Neuron	E-NCAM[a]	DSHB
Neuron	βtubulin 3	Sigma
Neuron	Neurofilaments (NFL, NFH)	Sigma
Neuron	Microtuble-associated protein 2	Sigma
Neuron	NeuN	Chemicon
Neuron/synapses	Synapsin I	Chemicon
Neuron/synapses	Synaptophysin	Sigma
Neuron	Islet1	DSHB
Astrocyte	GFAP	Dako
Oligodendrocyte	O4[a]	Chemicon
Oligodendrocyte	S100β	Sigma

[a]Surface antigen. Should not be permeabilized with Triton X-100.

[b]Developmental Studies Hybridoma Bank, Department of Biology Sciences, The University of Iowa, 007 Biology Building East, Iowa City, IA 52242, USA.

[c]Fix cells with MEMFA: 4% formaldehyde, 100 mM MOPS, pH 7.4, 2 mM EGTA, 1 mM MgSO$_4$.

[d]Division of Developmental Genetics, MRC National Institute for Medical Research, The Ridgeway, Mill Hill, London NW7 1AA, UK.

have derived ES cells that express either β-gal or green fluorescent protein (GFP) constitutively, and therefore enable the unambiguous visualization of donor cells following transplantation (*see* **Note 14**).

1. Preparation of cells: dissociate cells as described in **Subheading 3.1., steps 1–5**, spin cells down, and resuspend cell pellet in DMEM/F12 with N2 at 5–10×10^5 cells/μL. Keep cells on ice during surgery (*see* **Note 15**).
2. Anesthetize pregnant animal (E14-15) with Avertin at approximate dose of 0.2 mL/10 g body weight.
3. Lay the mouse on her back, and swab the abdomen with 70% alcohol to sterilize the skin.
4. Make a ventral midline incision, about 1.5-cm long staring at the level of the umbilicus.
5. Lift the body wall and make a similar incision to expose the contents of the abdomen.
6. With blunt forceps, gently pull out one uterus horn, cover with saline- or PBS-soaked muslin to maintain moist.
7. Place a fiber optic light cable to the side of a conceptus. Principle anatomical markers of the foetuses (specifically ventricles inside the translucent brain) can be readily visualized by transillumination of the embryonic sacs.
8. Hold the foetuses in place with fingers and inject cells into the telencephalic ventricle in 0.5–1 μL of cell suspension through a glass microcapillary controlled by mouth. Inject all the embryos in that horn if possible (*see* **Note 16**).
9. Replace the horn into the abdominal cavity, pull out the opposite horn, and repeat the procedure.
10. When finished, sew up the abdominal wall and skin using surgical sutures.

11. Keep the female warm (but not overheated) after surgery with a 37°C heated pad under the cage.
12. Let foetuses develop to term and deliver normally. Sacrifice the recipients at desired postgrafting time points by perfusion with 4% PFA.
13. Detect lacZ or GFP labeled cells on Vibratome or cryostat brain sections.

4. Notes

1. Serum quality is critical for ES viability. Good serum should neither be toxic nor induce ES cell spontaneous differentiation. We screen several batches before purchase. Once a good batch of serum is identified, we reserve enough of this batch to supply the laboratory's culture for a year.
2. LIF is required to maintain ES cells undifferentiated. A commercial source of recombinant LIF is a product known as ESGRO from Gibco BRL and Chemicon. We produce recombinant human LIF in-house by transfecting Cos7 cells with LIF-expression vector using standard techniques. Four days after transfection, the condition medium is harvested, and LIF activity is quantified by testing serial dilutions of the medium on ES cells plated at 10^4 cells per 16-mm diameter well *(9)*. This Cos cell supernatant is used at a dilution of 1 in 10^5 or greater.
3. Differentiation efficiency into neural cells appears to be higher from ES cells cultured on gelatin-coated plastic compared to the same ES cells maintained on feeder layers.
4. LIF is absent from the culture medium at all stages of the EB differentiation program.
5. The use of bacterial grade Pctri dishes for EB culture is necessary to prevent cell attachment. EBs that do attach to the dish appear to differentiate poorly into neural cells and, therefore, should be eliminated. This can be done by collecting only the floating EBs during medium change. Alternatively, precoating of Petri dishes with 0.1% ultrapure agarose can eliminate the occurrence of such phenomenon. We use a plastic pipet for handling EBs as they have a wider tip than a glass one.)
6. G418 is used to eliminate non-sox2 expressing neural cells from the mixed population of differentiated ES cell progeny *(5)*. The concentration to be used should be tested for each batch and is normally in the range of 100–200 µg/mL.
7. For the propose of expansion, we plate cells at $2–5 \times 10^4$ cells/well of a 4-well plate in 0.5 mL (diameter of 15 mm), or $4–8 \times 10^5$ cells in 5 mL in a 60-mm dish. For other purpose, i.e., for immunostaining straight after plating or to achieve terminal differentiation, plating density is increased by 5- to 10-fold in medium without growth factors and G418. In such condition, precursor cells differentiate into neurons and glial cells.
8. During culture under these conditions, proliferating cells may form clusters initially, which then merge with each other. This normally takes 3 to 4 d, and passaging is required at this point. Differentiation of precursors into neurons and glial cells becomes apparent when cell density is getting high.
9. After incubation with PBS for 10 min, cells round up. They still attach loosely to the plastics but can easily be harvested. The volume of PBS we use for this step is 0.2 mL for a 15-mm well or 2 mL for a 60-mm dish. After disassociation, cells are transferred to a conical tube, which has been rinsed with DMEM/F12 plus 1X N2 medium. This helps the recovery of cells following centrifugation, as cells are less likely to stick to the sidewall of the spin tube and subsequently be lost either during aspiration or resuspension in a small volume. Passaging can be also done by brief trypsinization (30 s).
10. Postmitotic neurons survive better in neurobasal medium supplemented with B27.
11. Most of the pan-neuronal and glial markers are expressed after 4–6 d of differentiation. However, maturation may be required for up to 2 wk or longer for markers, such as synaptic proteins and neurotransmitter transporters.

12. RA has been shown to promote neuronal differentiation of precursor cells *(10)*. The addition of FGF and FCS in the medium improves the survival of cells in low density culture and those that are clonally expanded. FGF2 at 1 ng/mL exhibits survival effect without mitogenic activity. The time required to express most of the terminal differentiated neuronal and glial markers are in the range of 7–10 d.

13. To minimize background staining, it is best to include 0.1% Triton in TBS at all time for intracellular antigens.

14. For analysis, recipient animals are perfused with 4% PFA in 0.1 *M* phosphate buffer (pH 7.4), embedded in OCT medium and cryosectioned. The sections could then be processed for direct visualization of GFP, histochemical staining for β-gal, or immuno-histochemical staining for GFP and/or β-gal.

15. Before the injection, trypan blue is added to the cell suspension at final concentration of 0.01% as a color marker to monitor the injection.

16. This procedure is best done by two people. One person performs the injection while the other holds the foetus in place. The surgery should be limited to 1 h.

Acknowledgments

The author wishes to thank Dr. Austin Smith for critical reading of the manuscript; and for his continuous support and encouragement. Thanks also to Dr. Ying, QL for active involvement in the development of intra-uterine injection technique in mice and the generation of β-gal and GFP expressing ES cells. This work was done in Dr. Smith's laboratory and was supported by the Medical Research Council of the U.K. and by the International Human Frontier Science Program. M. L. is a recipient of Career Development Award from the British Medical Research Council.

References

1. Doetschman, T. C. and Eistetter, H. (1985) The in vitro development of blastocyst-derived embryonic stem cell lines: formation of visceral yolk sac, blood islands and myocardium. *J. Embryol. Exp. Morphol.* **87,** 27–45.

2. Bain, G., Kitchens, D., Yao, M., Huettner, J. E., and Gottlieb, D. I. (1995) Embryonic stem cells express neuronal properties in vitro. *Dev. Biol.* **168,** 342–357.

3. Fraichard, A., Chassande, O., Bilbaut, G., Dehay, C., Savatier, P., and Samarut, J. (1995) In vitro differentiation of embryonic stem cells into glial cells and functional neurons. *J. Cell Sci.* **108,** 3181–3188.

4. Okabe, S., Forsberg-Nilsson, K., Sprio, A. C., Segal, M., and McKay, R. G. D. (1996) Development of neuronal precursor cells and functional postmitotic neurons from embryonic stem cells in vitro. *Mech. Dev.* **59,** 89–102.

5. Li, M., Pevny, L., Lovell-Badge, R., and Smith, A. (1998) Generation of purified neural precursors from embryonic stem cells by lineage selection. *Curr. Biol.* **8,** 971–974.

6. Bain, G., Ray, W. J., Yao, M., and Gottlieb, D. L. (1996) Retinoic acid promotes neural and represses mesodermal gene expression in mouse embryonic stem cells in culture. *Biochem. Biophys. Res. Commun.* **223,** 691–694.

7. Lendahl, U., Zimmerman, L. B., and McKay, R. G. B. (1990) CNS stem cells express a new class of intermediate filament protein. *Cell* **60,** 585–595.

8. Pevny, L. H., Sockanathan, S., Placzek, M., and Lovell-Badge, R. (1998) A role for sox1 in neural determination. *Development* **125,** 1967–1978.

9. Smith, A. G. (1991) Culture and differentiation of embryonic stem cells. *J. Tiss. Cult. Meth.* **13,** 89–94.

10. Takahashi, J., Palmer, T. D., and Gage, F. H. (1999) Retinoic acid and neurotrophins collaborate to regulate neurogenesis in adult-derived neural stem cell cultures. *J. Neurobiol.* **38,** 65–81.

11. Papaioannou, V. E. (1990) In utero manipulations, in *Postimplantation mammalian embryos. A practical approach* (Copp, A. J. and Cockroft, D. L., eds.), Oxford University Press, Oxford, pp. 61–80.

12. Cattaneo, E., Magrassi, L., Butti, G., Santti, L., Giavazzi, A., and Pezzotta, S. (1994) A short term analysis of the behaviour of conditionally immortalised neuronal progenitors and primary neuroepithelial cells implanted into the foetal rat brain. *Dev. Brain Res.* **83,** 197–208.

18

Selective Neural Induction from ES Cells by Stromal Cell-Derived Inducing Activity and Its Potential Therapeutic Application in Parkinson's Disease

Hiroshi Kawasaki, Kenji Mizuseki, and Yoshiki Sasai

1. Introduction

1.1. Mechanisms of Neural Induction in the Ectoderm of Vertebrate Embryos

In vertebrate embryogenesis, the primordia of the nervous systems arise from uncommitted ectoderm during gastrulation. Spemann and Mangold demonstrated that the dorsal lip of the amphibian blastopore, which gives rise mainly to axial mesoderm, emanates inductive factors that direct neural differentiation in the ectoderm. During the last decade, much progress has been made in the molecular understanding of early neural differentiation in *Xenopus*. Neural inducer molecules, such as chordin, noggin and follistatin, were identified, and several intracellular mediators of neural differentiation have been characterized (*1–3*). These neural inducers do not have their own receptors on target cells, instead they act by binding to and inactivating bone morphogenfic protein 4 (BMP4), which suppresses neural differentiation and promotes epidermogenesis. Noggin and chordin bind to BMP with dissociation constants comparable to that of BMP receptors to BMP (*4,5*). Thus, neural induction in *Xenopus* is controlled by a morphogenetic signaling involving a BMP activity gradient.

By contrast, relatively little is known about the regulatory factors involved in mammalian neural induction. One main reason for this is that good experimental systems for in vitro neural differentiation are still lacking in mice, that is, something comparable to the animal cap assay commonly used in *Xenopus* studies. Mammalian embryonic stem (ES) cells can differentiate into all embryonic cell types when injected into blastocyst-stage embryos. This pluripotency of ES cells can be partially recapitulated in vitro by a floating culture of ES cell aggregates or embryoid bodies (EBs). After a few weeks of culture without leukemia inhibitory factor (LIF), EBs frequently contain ectodermal, mesodermal, and endodermal derivatives. However, there had not been any good methods that can induce "selective" differentiation to a particular cell type such as neurons. Therefore, we attempted to develop such an experimental system by using colony assays.

From: *Methods in Molecular Biology, vol. 185: Embryonic Stem Cells: Methods and Protocols*
Edited by: K. Turksen © Humana Press Inc., Totowa, NJ

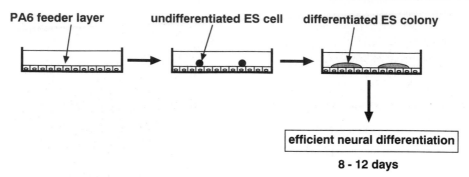

Fig. 1. Overview of the SDIA method.

1.2. Stromal Cell-Derived Inducing Activity Promotes Neural and Neuronal Differentiation of ES Cells

We first asked whether attenuation of BMP signaling is sufficient to induce neural differentiation of mouse ES cells by administering BMP antagonist molecules. Neither transfection of cytomegalovirus(CMV)-chordin plasmid into ES cells nor the addition of follistatin protein or neutralizing BMPR-Fc to culture medium induced significant neural differentiation of ES cells (**6**). On the other hand, the administration of BMP4 protein efficiently suppressed in vitro neural differentiation of mouse ES cells (e.g., EB + retinoic acid [RA] method) or of isolated mouse epiblasts even at a low concentration (0.5 nM) (H. K., K. M., and Y. S., unpublished observations). These results indicate that a blockade of BMP4 signaling is required but not sufficient for neural differentiation of undifferentiated mouse cells. This suggests that mouse ES cells, unlike *Xenopus* cells, require some unknown signals for initiating neural differentiation of ES cells, in addition to the attenuation of BMP signals.

By using a co-culture system, we screened various primary culture cells and cell lines for activities promoting neural differentiation of ES cells under serum-free conditions (**Fig. 1**) (**6**). Most of the cell types screened, including mouse embryonic fibroblasts (MEF), MDCK, and COS cells, did not significantly induce neural markers such as NCAM (pan-neural) in the overlying ES cells. However, some stromal (or mesenchymal) cells promoted neural differentiation when used as feeders. A low yet significant number of NCAM-positive colonies were observed in the presence of OP9 (stromal line derived from mouse calvaria) and NIH3T3 (an embryonic fibroblast line) cells. PA6 cells (stromal cells derived from skull bone marrow) (**7**) induced remarkably efficient neural differentiation when co-cultured with ES cells, resulting in 92% of colonies becoming NCAM-positive by d 12.

In these colonies, the majority of cells were stained with either the neuronal marker TuJ (class III β-tubulin) or the neural precursor marker nestin (**Fig. 2**). Very few colonies contained glial fibrillary acidic protein (GFAP)-positive cells (2%). The TuJ-positive neurons also expressed other neuronal markers such as MAP2 and neurofilament, and the presynaptic marker synaptophysin was detected on the induced neurons. To confirm that nestin-positive cells were neural precursors, cells were double-stained with nestin and RC2, a marker for neuroepithelium and radial glia. All of the nestin-positive cells

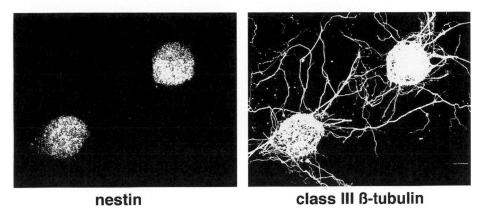

nestin   **class III ß-tubulin**

Fig. 2. Efficient induction of neural markers in SDIA-treated ES cells. Two colonies are double-stained with anti-nestin antibody and anti-class III β-tubulin antibody, which are a neural precursor cell marker and a postmitotic neuron marker, respectively.

were co-stained with the RC2 antibody. We next tested whether neural differentiation induced by PA6 was accompanied by mesodermal induction. In contrast to the high NCAM-positive percentage, very few colonies expressed mesodermal markers such as platelet-derived growth factor receptor-α (PDGFR α), Flk1, and MF20 (all <2% colonies). This is consistent with a previous report showing that PDGFR α and Flk-1 are induced in ES cells co-cultured with OP9 cells, but not in those co-cultured with PA6 cells (*8*). Thus, PA6 can promote neural differentiation of co-cultured ES cells without inducing mesodermal markers. We named the neural-inducing activity on the stromal cells "stromal cell-derived inducing activity" or "SDIA" (*6*).

1.3. Efficient Induction of Dopaminergic Neurons by the SDIA Method

Immunohistochemical analyses of the characteristics of SDIA-induced neurons revealed that 92% of colonies contained tyrosine hydroxylase (TH)-positive neurons (**Fig. 3A**). This value was much higher than those for GABAergic, cholinergic, and serotonergic neuron markers (43, 28, and 7%, respectively). TuJ-positive neurons and nestin-positive cells represented $52 \pm 9\%$ and $47 \pm 10\%$ of the total cells, respectively. TH neurons occupied $30 \pm 4\%$ of TuJ-positive neurons (i.e., approx 16% of the total cells). This value is again significantly higher than percentages of GABAergic, cholinergic and serotonergic neurons in TuJ-positive neurons ($18 \pm 8\%$, $9 \pm 5\%$, and $2 \pm 1\%$, respectively) at the cell level.

A time course study showed that TH-positive neurons appeared between d 6–8 of the induction period, following the appearance of nestin and tubulin markers (**Fig. 3B**). The cells remain negative for dopamine-β-hydroxylase (DBH), a marker for norepinephrine and epinephrine neurons, even after 13 d of induction. The mesencephalic dopaminergic neuron markers Nurr1 and Ptx3 were induced in SDIA-treated ES cells. High-pressure liquid chromatography (HPLC) analyses showed that ES cell-derived neurons released a significant amount of dopamine into the medium (7.7 pmol/10^6 cells) in response to a depolarizing stimulus. These data show that functional neurons producing dopamine were generated with this method.

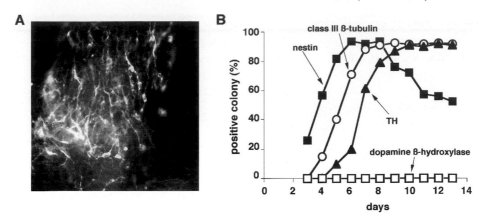

Fig. 3. Generation of dopaminergic neurons by the SDIA method. (**A**) TH-positive dopaminergic neurons induced by SDIA. (**B**) Time course of appearance of neural markers.

1.4. Epidermal Differentiation by Treating ES Cells with SDIA and BMP4

In *Xenopus*, BMP4 inhibits neural differentiation of animal cap ectoderm and promotes epidermogenesis *(1,2)*. We therefore tested whether BMP4 has a similar effect on the neural differentiation of mouse ES cells induced by SDIA *(6)*. The addition of 0.5 nM BMP4 suppressed neural differentiation and significantly increased the number of colonies positive for the non-neural ectoderm marker E-cadherin, without inducing mesodermal markers. Furthermore, after 11 d of culture, BMP4 induced keratin 14-positive colonies (0% without BMP4, 34% with BMP4). A time-course study showed that neural differentiation in ES cells was most sensitive to BMP4 during d 3–5 of the induction period.

Thus, with SDIA, neural and epidermal differentiation of ES cells are selectively induced in the absence and the presence of BMP signals, respectively.

1.5. Advantages of the SDIA Methods Compared to Previous Protocols

A superphysiological dose (up to 1 µM) of RA promotes neural differentiation in EBs *(9,10)*. Although this method can produce a good proportion of neural cells, it has two apparent problems. First, it is difficult to analyze and control each regulatory step of differentiation in this method, because EBs contain many different kinds of cells, including mesodermal and endodermal cells. Second, RA, a strong teratogen, is supposed to perturb neural patterning and neuronal identities in EBs, as it does in vivo. It is, therefore, preferable to avoid RA treatment unless RA induces the particular type of neurons of one's interest. In fact, we have noticed that RA treatment efficiently suppresses dopaminergic neuron differentiation in EB and SDIA methods *(6)*.

The SDIA method is technically simple, and the induction is efficient and speedy. Studies with various differentiation markers have demonstrated that time course features of neural differentiation by SDIA in vitro mimic well those observed in the developing embryo. The SDIA method does not involve EB formation or RA treatment, and each differentiating colony grows from a single ES cell in two dimensions under serum-free conditions. Because of these features, the SDIA method has advantages over the EB/RA methods when used for detailed analyses of differentiation, such as effects of exogenous growth factors.

Mesencephalic dopaminergic neurons can be efficiently produced by the SDIA method. Recently, it was reported that dopaminergic neurons could be generated from neural precursor cells amplified from EBs *(11)*. In contrast to our single-step method, they used the following 4 steps for the production of dopaminergic neurons. Since only small numbers of neural precursors are present in RA-untreated EBs, one needs to select and amplify nestin-positive cells for a long time (14 d) in this method before inducing differentiation (>24 d in total). This 4-step method produced TH-positive neurons at an efficiency of approx 7% of TuJ-positive neurons in conventional medium. When Shh, FGF8, and ascorbate were added, the production increased to about 30%, which is comparable to the efficiency of our method.

We infer that the quickness of dopaminergic neuron production could be crucial once this method is applied to primate–human ES cells, as primate–human ES cells seem to grow and differentiate much slower than mouse cells. Given that human ES cells take a 3 to 4 longer time period to become dopaminergic neurons than mouse cells in vitro, it seems that 8 d for mouse cells is a maximal tolerable length of time for the production of human cells by culturing cells in a practical sense.

1.6. Possible Mechanisms of Neural Induction by SDIA

At present, the molecular nature of SDIA remains to be identified. One important question that arises is whether PA6-derived factors act by antagonizing BMP4 in a similar manner as chordin and noggin. We actively blocked BMP signaling by administering neutralizing BMPR-Fc and found that the presence of BMPR-Fc did not affect the extent of neural differentiation in the ES cells under any conditions tested *(6)*. Thus, BMP antagonism is unlikely to explain the neuralizing activity of SDIA.

One possible explanation is that some additional factors such as SDIA are required for neural differentiation before mouse cells make the neural–epidermal binary decision in a BMP-dependent manner. Consistent with this idea, the SDIA-treated cells acquire their highest sensitivity to BMP4 subsequent to the onset of nestin expression (d 3) (**Fig. 3B**). This indicates that SDIA has already exerted some effects (nestin induction) before the cells react to BMP signals. The following scenario (**Fig. 4**) might be applicable to the mechanism of neuralization occurring in SDIA-treated ES cells. First, ES cells cultured on PA6 move in an ectodermal direction under the influence of SDIA. SDIA-treated ES cells then adopt a default neural status unless they receive a sufficient level of BMP4 signals. However, as the molecular nature of SDIA remains to be elucidated, we must await further study to judge this proposition and to understand the relevant roles of SDIA in the embryo.

1.7. Application to Cell Transplantation Therapies and Its Prospective

We tested whether SDIA-treated ES cells could be integrated into the mouse striatum after implantation. ES cell colonies were cultured on PA6 cells for 12 d and detached en bloc (approx 50 μm) from the feeders by a mild protease treatment without EDTA. In order to enrich for postmitotic neurons by eliminating mitotic cells, the SDIA-treated ES cells were treated with mitomycin C (MMC) before grafting. The isolated ES cell colonies were then implanted into the mouse striatum, which had been treated with 6-hydroxydopamine (6-OHDA). Ipsilateral implantation of SDIA-induced neurons significantly restored TH-positive areas in and around the DiI-positive graft. Two weeks

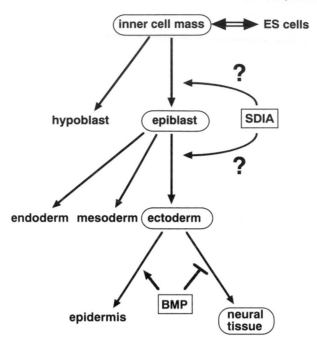

Fig. 4. A working model for the role of SDIA. SDIA may promote differentiation of ES cells into primary ectoderm, which becomes neural and epidermal tissues in the absence and the presence of BMP signals, respectively.

after implantation of 4×10^5 SDIA-treated ES cells, $3.9 \pm 0.6 \times 10^4$ grafted cells were found in the brain, and 74% of them were TuJ-positive neurons. The estimated survival rate of TH-positive neurons was approx 22% after these procedures. No teratoma formation was observed in the grafted tissue by histological analysis.

As the SDIA method produces a high yield of mesencephalic dopaminergic neurons, and once this method is successfully applied to human ES cells, SDIA-induced neurons may provide a noninvasive alternative to embryonic brain tissues and neural stem cells for neuronal replacement therapy of Parkinson's disease. Then, long-term survival of SDIA-induced neurons and its functional consequences, such as motor recovery, should be examined in Parkinsonian model monkeys before applied to patients. One advantage of using ES cells over neural stem cells is that genetic manipulations, such as modifying histocompatibility, are theoretically feasible by homologous recombination. In addition, we have to develop an even more efficient way to enrich for dopaminergic neurons from the total population of differentiated ES cells. Sorting by flow cytometry or separating with magnetic beads should be feasible once appropriate surface antigens for early dopaminergic neurons become available. We also need to test the safety of ES cell therapy in long-term implantation studies. It may be beneficial to select only postmitotic neurons by eliminating dividing cells with MMC (and/or cytosine β-D-arabinofuranoside [Ara-C]) before grafting to prevent tumor formation.

To explore further application possibilities, it is important to study how to regulate the differentiation of SDIA-treated ES cells into many specific types of central nervous system (CNS) neurons. For instance, as the generation of motor neurons from mouse

ES cells has been previously demonstrated *(12)*, the SDIA method may provide a protocol for more efficient production of dopaminergic neurons.

2. Materials

2.1. Cell Lines

1. PA6 cells: PA6 is a stromal cell line derived from newborn mouse calvaria *(7)*. Among various cell lines that we screened, PA6 cells have a marked activity to induce neural differentiation of ES cells (*see* **Note 1**).
2. ES cell lines: the ES cell line used in this study is EB5. EB5 cells carry the blasticidin S-resistant selection marker gene driven by the Oct3/4 promoter (active under the undifferentiated status) and are maintained in medium containing 20 µg/mL blasticidin S to eliminate spontaneously differentiated cells. EB5 is a subline derived from E14tg2a ES cells *(13)* and was generated by the targeted integration of the Oct-3/4-IRES-BSD-pA vector *(14)* into the Oct-3/4 allele. EB5 and CCE behaved similarly using our procedures.

2.2. Reagents (see Note 2)

1. PA6 culture medium: we use α minimum essential medium (αMEM, Gibco BRL, cat. no. 11900-024) supplemented with 50 U/mL penicillin and 50 µg/mL streptomycin. Fetal calf serum (FCS) (JRH, cat. no. 12103-78P) is added at a final concentration of 10%.
2. ES differentiation medium: Glasgow minimum essential medium (GMEM) (Gibco BRL, cat. no. 11710-035), KNOCKOUT serum replacement (KSR) (Gibco BRL, cat. no. 10828-028); 10% final concentration (*see* **Note 3**), nonessential amino acids (Gibco BRL, cat. no. 11140-050); 0.1 m*M* final concentration, sodium pyruvate (Sigma, cat. no. S-8636); 1 m*M* final concentration, 2-mercaptoethanol (WAKO, Japan, cat. no. 137-06862); 0.1 m*M* final concentration (*see* **Note 4**).
3. N-2 medium (optional): G-MEM (Gibco BRL, cat. no. 11710-035), N-2 supplement (100X, Gibco BRL, cat. no. 17502-048); 1X final concentration, nonessential amino acids (Gibco BRL, cat. no. 11140-050); 0.1 m*M* final concentration, sodium pyruvate (Sigma, cat. no. S-8636); 1 m*M* final concentration, 2-mercaptoethanol (WAKO, Japan, cat. no. 137-06862); 0.1 m*M* final concentration.

3. Methods

3.1. Maintenance of PA6 Cells

1. Remove PA6 culture medium and rinse with Ca^{2+} and Mg^{2+}-free phosphate-buffered saline (PBS).
2. Add 0.05% trypsin-EDTA (Gibco BRL, cat. no. 25300-054) and incubate for 5 min at 37°C.
3. Add PA6 culture medium and break cell aggregates by gentle pipetting.
4. Spin down the cells and resuspend the pellet in PA6 culture medium and plate one-fifth of the cells onto tissue culture dishes (*see* **Note 5**).
5. Incubate in a 37°C incubator with 5% CO_2 environment and passage cells every third day when the density reaches near the confluency.

3.2. Preparation of PA6 Feeder Layers

1. Into each well of a collagen type I-coated 8-well Biocoat culture slide (Becton Dickinson, cat. no. CBP40630), dispense 500 µL of PA6 cell suspension in PA6 culture medium. It is important to plate PA6 cells to ensure that a confluent uniform monolayer is produced (*see* **Note 6**).
2. Allow PA6 cells to attach overnight (*see* **Note 7**).

3.3. Induction of Neural Differentiation of ES Cells (see Note 8)

1. Rinse undifferentiated ES cells with PBS and collect the cells by trypsinization.
2. Add ES differentiation medium and disperse the cells into a single-cell suspension by pipetting (see **Note 9**).
3. Collect the cells by centrifugation at 250*g* for 5 min.
4. Resuspend the pellet with ES differentiation medium, count the cells, and dilute them into a final cell density of 2×10^3/mL (see **Note 9**).
5. Remove PA6 culture medium from the culture slide covered with a PA6 feeder monolayer and rinse with ES differentiation medium (see **Note 9**).
6. Plate 500 μL of ES cell suspension into each well (see **Note 10**) and incubate at 37°C under 5% CO_2.
7. Change the medium with fresh ES differentiation medium on d 4 and 6, and afterwards, everyday (see **Note 11**).
8. (Optional) Change the medium to N-2 medium on d 8. This medium promotes the differentiation of neural precursor cells to postmitotic neurons.
9. Fix the cells with an appropriate fixative and stain the cells with appropriate antibodies (see **Note 12**). Neural precursor cells, postmitotic neurons, and dopaminergic neurons first appear in culture after 3, 5, and 7 d, respectively, and 2 d later, the frequencies of these cells increase rapidly (**Fig. 3B**, see also **Note 13**). In general, more than 90% of colonies become NCAM-positive (see **Note 14**). In these colonies, the majority of the cells are neural precursor cells or postmitotic neurons. Dopaminergic neurons (tyrosine hydroxylase-positive and DBH-negative neurons) occupy about 30% of postmitotic neurons (see **Note 15**). Very few colonies contain GFAP-, PDGFRα-, Flk1-, and MF20-positive cells.

3.4. Mitomycin C Treatment of Differentiated ES Cells (see Note 16)

1. Prepare differentiated ES colonies in a 10-cm dish (see **Note 17**).
2. Remove the medium and add 6 mL of fresh ES differentiation medium containing 10 μg/mL MMC (Sigma, cat. no. M0503) onto the dish and then incubate at 37°C for 2 to 3 h.
3. Rinse the cells twice with PBS and use them for further experiments such as transplantation.

3.5. Isolation of Differentiated Colonies from PA6 Feeder Layer (see Note 18)

We routinely use the Papain Dissociation System (Worthington Biochemical Corp.) (see **Note 19**).

1. Prepare differentiated ES colonies in a 10-cm dish and, according to the manufacturer's protocol, reconstitute ovomucoid inhibitor, papain, and DNase.
2. Wash differentiated ES cells with PBS containing Ca^{2+} and Mg^{2+} (Gibco BRL, cat. no. 14040-133) (see **Note 20**). Discard PBS and pour 2 ml of the reconstituted papain solution.
3. Incubate at 37°C for 5 min (see **Note 21**).
4. Rock the dish gently. The colonies are detached from the feeder layer en bloc.
5. Collect the isolated colonies into a 15-mL tube. Spin down the cells briefly and resuspend the pellet in the reconstituted ovomucoid inhibitor solution.
6. Spin down briefly and replace the supernatant with appropriate medium.

4. Notes

1. We screened various primary culture cells and cell lines for activities promoting the neural differentiation of ES cells. While certain stromal cells such as OP9 cells and NIH3T3

cells generated neural cells to some degree, PA6 cells resulted in the highest yield and are therefore routinely used.

2. All tissue culture procedures described must be carried out under sterile conditions using sterile plasticware and detergent-free glassware. Water quality is very important. Medium or solutions should be warmed to 37°C before use. We recommend using medium less than 4 wk old to obtain a high yield of neuronal cells.

3. We noticed that the quality of KSR varies from lot to lot. We recommend lot checks to find appropriate KSR batches. It is essential to determine the optimal concentration of KSR to induce efficient neural differentiation. Optimal concentration of KSR may range from 5–15%. Inappropriate batches and concentration of KSR result in excessive E-cadherin-positive cells instead of neural cells.

4. Inappropriate concentration of 2-mercaptoethanol leads to low yields of neural cells.

5. Precoating of tissue culture dishes is not required. PA6 cells should be maintained carefully, so as not to lose their supportive activity. We recommend to passage the cells frequently (every 3 days) with a minimum dilution of the cells (1:5) using fresh medium.

6. Culture slides must be coated with collagen type I. Otherwise, PA6 cells dissociate during the differentiation period. Plasticware pretreated with a 0.1% solution of gelatin for 1 h is also useful. In the case that large numbers of the colonies are needed, we use a 10-cm tissue culture dish (Becton Dickinson, cat. no. 353003) coated with 0.1% gelatin and usually place a coverglass coated with 0.1% gelatin onto the dish. Plate 5×10^4 undifferentiated ES cells into a 10-cm dish covered with PA6 cells. After the differentiation period, the coverglass is used for immunocytochemistry to examine the efficiency of neuronal differentiation.

7. As long as PA6 cells are confluent, small differences in the cell density of PA6 cells do not significantly affect the efficiency of differentiation. MMC-treated PA6 cells can be used, although the treatment occasionally results in less efficient induction of neuronal cells. PA6 feeder slides may be used for up to 2 d after they become confluent.

8. ES cells are cultured to form a colony from a single cell. Within 8 d of differentiation, the cell number will increase more than 100-fold in a typical case.

9. Because serum strongly inhibits the neural differentiation of ES cells, it is essential to remove serum completely by washing cells with ES differentiation medium.

10. The addition of 0.5 nM BMP4 into ES differentiation medium markedly inhibits neural differentiation and promotes epidermogenesis.

11. In the case that PA6 cells dissociate during the differentiation period, the reasons may be as follows: *(1)* the cells and/or the media are old; *(2)* the plating density of PA6 cells in culture slides is extremely high; and *(3)* PA6 cells are incubated more than 3 d in culture slides before the differentiation.

12. In general, 8 d differentiation is enough to obtain high yields of postmitotic neurons, and neurites become visible by using a phase-contrast microscope. However, differentiation speed can vary depending on cell lines and other conditions. In the case that neuronal differentiation is not enough, 10–12 d differentiation should be tried. To obtain specific types of neurons that appear at later stage in vivo development, a longer incubation period should be considered.

13. Neural precursor cells, postmitotic neurons and dopaminergic neurons can be identified using anti-nestin antibody, anti-class III β-tubulin antibody, and anti-tyrosine hydroxylase antibody, respectively.

14. If yields of neural cells are low, the parameters that need to be considered are: (1) the cells and/or the media are old; (2) serum remains in the differentiation medium (*see* **Note 9**); and (3) the quality or concentration of KSR is inappropriate (*see* **Note 3**).

15. RA, which is commonly used in the EB method to induce the neural differentiation of ES cells, suppresses the induction of dopaminergic neurons without reducing the number

of postmitotic neurons. Thus, the absence of RA is required for the successful induction of dopaminergic neurons.

16. When we transplant differentiated ES cells into mouse brains, we treat the cells with MMC in advance to prevent teratoma formation.

17. If necessary, add Ara-C at a final concentration of 5 µg/mL for the last several days of the differentiation period.

18. The isolated colonies are used for transplantation, RNA analysis, and protein analysis. Although this method is useful to obtain large numbers of colonies, it is possible that a small number of PA6 cells remain on the isolated colonies. It should be noted that neurons, especially neurons whose neurites go out of colonies, might be damaged by the isolation.

19. Among several proteases that we have analyzed for their ability to isolate the colonies, papain resulted in the highest viability of the cells.

20. Removing Ca^{2+} from the medium weakens the binding between differentiated ES cells in the colonies and prevents the isolation of the colonies en bloc. Therefore, PBS containing Ca^{2+} and Mg^{2+} without EDTA is recommended.

21. After the papain treatment, attachment of ES and PA6 cell is weakened, while the integrities of a PA6 monolayer and ES colonies are not much affected. Do not incubate for more than 10 min, or a PA6 monolayer and ES colonies will be broken into pieces.

Acknowledgments

This method has been developed in a tight collaboration with the Nishikawa laboratory at Kyoto University. This work was supported by grants from the Ministry of Education, the Ministry of Health and Welfare, the Organization of Pharmaceutical Safety and Research, as well as HFSPO. We wish to thank Drs. Yoshio Yamaoka, Shigetada Nakanishi and Masatoshi Takeichi for their deep understanding and constant encouragement to our projects.

References

1. Hemmati-Brivanlou, A. and Melton, D. (1997) Vertebrate neural induction. *Annu. Rev. Neurosci.* **20,** 43–60.

2. Sasai, Y. and De Robertis, E. M. (1997) Ectodermal patterning in vertebrate embryos. *Dev. Biol.* **182,** 5–20.

3. Sasai, Y. (1998) Identifying the missing links: genes that connect neural induction and primary neurogenesis in vertebrate embryos. *Neuron* **21,** 455–458.

4. Piccolo, S., Sasai, Y., Lu, B., and De Robertis, E. M. (1996) Dorso-ventral patterning in *Xenopus:* inhibition of ventral signals by direct binding of chordin to BMP-4. *Cell* **86,** 589–598.

5. Zimmerman, L. B., De Jesus-Escobar, J. M., and Harland, R. M. (1996) The Spemann organizer signal noggin binds and inactivates bone morphogenetic protein 4. *Cell* **86,** 599–606.

6. Kawasaki, H., Mizuseki, K., Nishikawa, S., Kaneko, S., Kuwana, Y., Nakanishi, S., et al. (2000) Induction of midbrain dopaminergic neurons from ES cells by stromal cell-derived inducing activity. *Neuron* **28,** 31–40.

7. Kodama, H., Hagiwara, H., Sudo, H., Amagai, Y., Yokota, T., Arai, N., and Kitamura, Y. (1986) MC3T3-G2/PA6 preadipocytes support in vitro proliferation of hemopoietic stem cells through a mechanism different from that of interleukin 3. *J. Cell Physiol.* **129,** 20–26.

8. Kataoka, H., Takakura, N., Nishikawa, S., Tsuchida, K., Kodama, H., Kunisada, T., et al. (1997) Expressions of PDGF receptor alpha, c-Kit and Flk1 genes clustering in mouse

chromosome 5 define distinct subsets of nascent mesodermal cells. *Dev. Growth Differ.* **39,** 729–740.

9. Bain, G., Kitchens, D., Yao, M., Huettner, J. E., and Gottlieb, D. I. (1995) Embryonic stem cells express neuronal properties in vitro. *Dev. Biol.* **168,** 342–357.

10. Li, M., Pevny, L., Lovell-Badge, R., and Smith, A. (1998) Generation of purified neural precursors from embryonic stem cells by lineage selection. *Curr. Biol.* **8,** 971–974.

11. Lee, S.-H., Lumelsky, N., Studer, L. Auerbach, J. M., and McKay, R. D. (2000) Efficient generation of midbrain and hindbrain neurons from mouse embryonic stem cell. *Nat. Biotech.* **18,** 675–679.

12. Renoncourt, Y., Carroll, P., Filippi, P., Arce, V., and Alonso, S. (1998) Neurons derived in vitro from ES cells express homeoproteins characteristic of motoneurons and interneurons. *Mech. Dev.* **79,** 185–197

13. Hooper, M., Hardy, K., Handyside, A., Hunter, S., and Monk, M. (1987) HPRT-deficient (Lesch-Nyhan) mouse embryos derived from germline colonization by cultured cells. *Nature* **326,** 292–295.

14. Niwa, H., Miyazaki, J., and Smith, A.G. (2000) Quantitative expression of Oct-3/4 defines differentiation, dedifferentiation or self-renewal of ES cells. *Nat. Genet.* **24,** 372–376.

19

Epidermal Lineage

Tammy-Claire Troy and Kursad Turksen

1. Introduction

The epidermis is a stratified squamous epithelium that provides the protective layer of the skin. The mammalian epidermis is derived from embryonic ectoderm, and this layer eventually gives rise to a very early epithelial cell that further commits to become epidermal tissue (*for review see* **refs. *1,2***). A single layer of proliferating cuboidal cells (stratum germinatium) represents the putative epidermis in the mouse up to E8.5–E12.5 d of gestation. The stratum germinatium resides on a basement membrane and expresses markers characteristic of simple epithelial cells (i.e., keratins 8 and 18 [K8/K18]). At E12.5–E14.5, an intermediate layer (the stratum intermedium) develops, which is relatively undifferentiated and is able to proliferate. However at E15.5, the stratum intermedium begins to differentiate into the spinous layer and starts to lose its proliferative ability with all mitotic activity disappearing at about E16.5, at which time the granular layer appears followed by the stratum corneum at E16.5–E17.5. By E17.5–E18.5, the mouse epidermis is fully differentiated (*3–7*).

The role of keratins as markers of epidermal development and differentiation from mature K5/K14 positive basal cells to the terminally differentiated epidermal cells of the cornified layer has been widely researched (*for discussion see* **refs. *8,9***). Despite such intense study, there is still very little understood about the cell fate selection of stem cells towards the epidermal lineage (*10,11*). Equally so, there is a limited understanding of the developmental program of epidermal differentiation; such as, its regulation or the signals necessary to result in the differentiation of epidermal progenitors. While this is partly owing to a lack of appropriate markers, the most important stumbling block has been the absence of an appropriate in vitro model system in which to study the early events leading to commitment. The ability to generate differentiated epidermal progeny from a continuously growing stem cell population in vitro would provide a unique system for the study of stem cells and very early progenitors. Such a cell line would also make possible a comprehensive analysis of the underlying molecular mechanisms for the onset of embryonic epidermal commitment and differentiation. Similar in vitro approaches have yielded invaluable information on the mechanism of differentiation of other cell types. The development of stem cell models for hematopoietic, neuronal, and muscle lineages, for example, have led

From: *Methods in Molecular Biology, vol. 185: Embryonic Stem Cells: Methods and Protocols*
Edited by: K. Turksen © Humana Press Inc., Totowa, NJ

directly to important advances in our understanding of terminal differentiation. In the past, several attempts have been made to conduct similar studies on epidermal differentiation. However, the difficulties encountered, in maintaining a stable putative stem cell population in culture, limited the usefulness of these systems. Indeed, until relatively recently, even the maintenance of mature epithelial cells in culture has been problematic.

Embryonic stem cells (ES) have been shown to provide an excellent model system in which to study lineage commitment and progression in vitro for a number of lineages. ES cells are derived from the inner cell mass of 3.5-d mouse blastocysts *(12–14)*. When placed upon a suitable fibroblast feeder layer with leukemia inhibitory factor (LIF), ES cells proliferate and remain totipotent indefinitely. Removal of ES cells from their feeder layer induces aggregation and differentiation of the cells into simple or cystic embryoid bodies (EBs) *(14,15)*. Simple EBs consist of ES cells surrounded by a layer of endodermal cells, while cystic EBs develop an additional layer of columnar ectoderm-like cells around a fluid-filled cavity, morphologically similar to embryos at the 6- to 8-d egg cylinder stage *(12,16,17)*. The expression of markers for mesoderm, endoderm, and ectoderm indicates that cells derived from all three germ layers occur in cystic EBs *(18)*. We describe protocols here for culture conditions in which ES cells undergo epidermal differentiation in a highly reproducible fashion, providing a powerful system for investigating epidermal lineage, and a route to the isolation and characterization of epidermal stem cells.

2. Materials

2.1. Tissue Culture

1. 1X Phosphate-buffered saline (PBS): prepared from 10X PBS, then aliquoted into 100-mL bottles and autoclaved. To prepare 10X PBS (1 L) combine the following: 11.5 g sodium phosphate dibasic (Na_2HPO_4), 2.0 g potassium phosphate monobasic (KH_2PO_4), 80 g sodium chloride (NaCl), 2 g potassium chloride (KCl) in distilled water. It is important to dissolve the first two ingredients before adding the last two. Make the volume to 1 L. To make 1 L of 1X PBS, mix 100 mL of 10X PBS and 900 mL dH_2O.
2. 7X-OMATIC (4 gallons; Bellco Glass, cat. no. 4752-20001) (*see* **Note 11).**
3. Collagen Solution Type 1 from calf skin (20-mL bottle; Sigma, cat. no. C-8919).
4. Dimethyl sulfoxide (DMSO) (100-mL bottle; Sigma, cat. no. D-2650).
5. Dulbecco's modified Eagle medium (DMEM) 1X (500-mL bottle; Gibco BRL, cat. no. 11960-044) (*see* **Note 2**).
6. Fetal bovine serum (FBS) characterized and screened for ES cell growth (500-mL bottle; Hyclone, cat. no. SH30071.03) (*see* **Note 3**).
7. Growth factor reduced matrige basement membrane matrix (BM) (10-mL bottle; Collaborative Biomedical Products, cat. no. 35-4230) (*see* **Note 4**).
8. Insulin–transferrin–selenium (ITS) 100X (10-mL bottle; Gibco BRL, cat. no. 41400-045).
9. Keratinocyte Growth Media without Calcium (500-mL bottle; Clonetics, cat. no. CC-3112).
10. Minimum essential medium (MEM) nonessential amino acids (NEAA) solution 10 m*M*, 100X (100-mL bottle; Gibco BRL, cat. no. 11140-050).
11. MEM sodium pyruvate solution 100 m*M*, 100X (100-mL bottle; Gibco BRL, cat. no. 11360-070).
12. Mitomycin C (2-mg vial; Roche Diagnostics, cat. no. 107409) (*see* **Note 5**).
13. Penicillin–streptomycin 100X (100-mL bottle; Gibco BRL, cat. no. 15140-122).

14. Trypsin-EDTA: 0.05% trypsin, 0.53 m*M* EDTA•4Na 1X, (500-mL bottle; Gibco BRL, cat. no. 25300-062).
15. Trypsin-EDTA: 0.25% trypsin, 1 m*M* EDTA•4 Na 1X, (500-mL bottle; Gibco BRL, cat. no. 25200-072).
16. Corning polystyrene 100-mm tissue culture dishes (Corning, cat. no. 25020-100).
17. Corning polystyrene 60-mm tissue culture dishes (Corning, cat. no. 25010-60).
18. Corning polystyrene 35-mm tissue culture dishes (Corning, cat. no. 25000-35).
19. 100-mm Fisherbrand Petri dishes (Fisher, cat. no. 08-757-12).
20. 15-mL Polypropylene conical tubes (Beckton Dickenson, cat. no. 2097).
21. 5-cc Syringes (case of 100; Becton Dickenson, cat. no. 309603).
22. 0.2-µm Syringe filters (case of 50; Nalgene, cat. no. 190-2520).
23. Cryo tube vials (1.0-mL tubes, pack of 500; Nunc, cat. no. 375353).
24. StrataCooler cryo preservation module (Stratagene, cat. no. 400005).
25. Coverslips (22 × 22 mm; Bellco, cat. no.1916-12222; 12 mm; Bellco, cat. no. 1943-00012).

2.1.1. Media for Embryonic Fibroblast Cells

Embryonic fibroblast (EF) cells are maintained in supplemented DMEM. DMEM is supplemented with 10% heat-inactivated FBS, 1% sodium pyruvate, 1% NEAA, and 1% penicillin–streptomycin. This media is called 10% DMEM. To prepare 100 mL 10% DMEM, combine 10 mL heat-inactivated FBS, 1 mL penicillin–streptomycin, 1 mL NEAA, 1 mL sodium pyruvate, and 87 mL DMEM.

2.1.2. Media for ES Cells

R1 ES cells are maintained in supplemented DMEM. DMEM is supplemented with 15% heat-inactivated FBS, 1% sodium pyruvate, 1% NEAA and 1% penicillin–streptomycin. This media is called 15% DMEM (*see* **Note 6**). To prepare a bottle of 15% DMEM; to one 500-mL bottle of DMEM purchased from Gibco BRL add the following: 90 mL heat-inactivated FBS, 6 mL penicillin–streptomycin, 6 mL NEAA, and 6 mL sodium pyruvate.

2.1.3. Media for MKC Cells

Mouse keratinocyte cells (MKC) cells are maintained on collagen-coated 60-mm tissue culture dishes in supplemented keratinocyte growth media (KGM). KGM is supplemented with 1% ITS and 2% heat-inactivated FBS. The supplemented media has been termed KI2. To prepare 100 mL KI2 media combine 97 mL KGM, 1 mL ITS, and 2 mL heat-inactivated FBS.

2.1.4. General Comments and Required Equipment for Tissue Culturing

As a general rule, all tissue culture protocols must be performed using sterile techniques with great attention given to using clean and detergent-free glassware (*see* **Note 1**) and all media and solutions must be warmed to 37°C before use.

The tissue culture facility for ES cell culturing requires the following:

1. 37°C water bath.
2. Coulter Cell Counter Z2 series.
3. Glass pipets designated for tissue culture only (10 mL and 25 mL).
4. Humidified incubator at 37°C and 5% CO_2.

5. Inverted microscope with a range of phase contrast objectives (×10 to ×25) equipped with photographic capabilities.
6. Laminar flow cabinet.
7. Liquid nitrogen storage tank.
8. Pipetmen (2, 10, 20, 100, 200, and 1000 µL) designated for tissue culture use only.
9. Refrigerator (4°C) and freezer (–20°C).
10. Tabletop centrifuge.

2.2. In situ Hybridization

1. 5-Bromo-4-chloro-3-indolyl-phosphate 4-toluidine salt solution (BCIP) (3-mL vial; Roche, cat. no. 1 383 221).
2. Calf serum (100-mL bottle; Sigma, cat. no. C 6278).
3. Formamide, deionized (500-mL bottle; VWR, cat. no.4650).
4. Glycerol (1 L; Fisher, cat. no. BP229-1).
5. 4-Nitro blue tetrazolium chloride solution (NBT) (3 mL; Roche, cat. no.1 383 213).
6. Proteinase K (20 mg/mL stock) (50-mg vial; Fisher, cat. no. BP1700-50). To prepare a 20 mg/mL stock solution, add 2.5 mL water to a 50-mg vial, mix, and store at –20°C.
7. RNase A (10 mg/mL stock) (100-mg vial; Roche, cat. no. 109 169). To prepare 10 mg/mL stock solution, add 10 mL water to a 100-mg vial, boil for 5 min, and store at –20°C.
8. Digoxigenin (DIG) RNA labeling mix (Roche, cat. no. 1 277 073).
9. RNA polymerase (Roche, T3 cat. no. 1 031 163; T7 cat. no. 881 767).
10. 10X Transcription buffer (1-mL vial; Roche, cat. no. 1465384, supplied with RNA polymerase).
11. DIG-labeled control RNA (Roche, cat. no. 1 585 746).
12. QIAquick gel extraction kit (Qiagen, cat. no. 28704).
13. Heparin, porcine, sodium salt (Gibco, cat. no. 15077-019).
14. Glutaraldehyde: 8% aqueous solution (Sigma, cat. no. G 7526).
15. Levamisole hydrochloride (Sigma, cat. no. L9756).
16. 0.2 M EDTA: add 4 mL 0.5 M EDTA to 6 mL diethyl pyrocarbonate (DEPC) water.
17. Alkaline phosphatase (AP) buffer; for 10 mL, mix 1 mL 1 M Tris, pH 9.5, 500 µL 2 M NaCl, 500 µL 1 M MgCl$_2$, 10 µL 0.1 M levamisole, 100 µL 10% Tween-20, and 7.89 mL DEPC-H$_2$O.
18. Bovine serum albumin (BSA) (10 mg/mL).
19. DEPC-H$_2$O.
20. Glutaraldehyde (0.2%)/paraformaldehyde (4%) in PBS-Tween (PBS-T). For 4 mL add 100 µL of 8% glutaraldehyde from –80°C stock to 4 mL of 4% paraformaldehyde. (Glutaraldehyde solution is aliquoted upon receipt and stored at –80°C. Aliquots are thawed only once.)
21. Glycine 20 mg/mL (dissolve 0.04 g of glycine in 2 mL PBS-T and store at –20°C) and 2 mg/mL (add 100 µL of 20 mg/mL stock into 1 mL PBS-T).
22. Heparin 50 mg/mL; dissolve 50 mg (0.05 g) of heparin in 1 mL of 4X standard saline citrate (SSC). Store at –20°C.
23. Hybridization buffer. For 50 mL, combine 25 mL formamide (deionized), 12.5 mL 20X SSC, 50 µL yeast tRNA (50 mg/mL stock), 50 µL heparin (50 mg/mL stock), 500 µL Tween-20 (10% Tween-20 stock), and 11.9 mL DEPC-H$_2$O. Store at –20°C.
24. Levamisole hydrochloride (0.1 M).
25. MeOH/PBS-T series; 25% Methanol/PBS-T, 50% Methanol/PBS-T, and 75% Methanol/PBS-T.
26. MgCl$_2$ (1 M).
27. NaCl (2 M).

28. NBT plus BCIP staining solution; for 2 mL, mix 6.75 µL NBT, 7 µL BCIP, and 1.97 mL AP buffer.
29. Paraformaldehyde (4%); must be prepared fresh in the following manner: measure 2 g paraformaldehyde into 35 mL warmed DEPC water (55°–60°C), and add 8.5 µL 1 N NaOH to dissolve, and stir. Add 5 mL 10X PBS (RNase Free) and top volume to 50 mL. Check that the pH is 7.2–7.5 (using pH paper).
30. PBS (1X) (*see* **Subheading 2.1., step 1**).
31. PBS-T is 1X PBS plus 0.1% Tween-20.
32. PBS-T containing 5% calf serum. For 1 mL of solution, add 50 µL calf serum to 950 µL of PBS-T.
33. Proteinase K (10 µg/mL) must be made fresh. For 2 mL, add 1 µL 20 mg/mL proteinase K stock to 2 mL PBS-T.
34. SSC (20X); for 1 L, mix 175.25 g NaCl, 88.25 g sodium citrate, and adjust the final volume with DEPC-H_2O.
35. SSC (2X), 0.1% Tween-20; for 15 mL, mix 1.5 mL 20× SSC, 150 µL 10% Tween-20, and 13.35 mL DEPC-H_2O.
36. SSC (2X), 0.1% Tween-20 containing 20 µg/mL RNase A; for 3 mL, add 6 µL of 10 mg/mL RNaseA stock to 3 mL of 2X SSC, 0.1% Tween-20.
37. Tris (1 *M*), pH 9.5.
38. Yeast tRNA(50 mg/mL); dissolve 50 mg yeast tRNA in 1 mL of 4X SSC. Store at –20°C.

2.3. Ayoub Shklar Staining

1. 37% Formaldehyde (500 mL; Fisher, cat. no. F79-500).
2. Acid fuchsin (100 g; Sigma, cat. no. A 3908).
3. Aniline blue (25 g; Polysciences Inc., cat. no. 02570).
4. Orange G (100 g; Sigma, cat. no. O 7252).
5. Phosphotungstic acid (100 g; Sigma, cat. no. P-6395).
6. 95% Ethanol.
7. Filter paper and funnels.
8. 1X PBS (*see* **Subheading 2.1., step 1**).
9. 10% Neutral-buffered formalin; for 100 mL, combine 10 mL 10X PBS, 10 mL 37% formaldehyde, and 80 mL dH_2O, then store at 4°C.
10. 5% Acid fuchsin; for 100 mL, add 5 g acid fuchsin to 100 mL dH_2O, and stir at least 3 h or overnight. Filter and store at room temperature for 6 mo. Filter before use.
11. Aniline blue–orange G; for 100 mL, mix 0.5 g aniline blue, 2 g orange G, and 1 g phosphotungstic acid in 100 mL dH_2O. Stir at least 3 h or overnight. Filter and store at room temperature for 6 mo. Filter before use.

2.4. Immunofluorescence

1. Methanol (4 L; Fisher Scientific, cat. no. A452SK-4) with a quantity stored at –20°C.
2. Humidified chamber (*see* **Note 7**).
3. Nalgene centrifuge tubes.
4. 50°C Water bath.
5. Disposable pipets.
6. 1,4-Diazabicyclo[2. 2. 2]octane (DABCO) (25 g; Sigma, cat. no. D-2522).
7. 1X PBS (*see* **Subheading 2.1., step 1**).
8. 0.2 *M* Tris, pH 8.5.
9. Mowiol 4-88 (500 g; Polysciences, cat. no. 17951). For 20 mL, use the following procedure:
   a. Add 2.4 g Mowiol 4-88 to 6 g of glycerol and stir for 2 h at room temperature.
   b. Add 6 mL dH_2O and stir for several hours at room temperature.

    c. Add 12 mL 0.2 *M* Tris, pH 8.5, and heat to 50°C for 10 min with occasional mixing.

    d. Using a disposable pipet, transfer solution to a centrifuge tube and spin at 5000*g* for 15 min.

    e. Add 2.5% DABCO to reduce fading. Stir to dissolve.

    f. Aliquot into 1-mL tubes and store at –20°C. As needed, a tube is thawed and may be stored for several weeks at 4°C.

2.5. Reverse Transcription Polymerase Chain Reaction

1. Thermal cycler (Model 2400; Perkin Elmer, cat. no.0993-6057).
2. Agarose gel apparatus and reagents.
3. Trizol reagent (100 mL; Gibco BRL, cat. no. 15596-026).
4. Cell scrapers (Costar, cat. no. 3010).
5. 17 × 100-mm Polypropylene sterile culture tubes (Fisher, cat. no. 14-956-1J).
6. Chloroform (500 mL; Sigma, cat. no. C-2432).
7. Isopropanol (500 mL; Fisher, cat. no. A416-500).
8. 75% Ethanol.
9. RNase inhibitor (20 U/µL; Perkin Elmer, cat no. N808-0119).
10. DNase I (256 U/µL; Gibco, cat no. 18047-019).
11. Phenol, saturated (Intermedico, cat. no. AO/0945-100M).
12. 25 m*M* MgCl$_2$ (Perkin Elmer, cat. no. N808-0010).
13. 10X PCR buffer II (Perkin Elmer, cat. no. N808-0010).
14. 100 m*M* dNTP set (Gibco, cat. no. 10297-018).
15. RNase inhibitor (20 U/µL; Perkin Elmer, cat. no. N808-0119).
16. MuLV reverse transcriptase (50 U/µL; Perkin Elmer, cat. no. N808-0018).
17. Random hexamers (50 m*M*; Perkin Elmer, cat. no. N808-0127).
18. *Taq* DNA polymerase (5 U/µL; Perkin Elmer, cat. no. N808-0160).
19. DEPC-H$_2$O.
20. 10X DNase buffer (200 m*M* Tris-HCl, pH 8.4, 20 m*M* MgCl$_2$, 500 m*M* KCl); for 100 mL, combine 50 mL 1 *M* KCl, 1.5 mL 1 *M* MgCl$_2$, 10 µL gelatin, and 10 mL Tris-HCl (pH 8.4).
21. 10 m*M* dNTPs are required for reverse transcription (RT). Dilute stock dNTPs 1:10 with DEPC-H$_2$O.
22. 2.5 m*M* dNTPs are required for polymerase chain reaction (PCR). Combine 250 µL of each stock dNTP (100 m*M*) into a sterile microtube and mix (equals 25 m*M*). Aliquot 100 µL into each of 10 tubes and add 900 µL sterile H$_2$O and mix (equals 2.5 m*M*).
23. PCR primers diluted to 25 pmol/µL with sterile H$_2$O.

3. Methods

3.1. Tissue Culture

3.1.1. Freezing Cells

As a general rule, freeze cells slowly and thaw them quickly. For long-term storage (indefinitely), cells must be kept under liquid nitrogen.

To prepare 10 mL freezing medium, add 1 mL DMSO to 9 mL of appropriate media and proceed with the following protocol:

1. Trypsinize cells in the exponential phase of growth (approx 3 d in culture) from a 100-mm dish.
2. Pellet cells by centrifugation (700*g*, 2–3 min) and resuspend in an appropriate amount of freezing medium and gently resuspend (*see* **Note 8**).

3. Aliquot 1 mL of the cell suspension into freezing vials.
4. Immediately transfer the vials to a precooled (–20°C) StrataCooler and store it in a –70°C freezer for 24 h.
5. Transfer the tubes to liquid nitrogen.

3.1.2. Thawing Cells

When thawing vials of cells from the liquid nitrogen, one must work quickly and efficiently in order to maintain the integrity of the cells (*see* **Note 9**). Immediately after the cells are thawed, transfer the entire contents of the cryo vial into 10 mL of media in a 15-mL Falcon tube and collect the cells by centrifuging at 700*g* for 2 to 3 min. Remove the media and gently resuspend the cells with fresh growth media and transfer to a tissue culture dish. Allow the cells to adhere overnight in a 37°C incubator and change the media the next day.

3.1.3. Fibroblast Feeder Layers

ES cells require a feeder layer of mitotically inactivated fibroblast cells or gelatinized tissue culture dishes with the addition of LIF in order to remain pluripotent in culture. The ES cells that we routinely use, the R1 ES cell line *(19)*, were established on an EF feeder layer, and so we continue to use this substrata. Because of the short culturing life span of murine EFs, it is advisable to prepare stocks of frozen vials that are capable of supporting ES cells. Therefore, every new batch of EF cells must be tested before they may be used as feeder layers. We use the protocol outlined in **Subheading 3.1.3.1.** for the isolation of EF cells. However, there are a number of slight variations of this protocol that have been used very successfully by others.

3.1.3.1. METHOD FOR THE ISOLATION OF EF CELLS

1. Sacrifice pregnant mice at about 13–15 days postcoitum (dpc) (we routinely use 2 to 3 pregnant mice).
2. Clean the underside of the mouse with 70% ethanol and remove the uterus to a sterile Petri dish.
3. Remove the embryos from the uterus and transfer them to a sterile Petri dish.
4. Dissect away the heads and internal organs (liver, heart, kidney, lung, and intestine).
5. Wash 10–12 carcasses in a 50-mL Falcon tube with 50 mL 1X PBS, at least 3 times to remove as much blood as possible.
6. Cut the carcasses into small pieces with dissecting scissors. Use the following rule when cutting the embryos: (no. of embryos)^2 = the number of cuts; for example if there were 12 embryos, cut 144 times.
7. Transfer the tissue pieces into a sterile 50-mL Falcon tube and rinse once with 10% DMEM.
8. Rinse twice with 30 mL 1X PBS to remove blood cells and all traces of media.
9. Add 2 mL of 0.25% trypsin-EDTA per embryo (i.e., for 12 embryos add 24 mL trypsin-EDTA) and incubate at 37°C for 20 min.
10. Pipet up and down 24 times, remove the cells in suspension, and transfer 2 mL/100-mm tissue culture dish with 10 mL 10% DMEM and incubate overnight at 37°C.
11. Add 10 mL fresh 0.25% trypsin-EDTA to the remaining tissue and incubate for 20 min at 37°C.
12. Pipet up and down 10 times and plate 2 mL/100-mm tissue culture dish and add 10 mL 10% DMEM. Incubate cultures overnight at 37°C.

13. The next day, trypsinize and collect all cells to mix the populations. Plate on fresh tissue culture dishes.
14. The next day, change the media. When confluent (2 to 3 days), trypsinize, collect all cells, and freeze 1×100 mm = 1 vial, store in liquid nitrogen.

3.1.3.2. MAINTENANCE OF EF CELLS

When thawing EF cells, the contents of one cryo vial is transferred to one 100-mm tissue culture dish, and the cells are grown to confluency for 2 d and then subcultured 1:4 for use as feeder layers. Generally, a new vial of EF cells is opened on a Wednesday and subcultured 1:4 on Friday, in order to be ready for the following week's cultures (*see* **Table 1**). Once initial cultures have recovered, EF cells are maintained on 100-mm tissue culture dishes to confluency and then subcultured 1:2 every 3 to 4 d. To maintain a precise culturing schedule, EF cells are generally subcultured on Tuesday and Friday, in order for them to be on a comparable time course to the other cells (*see* **Table 1**). To subculture:

1. Remove media and rinse with 1X PBS.
2. Add 3 mL 0.05% trypsin-EDTA to each 100-mm dish.
3. Return plate to incubator for 2 to 3 min until the cells float with gentle agitation.
4. For 3×100-mm dishes, add 5 mL 10% DMEM to rinse the 3 dishes and collect all 14 mL into a 15-ml Falcon tube.
5. Centrifuge at $700g$ for 2 to 3 min to pellet the cells. While spinning, prepare 6×100-mm dishes by labeling the cell type and passage number and add 9 mL 10% DMEM to each plate.
6. After the spin has completed, suction off the media with a pasteur pipet, being cautious not to disturb the pellet.
7. Gently resuspend the pellet in 6 mL 10% DMEM, by carefully pipetting up and down 5 times.
8. Plate 1 mL (approx 2×10^6 cells) to each of the 6×100-mm dishes and gently agitate the plate back-and-forth and side-to-side to evenly distribute the cells.
9. Incubate at 37°C until they reach confluency (3 to 4 d). The media is changed every 2 d. EF cells are maintained only to the 7th passage. At this point, cell growth becomes retarded, and the cells exhibit signs of senescence.

3.1.4. Embryonic Stem Cells in Culture

3.1.4.1. MITOMYCIN C TREATMENT OF FEEDER LAYER FOR ES CELLS

ES cells are maintained on 100-mm Corning tissue culture dishes that have a mitomycin C-treated EF feeder layer. When the EF cells are confluent, they are treated with mitomycin C in order to halt the division of the cells while they are still able to condition the media.

1. Add 250 μL mitomycin C to a 100-mm confluent EF dish containing 10 mL of media, and agitate back-and-forth and side-to-side.
2. Incubate at 37°C for 2 to 3 h.
3. Rinse the dishes 3 times with 10 mL of 1X PBS each wash, then replace the media with 15% DMEM.

The EF cells are now ready to be used as a feeder layer for ES cells. Thaw ES cells according to **Subheading 3.1.2.** of this chapter. The content of one vial is thawed onto

Table 1.
Tissue Culture Time Table

Week #	Monday	Tuesday	Wednesday	Thursday	Friday	Saturday	Sunday
1			-thaw EF cells 1 vial onto 1x100mm dish = **EF1**	-change media **EF1**	-s/c EF 1:4 on 4x100mm dishes = **EF2**		
2		-s/c EF 1:2 on 6x100mm dishes = **EF3** -m/c treat 1x100mm EF2 and thaw 1 vial ES cells on it = **ESOF1**	-change media **ESOF1**	-change media **ESOF1**	-s/c EF 1:2 on 6x100mm dishes = **EF4** -m/c treat 2x100mm EF3 and s/c ESOF 1:10 on 2xm/c EF = **ESOF2** and 2xpetri = **EB1**		-change media **ESOF2**
3	-change media **EB1**	-s/c EF 1:2 on 6x100mm dishes = **EF5** -m/c treat 2x100mm EF4 and s/c ESOF 1:10 on 2xm/c EF = **ESOF3** and 2xpetri =**EB2**	-thaw EF cells 1 vial onto 1x100mm dish = **EF1**	-change media **EF1** -change media **ESOF3** -plate **dEB1** 1:10 -change media **EB2**	-s/c EF 1:2 on 6x100mm dishes = **EF6** -s/c new EF 1:4 on 4 x100mmdishes=**EF2** -m/c treat 2x100mm EF5 and s/c ESOF 1:10 on 2xm/c EF = **ESOF4** and 2xpetri =**EB3**	-change media **dEB1**	-change media **ESOF4**
4	-plate **dEB2** 1:10 -change media **EB3** -**dEB1** are ready for EPC generation -plate **EPC**	-s/c EF 1:2 on 6x100mm dishes = **EF7** (if required) -s/c EF 1:2 on 6x100mm dishes = **EF3** -m/c treat 2x100mm EF6 and s/c ESOF 1:10 on 2xm/c EF = **ESOF5** and 2xpetri =**EB4**	-change media **EPC**	-change media **ESOF5** -plate **dEB3** 1:10 -change media **EB4**	-s/c EF 1:2 on 6x100mm dishes = **EF4** -m/c treat 2x100mm EF3 and s/c ESOF 1:10 on 2xm/c EF = **ESOF6** and 2xpetri =**EB5** -**dEB2** are ready for EPC generation -plate **EPC**	-change media **EPC** -change media **dEB3**	-change media **ESOF6** -change media **EPC**
5	-plate **dEB4** 1:10 -change media **EB5** -**dEB3** are ready for EPC generation -plate **EPC**	-s/c EF 1:2 on 6x100mm dishes = **EF5** -m/c treat 1x100mm EF4 and thaw 1 vial ES cells on it = **ESOF1** -s/c ESOF 1:10 on 2xpetri =**EB6** -change media **EPC**	-thaw EF cells 1 vial onto 1x100mm dish = **EF1** -change med a **ESOF1** -change media **EPC**	-change media **EF1** -change media **ESOF1**	-s/c EF 1:2 on 6x100mm dishes = **EF6** -s/c new EF 1:4 on 4x100mm dishes = **EF2** -m/c treat 2x100mm EF5 and s/c ESOF 1:10 on 2xm/c EF = **ESOF2** and 2xpetri =**EB1** -**dEB4** are ready for EPC generation -plate **EPC** -change media **EPC**		-change media **ESOF2** -change media EPC and **EPC**

Legend:
- ● EF Cultures
- ● ES Cultures
- ● EB Cultures
- ● dEB Cultures
- ● EPC Cultures

one Mitomycin C-treated EF plate, and the media is changed 24 h after plating. At the first subculture, it is important to freeze some cells to assure that the cell line will not be depleted (*see* **Subheading 3.1.1.**).

3.1.4.2. MAINTENANCE OF ES CELLS

To subculture ES cells:

1. Gently remove media and rinse with 1X PBS.
2. Add exactly 2 mL 0.25% trypsin-EDTA to a 100-mm dish.
3. Return plate to incubator for 1 to 2 min until the cells float with gentle agitation. While waiting for trypsinization to be complete, prepare Fisherbrand 100-mm Petri dishes (*see* **Note 10**) by adding 10 mL 15% DMEM and labeling them to keep track of the passage number. These will be used for the generation of EBs.
4. When the ES cells are floating, gently pipet up and down 30 times with a 1-mL pipetman in order to make a single-cell suspension.
5. Distribute 200 µL (approx 2.5×10^6 cells) to each prepared EF cells and Petri dish, making a 1 : 10 dilution of the original culture.
6. Gently agitate up-and-down and side-to-side and incubate at 37°C.

The ES cell media must be changed every 2 d, and the cultures must be split every 3 to 4 d. For simplicity, ES cells are split every Tuesday and Friday, thereby producing EB's on the same days (*see* **Table 1**). It is important that one does not keep ES cells in culture for long periods in order to maintain pluripotency. Extensive culturing will result in abnormal karyotypes and inconsistent differentiation. It is for this reason that we generally keep ES cells in culture up to the 6th passage before new cells are thawed. Undifferentiated ES cells will stain positive for alkaline phosphatase (*see* **Subheading 3.1.4.3.** and **Fig. 1**).

3.1.4.3. AP HISTOCHEMISTRY

1. Rinse cells once with cold 1X PBS.
2. Fix for 15 min with cold formalin (on the bench).
3. Rinse with distilled water and incubate in fresh distilled water for 15 min.
4. Meanwhile, prepare the substrate (must be fresh). For 25 mL, combine 0.0025 g naphthol AS MX-PO$_4$, 100 µL N,N-dimethylformamide, 12.5 mL 0.2 M Tris-HCl (pH 8.3), and 12.5 mL distilled H$_2$O. Add 0.015 g Red Violet LB Salt and filter with Whatman's no. 1 paper directly onto dishes.
5. Incubate for 45 min at room temperature, then rinse with dH$_2$O.

3.1.5. Differentiation of ES Cells into the Epithelial Lineage

The three germ layers (ectoderm, mesoderm, and endoderm) are established during gastrulation. The ectoderm has the capacity to give rise to the epidermis and the nervous system. Specifically, the epidermis is derived from ectodermal precursor cells, which make epidermal cell fate selection as a result of inductive interactions during early embryogenesis. Following the fate selection, these cells give rise to putative multipotential epidermal stem cells capable of giving rise to the epidermal lineage. Within this lineage, one predicts, is contained progenitor cells with varying degrees of proliferative potential as well as committed–mature progenitors that are capable of further differentiation along the epidermal lineage (*see* **Fig. 2**). Based on this descriptive work, we envisioned that, in order to generate epidermal cells in culture, it is necessary to assay these three distinct stages to be successful in dissecting out the

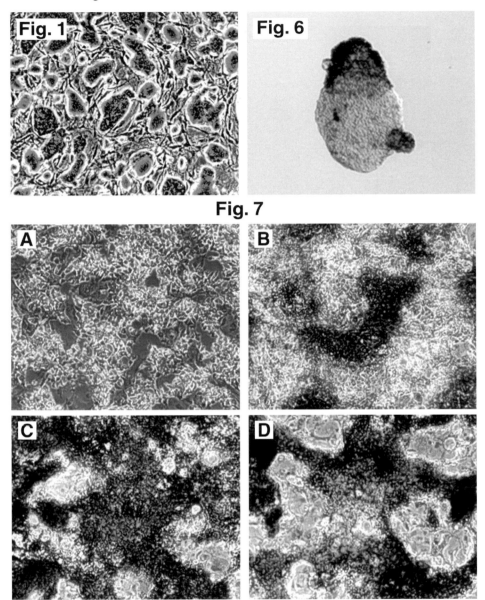

Fig. 1. *(top left)* AP staining of ES cells. ES cells growing on a fibroblast feeder layer after 3 d in culture stained with AP. All ES cells are AP positive, indicating that they are in an undifferentiated state.

Fig. 6. *(top right) In Situ* analysis of EBs. An EB after 6 d in culture showing the spatial localization of K8 in the developing EB. The expression of K8 in epithelial cells was detected using whole mount *in situ* with a K8-specific probe.

Fig. 7. *([A-D] bottom)* Ayoub Shklar staining for keratin. 2×10^6 EPCs were plated and stained with AS at four time points. **(A)** Day 2; the cells are relatively undifferentiated as indicated by the blue staining. **(B)** Day 4; the cells have begun to differentiate as indicated by the cells that have stained red. **(C)** Day 6; a large percentage of the cells have differentiated as determined by the expanding numbers of cells that have stained red. **(D)** Day 16; the cells have committed to further differentiation as indicated by a color change from orange to yellow to brown.

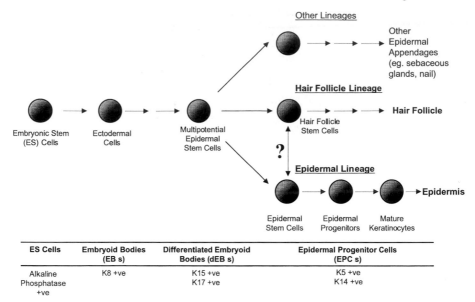

Fig. 2. A schematic representing the differentiation of committed epidermal cells from totipotent ES cells and the associated stage-specific markers.

stages of the lineage. In accordance with this rational, our culture protocol consists of three distinct stages:

1. Ectodermal cell fate selection; this can be achieved during EB formation.
2. The generation and differentiation of putatitive epidermal stem cells; during the epithelial sheet formation from EBs on matrigel-coated substrata.
3. The fate of early progenitors; differentiating epidermal stem cells presumably give rise to progenitors. These progenitors may potentially be maintained and expanded in vitro under appropriate conditions. Our secondary (epidermal progenitor cells cultures [EPC]) support this prediction.

These stages were assayed and followed for distinct markers. For example, EBs are cell aggregates that express K8 on their surface. K8 expression is assessed through whole mount *in situ* hybridization with a DIG-labeled probe for K8 (*see* **Subheading 3.2.1.**). Further differentiation is accomplished by the production of differentiated embryoid bodies (dEBs), which are adhered EBs that have an emerging spread area of epithelial cells that are K14 positive as determined by indirect immunofluorescence with antibodies against K14 (*see* **Subheading 3.2.2.**) as well as AS staining (*see* **Subheading 3.2.3.**). The culturing of secondary cultures from dEBs generates early epithelial cells called EPCs. These cells progress along the epithelial differentiation pathway to express the markers of differentiation as determined by indirect immunofluorescence and RT-PCR (*see* **Subheading 3.2.4.**).

3.1.5.1. EMBRYOID BODY (EB) FORMATION

EBs prepared in **Subheading 3.1.4.2., step 5**. remain in culture for 6 d. The media is changed on the third day (i.e., Monday for EBs produced on Friday, and Thursday for EBs produced on Monday) (*see* **Table 1**).

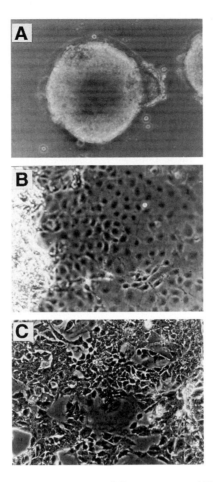

Fig. 3. Phase microscopy of cell cultures at different stages. (**A**) An EB after 6 d in culture. An ectodermal layer is easily discernible in the developing EB. (**B**) A dEB at 4 d in culture, showing the migration and formation of epithelial sheets from the differentiating EB. (**C**) EPCs after 2 d in culture generated from d 4 dEBs.

To change the media:

1. Using a 10-mL pipet, collect the entire contents of the Petri dish into a 15-mL Falcon tube.
2. Allow the EBs to settle to the bottom. Do not centrifuge, the settling process takes only a few minutes.
3. Suction off the media and gently resuspend the cells in 10 mL 15% DMEM.
4. Transfer to the original Petri dish. After 6 d in culture, EBs are ready to proceed to the next stage.

This procedure reproducibly generates EBs with an ectodermal layer that can easily be identified by microscopy (*see* **Fig. 3A**). The expression profile of K8, which is known to be associated with ectodermal cells during very early development, can be detected by whole mount *in situ* using a K8 riboprobe (*see* **Subheading 3.2.1.**). Other markers of ectoderm can be used to further characterize differentiation at this stage.

3.1.5.2. Differentiating Embryoid Body Formation

After 6 d, EBs are plated on BM-coated plates. This results in cell migration and the formation of epithelial sheets (*see* **Fig. 3B**).

1. Prepare 60-mm tissue culture dishes by coating them with BM (2 mL/60-mm dish) for 30 min at room temperature (*see* **Note 4**).
2. After 30 min, suction off the BM and gently replace it with 15% DMEM, being careful not to disturb the integrity of the matrix layer.
3. Collect the 6-d EBs (*see* **Table 1**) into a 15-mL Falcon tube, being sure to rinse the Petri dish well, with fresh 15% DMEM, in order to collect as much of the EBs as possible.
4. Allow the EBs to settle, do not centrifuge, for 2 to 3 min then suction off the media carefully as not to disturb the pelleted cells.
5. Gently resuspend the cells with 10 mL 15% DMEM and distribute 1 mL to each of the prepared 60-mm dishes (mixing frequently), thereby making a 1:10 dilution of the original culture. It is important to pipet gently up-and-down while plating the EBs, because they tend to sink quickly due to their mass.
6. Return the cells to the incubator and allow them to grow for 4 d, changing the media on the second day.

Once the cells attach and begin to spread on the plate, they are called dEBs. After a few days, dEBs have cells with epithelial morphology towards the edge of the spread area (*see* **Fig. 3B**) and, at this point, express keratin 8 (K8) (*see* **Fig. 4A**), keratin 17 (K17) (*see* **Fig. 4B**), and keratin 14 (K14) (*see* **Fig. 4C**) at different stages of the expanding dEB colony. The progression of differentiation is monitored by immunofluorescence, using epidermal-specific keratins (*see* **Subheading 3.2.2.**). With the onset of K14 expression (approx 4–6 d), secondary cultures may be produced. Detection of the mature epidermal marker K14 is indicative of the progression of differentiation along the epidermal lineage. K8 is also highly expressed in dEBs, indicating that the cells at this stage are "early" cells in the lineage. In addition, the epithelial sheets that are derived from dEBs also contain other early progenitors that express K15, K17, and K19 at varying frequencies during the differentiation process. These progenitors can be further expanded and characterized in secondary cultures by trypsinization.

3.1.5.3. Epithelial Progenitor Cells

The proliferative and differentiative capacity of EPCs in vitro can be studied through the establishment of EPC cultures from dEBs by trypsinization (*see* **Fig. 3C**). They are early epithelial cells that progress through differentiation along the epithelial pathway expressing K17 (*see* **Fig. 5A**), K14 (*see* **Fig. 5B,C**) and several other epithelial markers in a time- and density-dependent manner. These cultures are suitable for studying the role of growth factors and hormones to devise serum-free culture conditions. These studies are currently ongoing.

1. Prepare 35-mm tissue culture dishes by coating them with Matrigel (1 mL/35-mm dish, approx 0.1 mg) for 30 min at room temperature.
2. After 30 min, suction off the BM and gently replace it with 15% DMEM, being careful not to disturb the integrity of the matrix layer.
3. Trypsinize 4-d dEBs (*see* **Table 1**) with 2 mL 0.25% trypsin-EDTA per 60-mm dish for 2 to 3 min in the incubator.

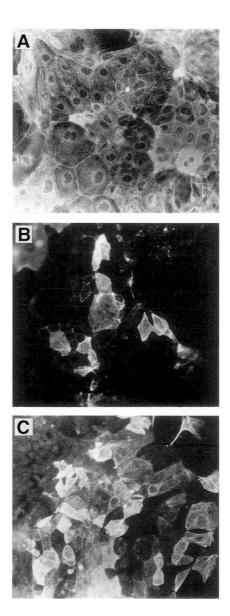

Fig. 4. Immunofluorescence of dEBs after 6 d in culture. (**A**) Almost all cells are K8 positive, indicating that they are "early" cells in epithelial lineage. (**B**) Emerging from the EB, the cells express K17 differentially, indicating that there are earlier progenitors close to the main body of the explant. (**C**) There are K14-positive cells towards the edge of the spreading epithelial cells of the dEB, indicating that the cells on the edge are much more differentiated.

4. Collect the cells, washing the dishes well with fresh 15% DMEM, into a 15-mL Falcon tube, then remove 100 μL to count the cells.
5. Centrifuge at 700*g* for 2 to 3 min for pellet formation.
6. While spinning, count the cells using a Coulter Counter.
7. Resuspend the pellet with 15% DMEM and dilute the cells appropriately to plate the desired number of cells.

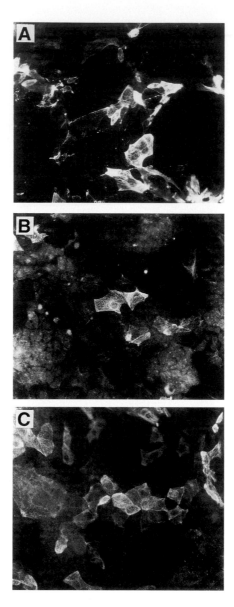

Fig. 5. Immunofluorescence of EPCs. (**A**) The detection of K17-positive cells in these cultures suggests the existence of early epithelial progenitors. (**B**) K14 expression after 2 d in culture indicates that our EPC cultures contain progenitors that are able to generate committed epithelial cells within a very short time period. (**C**) Shows increased expression of K14 after 4 d in culture, indicating that with our culture conditions, we are able to enrich and propagate epithelialization.

8. For immunofluorescence, we plate the cells on glass 22 × 22 mm coverslips in 35-mm dishes. EPCs may also be subcultured using a mouse keratinocyte co-culture, thereby enhancing their differentiation (*see* **Subheading 3.1.6.**).

3.1.6. Mouse Keratinocyte Feeder Layers for Low Density Progenitor Cultures

There are two different lines of MKCs that we use for co-culture with EPCs; they are MKC-5 and MKC-6. MKC-5 were isolated from the backskin of K14 knock-out mice, and MKC-6 were isolated from their normal counterpart. They are established cell lines that maintain their basal cell-like characteristics in vitro *(20)*. Our initial studies suggested that since MKCs are mature keratinocytes, they are conditioning the media, and they are providing factors that are enhancing the survival and the differentiation of EPCs towards the epithelial lineage (Turksen and Troy, in preparation).

3.1.6.1. THAWING MKC CELLS

When thawing MKCs, the contents of one vial are transferred to one 60-mm dish as described in **Subheading 3.1.2.** of this chapter. At the first subculture, extra dishes are seeded so that they may be frozen as not to deplete the cell line. Freezing is as per **Subheading 3.1.1.**, with the cells from one 60-mm dish being distributed to three cryo vials.

3.1.6.2. SUBCULTURING MKCs

1. Dishes are coated with 100 µL collagen-I solution in 2 mL of 1X PBS overnight in the 37°C incubator.
2. Before use, the dishes are washed 3 times with 1X PBS (2 mL/wash), then 4 mL KI2 media is added, and the dishes are appropriately labeled.
3. Trypsinize a 60-mm dish with 2 mL 0.25% trypsin-EDTA for 2 to 3 min in the 37°C incubator until the cells detach with gentle agitation.
4. Collect the cells with 3 mL KI2 into a 15-mL Falcon tube and centrifuge at 700g for 2 to 3 min for pellet formation.
5. Carefully suction off the media and resuspend the pellet in 5 mL K12.
6. Gently pipet up and down the entire volume to mix, and then distribute 1 mL to each of the pre-prepared plates, thereby making a 1:5 dilution of the original MKC-5 plate (approx 2×10^5 cells) *(20)*.

The plates are grown to confluency in about 3 to 4 d, and the media is changed every other day. The MKC-5 cells are subcultured twice a week, on Monday and Friday (*see* **Table 1**). At the point of subculturing, extra plates can be set up in order to use them as a feeder layer for EPCs. MKCs are mitomycin C-treated as previously described (**Subheading 3.1.4.1.**), using 50 µL for a 35-mm dish containing 2 mL of media.

3.2. Analysis of Differentiation

Three distinct differentiation stages, which were described previously, can be analyzed using the techniques that we describe in this section.

3.2.1. In situ Hybridization on EBs

This is a powerful technique that allows one to determine the spatial distribution of specific RNA within EBs without disrupting the morphology of the EBs. We have used this procedure to show the spatial distribution of K8 in the developing EB (*see* **Fig. 6**). This process is bone morphogenetic protein (BMP)-dependent, and in the presence of the BMP antagonist noggin, the expression of K8 and Scullin/Claud-6 are inhibited *(21)*. In addition, EBs are very suitable for the screening of newly identified genes for their role and association with ectoderm. It is our goal that a battery of new ectodermal genes will be identified.

3.2.1.1. Preparing a DIG-Labeled Ribo-Probe Using the In Vitro Transcription Method

1. Subclone a fragment of the gene of interest into a Bluescript plasmid that has a phage T7 and T3 promoter.
2. Linearize about 20 µg of the plasmid DNA with an appropriate restriction enzyme to produce a 5′ overhang at the far end of the fragment from the promoter and run an agarose gel to check for the absolute completion of the digestion.
3. Extract the linearized plasmid using the Qiagen gel extraction kit.
4. For a standard 20 µL of labeling reaction, add the following to a 1.5-mL microtube on ice:

13 µL	DEPC-water
1 µL	linearized plasmid DNA (approx 1 µg)
2 µL	DIG RNA labeling mixture
2 µL	10X transcription buffer
2 µL	RNA polymerase (T7/T3)

5. Mix and pulse the tube and incubate at 37°C for 2 to 3 h.
6. Add 2 µL of 0.2 *M* EDTA to stop the reaction, and add 30 µL DEPC-water to make the total volume 50 µL.
7. Use a Sephadex G-50 spin column to eliminate the unincorporated NTPs according to the manufacturer's directions:
   a. Vortex mix the column to resuspend the resin.
   b. Loosen the cap and snap off the bottom closure.
   c. Place the column in a 1.5-mL microfuge tube and pre-spin at 1500*g* for 1 min.
   d. Place the column in a fresh 1.5-mL microfuge tube and add the 50 µL sample to the top-center of the resin bed, being careful not to disturb the bed.
   e. Spin at 1500*g* for 2 min. This is the purified probe.
8. Estimate the yield of probe and dilute it to 100 µL, then store at −80°C. To estimate the probe concentration, load about 3 µL onto an agarose gel with 1 µL of linearized plasmid as a control and 1 µL DIG-labeled control RNA for yield estimation. Load DNA ladder to estimate the probe size. The 1.6-kb band on the ladder has a concentration of 10 ng/mL.

3.2.1.2. Whole Mount In Situ Procedure

Day One:
The following steps are performed on ice.

1. Collect EBs and fix with 10 mL 4% paraformaldehyde (4°C) in PBS on ice for 30 min, then change with fresh fixative for 3 h (or overnight), mixing occasionally.
2. Wash twice with PBS-T on ice. Routinely, we use a volume of 0.5 to 1 mL for all washes.
3. Wash with freshly made 25, 50, and 75% MeOH/PBS-T, then twice with 100% MeOH. Dehydrated samples may be stored at −20°C in 100% methanol for 1 mo.

Day Two:
The following are performed on the bench.

1. Take an aliquot of sample (about 100 µL), rehydrate by taking through a MeOH/PBS-T wash series in reverse (100 to 25%) and washing 3 times with PBS-T. Be sure to include a sample for antibody absorption.

2. Digest samples with 500 μL of 10 μg/mL proteinase K at room temperature for 15 min.
3. Wash with 2 mg/mL glycine in PBS-T to stop proteinase K digestion.
4. Wash twice with PBS-T.
5. Refix with fresh 0.2% glutaraldehyde/4% paraformaldehyde in PBS-T for 20 min.
6. Wash 3 times with PBS-T. At this step, one sample will be used for anti-DIG-AP antibody absorption. One absorption reaction is enough for 5 samples in the actual experiment.
7. For antibody absorption, add the following to the sample and incubate overnight at 4°C:

878 μL   PBS-T
100 μL   10 mg/mL BSA
 20 μL   calf serum (or normal sheep serum)
  2 μL   anti-DIG-AP

8. Wash the rest of the samples once with 50% hybridization buffer/50% PBS-T.
9. Wash once with 100% hybridization buffer at 50–55°C.
10. Prehybridize samples with hybridization buffer (500 μL) for 30 min in the water bath at 55°C, mixing occasionally.
11. Hybridize samples with hybridization buffer containing approx 50 ng/mL of probe. To dilute the probe, 20 μL of the probe reaction from **Subheading 3.2.1.1.** was diluted to a final vol of 100 μL. Add 5 μL of diluted probe into 500 μL of hybridization buffer. Boil it for 5 min. Cool it down to the hybridization temperature and add it to the sample. Hybridize samples (300 μL) in a 55°C water bath overnight.

Day Three:
1. Carefully take out the hybridization buffer containing the probe and store it at –20°C (*see* **Note 11**).
2. Wash with 100% hybridization buffer at 60°C for 30 min.
3. Wash with 50% hybridization buffer/50% PBS-T at 60°C for 30 min.
4. Wash with PBS-T at 60°C for 30 min.
5. Wash twice with 2X SSC, 0.1% Tween-20, 15 min each at 60°C.
6. Digest samples in 2X SSC, 0.1% Tween-20, containing 20 μg/mL RNase A at 37°C for 30 min.
7. Wash twice with PBS-T at room temperature, 5 min each.
8. Block for 30 min in PBS-T containing 5% calf serum at room temperature.
9. Incubate samples with absorbed anti-DIG antibody (*see* **Subheading 3.2.1.2**, from **Day 2, step 7**) overnight at 4°C. To prepare absorption, spin the absorbed sample for 1 min, take all the supernatant and add it to 3 mL of PBS-T and use it as absorbed anti-DIG-AP antibody. This will eliminate any nonspecific antibody binding.

Day Four:
1. Wash at least 5 times with PBS-T.
2. Wash twice with AP buffer.
3. Stain samples with NBT plus BCIP in AP buffer for 30 min to 2 h or until color development is completed. To stain the samples, place the sample tube on a rack wrapped with foil. Let it sit for about 30 min on a shaker, then check the color development (staining will be blue). Check the color development regularly every 10–15 min until the color turns very blue (up to 2 h or overnight). Be careful not to stain too long as the background increases (a dirty brown color).
4. Stop the reaction by washing twice with PBS-T.
5. Store samples in 100–200 μL of 50% glycerol/PBS-T (containing 1 to 2 μL sodium azide) at 4°C. Samples may be stored without losing staining intensity for extended periods of time at 4°C.
6. Mount the sample in glycerol and photograph.

3.2.2. Indirect Immunofluorescence for Differentiating Cells on Glass Coverslips

1. Cultures are set up and maintained on coverslips.
2. Fix and permeabilize the cells with –20°C methanol for 10 min at –20°C.
3. Rinse the coverslips 3 to 5 times with 1X PBS to remove the fixative.
4. Place coverslips in a humidified chamber and apply appropriately diluted primary antibody (*see* **Note 12**). Incubate for 30 min at room temperature.
5. Rinse each coverslip 3 times in 1X PBS to remove unbound primary antibody.
6. Apply secondary antibody to the coverslips. Again, make sure the cells do not dry out before the antibody is applied. Incubate for 30 min at room temperature.
7. Rinse each coverslip 3 times in 1X PBS.
8. Place Mowiol 4-88 (approx 20 µL for a 12-mm coverslip and 80 µL for a 22 × 22 mm coverslip) on each mounting slide, and carefully place the inverted coverslip on the slide.
9. Observe and photograph.

3.2.3. Ayoub Shklar Staining for Keratin

We routinely use Ayoub Shklar (AS) staining as a quick and reliable histological marker to determine the progression of epithelial differentiation of our cultures *(22,23)*. Undifferentiated epidermal cells are blue, differentiated epidermal cells are red, and further differentiation is indicated with a color change from orange to yellow to brown. Although the exact stages and extent of differentiation cannot be assessed via this method, AS staining is a good indication of epithelialization of EPC cultures (*see* **Fig. 7**) as a function of time.

3.2.3.1. STAINING PROCEDURE

1. Aspirate off the medium and rinse the dishes 3 times with 1X PBS.
2. Fix the plates with cold 10% formalin for 30 min at room temperature.
3. Rinse dishes 3 times with 1X PBS.
4. Stain for 3 min with filtered acid fuchsin solution.
5. Aspirate off excess stain, but do not let the dishes dry out.
6. Stain for 45 min with aniline blue–orange G solution.
7. Wash several times with 95% ethanol.
8. Allow the plates to dry inverted. View and photograph cultures in 95% ethanol. Stained plates may be stored indefinitely.

3.2.4. RT-PCR

3.2.4.1. TRIZOL RNA EXTRACTION

It is crucial that, during this extraction procedure, RNase-free instruments and solutions are used and that gloves are often changed, in order to minimize the risk of RNase contamination.

The following procedure has been derived from the manufacturer's protocol.

1. Lyse cells directly in the tissue culture dish by adding the appropriate amount of Trizol (i.e., for a 35-mm dish use 1 mL, for a 60-mm dish use 3 mL, and for a 100-mm dish use 6 mL).
2. Remove the cells with a cell scraper, then pipet the solution several times to collect all the cells, and transfer the volume to a sterile culture tube. Incubate the samples for 5 min at room temperature.

3. Add 0.2 mL of chloroform (RNase-free) per 1 mL of Trizol used (i.e., for a 35-mm dish use 200 μL, for a 60-mm dish use 600 μL, and for a 100-mm dish use 1.2 mL). Cap the tubes securely and shake vigorously for 15 s, then incubate at room temperature for 3 min.

4. Centrifuge at 12,000*g* for 15 min at 4°C. The mixture will separate into a lower phenol:chloroform phase, an interface, and a colorless upper aqueous phase. The RNA is in the aqueous phase and is about 60% of the original amount of Trizol added.

5. Transfer the aqueous phase into a fresh tube. Precipitate the RNA by adding 0.5 mL RNase-free isopropanol per 1 mL, Trizol originally added (i.e., for a 35-mm dish use 0.5 mL, for a 60-mm dish use 1.5 mL, and for a 100-mm dish use 3 mL). Incubate the samples for 10 min at room temperature.

6. Centrifuge at 12,000*g* for 10 min at 4°C. The RNA forms a pellet on the bottom and the side of the tube.

7. Remove the supernatant and wash the pellet with RNase-free 75% ethanol. Use 1 mL of ethanol for each 1 mL of Trizol used originally (i.e., for a 35-mm dish use 1 mL, for a 60-mm dish use 3 mL, and for a 100-mm dish use 6 mL). Vortex mix the sample and centrifuge at 7500*g* for 5 min at 4°C.

8. Air-dry the pellet and resuspend it in DEPC-H_2O. If it does not dissolve, try incubating for 10 min at 55–60°C.

9. Find the OD at A_{260} and A_{280}. The $A_{260}:A_{280}$ should be between 1.6–2.

10. Determine the concentration of the sample using the following equation:

$$[\ ]\ \mu g/\mu L = [A_{260} \times 40 \times \text{dilution factor}] \div 1000$$

3.2.4.2. To Remove Genomic DNA from RNA Samples

1. Digest 30–100 μg RNA so that RNA plus DEPC-H_2O = 87 μL
   Add: 10 μL 10X DNase buffer
             2 μL RNase Inhibitor
             1 μL DNase I

2. Mix and incubate for 15 min at room temperature.

3. Prepare chloroform:isoamyl alcohol (24:1): in a –50-mL Falcon tube put 12 mL RNase-free chloroform plus 0.5 mL RNase-free isoamyl alcohol.

4. Add an equal volume of phenol:chloroform to RNA sample, i.e., for a 100-μL sample, add 50 μL phenol and 50 μL chloroform:isoamyl alcohol (24:1).

5. Mix vigorously by inversion until an emulsion forms and spin at maximum speed at room temperature for 5 min.

6. Transfer the aqueous phase (the top phase) to a fresh RNase-free microtube.

7. Ethanol precipitate by adding a 1:10 dilution of RNase-free 3 *M* NaAc, i.e., for a 100-μL sample, add 10 μL 3 *M* NaAc.

8. Add 2X volume 100% EtOH (RNase-free, –20°C), i.e., for a 100-μL sample, add 200 μL 100% EtOH.

9. Mix vigorously and incubate tube at –80°C for 1 h.

10. Spin immediately at maximum speed for 20 min at 4°C.

11. Wash the pellet with 200 μL 70% EtOH (RNase-free, –20°C).

12. Spin again for 10 min at 4°C.

13. Air-dry the pellet (approx 15 min).

14. Resuspend the pellet in DEPC-H_2O.

15. Quantify the DNased RNA (dilute 1 μl in 500 μL DEPC-H_2O) using the following equation:

$$[\ ]\ \mu g/\mu L = [A_{260} \times 40 \times 500] \div 1000$$

16. Dilute samples to 1 μg/μL with DEPC-H_2O in order to be used for RT-PCR.

3.2.4.3. DENATURE RNA

1. Mix 1 µL RNA with 2 µL DEPC-H$_2$O in a 100-µL PCR tube. Vortex mix and pulse.
2. Place tubes in a Perkin Elmer thermal cycler and run the RNA at 65°C for 5 min, then hold the temperature at 4°C. The reaction must hold at 4°C for at least 5 min before proceeding to the RT reaction.

3.2.4.4. REVERSE TRANSCRIPTION

1. Prepare RT mixture:

4.0 µL	25 mM MgCl$_2$
2.0 µL	10X PCR buffer II
2.0 µL	10 mM dNTP
1.0 µL	RNase inhibitor
1.0 µL	MuLV reverse transcriptase
1.0 µL	Random hexamers

2. Add 17 µL of the RT mixture to the RNA and mix.
3. Place the tubes in the thermal cycler and run at 42°C for 15 min, 99°C for 5 min and then hold at 4°C. The reaction must hold at 4°C for at least 5 min before proceeding to PCR.

3.2.4.5. PCR

1. Prepare the PCR mixture (50 µL final volume):

0.5 µL	Template DNA (approx 10 ng)
5.0 µL	2.5 mM dNTPs
1.0 µL	10X polyelectrolyte complex (PEC) buffer
1.0 µL	3′ Primer (25 pmol/µL)
1.0 µL	5′ Primer (25 pmol/mL)
0.5 µL	*Taq* DNA polymerase
37.0 µL	dH$_2$0

2. Run samples as per the following schematic (*see* **Fig. 8** below):

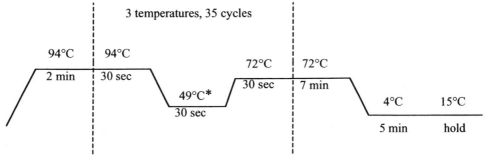

Fig. 8. The temperature of the extension step will depend upon the primers being used.

Various primers for transcription factors/signaling molecules involved in the formation of skin and hair follicles (LEF-1, SHH, and MSX-1; *see* **Table 2**) were used. The results indicate that our EPC cultures indeed follow in these lineages (*see* **Fig. 9A**). In addition, primers for the endodermal marker AFP (*see* **Table 2**) were used to prove

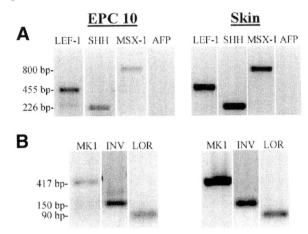

Fig. 9. RT-PCR comparison of EPCs after 10 d in culture and mouse backskin. (**A**) LEF-1, SHH, and MSX-1 are transcription factors–signaling molecules involved in the formation of skin and hair follicles. As indicated through RT-PCR analysis, our cultures are committed to this fate. In order to assure ourselves that our cultures are truly ectodermal and not endodermal in nature, an RT-PCR was conducted using primers specific for AFP (a marker of endodermal cells), and it was concluded that our cultures are not derived from the endodermal lineage. (**B**) MK1, involucrin, and loricrin are all markers of terminal differentiation, and their expression in our cultures, therefore, indicates that EPCs are progressing along terminal differentiation.

Table 2.
Primer Sequences

Gene	Forward Primer	Reverse Primer
LEF-1	5'-TTC AAG GAC GAA GGC GAT CCC CAG AAG G-3'	5'-TCT GAC GGG ATG TGT GAC GGG TGG GAT-3'
SHH	5'GAC AGC GCG GGG ACA GCT CAC-3'	5'-CCG CTG GCC CCT AGG GTC TTC-3'
MSX-1	5'-CGC TTC ACT CCT GCC CTT CA-3'	5'-ACT CCC GCT GCT CTG CTC AAA-3'
AFP	5'GCT CCG GCC TCT GTC CCA CC-3'	5'-GAT GTG AGC CAC ATC CAG GGC C-3'
MK1	5'-AAT TGC CAG AGG AGC AAG GC-3'	5'-TGG GAG TGC ACT CTC CAG AC-3'
Involucrin	5'-GGT GTA CAG AAG CTT CCA AGA TGT CC-3'	5'-GGC ATT GTG TAG GAT GTG GAG TTG G-3'
Loricrin	5'-GGT TCC CCT TCT CCT TAA AC-3'	5'-CTC CAC CAG AGG TCT TTC C-3'

that our EPC cultures were truly of epithelial and not endodermal lineage (*see* **Fig. 9A**). Primers for markers of epithelial terminal differentiation (MK1, involucrin, and loricrin; *see* **Table 2**) were used to show that our EPC cultures undergo this process (*see* **Fig. 9B**).

4. Notes

1. 7X-OMATIC is a special soap that we use for all washable tissue culture items. It is important that there is no soap residue on glassware in order to eliminate its potential effects on cultures.
2. This media is high in glucose and contains 4500 mg/L D-glucose and pyridoxine hydrochloride, but no L-glutamine or sodium pyruvate.
3. Serum is heat-inactivated in the following manner: thaw the bottle of serum overnight at 4°C, warm in a 57°C water bath for 7 min with constant mixing, then continue to incubate serum for 30 min, agitating every 10 min. The serum is then aliquoted into sterile 100-mL bottles in the tissue culture hood and allowed to cool before tightening the bottle and

freezing. It is important to cool the bottles in order to avoid breakage. Serum may be stored at –20°C for 6–12 mo.

4. Upon receipt of BM, the 10-mL bottle is thawed in a beaker of distilled water, and 1 mL is aliquoted into sterile tubes with a chilled pipet. The 1-mL aliquots are then stored at –20°C until required. When needed, thaw a tube in a beaker of distilled water and add 9 mL of unsupplemented DMEM. At this stage, it may be stored at 4°C. Before use, the diluted BM is further diluted another 1 : 10 with unsupplemented DMEM to a final working concentration of 0.1 mg/mL.

5. Prepare mitomycin C in the tissue culture hood. Add 5 mL of 1× PBS to a 2-mg vial, recap, and shake vigorously. Empty the vial into a clean weigh boat and syringe filter into a sterile 15-mL Falcon tube. Prepare 500-µL aliquots and store at –20°C until required to a maximum of 6 mo. To thaw for use, wrap the tube in tin foil to protect the mitomycin C from light and warm in a 37°C water bath.

6. Although the same serum is used for all our cultures, the quality of FBS is very important for the maintenance of ES cells. The ability of serum to support growth and differentiation of pluripotent ES cells is crucial. There are already number of suppliers that sell sera that are already tested for supporting ES cell growth. We use "ES cell tested" serum from Hyclone and have found that it is very reliable from batch to batch. Suitable sera batches should then be ordered in large quantities and then bottles may be stored at –20°C for up to 2 yr.

7. To prepare a humidified chamber for immunofluorescence, place a wet piece of filter paper in a 15-cm glass dish with a lid.

8. Usually when freezing cells in the exponential phase of growth, 5 cryo vials can be prepared from each 100-mm dish, and 3 cryo vials may be prepared from each 60-mm dish.

9. To protect oneself from the potential hazard of an exploding cryotube, a face shield and protective gloves should be worn. Prepare an ice bucket (with a lid) with a plastic beaker containing 37°C H_2O. Quickly remove the vial from the liquid nitrogen and place it in the water in the ice bucket and close the lid. Using a pair of forceps, gently invert the vial as to assure even thawing of the cells.

10. We have found that Fisherbrand Petri dishes (made in Canada) are the best dishes for the generation of EBs. In our experience, EBs formed on other plastics adhere to the plate or they aggregate. Fisherbrand Petri dishes reproducibly allow for the generation of EBs that never stick or aggregate.

11. This probe can be used 3 times. Therefore, after each hybridization, carefully take out the hybridization buffer and store it at –20°C. This probe does not need to be boiled again. Just warm it up to the hybridization temperature and add it to the sample right away.

12. Approximately 20–25 µL is required for each 12-mm coverslip and 100 µL for a 22 × 22 mm coverslip. Make sure the coverslips do not dry out before the antibody is applied.

References

1. Wolpert, L., Beddington, R., Brockes, J., Jessell, T., Lawrence, P., and Meyerowitz, E. (1998) *Principles of development.* Oxford University Press, Oxford.
2. Sengel, P. (1976) *Morphogenesis of skin.* Cambridge University Press, Cambridge.
3. DuBrul, E. F. (1968) Fine structure of epidermal differentiation in the mouse. *J. Exp. Zool.* **181,** 145–158.
4. Weiss, L. W. and Zelickson, A. S. (1975) Embryology of epidermis: ultrastructural aspects. I. Formation and early development in the mouse with mammalian comparisons. *Acta Derm. Venereol.* **55,** 161–168.

5. Weiss, L. W. and Zelickson, A. S. (1975) Embryology of epidermis: ultrastructural aspects II. Period of differentiation in the mouse with mammalian comparisons. *Acta Derm. Venereol.* **55,** 321–329.

6. Hanson, J. (1947) The histogenesis of the epidermis in the rat and mouse. *J. Anat.* **81,** 174–197.

7. Fuchs, E. and Byrne, C. (1994) The epidermis: rising to the surface. *Curr. Opin. Genet. Dev.* **4,** 725–736.

8. Fuchs, E. (1993) Epidermal differentiation and keratin gene expression. *J. Cell Sci.* **17,** 197–208.

9. Byrne, C. (1997) Regulation of gene expression in developing epidermal epithelia. *BioEssays* **19,** 691–698.

10. Lavker, R. M. and Sun, T. T. (2000) Epidermal stem cells: properties, markers, and location. *Proc. Natl. Acad. Sci. USA* **97,** 13,473–13,475.

11. Miller, S. J., Lavrek, R. M., and Sun, T. T. (1997) Keratinocyte stem cells of cornea, skin and hair follicles, in *Stem cells* (Pottten, C. S., ed.), Academic Press, London, pp. 331–362.

12. Martin, G. R. (1981) Isolation of a pluripotent cell line from early mouse embryos cultured in medium conditioned by teratocarcinoma stem cells. *Proc. Natl. Acad. Sci. USA* **78,** 7634–7638.

13. Evans, M. J. and Kaufman, H. M. (1981) Establishment in culture of pluripotent cells from mouse embryos. *Nature* **292,** 154–156.

14. Robertson, E. J. (1978) Embryo derived stem cell lines, in *Teratocarcinomas and embryonic stem cells: a practical approach* (Robertson, E. J., et al., eds.), IRL Press, Oxford, pp. 19–49.

15. Doetshchman, T. C., Eistetter, H., Katz, M., Schmidt, W., and Kemler, R. (1985) The in vitro development of blastocyt-derived embryonic stem cell lines: formation of visceral yolk sac, blood islands and myocardium. *J. Embryol. Exp. Morphol.* **87,** 27–45.

16. Coucouvanis, E. and Martin, G. R. (1995) Signals for death and survival: a two step mechanism for cavitation in the vertabrate embryo. *Cell* **83,** 279–287.

17. Keller, G. M. (1995) In vitro differentiation of embryonic stem cells. *Curr. Opin. Cell Biol.* **7,** 862–869.

18. Leahy, A., Xiong, J.-W., Kuhnert, F., and Stuhlmann, H. (1999) Use of developmental marker genes to define temporal and spacial patterns of differentiation during embryoid body formation. *J. Exp. Zool.* **284,** 67–81.

19. Nagy, A., Gocza, E., Diaz, E. M., Prideaux, V. R., Ivanyi, E., Markulla, M., and Rossant, J. (1990) Embryonic stem cells alone are able to support fetal development in the mouse. *Development* **110,** 815–821.

20. Troy, T.-C. and Turksen, K. (1999) In vitro characteristics of early epidermal progenitors isolated from keratin 14 (K14)—deficient mice: insights into the role of keratin 17 in mouse keratinocytes. *J. Cell. Physiol.* **180,** 409–421.

21. Turksen, K. and Troy, T.-C. (2001) Claudin 6: a novel tight junction gene is developmentally regulated during early commitment and differentiation of mouse embryonic epithelium. *Mech. Dev., submitted.*

22. Ayoub, P. and Shklar, G. (1963) A modification of the Mallory connective tissue stain as a stain for keratin. *Oral Surg.* **16,** 580–581.

23. Turksen, K. and Troy, T.-C. (1998) Epidermal cell lineage. *Biochem. Cell Biol.* **76,** 1–10.

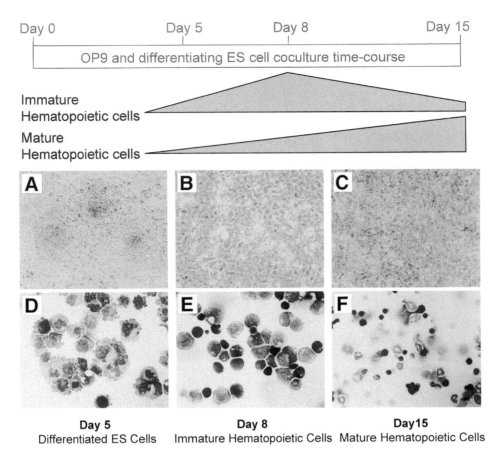

Day 5
Differentiated ES Cells

Day 8
Immature Hematopoietic Cells

Day15
Mature Hematopoietic Cells

Color Plate 1, Figure 1 (*see* full caption and discussion on p. 84).

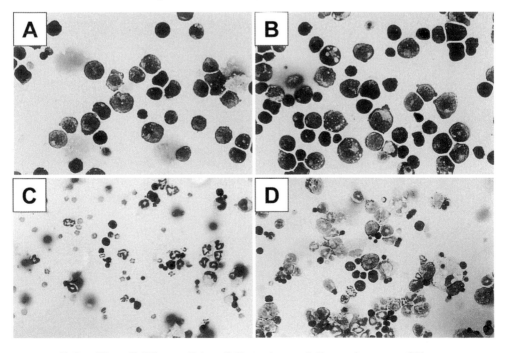

Color Plate 2, Figure 5 (*see* full caption and discussion on p. 92).

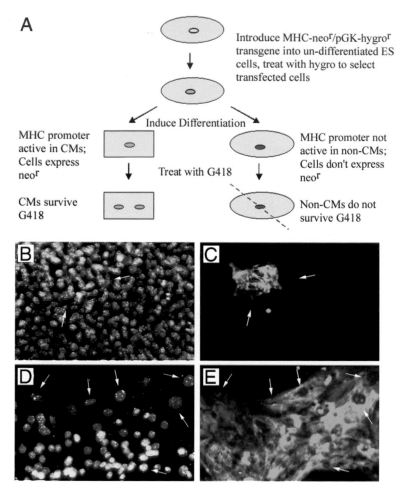

Color Plate 3, Figure 2 (*see* full caption and discussion on p. 160).

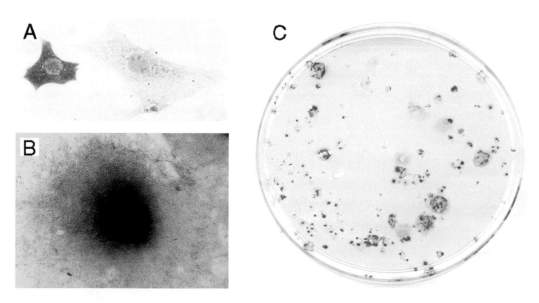

Color Plate 4, Figure 3 (*see* full caption and discussion on p. 163).

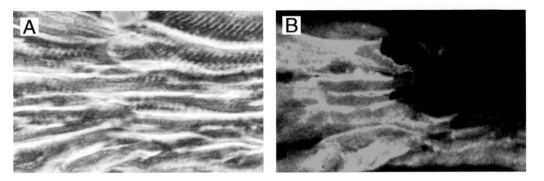

Color Plate 5, Figure 4 (*see* full caption and discussion on p. 164).

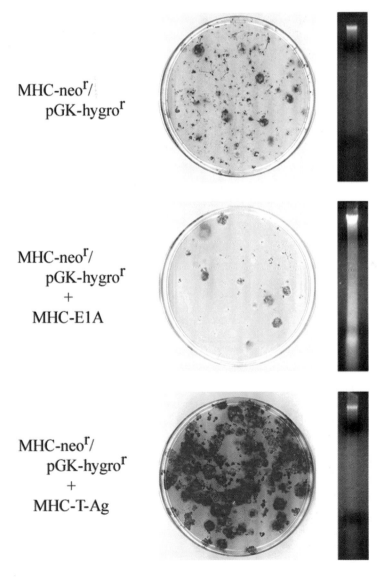

MHC-neo^r/
pGK-hygro^r

MHC-neo^r/
pGK-hygro^r
+
MHC-E1A

MHC-neo^r/
pGK-hygro^r
+
MHC-T-Ag

Color Plate 6, Figure 5 (*see* full caption and discussion on p. 166).

Mixed population
of cells

Magnetic stand to
retain labeled cells

Flow through–
unlabeled cells

Flow through–
labeled cells

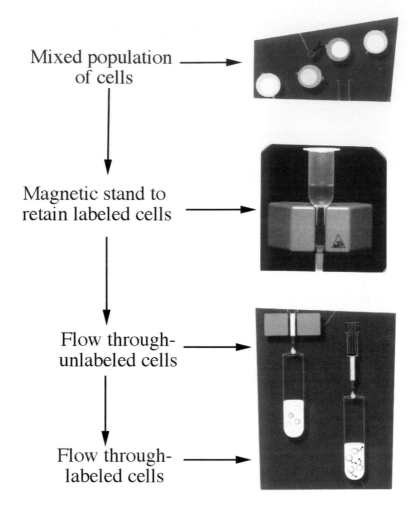

Color Plate 7, Figure 3 (*see* full caption and discussion on p. 198).

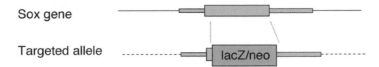

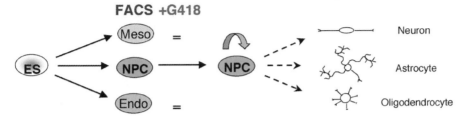

Color Plate 8, Figure 1 (*see* full caption and discussion on p. 206).

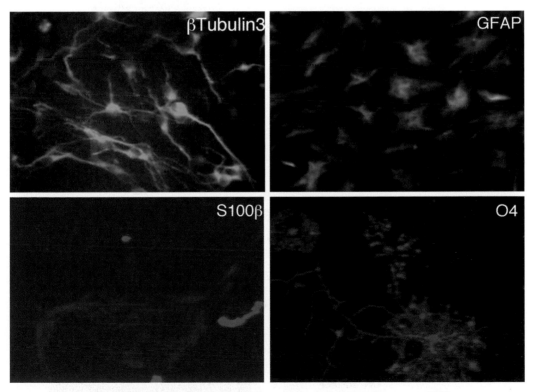

Color Plate 9, Figure 2 (*see* full caption and discussion on p. 207).

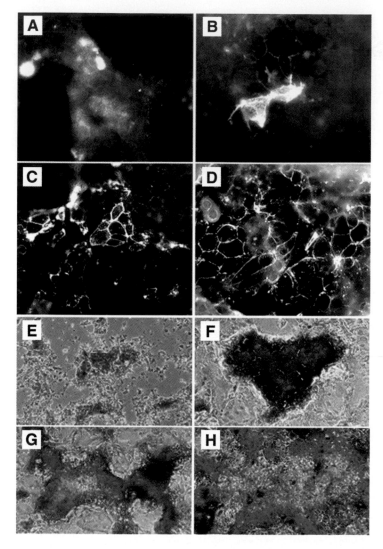

Color Plate 10, Figure 1 (*see* full caption and discussion on p. 258).

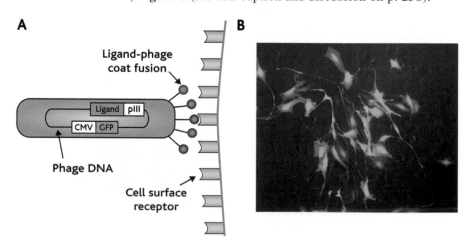

A

Ligand-phage
coat fusion

Ligand | pIII

CMV | GFP

Phage DNA

Cell surface
receptor

B

Color Plate 11, Figure 1 (*see* full caption and discussion on p. 394).

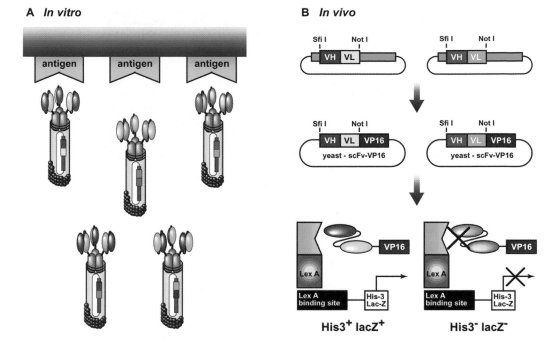

A *In vitro*

antigen   antigen   antigen

B *In vivo*

Sfi I   Not I
VH VL

Sfi I   Not I
VH VL

Sfi I   Not I
VH VL VP16
yeast - scFv-VP16

Sfi I   Not I
VH VL VP16
yeast - scFv-VP16

VP16

Lex A

Lex A
binding site

His-3
Lac-Z

VP16

Lex A

Lex A
binding site

His-3
Lac-Z

His3⁺ lacZ⁺

His3⁻ lacZ⁻

Color Plate 12, Figure 1 (*see* full caption and discussion on p. 435).

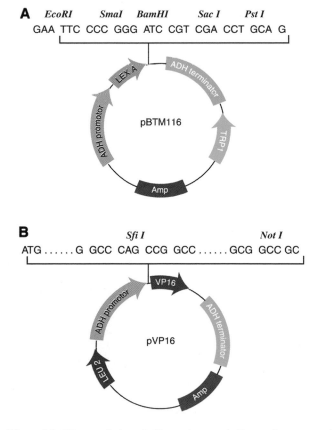

A   *EcoRI*   *SmaI*   *BamHI*   *Sac I*   *Pst I*
GAA TTC CCC GGG ATC CGT CGA CCT GCA G

LEX A
ADH terminator
ADH promotor
pBTM116
TRP1
Amp

B   *Sfi I*   *Not I*
ATG . . . . . . G GCC CAG CCG GCC . . . . . . GCG GCC GC

VP16
ADH promotor
ADH terminator
pVP16
LEU 2
Amp

Color Plate 13, Figure 2 (*see* full caption and discussion on p. 437).

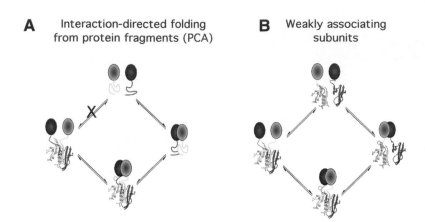

A Interaction-directed folding from protein fragments (PCA)

B Weakly associating subunits

Color Plate 14, Figure 1 (*see* full caption and discussion on p. 448).

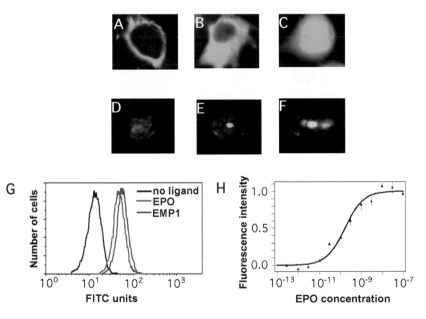

Color Plate 15, Figure 2 (*see* full caption and discussion on p. 451).

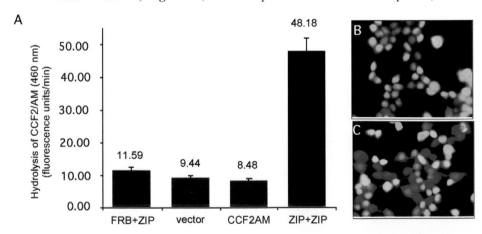

Color Plate 16, Figure 3 (*see* full caption and discussion on p. 455).

20

ES Cell Differentiation Into the Hair Follicle Lineage In Vitro

Tammy-Claire Troy and Kursad Turksen

1. Introduction

Hair is arguably the most crucial skin appendage that shapes one's self-image and acceptance in our society. However, despite this critical function, the precise regulatory steps governing the formation of hair follicles remains poorly understood. It is known, though, that hair follicle formation involves cell commitment, programmed cell death, mesenchymal–epithelial interactions and differentiation (1,2), and that it is a continuously renewing system. It cycles, from stages of resting (telogen), to stages of growth (anagen), as well as a short regression phase (catagen), before entering back to a new cycle (3–7); suggesting that there is a reservoir of cells in the hair follicle capable of hair follicle formation throughout life. However, the identity of the stem cells, as well as the signals and the steps from stem cells to committed epidermal progenitor cells to differentiated hair follicle cells, has not been elucidated. This has been owing to the lack of an in vitro model system that would enable the study of the steps and stages of hair follicle cell differentiation. Embryonic stem (ES) cells provide an excellent cell culture system with a capacity to give rise to various lineages in vitro. Using the R1 ES cell line (8), we have recently developed a culture system that allows ES cells to differentiate into hair keratin expressing mature hair follicle cells. It is predicted that this culture system will be invaluable to delineate the mechanisms involved in hair follicle formation, thereby allowing for the development of treatments and therapies for hair growth and regeneration.

2. Materials

2.1. Tissue Culture

1. Phosphpate-buffered saline (PBS), 1X.
2. Dulbecco's modified Eagle medium (DMEM), 1X (500-mL bottle; Gibco BRL, cat. no. 11960-044).
3. Fetal Bovine Serum (FBS) (500 mL bottle; Hyclone, cat. no. SH30071.03) (*see* **Note 1**).
4. Growth factor-reduced Matrigel basement membrane matrix (BM) (10-mL bottle; Collaborative Biomedical Products, cat. no. 35-4230) (*see* **Note 2**).

From: *Methods in Molecular Biology, vol. 185: Embryonic Stem Cells: Methods and Protocols*
Edited by: K. Turksen © Humana Press Inc., Totowa, NJ

5. MEM nonessential amino acids solution (NEAA), 10 m*M*, 100X (100-mL bottle; Gibco BRL, cat. no. 11140-050).
6. MEM sodium pyruvate solution, 100 m*M*, 100X (100-mL bottle; Gibco BRL, cat. no. 11360-070).
7. Penicillin–streptomycin, 100X (100-mL bottle; Gibco BRL, cat. no. 15140-122).
8. 0.25% Trypsin, 1 m*M* EDTA•4 Na, 1X, (500-mL bottle; Gibco BRL, cat. no. 25200-072).
9. Corning polystyrene 100-mm tissue culture dishes (Corning, cat. no. 25020-100).
10. Corning polystyrene 60-mm tissue culture dishes (Corning, cat. no. 25010-60).
11. Corning polystyrene 35-mm tissue culture dishes (Corning, cat. no. 25000-35).
12. 100-mm Fisherbrand Petri dishes (Fisher, cat. no. 08-757-12).
13. 15-mL polypropylene conical tubes (Beckton Dickenson, cat. no. 2097).
14. Coverslips (22 × 22 mm Bellco, cat. no.1916-12222).

2.2. Ayoub Shklar Staining

1. 37% Formaldehyde (500 mL; Fisher, cat. no. F79-500).
2. Acid fuchsin (100 g; Sigma, cat. no. A 3908).
3. Aniline blue (25 g; Polysciences, cat. no. 02570).
4. Orange G (100 g; Sigma, cat. no. O-7252).
5. Phosphotungstic acid (100 g; Sigma, cat. no. P-6395).
6. 95% Ethanol.
7. Filter paper and funnels.
8. 1X PBS.
9. 10% Neutral-buffered formalin; for 100 mL, combine 10 mL 10X PBS, 10 mL 37% formaldehyde, and 80 mL dH$_2$O, then store at 4°C.
10. 5% Acid fuchsin; for 100 mL, add 5 g acid fuchsin to 100 mL dH$_2$O and stir for at least 3 h or overnight. Filter and store at room temperature for 6 mo. Filter before use.
11. Aniline blue–orange G; for 100 mL, mix 0.5 g Aniline blue, 2 g Orange G, and 1 g phosphotungstic acid in 100 mL dH$_2$O. Stir at least 3 h or overnight. Filter and store at room temperature for 6 mo. Filter before use.

2.3. Immunofluorescence

1. Methanol (4 L; Fisher Scientific, cat. no. A452SK-4) at –20°C.
2. Humidified chamber (*see* **Note 3**).
3. 50°C Water bath.
4. Disposable pipets.
5. 1,4-diazabicyclo [2,2,2] octane (DABCO) (25 g; Sigma, cat. no. D-2522).
6. 1X PBS.
7. 0.2 *M* Tris, pH 8.5.
8. Mowiol 4-88 (500 g; Polysciences, cat. no. 17951).

3. Methods
3.1. Embryonic Stem Cells in Culture

ES cells are maintained on a mitomycin C-treated embryonic fibroblast (EF) feeder layer. The media must routinely be changed, and the cultures must be split according to a strict schedule. ES cells are maintained in culture only to the 6th passage in order to maintain an authentic undifferentiated stem cell culture system. ES cells were routinely tested for their stem cell characteristics by histochemical analysis of alkaline phosphatase activity. The conditions and scheduling of these cultures were followed exactly as described in Chapter 19, *Epidermal Lineage*, by Troy and Turksen, in this volume.

3.2. Maintenance of ES Cells

To subculture ES cells:

1. Gently remove the media, rinse with 1X PBS, and then incubate the cells with 2 mL 0.25% trypsin-EDTA to a 100-mm dish until the cells float with gentle agitation.
2. Gently pipet up and down 30 times in order to make a single-cell suspension and distribute 200 μL (approx 2.5×10^6 cells) to each mitomycin C-treated EF plate (to maintain stocks) and Petri dish (to proceed with embryoid body [EB] formation and differentiation).
3. Gently agitate and incubate at 37°C.

3.3. High Density Cultures of Epithelial Progenitor Cells Differentiate into Hair Follicle Cells

Our studies indicated that when EBs were plated onto BM-coated dishes, they give rise to epithelial sheets that are capable of differentiating into keratin 14 (K14)-positive cells. These studies also suggested that these cultures contain epidermal progenitor cells that are maintained in secondary cultures called epithelial progenitor cells (EPCs). EPCs progress along the hair follicle differentiation pathway, when cultured at high density, to express hair keratins, as determined by indirect immunofluorescence with antibodies against AE13 and AE15 *(9,10)* (**Fig. 1A–D**). It was shown that these hair keratin markers were expressed in a time- and density-dependent manner. It can, therefore, be projected that these cultures are suitable for studying the role of growth factors and hormones on hair follicle formation. Our goal is to devise a "serum-free" culture system in order to assess specific molecules and their affects on hair follicle morphogenesis.

1. Coat 35-mm tissue culture dishes with Matrigel (1 mL/35-mm dish, approx 0.1 mg) for 30 min at room temperature, then gently replace it with 15% DMEM.
2. Trypsinize 4-d-old differentiated embryoid bodies (dEBs) with 0.25% trypsin-EDTA for 2 to 3 min in the incubator. Collect the cells into a 15-mL Falcon tube and remove an aliquot to count the total available cells.
3. Centrifuge at 700*g* for 2 to 3 min for pellet formation. While spinning, count the cells using a Coulter Counter.
4. Resuspend the pellet with 15% DMEM and dilute the cells appropriately in order to plate 10^6 cells/35-mm dish. For immunofluorescence, plate the cells on glass 22 × 22 mm coverslips.

3.3.1. Indirect Immunofluorescence for Differentiating Cells on Glass Coverslips

Cells on coverslips were processed and labeled using indirect immunofluoresence *(11)* as described below.

1. Cultures are set up and maintained on coverslips.
2. Fix and permeabilize the cells with –20°C methanol for 10 min at –20°C. Then rinse the coverslips three times with 1X PBS.
3. Place the coverslips in a humidified chamber and apply the appropriately diluted primary antibody (*see* **Note 4**).
4. Incubate for 30 min at room temperature and rinse with 1X PBS to remove unbound primary antibody.

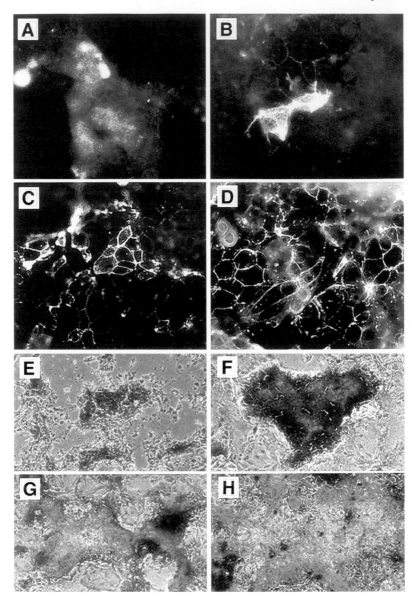

Fig. 1. EPC were plated at high density (10^6 cells/35-mm dish) and assayed after 10 and 12 d for hair follicle markers. Immunofluorescence analysis (**A–D**) (10× magnification) of these cultures revealed that expression of AE15 (antitrichlorohyalin granule) was first detected after 10 d in some cells (**A**) and was increased after 12 d (**B**). AE13 (antiacidic hair keratin) was expressed at d 10 (**C**) and was increased at d 12 (**D**). AS staining (**E–H**) (40× magnification) of EPC cultures plated at the same density was performed after 2 (**E**), 6 (**F**), 10 (**G**), and 12 (**H**) d to assay for the expression of epidermal and hair follicle markers. The staining revealed that d 12 cultures are sufficient for hair follicle precursor development, as shown by the brown, dark orange, and yellow staining. Together, immunofluorescence and AS staining analysis reveal that, with time, hair follicle differentiation is increased under our culture conditions. (*See* color plate 10, following p. 254).

5. Apply secondary antibody to the coverslips and incubate for 30 min at room temperature. Then rinse three times in 1× PBS.
6. Apply Mowiol 4–88 (approx 80 µL) on each mounting slide and carefully place inverted coverslip on the slide. Observe and photograph.

3.3.2. Ayoub Shklar Staining for Keratin

We routinely use Ayoub Shklar (AS) staining as a quick and reliable histological marker to determine the progression of epidermal differentiation and hair follicle formation of our cultures *(11,12)*. Undifferentiated epidermal cells are blue, differentiated epidermal cells are red, and further differentiation is indicated with a color change from orange to yellow to brown. It is believed that differentiated hair follicle cells can be assessed via this method with the higher differentiated colors, namely orange and yellow (**Fig. 1E–H**). Although the exact stages and extent of differentiation cannot be assessed via this method, AS staining is a good indication of hair follicle formation of EPC cultures as a function of time and density.

3.3.2.1. STAINING PROCEDURE

1. Aspirate off the medium and rinse the dishes 3 times with 1X PBS, then fix the plates with cold 10% formalin for 30 min at room temperature.
2. Rinse dishes three times with 1X PBS and stain for 3 min with filtered acid fuchsin solution.
3. Aspirate off excess stain, but do not let the dishes dry out.
4. Stain for 45 min with Aniline blue–orange G solution.
5. Wash several times with 95% ethanol.
6. Allow the plates to dry inverted. View and photograph cultures in 95% ethanol. Stained plates may be stored indefinitely.

4. Notes

1. Serum is heat-inactivated at 57°C waterbath for 30 min with constant mixing. Serum may be stored at –20°C for 6–12 mo.
2. The 1-mL aliquots of BM are then stored at –20°C until required. When needed, thaw a tube in a beaker of distilled water and add 9 mL of unsupplemented DMEM. At this stage, it may be stored at 4°C. Before use, the diluted BM is further diluted another 1:10 with nonsupplemented DMEM to a final working concentration of 0.1 mg/mL.
3. Incubation is performed in a humidified chamber to avoid drying of the coverslips, which results in increased background. To prepare, place a wet piece of filter paper in a 15-cm glass dish with a lid.
4. Approximately 100 µL is required for a 22 × 22 mm coverslip. Applying less may lead to insufficient coverage and subsequent drying of the sample.

Acknowledgments

We are very grateful to Mr. Ben C. Hulette for his having faith in our project, and we would like to thank Dr. Charles C. Bascom for his continuous support and encouragement throughout the course of these studies. This work was supported by a grant from the International Program for Animal Alternatives (IPAA) competition sponsored by the Procter & Gamble Company.

References

1. Hardy, M. H. (1992) The secret life of the hair follicle. *Trends Genet.* **8,** 55–61.
2. Phipott, M. and Paus, R. (1998) Principles of hair follicle morphogenesis, in *Molecular Basis of Epithelial Appendage Morphogenesis* (Chuong, C.-M., ed.), R. G. Landes Press, Austin, pp. 75–110.
3. Jones, P. H. (1997) Epidermal stem cells. *BioEssays* **19,** 683–690.
4. Miller, S. J., Lavker, R. M., and Sun, T. T. (1997) Keratinocyte stem cells of cornea, skin and hair follicles, in *Stem Cells* (Potten, C. S., ed.), Academic Press, London, pp. 331–362.
5. Taylor, G., Lehrer, M. S., Jensen, P. J., Sun, T. T., and Lavker, R. M. (2000) Involvement of follicular stem cells in forming not only the follicle but also the epidermis. *Cell* **102,** 451–461.
6. Fuchs, E. and Segre, J. A. (2000) Stem Cell: A new lease on life. *Cell* **100,** 143–155.
7. Stenn, K. S. and Paus, R. (2001) Controls of hair follicle cycling. *Physiol. Rev.* **81,** 449–494.
8. Nagy, A., Gocza, E., Diaz, E. M., Prideaux, V. R., Ivanyi, E., Markulla, M., and Rossant, J. (1990) Embryonic stem cells alone are able to support fetal development in the mouse. *Development* **110,** 815–821.
9. Lynch, M. H., O'Guin, W. M., Hardy, C., Mak, L., and Sun, T. T. (1986) Acidic and basic hair/nail ("hard") keratins: their colocalization in upper cortical and cuticle cells of the human hair follicle and their relationship to "soft" keratins. *J. Cell Biol.* **103,** 2593–2606.
10. Dhouailly, D., Xu, C., Manabe, M., Schermer, A., and Sun, T.-T. Expression of hair related keratins in a soft epithelium: Subpopulations of human and mouse dorsal toungue keratinocytes express keratin markers for hair-, skin- and esophageal-types of differentiation. *Exp. Cell Res.* **181,** 141–158.
11. Troy, T.-C. and Turksen, K. (1999) In vitro characteristics of early epidermal progenitors isolated from keratin 14 (K14)-deficient mice: insights into the role of keratin 17 in mouse keratinocytes. *J. Cell. Physiol.* **180,** 409–421.
12. Ayoub, P. and Shklar, G. (1963) A modification of the mallory connective tissue stain as a stain for keratin. *Oral Surg. Oral Med, Oral Pathol.* **16,** 580–581.

21

Embryonic Stem Cells as a Model for Studying Melanocyte Development

Toshiyuki Yamane, Shin-Ichi Hayashi, and Takahiro Kunisada

1. Introduction

Melanocytes residing in the skin, inner ear, and uveal tract are derived from neural crest cells. They are highly dendritic, heavily pigmented, and are generally located in the epidermal basal cell layer of these areas, including hair follicles *(1)*. The most convincing fact demonstrating the neural crest origin of these melanocytes came from mice mutated at the *c-Kit^W* and *Mgf^{Sl}* loci, in which all melanocytes were lost *(2)*. Retinal pigmented epithelial (RPE) cells, the melanin-producing polygonal cells forming epithelium in the retina, are directly induced from the neuroepithelium of the optic vesicle and are not affected by *c-Kit^W* and *Mgf^{Sl}* mutations. Another mutant, *Edn3^{ls}*, which lacks endothelin 3 (ET3), shows a severe defect in hair pigmentation *(3)*. Disruption of the receptor for ET3 is also known to cause a hair pigmentation defect. These genetic observations indicate that steel factor (also known as MGF, SCF) and ET3 are normally essential for melanocyte development (**Fig. 1**).

In this chapter, we describe a procedure for inducing the differentiation of the melanocyte lineage from embryonic stem (ES) cells *(4)*. We utilize a co-culture system of ES cells with a bone marrow-derived stromal cell line, ST2 *(5)*. Since ST2 cells are derived from an albino mouse *(4)*, any pigmented cells that appear in this system are derived from ES cells. This culture system is simple, as you only need to inoculate ES cells onto an ST2 layer and regularly change the culture medium. However, this system is highly dependent on the quality of the ES cells. Certain cell lines are found never to generate melanocytes even if the cell line differentiates into hematopoietic lineages. If appropriate ES cell lines are chosen, you can constantly and efficiently induce melanocytes from the cell line.

After inoculating undifferentiated ES cells onto plates pre-seeded with ST2 cells and culturing in the presence of dexamethasone (Dex) (and other supplements), colonies derived from single ES cells grow up, and melanocytes are generated within the colonies after 12 to 13 d of differentiation. The number of melanocytes continues to increase for up to 3 wk after plating the ES cells (**Fig. 2A–F**). It should be noted that these colonies are composed of highly heterogenous populations, which include cardiac muscle cells, hematopoietic cells, endothelial cells, and so on. Sometimes,

From: *Methods in Molecular Biology, vol. 185: Embryonic Stem Cells: Methods and Protocols*
Edited by: K. Turksen © Humana Press Inc., Totowa, NJ

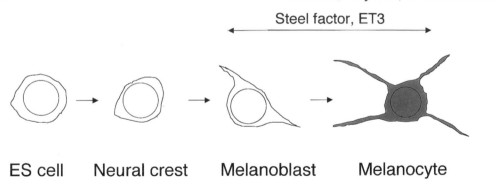

Fig. 1. Scheme describing the development of the melanocyte lineage. The stage at which steel factor and ET3 function is indicated. The melanin-containing stage is shaded.

highly polarized clusters of pigmented cells resembling RPE cells are observed (**Fig. 2G**, **H**). Markers for melanoblasts (committed melanocyte precursors) are first detected on d 6 of differentiation. The generation of melanocytes, but not RPE cells, from ES cells depends on steel factor and is promoted by ET3. More precise analysis suggested that steel factor-dependent melanoblasts are generated from d 6 to d 12 of differentiation. Provided that the ES cell lines are derived from the 3.5 d postcoitum blastocysts, the time course of in vitro differentiation of melanocytes from ES cells corresponds well to that of neural crest-derived melanocyte development in vivo *(4)*.

Instead of a conventional gene knock-out approach, the ES culture system allows for a rapid evaluation of whether or not the gene inactivated in the ES cell is indispensable for neural crest or melanocyte development. Applying this advantage, allows for the verification of genes, which could possibly affect melanocyte or neural crest development. In addition, factors that may affect neural crest cell lineages could be tested using this culture system.

2. Materials

1. 0.25% trypsin + 0.5 mM EDTA in phosphate-buffered saline (PBS).
2. 0.05% trypsin + 0.5 mM EDTA in PBS.
    For trypsin-EDTA solutions, we purchase 2.5% trypsin solution from a commercial supplier, for example, from Gibco BRL (Grand Island, NY, cat. no. 15090-046), dispense them into aliquots, and store at –20°C. Dilute them to appropriate concentrations with PBS and supplement with EDTA. Store at 4°C. Use within 1 mo.
3. Dulbecco's modified Eagle medium (DMEM) (Gibco BRL, cat. no. 12800-017).
4. RPMI-1640 (Gibco BRL, cat. no. 31800-022).
5. α-Minimum essential medium (α-MEM) (Gibco BRL, cat. no. 11900-024).
6. Medium for the maintenance of ES cells: DMEM supplemented with 15% fetal calf serum (FCS) (*see* **Note 1**), 2 mM L-glutamine (Gibco BRL, cat. no. 25030-081), 1X nonessential amino acids (Gibco BRL, cat. no. 11140-050), 0.1 mM 2-mercaptoethanol (2-ME), 1000 U/mL recombinant leukemia inhibitory factor (LIF) or equivalent amounts of a culture supernatant from Chinese hamster ovary (CHO) cells producing LIF, 50 µg/mL streptomycin, and 50 U/mL penicillin. Store at 4°C. Use within 1 mo.
7. Medium for embryonic fibroblasts: DMEM supplemented with 10% FCS, 50 µg/mL streptomycin, and 50 U/mL penicillin. Store at 4°C. Use within 2 mo.

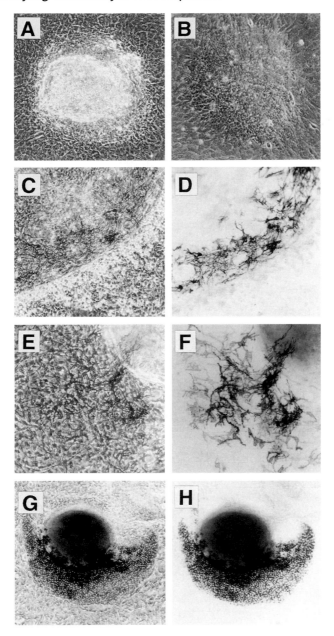

Fig. 2. Appearance of cultures. (**A**, **B**, **D**, **F**) Growing colonies on d 6 of differentiation. Variations exist in the appearance of the colonies. (**C–F**) Colonies that contain melanocytes on d 21 of differentiation. Melanocytes are observed both inside and outside of colonies. (**G**, **H**) A colony that generated RPE cells. Panels **D**, **F**, and **H** are bright-field images of **C**, **E**, and **G**, respectively.

8. 0.1% Gelatin solution: Dissolve 0.5 g of gelatin (Sigma, St. Louis, MO, cat. no. G2500) into 500 mL distilled water by autoclaving. Store at room temperature.
9. Mitomycin C (Sigma, cat. no. M0503). For a 100X solution, dissolve the powder into sterile distilled water and adjust the concentration to 1 mg/mL just before use.

10. Medium for the maintenance of ST2: RPMI 1640 supplemented with 5% FCS, 50 μM 2-ME, 50 μg/mL streptomycin, and 50 U/mL penicillin. Store at 4°C. Use within 2 mo.

11. Medium for the differentiation of ES cells on ST2: α-MEM supplemented with 10% FCS (*see* **Note 2**), 50 μg/mL streptomycin, and 50 U/mL penicillin. Store at 4°C. Use within 2 mo.

12. Dex (Sigma, cat. no. D4902). Store original stocks at 10^{-2} M in ethanol at –70°C. Prepare the stock at 10^{-3} M in ethanol at 4°C. Dilute from this stock for each use. Stable for at least 1 yr (stocks at –70°C) or 3 mo (stocks at 4°C).

13. Human recombinant basic fibroblast growth factor (bFGF) (R&D System, Minneapolis, MN, cat. no. 233-FB). Store the stocks at 200 nM in 0.1% bovine serum albumin (BSA)/PBS at –70°C. Store at 4°C after thawing. Stable for at least for 6 mo (stocks at –70°C) or 1 mo (stocks at 4°C).

14. Cholera toxin (CT) (Sigma, cat. no. C8052). Store the stock at 50 nM in distilled water at –70°C. Store at 4°C after thawing. Stable for at least for 1 yr (stocks at –70°C) or 1 mo (stocks at 4°C).

3. Methods

For each manipulation, prewarm the medium to 37°C. The cultures are maintained at 37°C with 5% CO_2 in a humidified incubator. The volume of culture medium is summarized in **Table 1**, for the culture vessels that appear in the protocol. Use trypsin-EDTA solution by one-tenth volume of medium for each vessel if not indicated.

3.1. Maintenance of ES Cells

We ordinarily maintain ES cells on mitomycin C-treated embryonic fibroblasts (*see* **Note 3**). ES cells are cultured in the medium described in the material section.

1. Prepare mitomycin C-treated confluent embryonic fibroblasts on gelatin-coated dishes (*see* **Note 4**).
2. Thaw ES cells on mitomycin C-treated embryonic fibroblasts.
3. Grow them up to a subconfluent state.
4. Replace the medium of the ES cell cultures with fresh medium 2–4 h before passage.
5. Wash three times with PBS and trypsinize the culture with 0.25% trypsin, 0.5 mM EDTA for 5 min at 37°C. Add medium and dissociate the cell clump by pipetting up and down. Centrifuge at 120g for 5 min at 4°C, aspirate the supernatant, and suspend the cell pellet in the medium. Count the number of ES cells.
6. Replate the cells onto the freshly prepared mitomycin C-treated embryonic fibroblasts. For 60- or 100-mm dishes, inoculate 10^6 or 3×10^6 ES cells, respectively.
7. Change the medium of the ES cell cultures with fresh medium the day after passage.
8. Two days after passage, the cultures must reach a subconfluent state. Maintain ES cells in the above-mentioned manner (*see* **Note 5**).

3.2. Maintenance and Preparation of ST2 Stromal Cells

ST2 cells (*see* **Notes 6,7**) are maintained in the RPMI 1640 media supplemented with 5% FCS and 50 μM 2-ME.

1. Trypsinize confluent cells in a dish or flask with 0.05% trypsin, 0.5 mM EDTA at 37°C for 1 to 2 min.
2. Add medium, dissociate the cells by pipetting, centrifuge them at 200g for 5 min, and then split them 1:4.
3. Maintain by regularly passing them every 3 or 4 d.

Table 1
Volume of Medium for Culture Vessels

Culture vessels	Volume of medium (mL)
25-cm^2 flask	6
60-mm dish	5
100-mm dish	10
6-well plate	2

4. To prepare feeder layers in 6-well plates, we ordinarily seed cells of a confluent T25 flask or 50% of a confluent 100-mm dish into a 6-well plate. Two days later, the cells reach confluency and are ready to use for the differentiation of ES cells (*see* **Note 8**). Neither irradiation nor treatment with mitomycin C is needed.

3.3. Induction of Differentiation into Melanocytes

1. Prepare confluent ST2 feeder layers in 6-well plates as described in **Subheading 3.2.**
2. Grow ES cells to a subconfluent state as described in **Subheading 3.1.** (*see* **Note 9**).
3. Replace the medium of the ES cell culture with fresh medium 2–4 h before inducing the differentiation.
4. Wash the ES cell cultures three times with PBS and trypsinize with 0.25% trypsin, 0.5 mM EDTA for 5 min at 37°C. Add α-MEM/10% FCS and dissociate the cell clump by pipetting up and down. Centrifuge at 120g for 5 min at 4°C, aspirate the supernatant, and suspend the cell pellet in α-MEM/10% FCS. Count the ES cell number (*see* **Note 10**).
5. Dilute to an appropriate cell density with α-MEM/10% FCS (*see* **Note 11**). Aspirate the medium from the plates of ST2 feeder layers. Dispense 2 mL of the cell suspension into each well. Supplement with 10^{-7} M Dex, 20 pM bFGF, and 10 pM CT (*see* **Note 12**).
6. Incubate cultures at 37°C with 5% CO$_2$ in an incubator.
7. Change the culture medium every 2 or 3 d (*see* **Notes 13–15**) and allow the cultures to differentiate up to 3 wk (*see* **Note 15**). Do not scratch the ST2 layer when you aspirate the medium, and do not leave the cells out of the incubator for a long time, since detachment of the ST2 feeder from the edge of the culture plates will occur.
8. To count the cells, see **Subheading 3.4.**

3.4. Counting of the Number of Melanocytes

As shown in **Fig. 2C–F**, dendritic mature melanocytes containing melanin gralules in their cytoplasm will be observed within and around the ES-derived colonies by the above-mentioned culture. An average of about 50,000 melanocytes would be generated per well. As the culture described here is not a two-dimensional one, and melanocytes overlap each other in the colony, it is necessary to dissociate colonies into single-cell suspensions as described below in order to count melanocytes. If the numbers of melanocytes are small enough to count without dissociation, score the number of melanocytes after the fixation of the culture with 10% formalin and PBS for 10 min on ice.

1. Aspirate the medium from the culture and wash three times with PBS.
2. Add 400 μL of 0.25% trypsin and 0.5 mM EDTA.

3. Incubate at 37°C for 10 min.

4. Dissociate the colonies by pipetting up and down vigorously, and then return the cultures to the incubator and incubate for a further 5 min.

5. Add 100 μL of FCS to inactivate the trypsin and dissociate the clumps into a single-cell suspension by pipetting vigorously.

6. Count the number of pigmented cells within a 3-mm^2 region of a hemocytometer. They will be easily detected by observation without phase-contrast. If the number of pigmented cells is small, count replicate aliquots so that reliable data is obtained.

4. Notes

1. Choose a good lot of serum that supports the growth of ES cells well, does not generate differentiated cells, and has a high plating efficiency. The culture at the clonal density without the addition of exogenous LIF would allow the selection of such a lot. The batches of serum suitable for the maintenance of ES cells are also available from several commercial suppliers.

2. There is batch-related variation of FCS. In our experience, about half the batches of serum are not good for the maintenance of ST2 for as long as 3 wk. Also, the plating efficiency of ES cells (*see* **Note 11**) and the number of melanocytes generated on ST2 varies according to the lot of FCS. Check the batches of serum by seeding ES cells in the range from 100–2000 cells/well of 6-well plates, as described in **Subheading 3.3.** Select serum that is good for the maintenance of ST2 and efficiently supports the generation of melanocytes. It should be noted that plating efficiency is not correlated with the ability to support melanocyte formation.

3. At least two ES cell lines, D3 *(6)* and J1 *(7)*, gave similar results using our protocol. It is also possible to use ES cells that are adapted to feeder independence.

4. Treat embryonic fibroblasts with 10 μg/mL mitomycin C for 2.5 h. After washing the cells' well, seed 10^6 or 3×10^6 cells into a 60- or 100-mm dish, respectively. It is also possible to freeze mitomycin C-treated cells. In this case, thaw the 6×10^6 cells of a stock vial into one 100-mm dish or three 60-mm dishes. Dishes should be coated with gelatin by covering them with a 0.1% gelatin solution at 37°C for 30 min or at room temperature for 2 h. Remove the gelatin solution just before the plating of embryonic fibroblasts.

5. The passage numbers of ES cells should be kept as low as possible. Prepare sufficient amounts of young stocks and recover them when needed.

6. ST2 cultures, cultured for long periods, sometimes change in appearance (i.e., changing to a more dendritic shape or senescent appearance). Discard such cultures and use freshly thawed young ST2 cell stocks. **Fig. 3** shows the appearance of normal ST2.

7. ST2 (RCB0224) and OP9 (RCB1124) cells have been registered in the RIKEN Cell Bank (http://www.rtc.riken.go.jp/CELL/HTML/RIKEN_Cell_Bank.html). If the cells do not function well, please contact us.

8. It is also possible to seed at higher or lower densities, and then culture the ST2 cells for a shorter or longer time, until the ST2 cells are in a confluent state when the ES cells are inoculated.

9. Do not use ES cells that have just been thawed from frozen stocks. Use ES cells that have been cultured for at least 1 d after thawing.

10. If you want to remove the embryonic fibroblasts, replate the ES cells on gelatin-coated dishes and incubate for 30 min in the culture medium, then collect the nonadherent cells by pipetting. However, the separation of the trypsinized cells is not necessary when plating on a fresh ST2 layer.

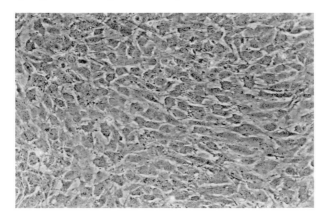

Fig. 3. Appearance of ST2 cells.

11. The efficiency with which ES cells form colonies on ST2 feeder layers changes according to the lot of serum from less than 1% to about 40%. It is recommended to seed ES cells so that approx 80 colonies are generated per well in the 6-well plate. Seeding at a higher density reduces the efficiency of melanocyte induction. Although inoculation at a lower density does not reduce the efficiency of melanocyte induction, the results of such experiments fluctuate from well to well. The plating efficiency also varies according to the ES line used. Carry out preliminary experiments to check the plating efficiency of specific cell lines and serum.

12. Dex is most important, while bFGF and CT are not so critical. If you remove Dex, the number of melanocytes generated is reduced by two orders of magnitude. You need not add steel factor, because ST2 cells produce a sufficient amount of this cytokine. Other factors, 50 nM 12-O-tetradecanoyl phorbol acetate (TPA), 10^{-8} M 1α, 25-dihydroxyvitamin D$_3$, and 100 ng/mL endothelin 3 also increase the number of melanocytes in this culture system *(4)*.

13. Just after the cultures are started, the growth of ST2 cells may be transiently promoted, and their appearance may become fibroblastic and more tightly packed due to the effects of bFGF and CT, but there is no need to worry, as this occurs normally, and ST2 cells will settle down after a while.

14. Adipocytes are frequently generated from ST2 layers. This occurs normally in well-maintained cultures.

15. After 2 or 3 d of differentiation, ES cell clusters derived from single ES cells are discernible. These colonies grow by piling up until around day 12 of differentiation. Pigmented melanocytes appear on d 12 to 13, and the number increases until 3 wk of differentiation. Further cultivation only augments the level of pigmentation of each cell.

Acknowledgments

We thank Dr. Hidetoshi Yamazaki (Tottori University) for collaborative work. Toshiyuki Yamane is a Research Fellow of the Japan Society for the Promotion of Science.

References

1. Nordlund, J. J., Boissy, R. E., Hearing, V. J., King, R. A., and Ortonne, J.-P. (eds.) (1998) *The Pigmentary System.* Oxford University Press, New York.

2. Galli, S. J., Zsebo, K.M., and Geissler, E. N. (1993) The kit ligand, stem cell factor. *Adv. Immunol.* **55,** 1–96.

3. Baynash, A. G., Hosoda, K., Giaid, A., Richardson, J. A., Emoto, N., Hammer, R. E., and Yanagisawa, M. (1994) Interaction of endothelin-3 with endothelin-B receptor is essential for development of epidermal melanocytes and enteric neurons. *Cell* **79,** 1277–1285.

4. Yamane, T., Hayashi, S.-I., Mizoguchi, M., Yamazaki, H., and Kunisada, T. (1999) Derivation of melanocytes from embryonic stem cells in culture. *Dev. Dyn.* **216,** 450–458.

5. Ogawa, M., Nishikawa, S., Ikuta, K., Yamamura, F., Naito, M., Takahashi, K., and Nishikawa, S.-I. (1988) B cell ontogeny in murine embryo studied by a culture system with the monolayer of a stromal cell clone, ST2: B cell progenitor develops first in the embryonal body rather than in the yolk sac. *EMBO J.* **7,** 1337–1343.

6. Doetschman, T. C., Eistetter, H., Katz, M., Schmidt, W., and Kemler, R. (1985) The in vitro development of blastocyst-derived embryonic stem cell lines: formation of visceral yolk sac, blood islands and myocardium. *J. Embryol. Exp. Morphol.* **87,** 27–45.

7. Li, E., Bestor, T. H., and Jaenisch, R. (1992) Targeted mutation of the DNA methyltransferase gene results in embryonic lethality. *Cell* **69,** 915–926.

22

Using Progenitor Cells and Gene Chips to Define Genetic Pathways

S. Steven Potter, M. Todd Valerius, and Eric W. Brunskill

1. Introduction

The zygote is a single cell, a fertilized egg, capable of giving rise to a complete organism. This is a truly remarkable transformation, with a single cell becoming trillions of cells or more, with a microscopic mass becoming in some cases tens or hundreds of kilograms, with a single amorphous cell making a number of complex organs, including the brain with its uncountable precise synaptic connections. For mammals, which generally appear to lack localized cytoplasmic determinants, this process is guided by a genetic program encoded in the DNA of the zygote.

The conceptual framework of the genetic regulatory circuitry controlling development was first presented by Britten and Davidson in a series of articles appearing about thirty years ago *(1,2)*. They proposed that each gene is regulated by DNA sequences, which we would today call *cis*-regulatory elements that respond to signals from upstream regulator genes. Some genes, in turn, will generate signals activating batteries of further downstream gene targets. They postulated the existence of complex genetic cascades. They also proposed the existence of sensor genes, which we would now call receptors, capable of responding to environmental cues. This broad framework of the genetic hierarchy of development has held true, but we are only now beginning to understand the details.

It is estimated that there are approx 30,000 to 50,000 genes in the mammalian genome. Furthermore, the genetic circuitry of development is not linear, but more neural in character. Each neuron of the brain receives input from multiple upstream neurons and projects to multiple downstream neurons. And so it is with the regulatory genes of development. Each gene regulated by multiple upstream genes. In turn, each regulator gene capable of modulating expression of multiple downstream target genes. Therefore, one can predict that the genetic regulatory pathways of development will be logarithmically more complex, for example, than the enzymatic pathways of intermediary metabolism, which tend to be more linear in nature, with one substrate generally converted to one product.

Despite the immense obstacle of complexity, there is cause for optimism. Considerable progress was made in defining important genetic processes of development. First, sequencing data now suggests a "core" genome of only a few thousand genes, with

From: *Methods in Molecular Biology, vol. 185: Embryonic Stem Cells: Methods and Protocols*
Edited by: K. Turksen © Humana Press Inc., Totowa, NJ

the total gene count high because of gene duplications. This suggests the presence of a limited number of fundamental principles and pathways to define. Second, again the result of massive cDNA and genomic sequencing efforts, essentially all of the genetic players in the process of development are now identified. It is clearly necessary to know what genes are involved before they can be positioned within a program. Third, the available tools for manipulating the mammalian genome have resulted in the generation of thousands of mice with targeted, and conditionally targeted, mutations allowing the characterization of developmental functions of genes. Fourth, new tools, in particular the advent of cDNA microarrays and Affymetrix GeneChips, allow the simultaneous analysis of expression patterns of thousands of genes. Properly used, these tools might allow the precise definition of the genetic hierarchies of development.

Understanding the genetic program of development may have profound practical implications. By understanding how cells are normally driven to develop and form organs, it may become possible to manipulate this process. We hope to improve our ability to regenerate and repair organs damaged by disease, old age, physical insult, or birth defect. Two possible strategies are envisaged. First, novel treatment regimens of an organ *in situ* may promote its repair. For instance, a specific combination or sequence of growth factors may trigger organ regeneration. Second, stem cells can be used. Stem cells are literally exploding on the scene, as illustrated in this volume. They are found in the marrow, liver, muscle, skin, and the brain. At low efficiency, these stem cells show remarkable developmental plasticity. Adult neural stem cells, for example, have been used to reconstitute the hematopoietic systems of irradiated mice *(3)*, and hepatocytes have derived from bone marrow stem cells *(4)*. With a more complete understanding of the genetic program of development, it may become possible to drive stem cells in desired directions to facilitate the repair, or replacement, of damaged organs. With recent advances in nuclear transplantation techniques, it becomes rational to consider taking a single-cell nucleus from a patient and transferring it to an enucleated stem cell or to an enucleated oocyte, which can then be used to make stem cells. The stem cells used will then be genetically identical to the recipient.

The characterization of genetic programs requires the identification of the downstream targets of transcription factor genes. This has been notoriously difficult. Nevertheless, a number of approaches have been used with some success. One of the most commonly used approaches to date is the *in situ* hybridization analysis of mutants. This so-called "molecular marker analysis" provides a molecular measure of the developmental perturbation caused by the mutation and defines altered expression patterns of downstream genes. Although extremely useful, this approach has severe limitations. First, it only tests preselected candidate genes. One must suspect that a gene is a target for it to be selected as a potential candidate. Second, *in situ* hybridization is a relatively laborious technique that severely limits the number of genes that can be tested. Third, *in situ* hybridization is not a very quantitative technique, which means that targets must be greatly altered in expression level to be identified. This technique will continue to be very useful, but is not the ideal method for the characterization of batteries of downstream targets. A second method is immunoprecipitation of target genes that have been in vivo fixed to the transcription factor of interest. This approach has provided significant insight in the Drosophila system *(5,6)*, but has had more limited success in mammalian studies *(7,8)*. In theory, immunoprecipitation offers a universal screen for downstream targets, but in practice, it can be technically difficult. Another strategy has been to sequence scan the promoter regions of genes of interest for potential transcription factor protein binding sites. This approach has identified the β-amyloid *(9)*, cytotactin *(10)*, and neural cell adhesion *(11)* genes as candidate targets of homeobox genes. Limitations of this procedure include the relative lack of specificity of the defined binding sequences. By chance, these sequences occur quite frequently in the genome. Also, the cotransfection assays for function are often conducted in heterologous cell types, such as HeLa cells, which are lacking appropriate cofactors that are necessary for binding specificity.

To avoid some of the cell type heterogeneity, variable timing, precise dissection, and limited material challenges presented by tissue approaches, some investigators have worked with cell lines grown in culture. Cell line studies have generally used a candidate target gene approach, looking for transcript level differences for specific genes in cells that do or do not express the transcription factor gene of interest. In this manner it was shown that *Hoxb 7* up-regulates *bFGF* expression in melanoma cells *(12)*, and that the *Pax-2* gene altered expression of *vimentin, E-cadherin*, and the *Wilms' tumor* gene *(13)*. One interesting target identification approach uses gene trap vectors with reporter or selectable marker genes. These vectors can integrate into multiple random positions in the genome. The cells are also stably transfected with an inducible version of the transcription factor gene, (or the cells are treated with growth factors or putative morphogens such as retinoic acid). Flourescent-activated cell sorting, replica plating, or antibiotics are used to select cells that show inducible reporter expression. These cells presumably have the reporter trap construct inserted into a responsive target gene. This approach offers the advantages of a universal screen and powerful selection techniques to identify cells of interest *(14,15)*. Disadvantages include reporters that are apparently inactivated at rather high rates by DNA methylation. Nevertheless, this strategy may make significant contributions to our understanding of genetic regulatory networks. Another approach that has been used is to identify transcript differences between wild-types and mutants (or, for hormones and growth factors, treated and untreated). If the transcription factor gene of interest were mutated, then presumably the downstream target genes would be altered in expression. A polymerase chain reaction (PCR)-based difference cloning strategy has been used to identify genes altered in expression following thyroid treatment of tadpoles *(16)*. The use of whole tissues can present signal-to-noise problems, as many tissues have considerable cell type heterogeneity. Nevertheless, Dulac and Axel *(17)* have described difference cloning procedures that work for abundant transcripts with little starting material, even from one cell. Coupled with laser capture microdissection methods, these techniques offer considerable future promise.

In this article, we describe a target identification strategy that uses a combination of developmentally appropriate progenitor cells, inducible gene expression systems, and gene chips.

2. Materials

The described target identification protocol uses a combination of several technologies. DNA constructs that allow cell-specific, developmental timing-specific expression of *SV40 T* in transgenic mice must be made. Transgenic mice are made and used to make cell lines. The cell lines are then stably transfected with inducible gene expression constructs, and the RNA from the cells used in conjunction with Affymetrix GeneChip probe arrays to identify transcript differences between induced and uninduced cells. In this article, we focus on the production of cell lines, the use of the tet inducible system to generate cells with and without expression of the transcription factor gene of interest, and the use of Affymetrix GeneChip probe arrays to identify gene expression profile differences representing downstream targets.

1. CO_2 incubator, Napco model 6000, (Fisher, cat. no. 11-686-103A).
2. Centrifuge (Fisher, cat. no. 05-112-116).
3. Zeiss inverted microscope (Fisher, cat. no. 12-070-310).
4. Dulbecco's modified Eagle medium (DMEM) (Gibco, cat. no. 10313-021).
5. Fetal bovine serum (FBS) (Gibco, cat. no. 26140-079).
6. Trypsin-EDTA (Gibco, cat. no. 25300-054).
7. L-Glutamine (HyClone, cat. no. SH30034.01).
8. Tet-Off gene expression system (Clontech, cat. no. K1620-1).
9. Superscript choice system for cDNA synthesis (Gibco, cat. no. 18090-019).
10. RNeasy mini kit (Qiagen, cat. no. 74104).
11. RNA transcript labeling kit (Enxo Diagnostic, cat. no. 42655).
12. Affymterix instrument system (Affymetrix, cat. no. 900227).
13. Murine gene chip probe array U74A (Affymetrix, 900321).
14. RNAzol (Tel-Test, cat. no. cs105).
15. GeneScreen Plus hybridization membrane (NEN, cat. no. NEF986).
16. Tetracycline system approved bovine serum (Clontech, cat. no. 8630-1).
17. Affymetrix eukaryotic hybridization control Kit (Affymetrix, cat. no. 900299).
18. 10X HEPES-EDTA buffer: 0.5 M HEPES, 0.01 M Na_2EDTA. Make with diethyl pyrocarbonate (DEPC)-treated water and RNA only dedicated reagents. Adjust to pH 7.8 and autoclave before use.
19. Buffer A: 294 μL 10X HEPES-EDTA buffer, 706 μL DEPC H_2O.
20. Formaldehyde–Formamide: 89 μL formaldehyde, 250 μL formamide.
21. Gel loading buffer: 322 μL buffer A, 5 mg xylene cyanol, 5 mg bromophemol blue, 400 mg sucrose. Mix to dissolve. Add 178 μL formaldehyde and 500 μL formamide.
22. 20X Standard saline citrate (SSC): 3 M NaCl, 0.3 M sodium citrate, pH 7.0.

3. Methods

3.1. A Novel Strategy

We devised a strategy to identify downstream targets of transcription factors by combining certain aspects of previous approaches, described above, with some novel features *(18)*. For several reasons, we elected to use developmentally appropriate cell lines rather than tissues. First, clonal cell lines offer great system homogeneity, not available with tissues. Cell lines can, therefore, reduce the noise problems associated with tissues resulting from the presence of diverse cell types, from dissection errors, and from developmental timing variations in different samples used for comparisons. Cell lines also provide an abundant amount of renewable starting material. This can be extremely important when examining early developmental processes, when extremely limited amounts of tissues are available. Finally, there is considerable evidence, presented later, suggesting it is possible to use the *SV40 T* gene to make cell lines that closely represent embryonic cells.

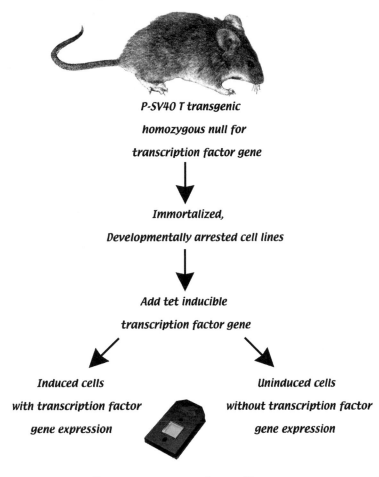

P-SV40 T transgenic

homozygous null for

transcription factor gene

Immortalized,

Developmentally arrested cell lines

Add tet inducible

transcription factor gene

Induced cells   *Uninduced cells*

with transcription factor   *without transcription factor*

gene expression   *gene expression*

Compare gene expression profiles

using GeneChip probe arrays

Fig. 1. Downstream target identification strategy. A transgenic mouse or embryo is made carrying a tissue-specific developmental timing-specific promoter driving the expression of the *SV40 T* gene. This mouse also is homozygous null for a targeted mutation in the transcription factor gene being studied. Activation of *SV40 T* immortalizes and developmentally arrests the cells of interest, which are grown as cell lines. A tet inducible version of the transcription factor gene is stably introduced, and a clone showing high levels of induction is identified. Induced and uninduced cells, identical except for the expression of the transcription factor gene in the induced cells, are grown. GeneChip probe arrays are used to identify the resulting expression level differences in downstream targets.

An outline of the strategy is shown in **Fig. 1**. Appropriate cell lines are made from mice homozygous for a targeted mutation in the transcription factor gene of interest. These cells are then stably transfected with a DNA construct, allowing inducible expression of the transcription factor gene. A single clone of cells showing a high level of induction and a low basal expression level is selected. The cells are then grown in duplicate, with and without inducer. It is important to emphasize that the two cell populations being compared are identical, except for the presence or absence of expression of the transcription factor gene of interest. This keeps noise levels to a minimum. The next step is to compare the gene expression profiles of the induced and uninduced cells. Expression of the transcription factor gene will, in turn, result in altered expression of downstream gene targets. Genes showing different expression levels in the two cell populations are, therefore, excellent target candidates. In our experience, Affymetrix GeneChip probe arrays offer a rapid and effective means of gene expression profile analysis. They are capable of comparing levels of gene expression for tens of thousand of genes simultaneously.

This strategy offers several advantages. As emphasized above, the close similarity of the two cell populations being compared eliminates cell heterogeneity, thus reducing noise levels. This provides an unbiased universal screen. No prior knowledge of possible targets is required. Even expressed seqeuence tags (ESTs) of unknown function can be found as targets. This strategy can find downstream targets at all levels and is not restricted to direct targets. Multiple members of a pathway at multiple levels can, therefore, be identified. By using clonal cell populations, one can make large amounts of starting materials. This strategy also allows the identification of targets showing modest quantitative expression responses, which would be missed by other techniques. Finally, by using developmentally appropriate cell lines, one gains genuine insight into the program of development. Transcription factors require an appropriate milieu of cofactors to bind to correct targets. Therefore, it is important to use cells that mimic the in vivo situation as closely as possible.

3.2. SV40 T Cell Lines

To generate embryologically appropriate cell lines, we have used the *SV40 large T* antigen gene (*SV40 T*). *SV40 T* has been shown previously to be remarkably effective at immortalizing and, in many cases, developmentally freezing cells. Use of an appropriate promoter in transgenic mice can drive precise spatiotemporal oncogene expression, targeting immortalization to the cells at the correct developmental stage. In culture only, the immortalized cells are selected by their continued growth. Pam Mellon and her colleagues have made powerful use of this strategy, generating a number of pituitary cell lines, using a variety of pituitary gene promoters. For example, a 1.8-kb promoter for the glycoprotein hormone α-subunit was used to immortalize and developmentally arrest cells that expressed the α-subunit and responded to gonadotrophin-releasing hormone *(19)*. A longer version of this promoter (5.5 kb), which expressed earlier, was subsequently used to establish more primitive α-subunit expressing cells, and the β-subunit leutinizing hormone promoter was used to make cell lines from later stages in the gonadotrope lineage *(20)*. These and other cell lines generated in similar fashion have made significant contributions to our knowledge of the genetic pathways of pituitary lineage-specific development. For example, studies of the upstream regulation in these cells of lineage-specific genes led to the discovery of the homeobox gene *Pit-1* *(21,22)* and the helix-loop-helix transcription factor gene CUTE *(23)*.

It is essential to use a promoter driving appropriate tissue or cell-specific expression of *SV40 T*. We have used promoters from the two homeobox genes, *Gsh-1* and *Hoxa 11*. With *Gsh-1-SV40 T*, we were able to make cell lines from the hypothalamus *(18)*. With *Hoxa 11-SV40 T*, we have made cell lines representing early metanephric mesenchyme of the developing kidney (Valerius et al., in preparation). In each case, it was possible to make cell lines from animals homozygous for a targeted null mutation in the homeobox gene used for the promoter. We found both *Gsh-1-SV40 T* and *Hoxa 11-SV40 T* transgenic mice survived long enough to allow breeding. It was possible to transfer the transgene onto the targeted null background by breeding. The cell lines were then made from mice a few weeks old. The resulting cell lines were characterized extensively by gene chips and were shown to retain early embryonic gene expression profiles and biological properties. Remarkably, some of the embryonic metanephric mesenchyme kidney cell lines retained the ability to induce growth and branching of the ureteric bud in co-culture experiments. This is consistent with the conclusion that certain cells were developmentally frozen following the onset of *SV40 T* expression and maintained their embryonic properties in mice a few weeks old and in the subsequent cell culture.

Cell lines are established as described previously *(19)*. The appropriate tissue is removed by dissection, minced with scissors, enzymatically dissociated with trypsin (and if necessary, collagenase), and cultured under relatively standard conditions with DMEM and 10% FBS. The *SV40 T* expressing cells are immortalized and grow indefinitely, while the nonexpressing cells senesce and die after repeated passage (10 to 30, depending on the cell type).

In characterizing a number of resulting cell line clones, we have observed that, even when using the same *SV40 T* transgene, not all cells appear developmentally frozen at exactly the same developmental stage. For example, some of the kidney cell lines made from *Hoxa 11-SV40 T* mice appear to represent early metanephric mesenchyme, while other cell lines represent metanephric mesenchyme at a somewhat (hours, not days) later developmental stage, undergoing epithelial conversion (Valerius et al., in preparation). This stochastic element can be useful, allowing a single transgene to generate cell lines with distinct developmental properties.

The use of *SV40 T* connected to a promoter driving expression in a cell-specific and developmental timing-specific manner allows a number of advantages. A single cell type can be immortalized, perhaps from within a developing organ consisting of multiple cell types. The cells are developmentally arrested at around the time of *SV40 T* expression and can be recovered later, even in the adult animal. Nevertheless, the desired promoter is not always available. An alternative strategy, therefore, is to use the "immortomouse" commercially available from Charles River. This mouse carries a temperature-sensitive, inducible form of *SV40 T* as a transgene. Tissues can be removed from this mouse, enzymatically dispersed, grown under conditions that activate *SV40 T*, and used to make cell lines. Embryos can be dissected to generate cell lines representing early stages of development. This *SV40 T* transgene can be bred onto a gene targeted background to make cell lines with null mutations in transcription factor genes. The use of a temperature-sensitive inducible *SV40 T* offers an additional advantage. The *SV40 T* can be inactivated, allowing the cells to proceed down their developmental pathway in culture. The immortomouse, therefore, can provide a very useful tool in the generation of progenitor cells for genetic circuitry analysis.

3.3. Generation of Cell Lines

1. Using sterile technique, dissect the desired tissue from the *SV40 T*-expressing transgenic mouse. Place the tissue on a plastic Petri dish and finely mince with scissors.

2. Add approx 30 vol of trypsin. Pipet the trypsin and minced tissue into a sterile plastic tube and seal the cap.

3. Incubate at 37°C, inverting the tube every few minutes to mix. Normally 5–15 min is sufficient. (*See* **Note 1**).

4. Invert the tube with tissue and trypsin to mix, then allow remaining tissue clumps to settle to the bottom for 30 s. Remove the solution above the clumps, add an equal volume of DMEM media with glutamine and 10% FBS to inactivate the trypsin.

5. Spin down the cells by centrifugation at 300*g* for 5 min.

6. Resuspend the cells in complete DMEM media with glutamine and 10% FBS, and pipet onto tissue culture plastic plates, with a final of 7 mL of media/100-mm plate. Each starting gram of tissue should be placed on 4 plates.

7. Media should be changed every other day. When the cells are confluent, the plates should be rinsed with 5 mL phosphate-buffered saline (PBS), treated with 4 mL trypsin for 5 min at 37°C, and the cells collected and pelleted by centrifugation for 5 min at 300*g*. The supernatant is discarded, and the cells are resuspended in complete DMEM media. The cells from each original plate should be passaged, or transferred, to 5 new plates. After 20–30 such passages, the surviving cells are likely transformed by *SV40 T*.

8. Clonal cell lines can be made by several different procedures. Serial dilutions of the cells can be placed in the wells of a 96-well plate. After dilutions, few wells will have cell growth after several days in culture, and the cells that do grow are likely of clonal origin. Alternatively, single cells of a trypsinized cell suspension can be drawn up with a finely drawn capillary, under the microscope, and individually placed in wells of a 96-well plate.

3.4. Expanding the Usefulness of the Cell Lines

As described in **Subheading 3.3.** above, the generation of the cell lines involves considerable work. According to the original strategy, each cell line is only useful for identification of the targets of the transcription factor targeted in the mice used to make the cells. We, therefore, expanded the scope of the strategy to allow target identification for additional transcription factor genes.

We found that the cell lines we made inevitably expressed a large number of transcription factor genes. It would be useful to understand the functions of these genes in the cell lines. By adding to the cells inducible dominant negative constructs and allowing removal of specific transcription factor gene function, it would again be possible to make duplicate sets of cells, with and without the transcription factor, to allow comparison of resulting gene expression profiles. It would then be possible to define the downstream targets of any transcription factor expressed in the cells.

The Drosophila "Engrailed" protein includes a very powerful transcription silencer domain that can be used to remove transcription factor gene function. This domain can be fused to the DNA binding domains of other transcription factors, generating a novel protein that will then selectively repress transcription factor downstream targets. Such Engrailed repressor domain-DNA binding domain fusions have shown effective gene-specific repressors in a number of previous studies *(24–26)*. It is expected that proper use of the Engrailed repressor will allow inducible silencing of downstream targets of specific transcription factors. Each cell line will then be useful for examining the downstream targets of each of the transcription factors it expresses. Each cell line provides an experimental system focused on a single cell type from single time point in development. By combining data from multiple cell lines, representing different time points and cell types of a single organ, a complex dynamic of the genetic program of organogenesis emerges. Such data is more powerful than a simple description of the changing gene expression profiles during organogenesis, because it includes regulatory information, describing how various transcription factors control different aspects of the observed gene expression patterns.

As mentioned previously, we observed different cell lines, even when resulting from the same transgene, often represent slightly different stages of development. For example, in making kidney metanephric mesenchyme cell lines, we found some cell lines exhibited gene expression profiles of earlier cells, prior to induction by the ureteric bud, while one cell line showed *Pax-2* expression, which is activated following bud induction. To identify *Pax-2* targets and better understand the function of this gene in mesenchyme induction, we introduced an inducible version of the *Pax-2* gene into the early mesenchyme cells. This strategy again expands the usefulness of the cell lines, allowing them to be used to study targets of genes that the cells would normally express slightly later in development.

3.4.1. Inducible Gene Expression

The "tet-off system," commercially available from Clontech is excellent for generating inducible gene expression in cell lines *(27)*. (*See* **Note 2**).

3.5. Preparation of Biotinylated RNA for Hybridization to Affymetrix GeneChip Probe Arrays (See Note 3)

3.5.1. First Strand Synthesis

Use reagents, including enzymes and solutions, from Superscript Choice System for cDNA synthesis kit from Gibco BRL.

1. Combine 10 µg of total RNA plus enough nuclease-free water to make to 10 µL total.
2. Use 0.7 µL T7 promoter sequence–(dT)24 primer (137 pM/µL).
3. Then incubate at 70°C in water bath for 10 min, quick spin to move contents to the bottom of the tube, and place on ice.
4. Add 4 µL of 5X *First* Strand cDNA buffer, 2 µL of 0.1 *M* dithiothreitol (DTT), 1 µL of 10 m*M* dNTP mixture.
5. Mix and incubate in 42°C water bath for 2 min, add 2 µL of SSII RT. Mix well.
6. Incubate 42°C for 1 h, then quick spin to the bottom of the tube and place on ice.

3.5.2. Second Strand Synthesis

7. Add 91 µL DEPC-treated water.
8. Add 30 µL of *Second* strand reaction buffer.
9. Add 3 µL of 10 m*M* dNTP mixture.
10. Add 1 µL of DNA ligase.
11. Add 4 µL of *Escherichia coli* DNA polymerase (not T4 DNA polymerase).
12. Add 1 µL of RNase H.
13. Mix gently by pipeting or giving the tube some taps.
14. Incubate for 2 h at 16°C.
15. Then add 2 µL of T4 DNA polymerase.
16. Incubate for another 5 min at 16°C.
17. Add 5 µL of 0.5 *M* EDTA to stop reaction.
    Clean up.

18. Add 30 μL of 3 *M* ammonium-acetate.
19. Add 150 μL of phenol:chloro:isoamyl alcohol (25:24:1).
20. Mix thoroughly, then spin in microfuge at full speed (15,000*g*) for 5 min.
21. Take top aqueous phase, add 2.5 vol of 95% ethanol, place on ice for 30 min.
22. Spin in refrigerated microfuge at full speed at 4°C for 30 min.
23. Pour off the supernatant, add 400 μL of ice-cold 80% ethanol, spin for 2 min at full speed, discard supernatant, and rinse again with 80% ethanol. One final rinse with 95% ethanol.
24. Dry in speed vac.
25. Dissolve in 12 μL of nuclease-free water.

3.5.3. In Vitro Transcription

Use reagents from the Enzo in vitro labeling kit from Enzo Diagnostics.

26. Add to a new 1.5-mL tube, 5 μL of the above cDNA (of the 12 μL from **step 30**).
27. Add 19 μL of nuclease-free water.
28. In setting up the in vitro transcription reaction, ingredients (except enzymes, which should be kept on ice until added) and reaction tube should be kept at room temperature to avoid precipitation.
29. Add 4 μL of 10X reaction buffer.
30. Add 4 μL of 10X biotin nucleotides.
31. Add 4 μL of 10X DTT.
32. Add 4 μL of RNase inhibitor.
33. Add 2 μL of T7 RNA polymerase.
34. Mix and incubate 37°C for 4 to 5 h.
35. Can store on ice briefly or freeze, then go to Qiagen clean up procedure.

3.5.4. Qiagen RNeasy Clean Up Procedure

Follow recipe on p. 48 of RNeasy Mini Handbook. Reagents are from Qiagen RNeasy mini kit.

36. Add 60 μL of nuclease-free water to the 40 μL in vitro transcription reaction.
37. Add 350 μL of buffer RLT. This buffer must have been made up fresh within the last month, by adding 10 μL of 2-mercaptoethanol (on shelf in tissue culture room) to 1 mL of RLT. Mix thoroughly.
38. Add 250 μL of ethanol (96–100%) and mix well by pipetting. Do not centrifuge.
39. Apply sample to labeled RNeasy spin column sitting in a collection tube. Centrifuge 15 s at 12,000*g*.
40. Transfer the column to a new collection tube. Add 500 μL of RPE buffer and centrifuge 15 s at 12,000*g*. Discard flow through.
41. Add another 500 μL of RPE to the column and spin at full speed for 2 min. Discard flow through.
42. Transfer column to a 1.5-mL centrifuge tube and spin again at full speed for 1 min, to remove all traces of residual RPE.
43. Add 35 μL of Rnase free water directly to the center of the Rneasy membrane. Transfer to a fresh 1.5-mL tube. Centrifuge for 1 min at full speed to elute. Keep the flow through as it has the RNA product.
44. Add another 35 μL of water directly to the RNeasy membrane again. Spin at full speed into the same collection tube used in the previous step.

3.5.5. Quantitate the Biotinylated RNA Product

45. Dilute 2 μL of the RNA with 98 μL of nuclease-free water.
46. Measure the absorbance (or OD) of the diluted sample at a wavelength of 260 using a spectrophotometer.
47. Multiply the absorbance reading times 50 (dilution factor) and multiply the product by 40 (an absorbance of one means the RNA is 40 μg/mL) to get the RNA concentration in μg/mL. In other words, multiply the absorbance reading by 2000 to find the RNA concentration.
48. This RNA can be used directly, without ethanol precipitation.

3.5.6. Fragmentation of Biotinylated RNA

49. Mix 8 μL of 5X fragmentation buffer (200 m*M* Tris-acetate, pH 8.1, 500 m*M* potassium-acetate, 150 m*M* magnesium-acetate) with 20 μg of biotinylated RNA, and bring to a final vol of 40 μL with water, in a 1.5-mL conical tube.
50. Incubate with sealed cap at 94°C for 25 min. Can store at –20°C until ready to make hybridization solution.

3.5.7. Making Hybridization Solution

The hybridization solution contains spikes of control RNAs, which provide borders read by the scanner, and measures of hybridization sensitivity. These spikes are most easily obtained through the Affymetrix eukaryotic hybridization control kit. The following assumes this kit is used.

51. To the 40 μg of fragmented RNA in 80 μL, add 6 μL of herring sperm DNA (10 mg/mL).
52. Add 6 μL of acetylated BSA.
53. Add 300 μL of 2X hybridization buffer (final 1× concentration is 100 m*M* 4-morpholinee-thansulfonic acid [MES], pH 6.6, 1 *M* NaCl, 20 m*M* EDTA, 0.01% Tween-20).
54. Add 30 μL of 20X eukaryotic hybridization controls.
55. Add 10 μL of control oligonucleotide B2, also from the Affymetrix eukaryotic hybridization control kit.
56. Add 168 μL of water.

3.5.8. GeneChip Probe Array Hybridization

57. The GeneChip probe array is allowed to equilibrate to room temperature, and the hybridization chamber is then filled with 1X hybridization buffer.
58. Prehybridization of the GeneChip probe array is carried out in the Affymetrix hybridization oven at 45°C for 15 min.
59. Meanwhile the hybridization cocktail with biotinylated RNA is heated at 99°C for 5 min in a heat block, followed by 45°C for 5 min.
60. The hybridization cocktail is centrifuged at full speed in a microfuge for 5 min to pellet particulate matter.
61. The 1X hybridization solution is removed from the chip and replaced with the above hybridization cocktail.
62. The chip is hybridized at 45°C overnight in the Affymetrix hybridization oven.

3.5.9. Washing, Staining, Scanning, and Analysis

63. The GeneChip probe array is then washed and stained in the Affymetrix automated Fluidics Station. This takes about 1.5 h.

64. The GeneChip probe array is then scanned by the Affymetrix GeneArray scanner, which takes about 5 min.
65. The Affymetrix GeneChip software then identifies transcript level differences in candidate downstream targets by comparing gene expression profiles of GeneChips hybridized to RNAs made from cells expressing or not expressing the transcription factor of interest.

3.6. Target Confirmation

The procedures listed above serve as a universal screen to identify candidate downstream targets, which then need to be confirmed. The first step in confirmation is to perform a Northern blot, using RNAs from the induced and uninduced cells, to verify that the candidate target alters its expression level in response to expression of the transcription factor gene being studied. In our experience, it is necessary to repeat this confirmation 3 times, with 3 separate RNA preparations. Each experiment, of course, should include a loading control hybridization, such as glyceraldehyde-3-phosphate dehydrogenase (GAPDH). We have found there is some "biological noise" of unknown origin. A few genes show altered expression levels that are not reproducible. These could represent cell cycle genes and result from harvesting cells at different levels of confluency.

3.6.1. Northern Blot Protocol

Poly(A) RNA (mRNA) (1–5 μg) is subjected to electrophoresis in a 1.5% agarose gel containing 0.6% formaldehyde.

3.6.2. Gel Preparation

To make 50 mL of a 1.5% formaldehyde gel:

1. Mix 0.75 g agarose and 5 mL 10X HEPES-EDTA and use DEPC-H_2O to bring up to approx 49.2 mL.
2. Boil to dissolve agarose, then cool to about 60°C.
3. Add 82 mL formaldehyde (37% stock) and 2.5 μL ethidium bromide (10 mg/mL stock).
4. Pour the gel and allow it to set for up to at least 45 min. The gel is run in 1X HEPES-EDTA buffer.

3.6.3. Sample Preparation

1. mRNA (1–5 μg in a vol of 4 μL is loaded per lane. If the volume is too large, dry down the sample and resuspend in 4 μL.

RNA in total volume of	4.0 μL
Buffer A	2.4 μL
Formaldehyde-formamide	4.6 μL
(Total vol	11.0 μL)

2. Heat at 70°C for 10 min.
3. Place on ice and spin down to collect.
4. Add 1.5 μL of gel loading buffer and load on gel.
5. Run gel at 50 V/cm length for 2 to 3 h.

3.6.4. Blotting

After electrophoresis, the gel is equilibrated in 50 m*M* NaOH for exactly 10 min. The GeneScreen Plus membrane is also equilibrated in 50 m*M* NaOH at this time. Also during this time, a downward transfer is prepared using 50 m*M* NaOH as the transfer buffer and GeneScreen Plus membrane. The setup starts with a stack of paper towels on the bottom, three layers of blotting paper, the membrane, the gel, three more layers of blotting paper, then a double-layered wick is placed over the top and into buffer tanks on the sides. A plastic or glass cover is used on top to supply light pressure on the stack and reduce evaporation.

After 2.5 h, the transfer is stopped, and the membrane is briefly neutralized in 2X SSC and then dried completely in a vacuum oven (15 min at 80°C).

The blot is now ready for hybridization by your method of choice.

Consistent response to induction of the transcription factor gene, as measured by the above Northern blot analysis, represents very strong evidence that the candidate gene is a genuine downstream target. In principle, some genes might be altered in expression as a result of the treatment with the inducer itself and not as a result of the altered expression of the transcription factor gene. This can be tested by inducer treatment of the original cell line, prior to transformation with the inducible DNA expression constructs.

Additional confirmation can include the following. *In situ* hybridization can be used to examine expression of the candidate target in mice mutant for the transcription factor gene. This represents an in vivo confirmation, but can only be applied to targets that show several-fold expression level differences, in response to the transcription factor. The target promoter can also be subjected to gel shift analysis and DNA footprinting, to determine if the transcription factor binds directly. Cotransfection studies, with one construct driving transcription factor expression and the other carrying the target promoter connected to a reporter gene such as luciferase, can be used to further dissect the transcription factor downstream target relationship. Specific mutation of the factor binding site should eliminate expression response of the putative target.

3.7. The Future

The application of these and other strategies will allow the definition of the elaborate genetic regulatory networks controlling development. The use of cell lines in conjunction with inducible gene expression systems and gene chips promises to dramatically accelerate our understanding of the genetic programs of organogenesis. The dominant negative strategy will be particularly useful, as it allows the removal of transcription factor gene function. A host of cell lines expressing a large number of interesting transcription factors are already available. This approach does not require the time-consuming production of new cell lines from mutant mice. Soon the entire sets of mouse and human genes will be represented on gene chips. Vast collections of data connecting transcription factors to their targets will be generated in the near future. Continued improvements in bioinformatics will be required to assimilate this data and to convert it into an improved understanding of the molecular mechanisms of development.

This deeper understanding of the genetic basis of development, in turn, will undoubtedly have remarkable practical consequences. Birth defects, for example, account for half of pediatric hospital admissions and are the second leading cause of death among children, only behind accidents. An improved understanding of normal development will lead to an better ability to regenerate and repair damage due to birth defect.

4. Notes

1. When making cell lines, the length of time required for trypsinization will vary with the tissue. One wants to digest as much tissue as possible without causing cell lysis. Over digestion will result in viscosity, as DNA is released from cells. Under digestion will release few cells from the minced tissue and give a low yield. Defining the proper digestion time may require some trial and error.

2. The protocols provided with the tet-off system from Clontech work well, but the following advice must be heeded. First, be aware that many lots of FBS are contaminated with tetracycline, presumably from feed given cows. This can result in low levels of expression in the induced state. Tetracycline-free serum can be purchased from Clontech. Second, the constructs used to generate inducible expression must be stably transfected separately. We have found cotransfection results in poor levels of induction. The constitutive expression construct apparently co-integrates with the inducible construct, driving low levels of expression even in the absence of induction. The preferred approach, then, is to first introduce the constitutive construct and to test clones for good expression by transient transfection with a tet operator-regulated luciferase reporter in the presence and absence of inducer. We recommend testing about 30 clones. A clone with good expression is then selected for subsequent stable transfection with the inducible construct. Clones are again selected and tested, generally by Northern blot, to identify those showing excellent levels of induction. It is generally wise to test at least ten clones.

3. The Affymetrix GeneChip system works well for comparing gene expression patterns in the induced and uninduced cells *(18)*. The Affymetrix Gene Chips consist of thousands of tiny squares or features, with different features containing millions of copies of unique oligonucleotide sequences. For each gene examined, there are 32–40 features, with half having centrally positioned single base mismatches to serve as hybridization controls. Single chips currently examine the expression of about 12,000 genes. As the features continue to become smaller, this number will increase. Affymetrix chips are available for analysis of expression of about 36,000 mouse genes and about 62,000 human genes. These are significant fractions of the genome, suggesting the chips are even now quite effective in finding most downstream targets. Further, the chips are very sensitive, able to detect hybridization signal from RNAs as rare as approximately one copy per cell. In our experience, the gene chips were much more effective in finding genuine targets than differential display or membrane-based hybridization approaches.

   The Affymetrix GeneChip probe arrays do, however, have the disadvantage of high cost. The Affymetrix GeneChip system is expensive, and then the GeneChips to use for a single experiment are also costly. Unfortunately, there is also a small but significant chip noise problem. To reduce the number of false positives, it is therefore best to do experiments in duplicate.

   Microarrays using spotted cDNA sequences are available from commercial vendors, such as Incyte, or can be made by the investigator, as per the Patrick Brown Web Site (http://cmgm.stanford.edu/pbrown/index.html). These systems, particularly when homemade, can dramatically reduce cost. The disadvantage of the homemade system is the difficulty in the preparation of the thousands of DNA preparations to be spotted.

References

1. Davidson, E. H. and Britten, R. J. (1973) Organization, transcription, and regulation of the animal genome. *Q. Rev. Biol.* **48,** 565–613.
2. Britten, R. J. and Davidson, E. H. (1969) Gene regulation for higher cells: a theory. *Science* **165,** 349–357.

3. Bjornson, C. R. B., Rietze, R. L., Reynolds, B. A., Magli, M. C., and Vescovi, A. L. (1999) Turning brain into blood: a hematopoietic fate adopted by adult neural stem cells in vivo. *Science* **283,** 534–537.

4. Thiese, N. D., Badve, S., Saxena, R., Henegariu, O., Sell, S., Crawford, J. M., and Krause, D. S. (2000) Derivation of Hepatocystes from bone marrow cells in mice after radiation-induced myeloblation. *Hepatology* **31,** 235–240.

5. Gould, A. P., Brookman, J. J., Strutt, D. I., and White, R. A. (1990) Targets of homeotic gene control in *Drosophila. Nature* **348,** 308–312.

6. Serrano, N., Brock, H. W., Demeret, C., Dura, J. M., Randsholt, N. B., Kornberg, T. B., and Maschat, F. (1995) Polyhomeotic appears to be a target of engrailed regulation in *Drosophila. Development* **121,** 1691–1703.

7. Bigler, J. and Eisenman, R. N. (1994) Isolation of a thyroid hormone-responsive gene by immunoprecipitation of thyroid hormone receptor-DNA complexes. *Mol. Cell. Biol.* **14,** 7621–7632.

8. Tomotsune, D., Shoji, H., Wakamatsu, Y., Kondoh, H., and Takahashi, N. (1993) A mouse homologue of the *Drosophila* tumour-suppressor gene l(2)gl controlled by Hox-C8 in vivo. *Nature* **365,** 69–72.

9. Violette, S. M., Shashikant, C. S., Salbaum, J. M., Belting, H. G., Wang, J. C., and Ruddle, F. H. (1992) Repression of the beta-amyloid gene in a Hox-3.1-producing cell line. *Proc. Natl. Acad. Sci. USA* **89,** 3805–3809.

10. Jones, F. S., Chalepakis, G., Gruss, P., and Edelman, G. M. (1992) Activation of the cytotactin promoter by the homeobox-containing gene Evx-1. *Proc. Natl. Acad. Sci. USA* **89,** 2091–2095.

11. Jones, F. S., Prediger, E. A., Bittner, D. A., De Robertis, E. M., and Edelman, G. M. (1992) Cell adhesion molecules as targets for Hox genes: neural cell adhesion molecule promoter activity is modulated by cotransfection with Hox-2.5 and -2.4. *Proc. Natl. Acad. Sci. USA* **89,** 2086–2090.

12. Care, A., Silvani, A., Meccia, E., Mattia, G., Stoppacciaro, A., Parmiani, G., et al. (1996) HOXB7 constitutively activates basic fibroblast growth factor in melanomas. *Mol. Cell. Biol.* **16,** 4842–4851.

13. Torban, E. and Goodyer, P. R. (1998) Effects of PAX2 expression in a human fetal kidney (HEK293) cell line. *Biochim. Biophys. Acta.* **1401,** 53–62.

14. Forrester, L. M., Nagy, A., Sam, M., Watt, A., Stevenson, L., Bernstein, A., et al. (1996) An induction gene trap screen in embryonic stem cells: identification of genes that respond to retinoic acid in vitro. *Proc. Natl. Acad. Sci. USA* **93,** 1677–1682.

15. Klucher, K. M., Gerlach, M. J., and Daley, G. Q. (1997) A novel method to isolate cells with conditional gene expression using fluorescence activated cell sorting. *Nucleic Acids Res.* **25,** 4858–4860.

16. Wang, Z. and Brown, D. D. (1991) A gene expression screen. *Proc. Natl. Acad. Sci. U.S.A.* **88,** 11,505–11,509.

17. Dulac, C. and Axel, R. (1995) A novel family of genes encoding putative pheromone receptors in mammals. *Cell* **83,** 195–206.

18. Li H., S. J. J., Fewell, G.D., MacFarlarnd K.L., Witte, D.P., Bodenmiller, D.M., Hsieh-Li, H.-M., et al. (1999) Novel strategy yields candidate Gsh-1 homeobox gene targets using hypothalamus progenitor cell lines. *Dev. Biol.* **211,** 64–76.

19. Windle, J. J., Weiner, R. I., and Mellon, P. L. (1990) Cell lines of the pituitary gonadotrope lineage derived by targeted oncogenesis in transgenic mice. *Mol. Endocrinol.* **4,** 597–603.

20. Alarid, E. T., Windle, J. J., Whyte, D. B., and Mellon, P. L. (1996) Immortalization of pituitary cells at discrete stages of development by directed oncogenesis in transgenic mice. *Development* **122,** 3319–3329.

21. Bodner, M., Castrillo, J. L., Theill, L. E., Deerinck, T., Ellisman, M., and Karin, M. (1988) The pituitary-specific transcription factor GHF-1 is a homeobox- containing protein. *Cell* **55,** 505–518.
22. Ingraham, H. A., Chen, R. P., Mangalam, H. J., Elsholtz, H. P., Flynn, S. E., Lin, C. R., et al. (1988) A tissue-specific transcription factor containing a homeodomain specifies a pituitary phenotype. *Cell* **55,** 519–529.
23. Therrien, M. and Drouin, J. (1993) Cell-specific helix-loop-helix factor required for pituitary expression of the pro-opiomelanocortin gene. *Mol. Cell. Biol.* **13,** 2342–2353.
24. Taylor, D., Badiani, P., and Weston, K. (1996) A dominant interfering Myb mutant causes apoptosis in T cells. *Genes Dev.* **21,** 2732–2744.
25. Kessler, D. S. (1997) Siamois is required for formation of Spemann's organizer. *Proc. Natl. Acad. Sci. U.S.A.* **24,** 13,017–13,022.
26. Mariani, F. V. and Harland, R. M. (1998) XBF-2 is a transcriptional repressor that converts ectoderm into neural tissue. *Development* **24,** 5019–5031.
27. Bujard, H. (1999) Controlling genes with tetracyclines. *J. Gene Med.* **5,** 372–374.

23

ES Cell-Mediated Conditional Transgenesis

Marina Gertsenstein, Corrinne Lobe, and Andras Nagy

1. Introduction

The use of mouse embryonic stem (ES) cells to alter the mouse genome has become a routine method to study gene function alongside classical transgenesis. Gene targeting by homologous recombination is the most widely used method of ES cell-mediated genome alterations. It allows the introduction of specific mutations of a gene of interest into the mouse germ line. ES cells can be used to mediate random insertional transgenesis as well, which is perhaps an alternative to the pronuclear DNA injection method. However, the combination of recently developed different site-specific recombinase systems combined with ES cell technology makes it possible to create more sophisticated conditional transgenic and conditional gene knock-out mouse models.

The site-specific recombinases catalyze the recombination between two consensus DNA sequences. If these sites are properly designed into a transgene or a gene-targeting vector, the site-specific recombination event can trigger the expression of a transgene. It can also create a loss-of-function allele of the targeted gene in a specific cell lineage and/or in a specific time of development, conditional to the expression of the recombinase. Cre recombinase of the bacteriophage P1 is the most widely used in mouse genetics (see detailed review in **ref. *1***). This prokaryotic enzyme was first shown to work in the mouse by Lakso et al. *(2)*. Cre protein recombines DNA between two 34-bp long *loxP* recognition sites. If these sites are placed in the same DNA strand, in the same orientation, the recombination results in the excision of the intervening sequence, leaving a single *loxP* site behind.

The second most popular is Flp recombinase from yeast. Concerning its mechanism of action, it is similar to *Cre/loxP*, but so far appears to be less efficient. However, recently developed enhanced Flp (Flpe) might change this situation *(3,4)*. The 34-bp consensus recombination site of Flp is called FRT.

The third potential enzyme is still in the earliest stage of its development for use in mammalian genome alterations, but certainly holds promise that it could become an additional recombinase system. This is an integrase from Streptomyces phage φC31, which was recently demonstrated to function in human cells *(5)*. It carries out an efficient site-specific unidirectional recombination between attP and attB sites. It presently seems as if this integrase can be used to delete sequences flanked by attP

From: *Methods in Molecular Biology, vol. 185: Embryonic Stem Cells: Methods and Protocols*
Edited by: K. Turksen © Humana Press Inc., Totowa, NJ

and attB sites. It is also possible that this could become the system of choice for recombinase-mediated site-specific insertion into the genome.

After mediating an insertion into the genome, both Cre and Flp, create two functional *loxP* or FRT sites, respectively, that flank the inserted sequence. Obviously, these sites could be the target of a second recombination, which will excise the inserted sequence. In contrast, φC31-mediated recombination between the attP and attB sites does not recreate either site, and therefore, no further recombination that could remove the inserted sequence will occur.

In the ideal but not at all unrealistic picture for the near future, if two recombinase systems are working in addition to the *Cre/loxP* system, there will be more freedom for the Cre system to be used as a postintegration switch. There is already a large collection of transgenic mouse lines expressing Cre recombinase at a high level and specificity, and there are more to come. The database presenting a collection of Cre transgenic lines created through the effort of many researchers can be found at (http://www.mshri.on.ca/nagy/Cre.html). Such transgenic lines expressing Cre recombinase driven by lineage–tissue-specific promoters can be derived by classical transgenesis as well as by the relatively new approach of gene-targeted knock-in of the Cre recombinase into an endogenous gene for a sometimes more reliable expression.

To be able to control the time of the switch, independently of the endogenous regulatory elements and lineage specificity, recombinase systems are combined with inducible gene expression systems. The recombinase gene can be placed under the control of an inducible promoter (ubiquitous or lineage-specific) or constructed as an inducible fusion protein.

Tamoxifen *(6)* and RU-486 *(7)* inducible systems use the nuclear localization capability of estrogen or a progesterone receptor ligand-binding domain in the presence of the ligand. The Cre recombinase is fused to a mutant ligand-binding domain, which has lost its ability to bind endogenous estrogen or progesterone, but still binds tamoxifen (an estrogen antagonist) or RU-486 (a synthetic steroid), respectively. In the presence of the synthetic ligand, the Cre fusion protein translocates into the nucleus and executes its function. So far, only partial tamoxifen-inducible Cre-mediated excision has been obtained *(8)*.

The tetracycline-inducible gene expression system uses the DNA-binding domain of the bacterial tet-repressor protein and a strong transcriptional activator domain (VP16 from the herpes virus), which are fused together. Such a heterologous protein can bind to the tetracycline operator element and activate transcription depending on the presence of tetracycline *(9)*. The combination of *Cre/loxP* and the doxycycline (Dox) system has proven promising *(10)*. However, problems with mosaic expression, toxic effects of the trasnsactivator, and a high background recombination level have been encountered. The fine tuning of that system has required quite a bit of time. Recently, however, ubiquitous Cre recombination was achieved upon per oral administration of antibiotic using a Dox inducible single-construct Cre transgene *(11)*. The authors placed both the Cre recombinase and the reverse tetracycline-dependent transactivator (rtTA) under the control of the same bidirectional Dox responsive promoter. In this arrangement, the transcription is auto-inducible depending on the availability of Dox and minimal amounts of rtTA. Both rtTA toxicity and background Cre recombination were found minimal in the absence of Dox.

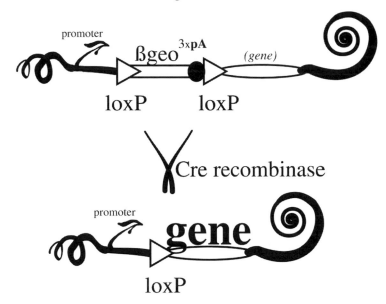

Fig. 1. Prototype construct for conditional transgenic expression of a gene of interest mediated by a Cre recombinase.

The efficiency and specificity of any recombinase system needs to be tested at the cellular level using a transgenic reporter mouse line (i.e., a transgenic line expressing a reporter gcnc in response to Cre-mediated recombination). For that, a single-copy transgene should have a ubiquitous promoter, followed by a *loxP*-flanked transcriptional STOP region, and then the coding region of a reporter gene. The reporter initiates expression under the control of the promoter only after Cre excision removes the STOP region. Such a conditional *lacZ* reporter construct was introduced into the ROSA26 gene-trap integration site *(12)*. The other recently developed system (Z/AP) uses two reporters: cells express *lacZ* before Cre excision and heat-resistant human alkaline phosphatase after excision. This was achieved by random insertion of the transgene and single-copy integration by ES cell-mediated transgenesis *(13)*.

ES cell-mediated as opposed to classical transgenesis has an advantage of a higher frequency of single-copy integration. It is an important issue, since multiple-copy integration can create more than two *loxP* sites and a potentially unpredictable outcome of the Cre excision and chromosome instability *(14)*.

Conditional transgenesis is based on a similar strategy to that behind the Cre reporter lines: the promoter and the coding region for the gene of interest are separated by a *loxP*-flanked STOP region that does not allow any transcription or translation of the gene of interest. The gene is expressed when this region is removed by Cre-mediated excision (**Fig. 1**). A ubiquitously expressed conditional transgenic line with single-copy integration is produced using ES cells. The transgene is activated when crossed with various Cre transgenic lines of different lineage specificity. Cre-mediated recombination, and therefore transgene activation, can also be inducible as discussed above.

Here, we describe our example applied in the design of the Z/AP reporter, which can be followed for any conditional transgenesis (**Fig.1**). The βgeo-*lacZ*/neo[R] fusion coding sequence *(15)* is placed upstream of a triple repeat of the simian virus 40 (SV40) polyadenylation signal (3xpA). The βgeo/3xpA is inserted between two *loxP* sites that are placed in a pCAGG vector *(16)* containing the cytomegalovirus (CMV) enhancer/chicken β-actin promoter in front of it. This vector is referred as pCALL *(13)*. The coding sequence of the transgene can be inserted downstream of *loxP*-flanked βgeo/3xpA to produce an expression vector of interest.

There are several variations of the construct for random integration into ES cells to generate a conditional transgenic mouse line. Alterations of the Z/AP construct can be customized for specific needs. Different positive selectable markers, such as hygromycin B and puromycin, allowing the survival of cells integrating the marker into the genome, can be used instead of neomycin. Among the possible reporter systems, bacterial β-galactosidase (β-gal) and human alkaline phosphatase allow the detection at a single cell level in histological sections. Green fluorescent protein (GFP) cloned from jellyfish, *Aequorea victoria* as well as its variants, yellow and cyan fluorescent proteins, permit the detection of gene expression and protein localization in living cells.

Perhaps the same principles are used for targeted insertion of conditional transgenes into an endogenous gene, but it requires significant target vector building and screening for targeted events. One of the strategies for the creation of a conditional gene knock-out is to place two *loxP* sites around a functionally essential part of the gene of interest, using gene targeting in ES cells (see review in **ref. *1***). Such minimal modification should leave the gene functional until lineage specific and/or inducible Cre recombinase is applied. This strategy requires a sophisticated gene targeting design, involving the proper insertion of *loxP* sites around the functional part of the gene to create the null allele. The selectable marker needs to be flanked with target sites for its later removal, preferably by a different recombinase system (Flp/FRT) to avoid multiple *loxP* sites. General gene targeting strategies and the use of site-specific recombination are described in great detail in many publications (e.g., **ref. *17***) and are out of the scope of the current chapter.

Here, we give a brief description of the steps involved in the creation of conditional transgenic animals.

2. Materials

2.1. ES Cell Tissue Culture

2.1.1. Equipment

1. Tissue culture facility, preferably used only for ES cells, including a laminar flow cabinet, humidified incubator (37°C, 5% CO_2), inverted phase-contrast microscope with 4×,10×, 20–25× objectives, stereomicroscope (dissecting) with transmitted light base, table-top centrifuge, water bath (optional), –70°C freezer, liquid nitrogen tanks. Microscope equipped with fluorescence, appropriate filters, and camera (e.g., Leica MZFLIII or universal fluorescence light source from BLS Ltd, Hungary [www.bls-ltd.com], e-mail: bls@euroweb.hu) is necessary if fluorescent proteins are to be used as reporters.
2. Sterile disposable cell culture plasticware (100-, 60-, 35-mm dishes, 6-well, 24-well, 4-well, flat and V-bottom 96-well plates, centrifuge tubes, cryovials). Various sterile disposable or reusable detergent-free glass pipets.
3. Electroporation apparatus (e.g., Bio-Rad Gene Pulser®, cat. no. 165-2106), and capacitance extender (cat. no. 165-21080). 4-cm electrode gap electroporation cuvettes (e.g., Bio-Rad, cat. no. 165-2088).

4. Multichannel pipettor with volume adjustable up to 200 μL, sterile disposable reagent reservoirs for multichannel pipettor (e.g., Costar, cat. no. 4870), and sterile pipet tips.
5. Multichannel aspirator system (optional) (e.g., Inotech Biosystems Vacuset®).
6. Isopropanol freezing container (optional) and/or styrofoam box with lid.

2.1.2. ES Cells and Feeders

Several ES cell lines allowing efficient germline transmission have been developed (*see* **ref.** *17* for review). To maintain pluripotency, ES cells are cultured on feeder cells: primary embryonic fibroblasts (Emfi) or STO fibroblast cell line; or on gelatinized plates in the presence of leukemia inhibitory factor (LIF). Growth media and culture conditions should be used as suggested for each ES cell line. Here we describe the protocols established and currently used for R1 ES cells *(18)*. Mitotically inactivated Emfi are used as feeders only for a long-term culture of R1 ES cells (typically before and after cryopreservation). Otherwise, culture of R1 ES cells for electroporation and during the selection is done on gelatinized plates. Emfi cells can be made from any strain of mice including transgenic mice that express bacterial neomycin or hygromycin genes depending on the choice of selectable markers used for altering the ES cell genome. Neo^R and Hygro^R mice are available from Jackson Laboratories (cat. nos. JR2354 and JR2356). Detailed protocols for preparation of Emfi stocks and mitomycin C-treated feeders for ES cell culture are presented in many publications (e.g., **ref.** *17*). Our brief protocols are given in **Tables 1** and **2**. Emfi are also commercially available from Specialty Media (cat. no. PMEF).

2.1.3. ES Cell Culture Media and Reagents

It is recommended to use only tissue culture grade or, preferably, ES cell-qualified reagents (e.g., from Life Technologies/Gibco or Specialty Media) whenever available. The water quality is a critical factor for optimal culture. It should be from a regularly maintained Milli-Q (Millipore) filtration system, preferably pretreated by deionization. Commercially available ultrapure water (e.g., Gibco or Sigma) can be used as an alternative. If large quantities of media and solutions are necessary, it can be prepared from powder and filter-sterilized. Otherwise, commercially available ES cell-qualified solutions are recommended.

1. *Complete ES Medium (ES-DMEM):*
   a. Dulbecco's modified Eagle's medium (DMEM) with 4.5 g/L D-glucose (high glucose), buffered with 2.2 g/L sodium bicarbonate (Life Technologies, cat. no. 1400-061 [powder], Gibco KO-DMEM, cat. no. 10829; Specialty Media, cat. no. SLM-220B). Store in the dark at 4°C.
   Prior to use, DMEM should be supplemented with the components listed below. If the complete media is stored for longer than 2 wk, it should be supplemented with additional 2 m*M* L-glutamine from 100X stock, as L-glutamine is unstable. GlutaMAX™ media from Gibco (cat. no. 10566) contains L-glutamine in a stabilized form of the dipeptide and can be used without extra supplements of L-glutamine.
   b. 15% ES cell-qualified fetal bovine serum (FBS) (HyClone; Gibco; Specialty Media; Gemini; Wisent, Quebec). Store main stock in the dark at –20°C. Heat-inactivate at 56°C for 30 min (optional) (*see* **Note 1**).
   c. 0.1 m*M* Nonessential amino acids (100X stock; Gibco, cat. no. 11140; or Specialty Media, cat. no. TMS-001-C) stored at 4°C.

Table 1
Preparation of Primary Embryonic Fibroblast Cell (Emfi) Stocks

Materials	Procedure
1. 15.5–17.5 dpc pregnant mice	1. Aseptically dissect fetuses from 1 to 2 mice in 100-mm Petri dish containing PBS.
2. Sterile dissecting instruments (scissors, forceps).	2. Transfer dissected embryos into a new dish with PBS; remove heads and all internal organs.
3. Autoclaved 3–5 mm diameter glass beads and 1 to 2 inch stir bars.	3. Remove as much blood as possible by washing carcasses at least twice with 50 ml of PBS using 50-mL tubes.
4. 50-mL tubes (Falcon, cat. no. 2070).	4. Mince carcasses into small pieces with sterile scissors in minimal volume of PBS using 50-mL tube with the top cut off.
5. 100-mm Petri dishes, 150-mm cell culture plates.	5. Add 10 mL of Trypsin-EDTA; transfer into 50-mL tube. Put around 5 mL of glass beads and stir bar inside the tube. If solution becomes viscous, add 100 µL of DNase per 10 mL.
6. DMEM plus 10% FBS.	6. Incubate at 37°C for 30 min with stirring.
7. 0.05% trypsin, 0.53 m*M* EDTA (e.g., Gibco, cat. no. 25300).	7. Add 10 mL of trypsin-EDTA, incubate at 37°C for another 30 min with stirring.
8. PBS without calcium and magnesium (e.g., Specialty Media, cat. no. BSS-1006-B).	8. Repeat step 7 one more time (final volume of 30 mL).
	9. Decant cell suspension into two 50-mL tubes each containing 3 mL of FBS.
	10. Wash the tube twice with DMEM plus 10% FBS and add to the tubes with cell suspension.
	11. Centrifuge at 270g for 5 min.
	12. Resuspend the pellet in DMEM plus 10% FBS.
9. DNase I (Sigma, cat. no. D4527) 10 mg/mL stock, approx 100 µg/mL final concentration (optional).	13. Count viable nucleated cells using trypan blue (Flow Labs, cat. no. 16-910-49). Approximately 5×10^{7}–10^{8} cells could be expected from 10 fetuses.
	14. Plate 5×10^{6} nucleated cells (approx from 1.25 embryo) per 150-mm dish, incubate at 37°C.
10. 1X freezing medium: 80% ES-DMEM, 10% FBS, 10% DMSO (prepared fresh prior to use and kept on ice).	15. Change the medium the next day.
	16. When confluent (in 2 to 3 days) trypsinize cells and split each plate onto 6 further plates.
	17. When these plates reach confluency, they can be frozen using 1X freezing media. Cells from each 150-mm plate are frozen in one cryovial in 1 mL of freezing media.
11. Cryovials (e.g., Nalgene, cat. no. 5000-0012).	18. Prepared Emfi stocks should be tested for mycoplasma, and mouse pathogens.

Table 2
Preparation of Mitomycin C-Treated Feeder Layers for ES Cell Culture

Materials	Procedure
1. Frozen vials of Emfi stocks.	1. Thaw a frozen vial quickly at 37°C.
2. Cell culture plates.	2. When ice crystals almost disappear, aseptically transfer cell suspension into a tube with DMEM plus 10% FBS. Centrifuge at 270g for 5 min, aspirate the supernatant.
3. DMEM plus 10% FBS.	
4. 0.05% trypsin, 0.53 mM EDTA (e.g., Gibco cat. no. 25300).	3. Resuspend the pellet in DMEM plus 10% FBS, seed onto 5 to 6 150-mm plates (25 mL per plate) and incubate at 37°C.
5. PBS without calcium and magnesium (e.g., Specialty Media, cat. no. BSS-1006-B).	4. When cells reach confluency (in about 3 d) they can either be:
	• split one more time before being treated with Mitomycin C, or
	• treated with mitomycin C and used directly as feeders for ES cell culture, or
6. Mitomycin C (Sigma, cat. no. M0503). 1 mg/mL stock solution in PBS is stored light protected at 4°C for no longer than 2 wk.	• treated with mitomycin C, frozen in cryovials and used as feeders later (alternative cost-efficient way for laboratories with small volume).
	Mitomycin C treatment
	• Remove the medium from the confluent plate; add 10 mL DMEM plus 10% FBS and 100 µL of mitomycin C stock solution (1 mg/mL). Incubate the plate for 2 h at 37°C.
As an alternative to mitomycin C treatment, γ irradiation at 6000–10,000 rads can be used for inhibition of cell growth.	• Rinse twice with PBS, trypsinize as usual, and either freeze cells from each 150-mm plate in one cryovial for later use, or seed cells onto tissue culture plates for immediate use at appropriate cell densities and volumes (2×10^5 cells/mL).
	One confluent 150-mm feeder plate can generate approximately the following number of feeder plates:
	• 5×100-mm plates (10 mL each);
	• 12×60-mm plates (5 mL each);
	• 25×35-mm plates (2 mL each);
	• 25×4-well plates or 4 to 5×24-well plates (0.5 mL/well);
	• 6×96-well plates (200 µL /well).
	Preferably, feeders are incubated overnight, or at least for a few hours before plating ES cells. The medium is changed to ES-DMEM prior to use. Mitomycin C treated feeders can be used within 7–10 d (medium is changed every 3 to 4 days).

    d. 1 m*M* Sodium pyruvate (100X stock; Gibco, cat. no. 11360) stored at 4°C.

    e. 0.1 m*M* β-mercaptoethanol (100X stock; Sigma, cat. no. M7522; or Specialty Media cat. no. ES-007-E) stored as aliquots at –20°C.

    f. 2 m*M* L-Glutamine (100X stock; Gibco, cat. no. 25030; or Specialty Media, cat. no. TMS-002-C) stored as aliquots at –20°C.

    g. 50 U/mL Penicillin and 50 μg/mL streptomycin (100X Pen–Strep stock; Gibco, cat. no. 15140; or Specialty Media, cat. no. TMS-AB2-C) stored as aliquots at –20°C. Alternatively, 100X Pen–Strep–L-Glu combo available from Gibco (cat. no. 10378).

    h. LIF (e.g., Chemicon International, ESCRO™ cat. no. ESG1107), stored as aliquots at –20°C (*see* **Note 2**).

2. *Feeder medium:* DMEM, 1X Pen–Strep, 10% FBS.
3. *ES cell freezing medium:* freezing medium should be prepared fresh prior to use and kept on ice. We commonly use 22% FBS as a final concentration. It is possible to increase the concentration of FBS in a freezing media up to 40% (e.g., 2X ES cell-culture freezing medium from Specialty Media) for better recovery of small amount of cells in 96-well plates. 2X: 60% ES-DMEM, 20% FBS, 20% dimethyl sulfoxide (DMSO) (Sigma, cat. no. D5879). 1X: 80% ES-DMEM, 10% FBS, 10% DMSO.
4. *0.1% Gelatin:* (Sigma, cat. no. G2500) 0.1% solution in water, autoclaved, and stored at 4°C. Alternatively ES cell-qualified 0.1% gelatin (Specialty Media, cat. no. ES-006-B).
5. *Trypsin-EDTA:* 0.05% trypsin, 0.53 m*M* EDTA (Gibco, cat. no. 25300; Specialty Media, cat. no. SM-2002-C). 0.25% Trypsin, EDTA (Gibco cat. no. 25200 1 m*M*). Store at 4°C (main stock at –20°C).
6. *Phosphate-buffered saline (PBS)* without Ca^{2+} and Mg^{2+} (e.g., Gibco cat. no. 10010; Specialty Media, cat. no. BSS-1006-B).
7. *Selection reagents:* Geneticin® (G418) (Gibco, in liquid, cat. no. 10131, in dry powder form, cat. no. 11811; Sigma, cat. no. G9516), puromycin (Sigma, cat. no. P8833), hygromycin B (Calbiochem, cat. no. 400051). Working concentration for selection agents must be determined by killing curves. We routinely use 150–200 μg/mL G418 for R1 cells.
8. *Reagents for the lipofection:* LipofectAMINE™ (Gibco, cat. no. 18324-012 or Lipo-fectAMINE™ 2000, cat. no. 11668-027); Opti-MEM® I reduced serum medium (Gibco, cat. no. 31985-062).

2.2. Isolation of DNA from 96-Well ES Cell Colonies

1. Lysis buffer: 10 m*M* Tris-HCl, pH 7.5, 10 m*M* EDTA, 10 m*M* NaCl, 0.5% sarcosyl, 1 mg/mL proteinase K added before use.
2. NaCl-ethanol mixture: 150 μL of 5 *M* NaCl per 10 mL of cold 100% ethanol (prepared fresh).
3. Restriction digestion mixture (per well): 1X appropriate restriction buffer, 1 m*M* spermidine, 100 μg/mL bovine serum albumin (BSA), 50–100 μg/mL RNase A, 10–20 U of enzyme. Use 35–40 μL per sample.

2.3. β-Gal Staining of ES Cells and Postimplantation Stage Embryos

1. PBS without Ca^{2+} and Mg^{2+} (e.g., Specialty Media, cat. no. BSS-1006-B).
2. Stock solutions: 10% Nonidet P-40 (NP40); 1% Na deoxycholate; 0.5 *M* EGTA; 1 *M* $MgCl_2$; 0.5 *M* K_3 [Fe(CN)$_6$], 0.5 *M* K_4 [Fe(CN)$_6$] (kept light protected at room temperature); 25–50 mg/mL 5-bromo-4-chloro-3-indolyl-β-o-galactopyranoside (X-gal) in DMSO or dimethyl formamide (DMF) (e.g., Specialty Media, cat. no. BG-3-G) kept light-protected at –20°C.

3. Fix solution: 0.2% glutaraldehyde in PBS (for cells); 0.2% glutaraldehyde in PBS, containing 5 mM EGTA and 2 mM MgCl$_2$ (for embryos). Glutaraldehyde (e.g., Sigma, cat. no. G6257) is added fresh prior to use. 2% Glutaraldehyde, 2 or 4% paraformaldehyde (PFA) are also used for fixation of embryos older than E12.5.
4. Wash solution: PBS for cells; PBS, containing 0.01% Na-deoxicholate, 0.02% NP40, 5 mM EGTA, 2 mM MgCl$_2$ for embryos.
5. Stain solution: 1 mg/mL X-gal, 5 mM K$_3$ [Fe(CN)$_6$] , 5 mM K$_4$ [Fe(CN)$_6$] in wash solution (prepared fresh prior to use). Stain solution can be used several times if filtered after use and stored at –20°C.
   Ready to use reagents for β-gal expression are available from Specialty Media (cat. nos. BG-1-C, BG-2-C, BG-4-C, BG-5-C, BG-6-B, BG-7-B, and BG-8-C).

2.4. Preimplantation Embryo Culture

2.4.1. Equipment

1. Stereomicroscope(s) (dissecting) with both transmitted and reflected lights or fiber optics light source with gooseneck. The use of two microscopes is convenient for 2-cell stage embryo fusion and embryo transfer into pseudopregnant females, however, one microscope is also enough. We find that the frosted glass, instead of the more common, transparent glass in the base of the microscope, gives better view of the zonae pellucidae of preimplantation stage embryos necessary for zona removal.
2. Humidified incubator (37°C, 5% CO$_2$).
3. Sterile Petri dishes of different size (60, 100 mm), organ culture dishes (Falcon, cat. no. 3037). We find the plastic of 35-mm Easy Grip Falcon (cat. no. 3001) tissue culture plates to be best suited for making depressions for aggregations.
4. Sterile 1-mL syringes, 26-gauge 1/2 inch long needles (26G1/2), 30G1/2 needles. To make a flushing needle, the sharp tip of 30G1/2 needle is first cut off and then polished on a sharpening stone or sand paper. The flushing needle is flushed with 70% ethanol before and after use.
5. Bunsen or alcohol burner.
6. Pasteur pipets that are drawn by hand over a flame and broken to produce pipets for embryo manipulation, with a diameter of a capillary slightly larger than an embryo. It is important to flame-polish the tip of the pipet as the embryos without zonae are easily damaged. Embryo manipulating pipets are connected through a latex tubing to an aspirator mouthpiece (HPI Hospital Products Med. Tech., cat. no. 1501P-B4036-2). Such a mouth-controlled pipet is used for all embryo manipulations and embryo transfer. Alternatively, a finger-controlled pipet (small piece of tubing closed at one end or small bulb connected to the drawn capillary) is used for embryo manipulations.
7. Surgical instruments (e.g., Fine Scientific Tools [FST]): sharp fine-pointed scissors, fine forceps (e.g., Dumont cat. no. 5 or ss/mc), straight or curved blunt forceps with serrated tips, forceps with 1 × 2 teeth, serrefine (FST, cat. no. 18050-28); wound clips and Autoclip applier (Clay Adams B-D 7631 and B-D 763007).
8. Cell-fusion instrument CF-150B, aggregation needle DN-09 (BLS Ltd, Hungary, [www.bls-ltd.com], e-mail: bls@euroweb.hu).

2.4.2. Mouse Stock

C57 Bl/6 mice are the most common strain used as donors of host embryos for chimera production by blastocyst injection of different ES cell lines. In our facility, random-bred ICR (CD-1) mice (available from Charles River Laboratories, Harlan Sprague Dawley, or Taconic) have been successfully used for many years as both donors of host embryos and recipients of manipulated embryos. The mouse colony necessary for the creation of chimeras should contain a stock of females, stud males, and vasectomized males. The details of maintenance of such a colony as well as all procedures involved in the production of superovulated and pseudopregnant animals are described in multiple publications *(17,19)*.

2.4.3. Preimplantation Embryo Culture Media and Reagents

Since embryos are cultured for 24 or even 48 h in aggregation experiments, the quality of culture conditions is more critical than in blastocyst injection. Embryo culture media is commercially available (e.g., from Specialty Media: KSOM medium, cat. no. MR-023-D, M2 medium, cat. no. MR-015-D). However, as culture media can not be used for longer than 2, or a maximum of 3 wk, it is often necessary to prepare media from scratch or from stocks, as described previously *(19)*. We keep all our concentrated stocks (*see* **Tables 3** and **4**) at –70°C for a few months. BSA (Sigma, cat. no. A3311) is kept desiccated at 4°C (*see* **Note 3**).

1. M2 is a HEPES-buffered media that is used during embryo collection and other manipulations in room atmosphere (fusion, zonae removal). Filtered aliquots are stored at 4°C and brought up to room temperature prior to use. Embryos should not be kept in M2 for prolonged periods of time and are rinsed well with a few drops of equilibrated KSOM media before being placed into the incubator.
2. KSOM is a bicarbonate-buffered media used for embryo culture and was developed through a simplex optimization procedure by Lawitts and Biggers *(20)*. It may also be supplemented with both essential and nonessential amino acids, which were shown to improve the development in vitro *(21)*. Filtered aliquots are stored at 4°C and equilibrated by placing the tube, a 1-mL syringe, or the culture dish containing KSOM in the incubator well in advance before use.
3. Embryo-tested light mineral oil (eg., Sigma, cat. no. M8410).
4. Acidic tyrode solution (Sigma, cat. no. T1788) for removing zonae. Main stock aliquots are stored at –20°C. An aliquot is thawed and kept at 4°C when needed, and it is brought to room temperature prior to use.
5. 0.3 mol/L Mannitol (Sigma, cat. no. M4125) in ultrapure water containing 0.3% BSA for embryo fusion. Aliquots are stored at –20°C. A freshly thawed aliquot is used for fusion, and the unused portion is discarded.

3. Methods
3.1. ES Cell Culture
3.1.1. Passage of ES Cells

See **Note 4** for the general considerations on ES cell maintenance.

1. Prepare the necessary number of gelatinized plates by coating them with 0.1% gelatin: rinse the plate with 0.1% gelatin solution covering the surface (5 mL/100-mm plate), leave for a few minutes, aspirate, and allow to dry for a few minutes, add ES-DMEM, and place in the incubator. Alternatively, replace the media on the appropriate number of prepared feeder plates to ES-DMEM.
2. Aspirate the growth medium, rinse twice with PBS, add trypsin (1.5–2 mL/100-mm, 1–1.5 mL/60-mm, 0.5 mL/35-mm dish, 0.25–0.3 mL/well for 4- or 24-well plates), place in the incubator for 5 min.

Table 3
Preparation of M2 Medium from Concentrated Stocks

| Component | M2 Medium | | | |
| | Final concentration | | Concentrated stock | Stock volume for 100 mL |
	mM	g/L		
			A (10X)	
			g/100 mL	10 mL
NaCl	94.66	5.534	5.534	
KCl	4.78	0.356	0.356	
KH$_2$PO$_4$	1.19	0.162	0.162	
MgSO$_4 \times 7$ H$_2$O	1.19	0.293	0.293	
Glucose	5.56	1.000	1.000	
Penicillin G		0.060	0.060	
Streptomycin		0.050	0.050	
Sodium lactate	23.28	2.610 or 4.349 g of 60% syrup	2.610 or 4.349 g of 60% syrup	
			B (10X)	
			g/100 mL	*1.6 mL*
NaHCO$_3$	4.15	0.349	2.101	
Phenol Red			0.001 or 0.01 mL of 0.5% solution	
			C (100X)	
			g/10mL	*1 mL*
Na pyruvate	0.33	0.036	0.036	
			D (100X)	
			g/10 mL	*1 mL*
CaCl$_2 \times 2$H$_2$O	1.71	0.25	0.25	
			E (10X)	
			g/100 mL	*8.4 mL*
HEPES	20.85	4.969	5.958	
Phenol Red			0.001 or 0.01 mL of 0.5% solution	
BSA (Sigma, cat. no. A3311)		4.000		400 mg

If necessary, pH of M2 is adjusted to 7.2–7.4 with 0.2 N NaOH. Osmolarity of M2 should be 285–287 mOsm. Media is filter-sterilized, aliquoted in polypropylene tubes ,and stored at 4°C.

3. Swirl the plate to detach clumps from the bottom of the plate, pipet the cells gently to break the clumps (optional). Add an equal volume of ES-DMEM to neutralize trypsin, pipet up and down several times, transfer the suspension into a 12-mL tube. Pellet the cells by low-speed centrifugation ($270g$) for 5 min at room temperature.
4. Aspirate the supernatant, add 1 drop of PBS or ES-DMEM to the pellet, flick the tube to resuspend the cells before adding ES-DMEM (optional).
5. Add 5–7 mL of ES-DMEM to the tube, pipet gently to mix well, and split the contents at a 1:5 or 1:7 ratio into the new plates containing a sufficient volume of medium (5 mL/60-mm, 10 mL/100-mm plate). About 1×10^6 cells onto 60-mm plate, 2×10^6 cells onto 100-mm plate.
6. Change the medium the next day and split every second day as described.

Table 4
Preparation of KSOM Medium from Concentrated Stocks

| | KSOM | Medium | | |
Component	Final concentration mM	g/L	Concentrated stock	Stock volume for 100 mL
			A' (10X) *g/100 mL*	*10 mL*
NaCl	95.00	5.55	5.55	
KCl	2.50	0.186	0.186	
KH_2PO_4	0.35	0.0476	0.0476	
$MgSO_4 \times 7\,H_2O$	0.20	0.0493	0.0493	
Glucose	0.20	0.036	0.036	
Penicillin G		0.060	0.060	
Streptomycin		0.050	0.050	
Sodium lactate	10.0	1.12 or 1.87 g of 60% syrup	1.12 or 1.87 g of 60% syrup	
			B' (10X) *g/100 mL*	*10 mL*
$NaHCO_3$	25.00	2.10	2.10	
Phenol Red			0.001 or 0.01 mL of 0.5% solution	
			C' (100X) *g/10 mL*	*1 mL*
Na pyruvate	0.20	0.022	0.022	
			D (100X) *g/10 mL*	*1 mL*
$CaCl_2 \times 2\,H_2O$	1.71	0.25	0.25	
			F (1000X) *g/10mL*	*0.1 ml*
EDTA (Na disodium salt)	0.01	0.0038	0.038	
			G (200X) *200 mM*	*0.5 mL*
L-Glutamine (Gibco, cat. no. 25030)	1.00	0.146	Liquid form	
BSA (Sigma, cat. no. A3311)		1.000		100 mg

Media is filter-sterilized, aliquoted in polypropylene tubes and stored at 4°C. Osmolarity of KSOM should be 256 mOsm. Optimally, KSOM media is gassed with 5% CO_2 in air before storage. KSOM media needs to be equilibrated in 5% CO_2 prior to use for embryo culture.

3.1.2. Freezing and Thawing of ES Cells in Cryovials

Usually ES cells are frozen at about 5×10^6 cells/mL of 1X freezing media (approx 4 vials from 100-mm dish).

1. Change growth media 2 to 3 h before freezing the cells (optional).
2. Freshly prepare 2X or 1X freezing media (*see* **Subheading 2.1.3.**).

3. Harvest the cells in a 12-mL tube containing ES-DMEM as described in **Subheading 3.1.1.** Pellet the cells at 270*g* for 5 min at room temperature.

4. Remove the supernatant, resuspend the cells gently in half of the final volume required using ES-DMEM, gradually add an equal volume of 2X freezing medium while shaking the tube, and mix by pipetting up and down several times. Alternatively, gently resuspend the pellet in 1X freezing media.

5. Quickly aliquot 1 mL of the cell suspension into labeled cryovials and immediately place them in a precooled styrofoam box that will allow them to cool down gradually. Alternatively, isopropanol containers purchased from a number of manufacturers (e.g., Nalgene, cat. no. 5100-0001) can be used.

6. Immediately place the container in a –70°C freezer for 1 to 2 d, then transfer cryovials into a liquid nitrogen tank for long-term storage.

A vial, frozen in such a way, can be thawed onto a 60-mm plate. Freezing and thawing are usually counted as one passage.

1. Thaw the vial by quickly warming it at 37°C.
2. When ice crystals almost disappear, aseptically transfer the cell suspension into a 12-mL tube using the pipet filled with ES-DMEM to slowly dilute DMSO.
3. Pellet the cells at 270*g* for 5 min and aspirate the supernatant.
4. Resuspend the pellet in fresh ES-DMEM, plate on a gelatinized or feeder plate, gently swirl the plate bidirectionally to evenly distribute the cells, and place in the incubator.
5. The next day, remove floating dead cells and change the media. If the correct procedure was used, cells should be ready for passage in 2 to 3 d.

3.2. Introduction of DNA into ES Cells

Electroporation is the most common way of in vitro introduction for both stable integration and transient expression in ES cells. Recently, we have successfully used lipofection for transient transfection of Cre recombinase into single-copy integrants. It is important to test the excision of the STOP region in vitro before in vivo experiments. In our example, ES cells become neo-sensitive and lose *lacZ* activity after Cre excision (*see* **Fig. 1**).

3.2.1. Electroporation of ES Cells

Cells are routinely passaged 2 d prior to electroporation. Usually one 10-cm plate at approx 80% confluency ($15–20 \times 10^6$ cells/mL) will provide enough cells for 1 to 2 electroporations. We regularly electroporate 20–40 µg of DNA into cells at the density of 7×10^6 cells/mL. Vector DNA is linearized by restriction enzyme digestion, extracted twice with phenol:chloroform (optional), ethanol-precipitated, washed twice in 0.5–1 mL of 70% ethanol, and resuspended in sterile 0.1 × Tris-EDTA (TE), PBS, or water at a concentration of 1 µg/µL. For transient expression of Cre recombinase, circular plasmid is used for electroporation. Electroporated cells are then plated very sparsely (approx 1000 cells/100-mm dish).

1. Change medium on ES cells at least 2 h prior to electroporation.
2. Switch on the electroporation apparatus and set up conditions in advance. We routinely use 250 *V*, 500 µF for the Bio-Rad GenePulser® (cat. no. 165-2106) and capacitance extender (cat. no. 165-2108).
3. Harvest cells as described in **Subheading 3.1.1.** It is critical to get a single-cell suspension. Pool the cells from all dishes into one tube.

4. Resuspend the pellet in a minimal volume of ice-cold PBS (about 1 mL/100-mm plate). Determine the cell density with a hemocytometer and dilute with a volume of PBS that will give the required density of a cell suspension (see above). Keep the suspension on ice. Recently, we found that ES cell electroporation buffer available from Specialty Media (cat. no. ES-003-D), used instead of PBS in this step, gives significantly better recovery of electroporated ES cells.

5. Gently mix 0.8 mL of the ES cell suspension and 20–40 µg of DNA in a precooled electroporation cuvette.

6. Electroporate the cells, then place the cuvettes on ice for 20 min (optional).

7. Prepare the appropriate number of gelatinized plates. The number of plates will depend on the cells' survival, specific vector, selection approach, and desired density of colonies. We routinely plate cells from one cuvette onto 1.5–2 × 100-mm plates.

8. Transfer electroporated cells from the cuvette into the tube with the appropriate volume of ES-DMEM (15–20 mL/cuvette). Cells from several cuvettes can be pooled into one tube and gently mixed by pipetting into a uniform suspension. Transfer cell suspension onto gelatinized plates, swirl bidirectionally to evenly distribute cells across the surface. Incubate overnight and change the medium the next day.

9. Start drug selection 24–48 h after electroporation. Change the selection media every day for the first few days, then every other day. Continue the selection until colonies become apparent and ready to pick (*see* **Subheading 3.3.1.**) i.e., visible to the naked eye, it usually takes 6–10 d.

3.2.2. Lipofection for Transient Expression of Transgenes

Based on Gibco protocols for LipofectAMINE, cat. nos. 18324-012 and 11668-027.

1. Plate ES cells to a 35-mm or 6-well gelatinized dish (2×10^5 cells/dish/well) using ES-DMEM.

2. Incubate overnight, the cells should reach 30–40% confluency.

3. Prepare the following solutions in 5-mL polystyrene tubes (Falcon, cat. no. 352058). It is possible to use serum-free or reduced serum ES-DMEM instead of OPTI-MEM. Antibacterial agents should not be present in either media during the transfection.
   Solution A: For each well of a 6-well plate or 35-mm dish, mix 1 µg of circular plasmid containing an expression vector DNA with 300 µL of OPTI-MEM reduced serum medium (Gibco, cat. no. 31985).
   Solution B: For each well of a 6-well plate or 35-mm dish, mix 8 µL of Lipofectamine reagent (**LF**) with 300 µL of OPTI-MEM (*see* **Note 5**).

4. Combine necessary volumes of solutions A and B, mix gently, and incubate at room temperature for 20 min.

5. Place the solution (500 µL/well/dish) onto ES cells rinsed with OPTI-MEM. Incubate for 5 h at 37°C.

6. Replace the media with drug-free ES-DMEM or proceed with selection.

3.3. Growing Drug-Resistant ES Cell Clones

3.3.1. Picking ES Cell Clones into 96-Well Plates

Well-separated colonies of similar size with a defined perimeter and a compact center with undistinguishable individual cells should be picked. Large colonies and those with large distinguishable cells are differentiating and should be avoided if possible. Colonies can either be picked with the naked eye or by using a dissecting microscope with transmitted light. They can also be circled on the bottom of the plate with a marker for easier visualization.

1. Using the multichannel pipettor, prepare the appropriate number of gelatinized flat-bottom 96-well plates. Aliquot 30–50 μL of trypsin into the wells of a V-bottom 96-well plate.
2. Aspirate growth media from the plate with ES colonies, rinse it twice with PBS, add 6–8 mL of PBS to completely cover the dish
3. Using a drawn Pasteur pipet or a Gilson P20 or P200 set at 15 μL, carefully dislodge the colony from the dish and pull it into the pipet tip with as little volume of PBS as possible (usually 3–5 μL).
4. Transfer each individual colony in a minimal volume of PBS into one of the wells of a V-bottom 96-well plate containing trypsin.
5. Using a new tip for each colony, proceed with the rest. This process should not take longer than 30–60 min, 48 or 96 colonies should be picked at a time, depending on the picking speed.
6. Place a 96-well plate in a 37°C incubator for 10 min.
7. Working row by row with a multichannel pipettor, add 50–70 μL of selective ES-DMEM to each well with trypsin (to 100 μL/well). Gently pipet up and down several times to disaggregate the cells. Transfer the suspension to the equivalent row of gelatinized plates. Alternatively, add ES-DMEM to all the wells to neutralize the trypsin first, and then proceed with pipetting and transfer.
8. Using the multichannel pipettor, wash each well of the V-bottom plate with another 100 μL of medium, and transfer the volume to the equivalent row in a flat-bottom 96-well gelatinized plate (to total vol of 200 μL/well). Place the plate into the 37°C incubator.
9. Change the media daily until the cells are ready for passage (80% confluency).

3.3.2. Passage of ES Cells in 96-Well Plates

Optimally, 3 or 4 d after picking colonies into the 96-well plates, the cells reach the density required for passage (*see* **Note 6**). Replica plates are used for the preparation of DNA for screening, X-gal staining, and for creating frozen stocks. It is generally recommended to have two confluent replica plates for DNA preparation and one or two replica plates to keep as a frozen stock. It is also important to identify strong overall expresser clones using the reporter system of choice (X-gal staining in our example). The wells equivalent to such identified strong expresser clones are used for later isolation of DNA for Southern blotting to test single site–copy integration. Confirmed single-copy integrants are thawed and expanded for further in vitro and in vivo analysis.

1. Prepare the required number of gelatinized flat 96-well plates. Add 150 μL of ES-DMEM per well and place in a 37°C incubator until necessary.
2. Aspirate the medium from the plate to be split and wash twice with 200 μL of PBS using the multichannel pipettor.
3. Add 50 μL of trypsin per well. Incubate at 37°C for 10 min. The cells should detach with gentle tapping on the plate.
4. Add 50 μL of medium into each of the wells to stop trypsinization. Pipet up and down at least five times to mix well. Working row by row, transfer the cell suspension into 2 or 3 new gelatinized plates containing ES-DMEM. Place in the 37°C incubator and change the media the next day. Alternatively, one half of such a cell suspension can be frozen right away in one new V-bottom or flat 96-well plate containing 2X freezing media (*see*

Subheading 3.3.3.). At least one more passage is required to create two replica plates for DNA preparation and one for X-gal staining in addition to one or two frozen plates.

3.3.3. Freezing and Thawing of ES Cells in 96-Well Plates

1. Freshly prepare 2X cell freezing media and keep it on ice.
2. Trypsinize the cells of an 80% confluent 96-well plate as described in **Subheading 3.3.2.** The final volume of the cell suspension is 100 µL/well.
3. Working quickly on ice, aliquot 100 µL of 2X freezing media into each well. Pipet the cells up and down several times to get a homogeneous suspension. Alternatively, transfer the cell suspension into the new V-bottom 96-well plate, containing cold 2X freezing media.
4. Add 50 µL of cold sterile mineral oil (e.g., Sigma, cat. no. M8410; or Specialty Media, cat. no. ES-005-C) to each well on top of the cell suspension.
5. Wrap the plates in parafilm and foil (the latter is optional). Place in a precooled styrofoam box, and store in a –70°C freezer, preferably not longer than 2 mo, until ready for thawing and expansion.

The first frozen plates are considered as master plates. The unfrozen replica plates are used for further characterization of the clones. After strong expresser and single-copy integrant clones have been identified, the frozen stock can be thawed and expanded for further analysis as described below.

1. Prepare the necessary number of 4- or 24-well feeder plates containing ES-DMEM.
2. Remove the plate containing the identified clones from the freezer. Unwrap the plate and place in the incubator to thaw.
3. When ice crystals almost disappear, wipe the outside of the plate with 70% ethanol.
4. Transfer the content of the well into the wells of prepared feeders plate.
5. Rinse the original wells of the 96-well plate with more ES-DMEM and transfer to the same wells. Change the media after overnight culture and daily.
6. Passage the cells when they reach 70–80% confluency to a larger plate (i.e., 35 mm). If cells do not reach confluency in a few days, but form few colonies in a well, they can be trypsinized, broken into smaller cell clumps, and plated on the same plate (*see* **Note 6**).
7. Passage every other day, freeze the cells in vials as described (*see* **Subheading 3.1.1.** and **3.1.2.**). Use them for in vitro tests (i.e., transient expression of Cre recombinase by electroporation or lipofection).

3.4. β-*Gal Staining of ES Cells and Postimplantation Stage Embryos*

lacZ gene expression is detected by the enzymatic activity of the gene product β-gal in both embryos and cells. Generally, the staining becomes visible after a 15-min or an overnight incubation, depending on the level of expression.

1. Wash cells or embryos twice in PBS.
2. Add freshly prepared fixative solution to completely cover the cells or embryos, incubate at room temperature for 5–7 min (for cells), 5 min (for up to E9.5 embryos), and 15–20 min (for up to E12.5 embryos) (*see* **Note 7**).
3. Wash three times with PBS (cells) or wash buffer (embryos).
4. Replace the wash buffer with X-gal stain to completely cover the cells or embryos and incubate at 37°C light protected with optional shaking.
5. Replace the X-gal stain with PBS (cells) or wash buffer (embryos) and store at 4°C. For longer storage, embryos can be refixed in a fresh 4% solution of PFA or formaldehyde.

3.5. DNA Isolation from ES Clones in 96-Well Plates

This procedure was established by Ramirez-Solis et al. *(22)* and allows cell lysis, DNA precipitation, and the restriction digestion in the original 96-well plate in which ES cells were growing. Usually two replica plates are used for DNA isolation, then one is processed for Southern blot analysis, leaving the second plate as a back up. The cells should be lysed, and genomic DNA isolated from them when the majority of clones are confluent. The yellow color of the media within 24 h of its change indicates cell confluency.

1. Aspirate media from each well, wash twice with PBS.
2. Add 50 µL of lysis buffer to each well.
3. Incubate the plates overnight at 55°C in a humid atmosphere (wrap the plates sealed with the parafilm with wet paper towels and place in a plastic container or sealed plastic bag).
4. The next day, add 100 µL of cold NaCl-ethanol mixture to each well.
5. Leave the plate undisturbed at room temperature for 30–60 min (or more if necessary) until the precipitated DNA attached to the dish is visible against a dark background.
6. Gently invert the plate on the paper towel to drain the liquid.
7. Rinse 3 times with 150–200 µL of 70% ethanol per well, inverting the plate each time. After this step, DNA can be stored in 70% ethanol at –20°C.
8. Invert the plate after the final wash and allow to air-dry for 10–15 min (it is important for all the ethanol to dry out).
9. Add 34–40 µL of restriction digest mixture per well, mix, seal the plate, and incubate overnight at 37°C in a humid atmosphere.
10. Proceed with Southern blot analysis.

3.6. Introduction of ES Cells into Mice

As our ES cell line of choice is R1*(18)*, we use aggregation of ES cells with cleavage stage host ICR embryos to create the chimeras *(23)*, although blastocyst injections may be used as well. Concerning the conditional transgenic lines, a faster way to test the in vivo expression of transgenes introduced into several candidate ES cell clones is by analyzing chimeras produced by aggregation with tetraploid or diploid embryos *(17)*. Clones that give the expected results are then chosen for germline transmission. After successful germline transmission, conditional transgenic animals can be crossed with switch transgenic lines, where Cre recombinase is driven by lineage-specific promoters.

3.6.1. Preparation of ES Cells for Aggregation

It is important to maintain optimal culture conditions for all ES cell cultures, but particularly for ES cell clones to be introduced into mice (*see* **Note 4**). At least one passage on a gelatinized plate is required before aggregation. Sparser than usual, passage 1 or 2 d before aggregation produces the colonies of 8–15 cells that are lifted by gentle trypsinization.

1. Three or four days prior to aggregation, thaw a vial of ES cells or passage as described above. Change the media the next day.
2. One or two days prior to aggregation, trypsinize ES cells as described in **Subheading 3.1.1.** and ensure a single-cell suspension. Twenty-four hours growth is enough for most clones, but 48 h are necessary for slower growing clones.

3. Resuspend the pellet in ES-DMEM. Leave the tube undisturbed for a few minutes to allow for large clumps and feeders to settle.
4. Seed the cell suspension from the top portion onto a few gelatinized plates using different dilutions (e.g., 1:10 to 1:50).
5. On the day of aggregation, after preparation of the embryos (*see* **Subheading 3.6.3.**), wash the cells first with PBS, then with trypsin (optional).
6. Add minimal amount of trypsin to just cover cells (0.5 mL/60-mm plate) and place in the incubator for 1 to 2 min or leave at room temperature until colonies start to detach from the plate.
7. Watch under the microscope for colonies to detach from the plate, gently swirl the plate, or tap at the microscope stage. Do not over-trypsinize, as cells will become sticky and hard to manipulate. Add the required volume of ES-DMEM to the plate. Do not pipet. ES cells are now ready for aggregation during the next hour or two. Keep the plate at room temperature, as cells will start attaching to the plate in the incubator (*see* **Note 8**).

3.6.2. Recovery of 8- and 2-Cell Stage Embryos

It is preferable to use only noncompact 8-cell stage embryos for aggregation. However, we routinely use all embryos with intact blastomeres, from the 8-cell to morula stage, collected at 2.5 d postcoitum (dpc). Two-cell stage embryos, collected at 1.5 dpc, are used for production of tetraploid embryos (*see* **Subheading 3.6.4.**). The recovery procedure for 2-cell stage embryos is essentially the same.

1. Prepare the culture plates using KSOM media (organ culture or microdrops overlayed with mineral oil). Place the tube or the syringe with KSOM into the incubator.
2. Dissect the oviducts from 2.5 dpc pregnant females, leaving the upper part of the uterus attached, and place in the drop of M2.
3. Transfer one oviduct into a small drop of M2.
4. Using the dissecting microscope, insert the flushing needle attached to a 1-mL syringe filled with M2 into the infundibulum. The use of fine forceps helps to place the needle in the right position. Flush M2 media through the oviduct and observe its swelling.
5. Proceed with the remaining oviducts, keeping the time of manipulations in M2 minimal.
6. Collect embryos into a fresh M2 drop using embryo-manipulating mouth- or finger-controlled drawn Pasteur pipet. Wash them through several M2 drops to get rid of debris.
7. Wash embryos through few drops of equilibrated KSOM and place them in the culture dish.

3.6.3. Aggregation of ES Cells with Cleavage Stage Embryos

3.6.3.1. PREPARATION OF THE AGGREGATION PLATE

1. Using a 1-mL syringe filled with KSOM media, place microdrops (approx 3 mm in diameter) on a 35-mm dish. We usually place 2 rows of 4 to 5 drops in the middle of the plate and 2 more rows of 3 drops on each side. Cover with mineral oil.
2. Wipe the aggregation needle with 70% ethanol. Press the needle into the plastic, make a slight circular movement (*see* **Note 9**). We usually make 6 depressions per KSOM drop. Three or 6 drops on the sides are left without depressions. They are used for the final selection of ES cell clumps. The plate holds 40–60 aggregates.
3. The plate is placed into the incubator to allow good equilibration of the media.

3.6.3.2. ZONA REMOVAL

In order to allow the attachment of ES cells to the blastomeres of the embryos as an obvious precondition of aggregation, the zonae pellucidae of the embryos need to be removed. We use acidic Tyrode solution to dissolve this glycoprotein membrane. Since the acid diluted with buffered solution will not work as efficiently and the acid transferred into culture media will damage the embryos, it is important to transfer a minimal amount of solutions between drops and use multiple washes. The number of embryos manipulated at a time depends on the speed of manipulations.

1. Place M2 and acid Tyrode's drops into a 100-mm Petri dish. The temperature of the acid tyrode should not exceed room temperature.
2. Pick 20–50 embryos with a minimal volume of media and wash them through one acid drop. Transfer embryos to a fresh drop of acid. Move them by pipetting and observe zonae dissolution.
3. As soon as the zonae dissolve, immediately transfer the embryos with a minimal volume of acid to a M2 drop. Rinse through several M2 drops. Do not allow the embryos to touch each other. Proceed with the remaining embryos.
4. Wash the embryos through several drops of equilibrated KSOM and place them into the aggregation plates.

3.6.3.3. ASSEMBLY OF AGGREGATES

We aggregate an ES cell clump of 8–15 cells with one diploid embryo. Two tetraploid embryos are also usually aggregated with such a clump of ES cells (*see* **Subheading 3.6.3.4.**). It is possible to use one tetraploid embryo for aggregation, but it might be necessary to culture aggregates for one more night to ensure blastocyst formation. Aggregates are assembled in either of two ways: *(1)* the embryo is first placed into the depression, and the clump of ES cells is placed next to it, or *(2)* the clumps of ES cells are first distributed into depressions, and then the embryos are placed next to the clumps. Both ways work equally well.

1. After zona removal, embryos are placed into the aggregation plates inside the depressions or beside them, making sure they are not touching each other. The plates are kept in the incubator until the ES cells are ready.
2. Prepare ES cells for aggregation as described in the **Subheading 3.6.1.**
3. Under the dissecting microscope, choose a number of clumps of required size and transfer them into the microdrops, not containing depressions, for final selection.
4. Select a few clumps of 8–15 cells and carefully transfer them individually into the depressions, next to the embryos. Alternatively, distribute clumps into all depressions in a plate, then place the embryos next to each clump.
5. Assemble all the aggregates in this manner, check the plate to ensure that all ES cell clumps are in contact with embryos, and incubate overnight. The next day, the majority of aggregates will be blastocysts or late morulae. They can be transferred into 2.5 dpc pseudopregnant females as described previously *(17,19)* (*see* **Note 10**).

3.6.4. Generation of Tetraploid Embryos

The electrofusion of blastomeres of 2-cell stage embryos for the generation of tetraploid embryos was first developed by Kubiak and Tarkowski *(24)*. Most tetraploid embryos die shortly after implantation *(25)*, but when aggregated with diploid embryos, they can contribute to the primitive endoderm and trophoectoderm lineages and are excluded from the primitive ectoderm lineage *(26)*. When the tetraploid embryos are aggregated with ES cells, the resulting fetuses are ES cell derived, as ES cells do not contribute to trophectoderm and primitive endoderm. Thus, ES cell and tetraploid components complement each other. Therefore, the method of ES cell ↔ tetraploid embryo aggregation can be used, for example, as a rapid test for the developmental potential as well as for the transgene expression of an ES cell clone. The application of ES cell ↔ tertaploid embryo chimeras is, however, much broader (*see* **ref. 27** for review).

1. Prepare the culture plate (microdrops of KSOM overlayed with mineral oil). Thaw a frozen aliquot of mannitol.
2. Collect 2-cell stage embryos as described in **Subheading 3.6.2.**
3. Turn on the cell-fusion instrument and set up the parameters. The fusion of blastomeres of 2-cell stage embryos occurs when a DC electric pulse is applied perpendicular to the plane of the blastomeres' contact. We apply one or two pulses of 30 V and 40 μs for the fusion in nonelectrolyte solution (mannitol) using the CF-150B cell-fusion instrument from BLS Ltd. The adjustable 1 MHz AC field (1 to 2 V) allow the orientation of 2-cell stage embryos in the electrode chamber (distance between electrodes is 250 μm). The actual parameters might vary and need to be determined in a pilot experiment. The goal is to reach 90% fusion in 30–60 min without embryo lysis.
4. Place an electrode chamber connected to the pulse generator into the 100-mm Petri dish. Use the same dish for embryo washes; alternatively, use another plate on a second microscope.
5. Place two large drops of M2 and a drop of mannitol solution in the dish. Place the mannitol drop over the electrode chamber.
6. Pick 25–30 embryos and rinse them well with mannitol by quickly pipetting up and down. Place the embryos between the electrodes, spacing them from each other (*see* **Note 11**). Manually move the embryos that are not oriented.
7. When all the embryos are properly oriented, push the trigger pulse.
8. Immediately transfer the embryos into an M2 drop. Wash the embryos through a few drops of equilibrated KSOM. Place them into the culture plate in the incubator.
9. Proceed with the rest of the embryos. The mannitol drop over the electrode chamber should not be used for longer than 15 min and should be replaced with a fresh one after that time. The number of embryos handled in 15 min depends on the speed of manipulations.
10. It is very important to select perfectly fused tetraploid embryos 30–60 min after application of the pulse and transfer them into a fresh drop. Since embryos are recovered at the late 2-cell stage, the second mitotic division is expected soon after fusion. If not checked in time, fused and cleaved tetraploid embryos could be confused with non-fused diploid 2-cell stage embryos. Under optimal conditions, around 90–95% of embryos should fuse.
11. After overnight incubation, 3- to 4-cell stage embryos are used for sandwich aggregation as described in the **Subheading 3.6.3.**

4. Notes

1. The quality of FBS is the most important factor for successful ES cell culture. ES cell qualified pretested serum is available from some suppliers (e.g., Gibco and Specialty Media). It is still recommended to test different lots of serum from different companies before ordering large quantities (*see* **ref. 17** for testing serum batches protocol). It is also possible to use Knockout Serum Replacement (SR)™ from Gibco (cat. no. 10828) in combination with regular DMEM or Knockout D-MEM (Gibco, cat. no. 10829) for ES cell culture. However, in this case, ES cells need to be grown on feeders all the time.
2. It is possible to purify LIF from bacteria transfected with recombinant LIF constructs (*28*). Each new batch of LIF should be tested for the final concentration necessary to maintain pluripotency of ES cells. For R1 ES cells, we use final concentration of 1000 U/mL.

3. The quality of water is possibly even more critical for embryo culture than for ES cell media (*see* **Subheading 2.1.3.**). Disposable plasticware is highly recommended, or if clean glassware is used, it should never be exposed to detergent or organic solvents. All chemicals should be of highest grade, or embryo-tested (available from Sigma) and used only for media preparation. Embryos are cultured in organ culture dishes or in microdrops of KSOM media covered with embryo-tested light mineral oil (e.g., Sigma, cat. no. M8410). A majority of E0.5 dpc embryos reaching blastocyst stage demonstrates the optimal culture conditions. The minimal time between sacrificing the embryo donors and putting the embryos in the culture dish and the proper removal of debris after flushing also contributes to successful embryo culture.

4. It is important to carefully follow the ES cell culture protocol to maintain their pluripotency and ability to contribute to the germline. Typically, ES cells are kept at relatively high density and should be passaged when they reach a subconfluent state of 70–80%, i.e., tightly packed colonies almost touch each other. Media should be changed every day. For regular maintenance ES cells should never be seeded too sparsely (when 4 to 5 d is required to reach subconfluency), nor should they grow past 90% confluency before passage. Both conditions induce cell differentiation. Usually ES cells are split at 1:5 to 1:7 dilution depending on their growth rate, ideally every other day (about 1×10^6 cells are seeded onto 60-mm plate, 2×10^6 cells onto 100-mm plate). Cells should be trypsinized to a single-cell suspension as large clumps might differentiate. The passage number of ES cells must be kept as low as possible with a stock of frozen vials kept in liquid nitrogen. It is recommended to create a frozen pool of low passage number of ES cells that serve as the only regular source of ES cells for many years of experiments.

5. For example, for 2 wells: $\times 2.5$, i.e., 2.5 µg DNA in 600 µL OPTI-MEM$^+$ 20 µL LF in 600 µL of OPTI-MEM.

6. Often cells in different wells might not grow at a synchronous rate. Some wells might not be subconfluent but have one or two colonies. Such wells can be trypsinized individually to allow more even growth after plating back the single-cell suspension. It is best to choose a time for the passage when the majority of the wells have reached 80% confluency. Ninety-six-well plates are split into 2 or 3 replica plates, which can then be passaged one more time if necessary.

7. For E12.5 and older embryos as well as tissues, the samples can be sectioned sagitally using a razor blade after 30 min of incubation at room temperature in freshly prepared prefix solution (2% PFA in PBS) and then fixed for an extra 30–60 min on ice.

8. If a plate has too many ES cell colonies that are also larger than necessary, they can be gently resuspended in trypsin. Aliquots of cell suspension can be transferred into a new plate with ES-DMEM after pipetting up and down 1–3 times to achieve clumps of the right size.

9. The goal is to create a small depression with a smooth surface, deep enough to hold the aggregate safely even when moving the plate to the incubator.

10. We usually aggregate 150–200 embryos (transferred into 8–10 recipients) per ES cell clone for germ line transmission. If chimeric embryos are dissected at mid-gestation to assess in vivo expression, two ES cell clones can be done in one experiment. As ES cells are derived from a pigmented mouse strain (in the case of R1 from an F1 hybrid of 129Sv-cp $\times$ 129 SvJ cross), and host embryos are derived from albino mice, the degree of chimerism can be estimated by eye pigmentation (from E11.5). Pigmented cells in the eyes are ES cell-derived and indicate overall ES cell contribution. By analyzing such chimeric embryos, a few candidate ES cell clones can be chosen for best contribution potential and gene expression for further aggregation for germ line transmission.

11. The AC field can be set up in advance; alternatively, slowly increase the AC field, so the embryos will orient properly.

References

1. Nagy, A. (2000) Cre recombinase: the universal reagent for genome tailoring. *Genesis* **26,** 99–109.
2. Lakso, M., Sauer, B., Mosinger, B., Jr., Lee, E. J., and Manning, R. W. (1992) Targeted oncogene activation by site-specific recombination in transgenic mice. *Proc. Natl. Acad. Sci. USA* **89,** 6232–6236.
3. Buchholz, F., Angrand, P. O., and Stewart, A. F. (1998) Improved properties of FLP recombinase evolved by cycling mutagenesis. *Nat. Biotechnol.* **16,** 657–662.
4. Rodriguez, C. I., Buchholz, F., Galloway, J., Sequerra, R., and Kasper, J. (2000) High-efficiency deleter mice show that FLPe is an alternative to Cre-loxP [letter]. *Nat. Genet.* **25,** 139–140.
5. Groth, A. C., Olivares, E. C., Thyagarajan, B., and Calos, M. P. (2000) A phage integrase directs efficient site-specific integration in human cells. *Proc. Natl. Acad. Sci. USA* **97,** 5995–6000.
6. Brocard, J., Feil, R., Chambon, P., and Metzger, D. (1998) A chimeric Cre recombinase inducible by synthetic, but not by natural ligands of the glucocorticoid receptor. *Nucleic Acids Res.* **26,** 4086–4090.
7. Kellendonk, C., Tronche, F., Casanova, E., Anlag, K., and Opherk, C. (1999) Inducible site-specific recombination in the brain. *J. Mol. Biol.* **285,** 175–182.
8. Danielian, P. S., Muccino, D., Rowitch, D. H., Michael, S. K., and McMahon, A. P. (1998) Modification of gene activity in mouse embryos in utero by a tamoxifen-inducible form of Cre recombinase. *Curr. Biol.* **8,** 1323–1326.
9. Gossen, M., Freundlieb, S., Bender, G., Muller, G., and Hillen, W. (1995) Transcriptional activation by tetracyclines in mammalian cells. *Science* **268,** 1766–1769.
10. Kistner, A., Gossen, M., Zimmermann, F., Jerecic, J., and Ullmer, C. (1996) Doxycycline-mediated quantitative and tissue-specific control of gene expression in transgenic mice. *Proc. Natl. Acad. Sci. USA* **93,** 10933–10938.
11. Holzenberger, M., Zaoui, R., Leneuve, P., Hamard, G., and Le Bouc, Y. (2000) Ubiquitous postnatal LoxP recombination using a doxycycline auto-inducible Cre transgene (DAI-Cre). *Genesis* **26,** 157–159.
12. Soriano, P. (1999) Generalized lacZ expression with the ROSA26 Cre reporter strain. *Nat. Genet.* **21,** 70–71.
13. Lobe, C. G., Koop, K. E., Kreppner, W., Lomeli, H., and Gertsenstein, M. (1999) Z/AP, a double reporter for cre-mediated recombination. *Dev. Biol.* **208,** 281–292.
14. Lewandoski, M. and Martin, G. R. (1997) Cre-mediated chromosome loss in mice. *Nat. Genet.* **17,** 223–225.
15. Friedrich, G. and Soriano, P. (1991) Promoter traps in embryonic stem cells: a genetic screen to identify and mutate developmental genes in mice. *Genes Dev.* **5,** 1513–1523.
16. Niwa, H., Yamamura, K., and Miyazaki, J. (1991) Efficient selection for high-expression transfectants with a novel eukaryotic vector. *Gene* **108,** 193–199.
17. Joyner, A. (ed.) (1999) *Gene targeting: a practical approach*, 2nd ed., Oxford University Press, New York.
18. Nagy, A., Rossant, J., Nagy, R., Abramow-Newerly, W., and Roder, J. C. (1993) Derivation of completely cell culture-derived mice from early-passage embryonic stem cells. *Proc. Natl. Acad. Sci. USA* **90,** 8424–8428.
19. Hogan, B., Beddington, R., Costantini, F., and Lacy, E. (1994) *Manipulating the mouse embryo: a laboratory manual*, 2nd ed. series. CHS Laboratory Press, Cold Spring Harbor, N.Y.

20. Lawitts, J. A. and Biggers, J. D. (1992) Joint effects of sodium chloride, glutamine, and glucose in mouse preimplantation embryo culture media. *Mol. Reprod. Dev.* **31,** 189–194.

21. Ho, Y., Wigglesworth, K., Eppig, J. J., and Schultz, R. M. (1995) Preimplantation development of mouse embryos in KSOM: augmentation by amino acids and analysis of gene expression. *Mol. Reprod. Dev.* **41,** 232–238.

22. Ramirez-Solis, R., Rivera-Perez, J., Wallace, J. D., Wims, M., and Zheng, H. (1992) Genomic DNA microextraction: a method to screen numerous samples. *Anal. Biochem.* **201,** 331–335.

23. Wood, S. A., Allen, N. D., Rossant, J., Auerbach, A., and Nagy, A. (1993) Non-injection methods for the production of embryonic stem cell-embryo chimaeras. *Nature* **365,** 87–89.

24. Kubiak, J. Z. and Tarkowski, A. K. (1985) Electrofusion of mouse blastomeres. *Exp. Cell Res.* **157,** 561–566.

25. Tarkowski, A. K., Witkowska, A., and Opas, J. (1977) Development of cytochalasin in B-induced tetraploid and diploid/tetraploid mosaic mouse embryos. *J. Embryol. Exp. Morphol.* **41,** 47–64.

26. Nagy, A., Gocza, E., Diaz, E. M., Prideaux, V. R., and Ivanyi, E. (1990) Embryonic stem cells alone are able to support fetal development in the mouse. *Development* **110,** 815–821.

27. Lobe, C. G. and Nagy, A. (1998) Conditional genome alteration in mice. *Bioessays* **20,** 200–208.

28. Mereau, A., Grey, L., Piquet-Pellorce, C., and Heath, J. K. (1993) Characterization of a binding protein for leukemia inhibitory factor localized in extracellular matrix. *J. Cell Biol.* **122,** 713–719.

24

Switching on Lineage Tracers
Using Site-Specific Recombination

Susan M. Dymecki, Carolyn I. Rodriguez, and Rajeshwar B. Awatramani

1. Introduction

Methods to study the establishment and distribution of embryonic cell lineages have increased our understanding of the diverse events that comprise normal development. The resultant fate maps have also provided an important framework for systematically analyzing genotype–phenotype relationships uncovered by mutagenesis. Until recently, vertebrate fate maps have been plotted principally in avian systems because of the ease of manipulating tissue in ovo. These studies, using methods comprised of injecting retroviral (1,2) or fluorescent lineage tracers (3) or of grafting quail cells into chick embryos (4), have provided the core of what we know about vertebrate development. In contrast to chicken embryos developing in eggs, mouse embryos developing in utero are much less accessible, making the established tracing methods significantly more difficult. To circumvent these difficulties, noninvasive methods have recently been developed to genetically activate lineage tracers in mice using site-specific recombination (5,6).

Two recombinase systems from the λ integrase family have been established in mice for catalyzing gene (in)activation: the Cre-*loxP* system from the bacteriophage P1 (7–9) and the FLP-*FRT* system (10) from the budding yeast *Saccharomyces cerevisiae*. Cre and FLP each cleave DNA at distinct 34 bp target sequences (designated *loxP* (11) or *FRT* sites (12–14), respectively) and then ligate it to the cleaved DNA of a second identical site to generate a contiguous strand (15–17). The orientation of *loxP* (or *FRT*) sites relative to each other on a segment of DNA determines the type of modification catalyzed by the recombinase, with directly oriented identical sites leading to excision of the intervening DNA. This excision reaction can be exploited to activate lineage tracers in vivo.

Current recombinase-based fate mapping systems (5,6,18) are comprised of two basic elements: (*i*) the recombinase mouse, expressing the site-specific recombinase Cre or FLP in a gene-specific fashion; and (*ii*) the indicator mouse, harboring a transgene that "indicates" a recombination event in a given cell and "remembers" or "provides a permanent record" of that recombination event in a heritable manner far after the time of recombinase expression. In embryos harboring both the recombinase and indicator

From: *Methods in Molecular Biology, vol. 185: Embryonic Stem Cells: Methods and Protocols*
Edited by: K. Turksen © Humana Press Inc., Totowa, NJ

transgenes, the recombinase will catalyze excisional recombination between two target sites in the indicator transgene, thereby activating the indicator (e.g., switching-on expression of the β-galactosidase (β-gal)-encoding *lacZ* to irreversibly mark a progenitor population and its descendant cells. Cell fates are then determined by colocalization of the indicator molecule (e.g., β-gal) with specific cell-identity markers (molecular and morphological). In this way, a fate map is generated by noninvasive means that not only links a progenitor population to its final array of fates, but also links progenitor- or subtype-specific gene expression to those fates. Applied to the developing nervous system, this approach has led to a significant breakthrough in being able to determine how embryonic cell identity (embryonic gene expression) links to the final location and properties of those neurons in the adult *(6,18,19)*. This link is made regardless of the transient nature of the expression profiles characterizing many genes, and holds irrespective of intervening cell migration or morphogenetic movement. In other words, transient gene expression patterns can be transformed by site-specific recombination to become neutral tracers of cell fate.

1.1. Recombinase-Expressing Mice

In recombinase-based fate mapping experiments, the progenitor population to be tracked is set by the pattern of recombinase expression; thus, it is critical to be able to restrict recombinase activity to a specific progenitor population during development. Such progenitor-specific recombinase transgenics can be generated in one of two ways: *(i)* zygote injection of a transgene that can express the recombinase; or *(ii)* targeted insertion of the recombinase transgene into an endogenous gene (a knock-in targeting experiment). Each method has both unique advantages as well as drawbacks.

The approach of zygote injection has at least two advantages. First, specific enhancer elements can be selected to drive recombinase expression in a very restricted and transient fashion, enabling both a spatially and temporally defined subgroup of progenitor cells to be traced *(6)*. Second, the method allows for the relatively rapid generation of recombinase transgenics. While this is an essential and widely exploited approach, there are two drawbacks worth mentioning. Constructing a recombinase transgene requires previous isolation and characterization of promoter and enhancer elements; these elements may not always be readily available (although the number is increasing significantly as genome analyses expand). Additionally, recombinase expression from an integrated transgene (which is typically comprised of a head-to-tail array of multiple transgene copies) can be mosaic, with only a subset of the expected cell population expressing the transgene. This variation is usually stochastic and can be counterbalanced in fate mapping studies by analyzing a larger number of recombinase–indicator double transgenics in order to achieve full coverage of the lineage(s) under investigation.

If the goal is to fate map every cell that at some point in its history expressed a given gene (versus just a subset of expressing-cells in which a discrete enhancer element is active, as described above), the recombinase knock-in approach is preferable. Moreover, by placing a single copy of the recombinase transgene under the transcriptional control of an endogenous locus, recombinase expression should be optimally regulated with minimal mosaicism. However, when interpreting fate maps generated through the use of recombinase knock-in strains, it is important to remember that recombinase knock-ins can have continuous and changing expression patterns, reflecting the dynamic nature of the wild-type locus. Thus, at a given developmental stage, indicator-positive ("mapped") cells may reflect either recombinase-expressing cells or recombinase-negative lineal descendants of earlier recombinase-expressing cells.

Constructing a recombinase transgene for zygote injection involves the following steps: *(1)* the appropriate promoter–enhancer sequences are coupled to Cre- *(20,21)* or FLP-encoding sequences *(22–24)*; *(2)* the translation start codon is bracketed with Kozak consensus nucleotides for efficient translation of the recombinase *(25)*; *(3)* splice donor–splice acceptor sequences and a polyadenylation signal sequence (pA) are incorporated to maximize transgene expression *(26,27)*; and *(4)* unique restriction sites are engineered into the construct so that the transgene can be isolated away from the bacterial plasmid backbone. Consideration should also be given to flanking the transgene with insulator sequences *(28,29)* to minimize the effects of integration site on proper transgene expression.

Essential components of a knock-in targeting vector have been well described *(30)* and include: *(1)* two arms of isogenic DNA homologous to the genomic target locus; *(2)* the Cre- or FLP-encoding transgene positioned between the homology arms in such a way that it will be under the transcriptional control of the endogenous locus following homologous recombination; *(3)* a positive selection gene placed between the homology arms; *(4)* a negative selection gene positioned at the end of a homology arm; *(5)* a unique restriction site to linearize the vector outside the homologous sequences; and *(6)* directly oriented recombinase target sites flanking the positive selection marker. Later recombinase-mediated removal of the selection marker *(31)* ensures that the strong regulatory elements driving expression of the selectable marker do not interfere with normal regulation of either the knock-in allele or neighboring endogenous genes *(32–35)*.

Whether recombinase strains are generated by zygote injection of a transgene or by gene replacement techniques, it is critical that the recombinase activity profile by adequately determined for each recombinase mouse strain. Expression analyses should be performed at cellular resolution and should include: *(1)* detection of recombinase mRNA by in situ hybridization or by the detection of a histochemical or fluorescent marker encoded on a bicistronic transcript that also encodes the recombinase; *(2)* visualization of recombinase protein by immunodetection *(36)* or by fluorescence using a version of the recombinase tagged with either green fluorescent protein (GFP) or one of its spectral variants (red, yellow, and cyan fluorescent proteins; Clonetech) *(37,38)*; and *(3)* detection of recombinase activity by crossing to an indicator mouse strain and analyzing recombinase–indicator double transgenic progeny for sites of reporter expression *(18,39–42)*. As a means to assess recombinase expression, it is important to emphasize that analyses of recombinase–indicator double transgenics, alone, are insufficient. This is because recombinase and indicator profiles will initially be coincident, but later will diverge, with indicator-positive cells reflecting either new sites of recombinase expression or recombinase-negative cells that are lineal descendants of the earlier recombinase-expressing cells. Distinguishing between these two possibilities requires analyzing recombinase expression by an independent method, such as in situ mRNA hybridization. Overall, expression analyses should be performed on a developmental series of staged embryos as well as on adult tissues. Especially critical is to rule out ectopic recombinase expression. Any unexpected recombinase expression would confound subsequent fate mapping studies by switching on the indicator (e.g., β-gal) in unrelated cells that would be erroneously interpreted as part of (or lumped into) a given lineage. Methods for profiling recombinase activity are detailed in **Subheading 3.1.**

1.2. Indicator Mice

In contrast to the transient expression of most recombinase transgenes, the critical feature for indicator transgenes is permanent reporter expression in all arms of a given lineage. Success, therefore, hinges on using a promoter with constitutive and ubiquitous activity to drive the indicator transgene such that following a recombination event in a given cell, that cell and all progeny cells will be marked regardless of subsequent fate. In constructing an indicator transgene *(7,18,22,40,42,43)*, such constitutive promoter sequences are placed upstream of the reporter gene, but separated from the reporter sequence by a "stop" segment of DNA flanked by either directly oriented *FRT* or *loxP* sites. One efficient and strong stop DNA segment is a head-to-tail array of four simian virus 40 (SV40) pA sequences coupled with translational stop codons in all reading frames *(42,43,44)*. Using such a *loxP*- or *FRT*-disrupted indicator transgene, the reporter molecule is not expressed unless the disrupting stop DNA is excised by the appropriate recombinase. Following such an excision event, the reporter is switched-on and cell marking becomes heritable and dependent on constitutive reporter expression only (reporter expression is now independent of the recombinase).Thus, in designing an indicator transgene, the first key decision involves choosing the most suitable promoter. The second critical step concerns choosing a reporter molecule that will best highlight the cell population under study, including specialized features of that cell type (e.g., dendrite or axon), and that will enable colocalization of the reporter molecule with informative cell identity markers (nuclear or cytoplasmic). As for any transgene, the indicator transgene should also contain Kozak consensus nucleotides *(25)* surrounding the initiator ATG for efficient translation of the reporter, and splice donor–splice acceptor sites and pA sequences *(26,27)* to maximize reporter expression following recombinase-mediated activation.

Ideally, the most suitable promoter is one capable of achieving robust and ubiquitous indicator expression following site-specific recombination, as this would enable mapping all cell types. The identification of such a *promoter*::*indicator* combination has not been a trivial task, having only been approached recently through two different strategies: either targeting indicator transgenes to the constitutive *ROSA26* locus *(42)* or by using the cytomegalovirus (CMV) enhancer/chicken β-*actin* (*CAG*) promoter *(39,45)* to drive single-copy indicator transgenes *(43,46)*. The *ROSA26* locus was identified through the gene-trap strain *ROSA* β*geo 26*, in which a βgeo reporter was constitutively expressed during embryonic development *(47)*. Multiple genes have now been targeted to the *ROSA26* locus *(42,48,49)*, with *ROSA26* regulatory sequences driving generalized transgene expression from preimplantation onward. The second approach, using the *CAG* promoter, involved an extensive screen for embryonic stem (ES) cell clones that both harbored a single integrated copy of a *CAG*::*indicator* transgene and robustly expressed the indicator following transient recombinase expression *(43,46)*. Such permissive ES clones were then used to generate transgenic mice. In addition to these two approaches, other promoters have been used to drive indicator trangenes. While of less universal use, these include promoter–enhancer sequences from the mouse *Hmgcr* gene (encoding hydroxy-methylglutaryl-coenzyme A reductase) *(18)*, the chicken β-actin gene *(6,40)*, and the mouse *BT5* gene-trap locus *(41)*.

To date, the most widely used reporter molecule in indicator transgenes is β-gal encoded by the bacterial *lacZ* gene. In situ β-gal activity can be visualized either by 5-bromo-4-chloro-3-indolyl-β-D-galactopyranoside (X-gal) histochemistry (described in **Subheading 3.2.1.**) or by immunodetection using anti-β-gal antibodies, the latter method enabling immunolocalization of β-gal with other cell-specific antigens *(18)*. As a cytoplasmic protein, β-gal typically gives good staining of cell bodies. In some cases, however, such staining can make it difficult to distinguish individual cells within a cluster of β-gal-positive cells. This potential limitation can be circumvented by using a form of β-gal that localizes to the cell nucleus. In addition to providing clear visualization of individual cells, nuclear β-gal both increases the sensitivity of the indicator system by concentrating the β-gal signal and is distinguished from any confounding endogenous β-gal activity that would be cytoplasmic.

In many fate mapping experiments, such as when studying neurons where the dendritic arborization and axon morphology reflect very specialized functions, additional insights could be obtained if entire cells were outlined by the reporter molecule. While β-gal serves as an excellent cell tracer, it does not completely fill cells, particularly long processes. Membrane-localized reporters, such as heat-resistant human placental alkaline phosphatase (PLAP) *(50–52)* and the farnesylated version (EGFPF) of the enhanced green fluorescent protein (EGFP) *(53)* are capable of outlining long cellular processes without seeming to compromise viability.

PLAP is anchored to the outer (noncytoplasmic) surface of the plasma membrane by a glycan-phosphatidyl inositol (GPI) moiety. Histochemical detection of PLAP activity is so sensitive that single axons can be traced to their terminals and individual growth cones and dendrites visualized quite readily *(52)*. In addition, monoclonal and polyclonal antibodies specific to PLAP are available (supplied by Biomedia or Dako) and can be used to immunolocalize PLAP expression with cell-identity markers. In situ detection of PLAP activity in mouse embryos and tissue cryosections is described in **Subheading 3.4.2.**

EGFPF is a fusion between EGFP and the carboxy-terminal 20 amino acids of c-Ha-Ras containing membrane-targeting farnesylation and palmitoylation sequences *(53)*. In *Drosophila* and mammalian cells, EGFPF provides a robust fluorescent signal highlighting long cellular processes such as dendrites and axons (*54*, and Awatramani and Dymecki, unpublished results). To date, no defects in neural connections or cells viability have been found associated with either EGFPF or PLAP, as have been associated with some other axon-targeted reporter molecules *(52)*. Membrane-localized EGFPF should allow for unequivocal marking of long cellular processes (such as axons) and should enable visualization of migrating cells in live tissue.

Before using an indicator strain for lineage studies, it is important to determine the expression profile of the activated reporter molecule. This profile determines the range of cell types that can be mapped reliably by the indicator strain, establishing its overall utility. If a given cell is incapable of expressing the indicator transgene, then it would not be possible to determine by reporter detection whether the recombinase was ever expressed in that particular lineage. To unambiguously determine the scope of the potential indicator profile, a derivative mouse strain should be generated in which every cell harbors a recombined copy of the indicator transgene while maintaining the chromosomal environment of the starting (unrecombined) transgene. Generating such a derivative ("activated") indicator strain is described in **Subheading 3.3** and involves crossing the indicator strain to a recombinase-expressing deleter strain (defined in **Subheading 2.3.1** and **ref. *24***), which will catalyze indicator recombination in the germ line and transmit this recombined indicator transgene to progeny. Since any chromosomal position effects on indicator expression will be identical between the derivative activated strain and the parental strain, the reporter profile for the activated strain will precisely reflect the expression potential of the parental indicator strain.

1.3. Expanding the Capabilities of Recombinase-Based Tools

Through the production of new FLP variants, unique capabilities have been added to recombinase-based fate mapping systems, allowing for an important degree of flexibility in experimental design. The three variants include: FLP-wild-type (FLP-wt), FLPe, and FLP-L. "Enhanced FLP" (FLPe) harbors four amino acid changes that collectively render FLPe more thermostable, thereby improving recombinase activity in vitro 4-fold at 37°C and 10-fold at 40°C as compared to FLP-wt *(55)*. Importantly, when expressed in mice, FLPe has been able to achieve maximal target gene recombination *(24,49)*. In contrast, FLP-L, containing an alternative point mutation, exhibits at least a 5-fold decrease in activity as compared to FLP-wt *(24,56)*. These activity differences can be exploited in vivo. For example, maximal target gene recombination, such as approached with FLPe, is desirable for mapping cell populations en masse when examining either gross morphogenetic movements or the development of an entire neural system. The more modest recombination frequencies achieved with FLP-L are advantageous when visualization of fewer cells improves the resolution of a developmental process, such as following discrete axon trajectories of lineally related neurons. In addition to the expanded flexibility afforded by the different variants, use of the FLP system for fate mapping in general has the advantage that mouse strains with FLP-marked lineages can be combined with Cre-induced mutations and deficiencies without concern for potentially confounding cross-reactions between the fate mapping (FLP) and mutagenesis (Cre) systems (which could occur should the same recombinase system be used for both purposes). It then becomes possible to analyze mutant gene activities directly for their effect on the FLP-marked lineage and its fate. Due to these advantages, the number of available FLP-expressing mouse strains is on the rise. Alternatively, if Cre is used as the fate mapping tool, the large number of already-generated Cre transgenics can be exploited. Regardless of the recombinase used, it is essential that the recombinase expression patterns be detailed both spatially and temporally for each strain prior to their use in order to make proper sense of the resultant fate maps.

In addition to using a single-recombinase system for fate mapping (e.g., FLP or Cre alone), it is also possible to exploit both FLP and Cre together for fate mapping purposes, enabling the generation of higher resolution maps. For example, indicator transgenes have been generated (R. Awatramani and S. Dymecki, unpublished data) in which cell lineages are marked only if they have expressed both FLPe and Cre at some point in their history. Such "intersectional" fate mapping can be used to determine the fate of more restricted progenitor populations, subsets demarcated by specific pair-wise combinations of expressed genes. Additional recombinase systems are being developed that will likely increase capabilities even further. These include the integrase from the Streptomyces phage ϕC31 *(57)* and the Cre-FLP hybrid enzymes, Clp and Fre, which recognize novel DNA target sites *(58)*. Moreover, all of these systems can gain finer temporal control of recombination events either by using inducible promoters (broadly active or lineage-specific) to express the recombinase *(59,60)* or by using a recombinase–steroid receptor fusion protein that can be activated at will by systemic administration of hormone *(61–66)*.

While the described Cre and FLP fate mapping techniques are limited by our current knowledge of early gene expression patterns, the number of well-characterized progenitor-specific genes is growing rapidly due to transcript profiling at the genome level through the use of such methods as DNA microarrays *(67)* and the SAGE method for serial analysis of gene expression *(68)*. Thus, the possibilities for highly sophisticated spatial and temporal genetic manipulations appear without limit.

Recombinase-based fate mapping approaches have been used principally in vivo; however, it should be feasible to exploit this basic strategy, or at least certain aspects of the method, in the in vitro study of lineages differentiated from cultured stem cells. The underlying principles and transgenes are the same. These methods, however, should be applied to stem cell cultures with some degree of caution because the efficiency of each recombinase variant has not been unequivocally established in ES cells.

In this chapter, we describe methods to assess recombinase expression patterns, to determine the overall utility of a given indicator strain, and to histochemically detect either β-gal or PLAP activity in tissue sections in order to track cell populations during development. These approaches will be presented within the context of in vivo analyses.

2. Materials

2.1. Profiling Recombinase Expression by In Situ mRNA Hybridization

2.1.1. Pretreatment and Cryosection of Tissues

1. Phosphate-buffered saline (PBS): 137 mM NaCl, 2.6 mM KCl, 10 mM Na_2HPO_4, and 1.8 mM KH_2PO_4. Adjust pH to 7.4 with HCl, autoclave, and store at room temperature.
2. 4% Fix: 4% paraformaldehyde in PBS. Prepare fresh from frozen 20% paraformaldehyde-PBS stock. Store at 4°C for up to 24 h.
3. 30% Sucrose solution: 30% sucrose (w/v) in PBS. Filter-sterilize (0.2 µm) and store at 4°C.
4. OCT: embedding resin (Sakura Fineteck, cat. no. 4583).
5. Equipment: dissecting instruments (forceps and scissors, Fine Science Tools), embedding molds (Polysciences, cat. no. 17177C), Superfrost/Plus glass slides (Fisher Scientific, cat. no. 12-550-15), freezer-safe elastic sealing tape (3M, cat. no. 471), freezer-safe slide boxes (Central Carolina Products, cat. no. 634200), and a cryostat (Leica, cat. no. CM3050).

2.1.2. Synthesis of Digoxigenin-Labeled Riboprobes for In Situ Detection of Recombinase mRNA

Precautions should be taken to avoid any contamination of reagents or equipment with ribonucleases. Sterile, disposable plasticware is essentially free of RNases and should be used whenever possible.

1. Linearized plasmid (1μg/μL) containing T3, T7, or SP6 promoter sequence and the cDNA of interest as template for riboprobe synthesis.
2. Diethyl pyrocarbonate (DEPC)-water: to inhibit any contaminating ribonucleases, add 1 mL of DEPC (Sigma, cat. no. D-5758) per liter of distilled water. Note that DEPC is suspected to be a carcinogen and should be handled with care. Stir the 0.1% DEPC in water overnight in a fume hood at room temperature. Autoclave to inactivate the DEPC. Store at room temperature.
3. Stock solutions (prepare in DEPC-water): 10 mg/mL proteinase K (store in aliquots at –20°C); 0.5 M EDTA, pH 8.0; 4 M LiCl; 10 mg/mL ethidium bromide.
4. 10X Transcription buffer: 400 mM Tris-HCl, pH 8.0, 60 mM MgCl$_2$, 100 mM dithioerythritol, 20 mM spermidine, 100 mM NaCl, and 1 U/mL RNase inhibitor (Roche, cat. no. 1465384).
5. 10X NTP labeling mixture. 10 mM ATP, 10 mM CTP, 10 mM GTP, 6.5 mM UTP, and 3.5 mM digoxigenin-UTP in Tris-HCl, pH 7.5 (Roche, cat. no. 1 277 073).
6. RNase inhibitor, 40 U/μL (Roche, cat. no. 799 017).
7. T3, T7, or SP6 RNA Polymerase, 20 U/μL (Roche, cat. nos. 1 031 163, 881 767, or 810 274, respectively).
8. DNase I (RNase-free), 10 U/μL (Roche, cat. no. 776 785).
9. TE$_{50}$: 50 mM Tris-HCl, pH 8.0, 1 mM EDTA, pH 8.0. Autoclave and store at room temperature.
10. TE: 10 mM Tris-HCl, pH 8.0, 1 mM EDTA, pH 8.0. Autoclave and store at room temperature.
11. 100 and 70% ethanol. Store at –20°C.
12. Electrophoresis-grade agarose.
13. Digoxigenin-labeled RNA standards, 0.1 μg/μL (Roche, cat. no. 1 585 746).
14. RNase-free microfuge tubes and pipet tips.

2.1.3. In Situ mRNA Detection

1. Stock solutions (prepare in DEPC-water, *see* **Subheading 2.1.2.**): 1 M triethanolamine (Sigma, cat. no. T-9534); 5 M NaCl; 1 M KCl; 1 M Na$_2$HPO$_4$; 1 M KH$_2$PO$_4$; 1 M levamisole (store in aliquots at –20°C); 1 M Tris-HCl, pH 7.5 and pH 9.5; 20% paraformaldehyde in PBS (store in aliquots at –20°C); heat-inactivated sheep serum (Gibco, cat. no. 16070-096) (*see* **Note 1**); 20% sodium dodecyl sulfate (SDS); polyoxyethylene-sorbitan monolaurate (Tween-20; Sigma, cat. no. P7949); acetic anhydride (Sigma, cat. no. A 6404); 10 mg/mL proteinase K (store in aliquots at –20°C); 10 mg/mL yeast tRNA (store in aliquots at –20°C; Roche, cat. no. 109 495); 10 mg/mL heparin (store in aliquots at –20°C; Sigma, cat. no. H-3149); DEPC-water.
2. PBS: 137 mM NaCl, 2.6 mM KCl, 10 mM Na$_2$HPO$_4$, and 1.8 mM KH$_2$PO$_4$. Adjust pH to 7.4 with HCl, autoclave, and store at room temperature.
3. PBS$_{DEPC}$; 137 mM NaCl, 2.6 mM KCl, 10 mM Na$_2$HPO$_4$, and 1.8 mM KH$_2$PO$_4$. Adjust pH to 7.4 with HCl. Add 1 mL DEPC/L PBS, stir overnight in fume hood, autoclave, and store at room temperature.
4. 4% Fix: 4% paraformaldehyde in PBS$_{DEPC}$. Prepare fresh from frozen 20% paraformaldehyde/PBS stock. Store at 4°C for up to 24 h.

5. Glycine-PBS$_{DEPC}$: 2 mg/mL glycine in PBS$_{DEPC}$, prepared just before use.
6. 20X standard saline citrate (SSC): 3 M NaCl, 0.3 M Na-citrate in DEPC-water. Adjust to pH 4.5 with 0.6 M citric acid.
7. Hybridization buffer: 50% formamide, 5X SSC, 50 μg/mL yeast tRNA, 1% SDS, and 50 μg/mL heparin in DEPC-water. Store at –20°C.
8. Riboprobe: 0.1 μg/μL in TE$_{50}$ (see **Subheading 2.1.2.**).
9. Wash I: 50% formamide, 5X SSC, 1% SDS in DEPC-water.
10. Wash II: 0.5 M NaCl, 10 mM Tris, pH7.5, 0.1%. Tween-20 in DEPC-water.
11. Wash III: 50% formamide, 2X SSC in DEPC-water.
12. Rnase A Wash: 25 μg/mL Rnase A in Wash II.
13. Anti-digoxigenin Fab fragments conjugated to alkaline phosphatase (αdigAP), 0.75 U/μL (Roche, cat. no. 1 093 274). Aliquot and store at 4°C.
14. 10X TBST: 1.4 M NaCl, 27 mM KCl, 250 mM Tris-HCl, pH 7.5, and 10% polyoxyethylene-sorbitan monolaurate (Tween-20). Store at room temperature.
15. TBST-L: 1X TBST, 2 mM levamisole. Prepare fresh prior to use.
16. Antibody solution: αdigAP at 1:5000 in TBST-L with 1% heat-inactivated sheep serum.
17. NTMT-L: 100 mM NaCl, 100 mM Tris-HCl, pH 9.5, 50 mM MgCl$_2$, 1% polyoxyethylene-sorbitan monolaurate (Tween 20), and 2 mM levamisole. Prepare immediately prior to use.
18. NBT: 4-nitro blue tetrazolium chloride, 100 mg/mL (Roche, cat. no. 1 383 213).
19. BCIP: X-phoshate-5-bromo-4-chloro-3-indolyl-phosphate, 50 mg/mL (Roche, cat. no. 1 383 221).
20. Alkaline phosphatase staining solution: 4.5 μL NBT and 3.5 μL BCIP/mL NTMT-L. Prepare immediately prior to use.
21. Gelvatol (*see* **Subheading 3.1.4.** for instructions on preparation) or other aqueous mounting media (e.g., Vectashield; Vector Laboratories, cat. no. H-1200).
22. Equipment: plastic staining dishes (capacity for 24 slides in 250 mL of solution; Tissue-Tek, cat. no. 4456), slide holders with handles (each holds 24 slides; Tissue-Tek, cat. no. 4465A), slide mailers (capacity for 5 slides in 15 mL of solution; Ted Pella, cat. no. 22518), 24 × 60 mm glass coverslips (Fisher Scientific, cat. no. 12-548-5P).

2.1.4. Preparation of Gelvatol Aqueous Mounting Media

1. Stock solutions: 10 mM KH$_2$PO$_4$; 10 mM Na$_2$HPO$_4$. Store each at room temperature.
2. Polyvinyl alcohol resin (grade 205) (Air Products & Chemicals).
3. NaCl.
4. Glycerol.
5. Equipment: a 2-L beaker, a 6-inch stir bar (VWR, cat. no. 58948-958), 250-mL polypropylene centrifuge bottles, and 50-mL polypropylene conical tubes.

2.2. Profiling Recombinase Expression by Crossing Recombinase Mice to a lacZ Indicator Strain

2.2.1. Detection of β-Gal Activity on Tissue Cryosections by X-Gal Histochemistry

1. Stock solutions: 10% Nonidet P-40 (NP40); 1% Na deoxycholate; 1 M Na-phosphate (combine 1 M NaH$_2$PO$_4$ and 1 M Na$_2$HPO$_4$ to pH 7.4); 0.25 M EGTA, pH 7.4; 0.5 M piperazine-N,N′-bis(2-ethanesulfonic acid) (PIPES) pH 6.9; 1 M MgCl$_2$; 20% paraformaldehyde in PBS (store in aliquots at –20°C); 0.5 M K$_3$(Fe[CN]$_6$); 0.5 M K$_4$(Fe[CN]$_6$); 25 mg/mL X-gal (Angus, cat. no. 7240-90-6) in N,N-dimethyl-formamide (DMF) (Sigma, cat. no. D8654) (store at –20°C protected from light).

2. 0.2% Fix solution: 0.2% paraformaldehyde (from frozen 20% paraformaldehyde-PBS stock), 0.1 *M* PIPES, pH 6.9, 2 m*M* MgCl$_2$, and 5 m*M* EGTA, pH 7.4. Prepare immediately prior to use. Store at 4°C for up to 24 h.
3. 2.0% Fix solution: 2.0% paraformaldehyde (from frozen 20% paraformaldehyde-PBS stock), 0.1 *M* PIPES, pH 6.9, 2 m*M* MgCl$_2$, and 5 m*M* EGTA, pH 7.4. Prepare immediately prior to use. Store at 4°C for up to 24 h.
4. Reaction buffer: 100 m*M* Na-phosphate, pH 7.4, 2 m*M* MgCl$_2$, 0.1% Na-deoxycholate, 0.2% NP40 (*see* **Note 2**).
5. X-gal staining solution: 5 m*M* K$_3$(Fe[CN]$_6$), 5 m*M* K$_4$(Fe[CN]$_6$), and 1mg/mL X-gal in rinse solution. X-gal staining solution can be reused several times if stored at 4°C protected from light.
6. PBS: 137 m*M* NaCl, 2.6 m*M* KCl, 10 m*M* Na$_2$HPO$_4$, and 1.8 m*M* KH$_2$PO$_4$. Adjust pH to 7.4 with HCl, autoclave, and store at room temperature.
7. PBS-M: 2 m*M* MgCl$_2$ in PBS.
8. Gelvatol (*see* **Subheading 3.1.4.** for instructions on preparation) or other aqueous mounting media (e.g., Vectashield; Vector Laboratories).
9. Equipment: glass staining dishes (Brain Research Laboratories, cat. no. 3125), slide-racks (Brain Research Laboratories, cat. no. 3003), and 24 × 60 mm glass coverslips (Fisher Scientific, cat. no. 12-548-5P).

2.3. Determining the Utility of an Indicator Strain

2.3.1. Germ Line Transmission of an Activated Indicator Transgene

1. Deleter mice: a recombinase (Cre or FLP)-expressing mouse strain that is capable of mediating site-specific excisional recombination in the germ line.
2. Indicator mice: a mouse strain harboring a *loxP*- or *FRT*-disrupted reporter transgene that is capable of indicating a recombination event and providing a permanent record of that recombination event in a heritable manner far after the time of recombinase expression.
3. Nontransgenic mice of the same genetic background as the indicator strain.

2.3.2. DNA Extraction from Tail Biopsies

To minimize contamination of samples with unwanted plasmid or genomic DNA, all solutions should be prepared using polymerase chain reaction (PCR)-grade reagents, dispensed using aerosol-resistant tips, and stored and/or incubated in autoclaved microfuge tubes or containers.

1. Stock solutions: proteinase K (store in 10 mg/mL aliquots at –20°C); phenol:chloroform (1:1); chloroform:isoamyl alcohol (24:1); 3 *M* NaOAc.
2. Phenol: phenol equilibrated with TE, pH 8.0 (USB, cat. no. 75829).
3. 100% Ethanol. Store at –20°C.
4. STES buffer: 100 m*M* NaCl, 50 m*M* Tris-HCl, pH 7.4., 10 m*M* EDTA, pH 7.5, and 1% SDS.
5. TE: 10 m*M* Tris-HCl, pH 8.0, 1 m*M* EDTA, pH 8.0. Autoclave and store at room temperature.
6. Equipment: polypropylene microfuge tubes (autoclaved), aerosol resistant tips (Rainin, cat. nos. GP-1000F, GP-200F, GP-20F), and glass capillary tubes (VWR Scientific, cat. no. 53432-728).

2.3.3. Identifying Recombinase and Indicator Transgenics by PCR-Genotyping DNA isolated from Tail Biopsies

1. 10X PCR buffer: 100 m*M* Tris-HCl, pH 9.0, 500 m*M* KCl, 1% Triton X-100 (Promega, cat. no. M190G).

2. 25 m*M* MgCl$_2$ (Promega, cat. no. A351H).
3. *Taq* DNA polymerase, 5 U/µL (Promega, cat. no. M2035).
4. 50X dNTP mixture: 10 m*M* dATP, dTTP, dGTP, and dCTP. Aliquot and store at –20°C.
5. PCR primers stored at –20°C as concentrated stocks ≥3 µg/µL) and as working stocks (20 µ*M*). For example, *FRT*7 specific indicator primers:

219 (5′-CTAGAGGATCCCCGGGTACCG-3′),
218 (5′-GCATCGTAACCGTGCATCTGCC-3′),
90 (5′-CAGTTCATTCAGGGCACCGGACAGG-3′),
204 (5′-CACGAGCATCATCCTCTGCATG-3′),
205 (5′-CAGCGACTGATCCACCCAGTCC-3′),
FLP-specific primers:
222 (5′-CCCATTCCATGCGGGGTATCG-3′),
223 (5′-GCATCTGGGAGATCACTGAG-3′) *(18,24,49)*.

2.3.4. Determining the β-Gal Profile of an Activated Indicator Strain by X-Gal Histochemistry on Tissue Cryosections (see **Subheadings 2.1.1.** and **2.2.1.**).

2.4. Generation of a Recombinase-Based Fate Map

2.4.1. Identifying Recombinase–Indicator Double Transgenic Embryos by PCR-Genotyping DNA from Yolk Sac Membranes

1. PBS: 137 m*M* NaCl, 2.6 m*M* KCl, 10 m*M* Na$_2$HPO$_4$, and 1.8 m*M* KH$_2$PO$_4$. Adjust pH to 7.4 with HCl, autoclave, and store at room temperature.
2. Yolk sac lysis buffer: 50 m*M* KCl, 1.5 m*M* MgCl$_2$, 10 m*M* Tris-HCl, pH 8.5, 0.01% gelatin, 0.45% NP40 (Sigma, cat. no. N-6507), 0.45% polyoxyethylene-sorbitan monolaurate (Tween-20). Store at room temperature.
3. 10 mg/mL Proteinase K (store in aliquots at –20°C).
4. Equipment: autoclaved microfuge tubes, aerosol-resistant tips (Rainin, cat. nos. GP-1000F, GP-200F, GP-20F), and microfuge tube covers (or screw cap tubes) to prevent splash contamination.

2.4.2. Detection of Progenitor Cells and Their Descendants by Staining for Heat-Resistant Alkaline Phosphatase Activity

1. Stock solutions: 1 *M* Tris-HCl, pH 9.5; 5 *M* NaCl; 1 *M* MgCl$_2$; 0.5 *M* EDTA, pH 8.0; 20% paraformaldehyde in PBS (store in aliquots at –20°C); 1 *M* levamisole (store in aliquots at –20°C).
2. PBS: 137 m*M* NaCl, 2.6 m*M* KCl, 10 m*M* Na$_2$HPO$_4$, and 1.8 m*M* KH$_2$PO$_4$. Adjust pH to 7.4 with HCl, autoclave, and store at room temperature.
3. PBS-M: 2 m*M* MgCl$_2$ in PBS.
4. 4% Fix: 4% paraformaldehyde in PBS. Prepare fresh from frozen 20% paraformaldehyde-PBS stock. Store at 4°C for up to 24 h.
5. Detection buffer: 100 m*M* Tris-HCl, pH 9.5, 100 m*M* NaCl, and 50 m*M* MgCl$_2$
6. NBT: 100 mg/mL.
7. BCIP: 50 mg/mL.
8. EDTA wash solution: 20 m*M* EDTA (from 0.5 *M* EDTA, pH 8.0 stock) in PBS.
9. Gelvatol or other aqueous mounting media (Vectashield; Vector Laboratories).
10. Equipment: glass staining dishes (Brain Research Laboratories, cat. no. 3125), slide-racks (Brain Research Laboratories, cat. no. 3003), and 24 × 60 mm glass coverslips (Fisher Scientific, cat. no. 12-548-5P).

3. Methods

3.1. Profiling Recombinase Expression by In Situ mRNA Hybridization

3.1.1. Pretreatment and Cryosection of Tissues

1. Dissect embryos or adult tissue in PBS, removing extra-embryonic membranes and opening any cavities to avoid trapping of reagents. Place tissue in cold PBS on ice until all dissections are complete. Wash tissue with cold PBS to remove any residual blood.
2. Replace PBS with 4% fix solution and gently rock at 4°C for 2 h to overnight.
3. Wash three times with PBS for 5 min each to eliminate residual fixative.
4. To preserve tissue architecture on later sectioning, soak tissues in 30% sucrose solution, rocking gently at 4°C for 6 h to overnight.
5. Carefully dip or roll tissue in an aliquot of OCT to remove residual sucrose solution prior to embedding.
6. Submerge and position tissue in an OCT-containing embedding mold. Gently "push out" air bubbles with forceps, and immediately freeze by floating mold in a dry ice/ethanol bath until OCT has turned from clear to white.
7. Carefully remove mold from ice bath, wiping off excess ethanol. Wrap mold tightly in aluminum foil and place in an airtight plastic bag to avoid specimen dehydration. Store indefinitely at –80°C.
8. Cryosection tissue (10–12 µm) and collect on glass slides. Air-dry 2 h to overnight or dry under vacuum with desiccant. Store sections in sealed box at –80°C.

3.1.2. Synthesis of Digoxigenin-Labeled Riboprobes for In Situ Detection of Recombinase mRNA

Digoxigenin-labeled complementary single-stranded (antisense) RNA probes (riboprobes) are transcribed from recombinant plasmids containing the cDNA sequence of interest downsteam from a promoter for either T3, T7, or SP6 RNA polymerase. Synthesis of such probes has been described previously in detail *(69,70)*. First, plasmid DNA is cut downstream of the cDNA sequence of interest by digestion with an appropriate restriction endonuclease. This ensures that the generated riboprobe is free of contaminating vector sequence. Second, riboprobe is transcribed from the truncated plasmid template using either T3, T7, or SP6 RNA polymerase, as determined by the upstream promoter sequence. Riboprobes ranging in size from 0.3 to 1.5 kb have been found to give good signal-to-noise results—short enough to enable thorough penetration to target mRNAs and long enough to give a strong hybridization signal. Probe fragments under 100 bp should be avoided since the melting temperature for hybridization decreases exponentially with fragment length, making it difficult to control the stringency of hybridization *(69,70)*. Sense strand probes provide useful controls on nonspecific background and are synthesized by transcribing in the opposite direction on the template DNA.

Use T3 RNA polymerase and *Eco*RV-digested pBS-FLP-L *(5)* to generate a FLP antisense riboprobe; use T7 RNA polymerase and *Eco*RV-digested pBS-FLP-L to generate control sense probes. Precautions should be taken to avoid any contamination of reagents or equipment with RNases. Sterile disposable plasticware is essentially free of RNases and should be used whenever possible.

1. To generate approx 10 µg of a given riboprobe, mix these reagents in the following order at room temperature: 13 µL DEPC-water, 2 µL 10X transcription buffer, 2 µL 10X NTP

labeling mixture, 1 μL linearized plasmid (1μg/μL), 1 μL RNase inhibitor (40 U/μL), and 1 μL T3, T7, or SP6 RNA polymerase (20 U/μL). Incubate at 37°C for 2 h.

2. Remove a 1-μL aliquot and electrophorese on a 1% agarose gel containing 0.5 μg/mL ethidium bromide (*see* **Note 3**). An RNA band approximately 10-fold more intense than the linearized plasmid band should be seen, indicating that about 10 μg of riboprobe has been synthesized. (Also *see* **Subheading 3.1.2., step 9** for estimating the final concentration of synthesized riboprobe.)

3. To digest plasmid from the riboprobe synthesis reaction (**Subheading 3.1.2., step 1**), add 1 μL DNase I (10 U/μL) and incubate 25 min at 37°C.

4. Add 2 μL 0.5 *M* EDTA.

5. Add 100 μL TE$_{50}$, 10 μL 4 *M* LiCl, and 300 μL 100% ethanol. Mix well and incubate at –20°C for 30 min.

6. Spin out precipitate at approx 10,000*g* at 4°C for 15 min in a microcentrifuge. Discard supernatant while ensuring that the riboprobe pellet remains in the microfuge tube. Wash pellet with 70% ethanol and air-dry.

7. Resuspend pellet in 120 μL TE$_{50}$. Precipitate again by adding 10 μL 4 *M* LiCl, and 300 μL 100% ethanol. Mix well and incubate at –20°C for 30 min. (*see* **Note 4**).

8. Spin out precipitate at approx 10,000*g* at 4°C for 15 min in a microcentrifuge. Discard supernatant while ensuring that the riboprobe pellet remains in the microfuge tube. Wash pellet with 70% ethanol and air-dry.

9. Resuspend pellet in 100 μL Te$_{50}$ to achieve a final concentration of ~0.1 μg/μL and store at –80°C. Use approx 0.5–1 μg of iboprobe per mL of hybridization mixture.

10. To more precisely estimate riboprobe concentration, remove 1- and 5-μL aliquots (approx 100 ng and 500 ng of riboprobe, respectively) from **Subheading 3.1.2., step 8** and electrophorese on a 1% agarose gel containing 0.5 μg/mL ethidium bromide. Next to the riboprobe, load digoxigenin-labeled RNA standards (100, 200, 400, and 800 ng). Compare band intensities to estimate the amount of riboprobe synthesized. **Subheading 3.1.2., steps 2** and **9** can be performed simultaneously using the same agarose gel.

3.1.3. In Situ mRNA Detection

The following protocol has been derived from previously published methods *(70–72)* and has been optimized for detection of *FLP* mRNA. Some steps may need to be altered to best detect other mRNAs of differing abundance and composition. The protocol is written for processing 24 slides (one rack) and can readily be expanded by processing additional slide-filled racks, each starting *see* **Subheading 3.1.3., step 2** at approx 30-min intervals. See **Table 1** for a quick-reference summary of the in situ hybridization protocol. In **Subheading 3.1.3.,** Incubations in **steps 2–10**, **13–24**, **26–28**, and **31–33** are done in plastic staining dishes containing 250 mL of solution; **steps 11–12**, **25**, and **29** and **30** are performed in slide mailers containing 15 mL of solution. To expedite this multistep protocol, label each plastic staining dish (total of 14) as either: PBS, 4% fix, proteinase K, glycine, acetic anhydride, wash I-1, wash I-2, wash I-3, wash II, wash II: wash I (1:1), Rnase A wash III-1, wash III-2, wash III-3, TBST-L, block, or NTMT-L.

In **Subheading 3.1.3., steps 1–12**, precautions should be taken to avoid any contamination of reagents or equipment with RNase. Sterile disposable plasticware is essentially free of RNase and should be used whenever possible; glassware should be baked. Posthybridization washes do not require these precautions, because hybrids formed in situ are resistant to RNase activity in the presence of high salt.

Table 1
In Situ Hybridization Quick-Reference
Summary of Incubations

In situ hybridization	
PBS_{DEPC}	5 min at RT
4% fix	15 min at RT
PBS_{DEPC}	5 min at RT
Proteinase K	5 min at RT
Glycine/PBS_{DEPC}	5 min at RT
PBS_{DEPC}	5 min at RT
4% fix	15 min at RT
Acetic anhydride	10 min at RT
PBS_{DEPC}	5 min at RT
Prehybridization	15 min at 65°C
Hybridization	O/N at 70°C
Prewash (Wash I)	dip at 65°C
Wash I-1	15 min at 65°C
Wash I-2	15 min at 65°C
Wash I-3	15 min at 65°C
Wash II: Wash I (1:1)	10 min at 65°C
Wash II	3×5 min at 65°C
RnaseA	30 min at 37°C
Wash II	5 min at 65°C
Wash III-1	15 min at 65°C
Wash III-2	15 min at 65°C
Wash III-3	15 min at 65°C
TBST-L	3×10 min at RT
Block	30 min at RT
Antibody solution	2 h at RT or O/N
TBST-L	Prewash dip
TBST-L	5 min at RT
TBST-L	4×15 min at RT
NTMT-L	3×5 min at RT
Alkaline phosphatase staining solution	30 min–4 d at RT
To mount	
PBS	10 min at RT
4% fix	15 min at RT
PBS	10 min at RT

RT, room temperature; o/n, overnight

1. Bring the sealed box containing tissue sections from –80°C to room temperature prior to opening; this prevents unwanted water condensation on the tissue sections.
2. Load slides into a rack and place into PBS_{DEPC}, in the "PBS"-labeled staining dish. Incubate 5 min at room temperature.

3. Transfer slides into 4% fix solution in the "4% fix"-labeled staining dish. Incubate 15 min at room temperature, after which discard used fix solution as hazardous waste.

4. Transfer slides to fresh PBS$_{DEPC}$ in the "PBS"-labeled staining dish. Incubate 5 min at room temperature.

5. Transfer slides into 1 µg/mL proteinase K in PBS$_{DEPC}$ (prepared from freshly thawed 10 mg/mL proteinase K stock) in the "proteinase K"-labeled staining dish (*see* **Note 5**). Incubate 5 min at room temperature.

6. To quench proteinase K activity, transfer slides into glycine/PBS$_{DEPC}$ solution in the "glycine"-labeled staining dish. Incubate 5 min at room temperature.

7. Transfer slides to fresh PBS$_{DEPC}$ in the "PBS"-labeled staining dish. Incubate 5 min at room temperature.

8. Refix tissue by transferring slides to fresh 4% fix solution. Incubate 15 min at room temperature.

9. Transfer slides into 0.25% acetic anhydride in 0.1 *M* triethanolamine in the "acetic anhydride"-labeled staining dish. The acetic anhydride should be added to the triethanolamine solution just prior to use. Incubate 10 min at room temperature. Discard acetic anhydride as hazardous waste.

10. Transfer slides to fresh PBS$_{DEPC}$ in the "PBS"-labeled staining dish. Incubate 5 min at room temperature.

11. Prehybridization of tissue sections in slide mailers: transfer slides into slide mailers containing prewarmed (70°C) hybridization buffer (slide mailers hold 5 slides and 15 mL of solution) (*see* **Note 6**). Incubate for 15 min at 70°C.

12. Hybridization: transfer slides into slide mailers containing prewarmed (70°C) hybridization solution with 0.5–1 µg/mL digoxigenin-labeled RNA probe. Seal slide mailers tightly and incubate overnight at 70°C.

13. Transfer slides from mailers to a rack and dip 3 times in prewarmed (65°C) wash buffer I in "wash-I-1" staining dish. Discard used wash buffer as hazardous waste and replace with fresh prewarmed (65°C) wash I-1. Incubate 15 min at 65°C. Discard used wash buffer as hazardous waste.

14. Transfer slides to fresh prewarmed (65°C) wash buffer I in the "washI-2"-labeled staining dish. Incubate 15 min at 65°C.

15. Transfer slides to fresh prewarmed (65°C) wash I in the "wash I-3"-labeled staining dish. Incubate 15 min at 65°C.

16. Transfer slides to fresh prewarmed (65°) Wash II: Wash I (1:1 ratio) in the "Wash II-Wash I"-labeled staining dish. Incubate 10 min at 65°C.

17. Transfer slides to fresh prewarmed (65°C) Wash II in the "Wash II"-labeled staining dish. Incubate 5 min at 65°C. Repeat twice using fresh Wash II.

18. Transfer slides to fresh Rnase A wash in the "Rnase"-labeled staining dish. Incubate 30 min at 37°C.

19. Transfer slides to fresh prewarmed (65°C) Wash II in the "Wash II"-labeled staining dish. Incubate 5 min at 65°C.

20. Transfer slides to fresh prewarmed (65°C) wash III in the "wash III-1"-labeled staining dish. Incubate 15 min at 65°C.

21. Transfer slides to fresh prewarmed (65°C) wash buffer III in the "wash III-2"-labeled staining dish. Incubate 15 min at 65°C.

22. Transfer slides to fresh prewarmed (65°C) wash buffer III in the "wash III-3"-labeled staining dish. Incubate 15 min at 65°C.

23. Transfer slides to TBST-L in the "TBST-L"-labeled staining dish. Incubate 10 min at room temperature. Repeat two additional times using fresh TBST-L.

24. Transfer slides to TBST-L containing 10% heat-inactivated lamb serum in the "block"-labeled staining dish. Incubate 30 min at room temperature.
25. Add 15 mL of antibody (αdigAP) solution into each of an appropriate number of slide mailers. Transfer slides from **step 20** into each mailer. Incubate at room temperature for 2 h or overnight at 4°C.
26. Transfer slides into rack and dip in fresh TBST-L in the "TBST-L"-labeled staining dish.
27. Replace with fresh TBST-L and incubate 5 min at room temperature.
28. Replace with fresh TBST-L and incubate 15 min at room temperature. Repeat 3 additional times using fresh TBST-L.
29. Transfer slides into fresh NTMT-L in the "NTMT-L"-labeled staining dish. Incubate 5 min at room temperature. Repeat two additional times.
30. Transfer slides into slide mailers containing freshly prepared alkaline phosphatase staining solution. Keep in the dark as much as possible by wrapping containers in aluminum foil. Incubate at room temperature until desired degree of color development has occurred (a few hours to a few days depending on the abundance of the mRNA under detection). Change alkaline phosphatase staining solution after the first 8 h and then every 24 h thereafter to avoid unwanted precipitate (*see* **Note 8**).
31. When the color has developed to the desired extent, wash twice in NTMT for 5 min each.
32. Transfer slides to rack and place in PBS for 10 min at room temperature.
33. Transfer slides to 4% fix solution and incubate 15 min at room temperature. Discard fix solution as hazardous waste.
34. Transfer slides to fresh PBS and incubate 10 min at room temperature.
35. Remove slides one at a time, dabbing sides with a Kimwipe to remove excess PBS. Mount slides in aqueous media such as prewarmed (37°C) (*see* **Note 9**) gelvatol (*see* **Subheading 3.1.4.** below) or Vectashield. Apply approximately three drops of the prewarmed mounting media per slide. Gently place glass coverslip on top, being careful not to introduce bubbles. Press down on the coverslip lightly to remove excess gelvatol and small air bubbles.
36. Air-dry slides until gelvatol is firm. Wipe slides clean of excess mounting media and examine by light microscopy. Slides can be stored at –80°C to prevent any unwanted additional color development.

3.1.4. Preparation of Gelvatol Aqueous Mounting Media

The following protocol, modified from Rodriguez and Deinhardt *(73)*, makes approximately 1.25 L of gelvatol.

1. Adjust a 10 mM KH$_2$PO$_4$ solution to pH 7.2 by adding 10 mM Na$_2$HPO$_4$. Prepare 500 mL.
2. Place a 6-inch stir bar into a 2-L beaker. Add 500 mL of 10 mM KH$_2$PO$_4$/Na$_2$HPO$_4$ solution and 4.1 g NaCl (0.14 M NaCl final concentration). Stir at room temperature.
3. Slowly add 125 g of polyvinyl alcohol resin. Stir at room temperature for 3 h.
4. Remove stir bar and microwave mixture until it reaches a boil (about 4 min).
5. Replace stir bar and mix at room temperature overnight.
6. While stirring, gradually add 525 mL glycerol. The solution will become very viscous. Continue stirring mixture at room temperature for 3 d.
7. To remove undissolved particles, aliquot into 250-mL polypropylene centrifuge bottles and spin in Sorvoll GSA rotor at 16,000g at room temperature for 30 min (*see* **Note 10**).
8. Pool gelvatol supernatants and readjust the pH to between 6.0 and 7.0 using either 10 mM KH$_2$PO$_4$ or 10 mM Na$_2$HPO$_4$ as needed
9. Aliquot gelvatol into polypropylene tubes (50 mL size) and store at 4°C.

3.2. Profiling Recombinase Expression by Crossing Recombinase Mice to a lacZ Indicator Strain

3.2.1. Detection of β-Gal Activity on Tissue Cryosections by X-Gal Histochemistry

Process staged embryos and adult tissues isolated from recombinase–indicator double transgenics as described in **Subheading 3.1.1.** with the following modifications:

A. In **step 2**, gently rock embryos or adult tissues in 0.2% fix solution (rather then 4% fix solution, *see* **Note 11**) at 4°C for 1 h to overnight. Fixation times will vary with the size and nature of the tissue. Early- to mid-gestation mouse embryos should be exposed to the fixative for shorter time periods (e.g., 1 h), while late-stage embryos and adult tissues should be exposed for longer time periods (e.g., 4 h to overnight). In general, shorter fixation times are preferable, as X-gal staining may be decreased by lengthy fixation.
B. Collecting thicker cryosections (60 μm as opposed to 10 μm) allows more tissue to be analyzed per given section.

All histochemistry is performed in glass staining dishes: one for the fixative, one for the reaction buffer, and one for the X-gal staining solution. The following protocol is for 30 slides (one rack) and can readily be expanded by processing additional racks at approximately 10-min intervals.

1. Bring a sealed box containing tissue sections from –80°C to room temperature prior to opening; this prevents unwanted water condensation on the tissue sections.
2. Load slides into rack and place into cold 2% fix solution for 10 min on ice. Do not discard used fix solution; store at 4°C until **step 6**.
3. Transfer slides into cold rinse solution for 10 min on ice.
4. Transfer slides to X-gal staining solution. Protect from light by wrapping staining dish with aluminum foil and incubate at 37°C until color has developed to the desired extent (typically 2–48 h). Optional: place slides in X-gal staining solution at 4°C for an additional 24 h to maximize precipitation of the blue X-gal product.
5. Place slides in PBS-M for 5 min at room temperature to remove any residual X-gal staining solution.
6. Refix slides by placing in the 2% fix solution from **step 2** for 5 min at room temperature. Discard fix solution as hazardous waste.
7. Transfer slides to PBS-M to remove any residual fix solution. Coverslip tissue using aqueous mounting media such as prewarmed gelvatol (37°C) (*see* **Note 10**) or Vectashield as in **Subheading 3.1.3.**
8. Air-dry until gelvatol is firm. Wipe slides clean of excess mounting media and examine by light microscopy.

As an alternative approach to X-gal detection of β-gal activity, antibody detection of β-gal can also be employed. We have had success using a rabbit polyclonal antibody to β-gal at a 1:500 dilution. As a secondary antibody, we use either lissamine rhodamine B sulfonyl chloride (LRSC)-conjugated goat anti-rabbit IgG (Jackson Laboratories, cat. no. 111-295-144) at 1:200 dilution or fluorescein isothiocyanate (FITC)-conjugated goat anti-rabbit IgG (Jackson Laboratories, cat. no. 111-095-144) at 1:200 *(74)*.

3.3. Determining the Utility of an Indicator Strain

Before using an indicator strain for lineage studies, it is important to determine the expression profile of the activated reporter molecule. This profile determines the range of cell types that can be mapped reliably by the indicator strain, establishing its overall utility. To unambiguously determine the scope of the potential indicator profile, it is necessary to generate a derivative mouse strain, in which every cell harbors a recombined copy of the indicator transgene, while maintaining the chromosomal environment of the starting (unrecombined) transgene. As described below in **Subheading 3.3.1.**, this derivative strain is generated by germ line transmission of the recombined indicator transgene. Means to genotype this derivative strain are presented in **Subheadings 3.3.2.** and **3.3.3.** Determination of the expression profile of the recombined indicator is described in **Subheading 3.3.4.**

The following protocols were worked out for assessing our *FRT*-disrupted *lacZ* indicator strain, *Hmgcr*::FRTZ *(18)*, and involve using the deleter strain *hACTB* :: FLPe.9205 *(18,24)* and standard X-gal detection assays. The same general strategy applies to either *FRT*- or *loxP*-disrupted indicator strains harboring other reporter genes such as PLAP or EGFPF.

3.3.1. Germ Line Transmission of an Activated Indicator Transgene

1. Cross deleter (e.g., *hACTB*::FLPe.9205 *(18)*) and indicator (e.g., *Hmgcr*::FRTZ *[18]*) mice.
2. Identify double transgenic *(hACTB::FLPe.9205/ Hmgcr::FRTZ)* F1 progeny by PCR-genotyping DNA isolated from tail biopsies (*see* **Subheadings 3.3.2.** and **3.3.3.** below). These mice will harbor recombined activated indicator transgenes (*Hmgcr*::*FRTZ-A*) in their germ cells.
3. To generate the control derivative mouse strain (*Hmgcr*::*FRTZ-A*), in which every cell harbors a recombined copy of the indicator transgene, outcross double transgenic F1 mice to nontransgenic wild-type mice. By PCR-genotyping DNA isolated from tail biopsies, identify progeny harboring the recombined indicator transgene, *Hmgcr*::*FRTZ-A*, in the absence of the recombinase transgene. These *Hmgcr*::*FRTZ-A* mice constitute the derivative control strain.

3.3.2. DNA Extraction from Tail Biopsies

This protocol yields high-quality genomic DNA, suitable for both PCR and Southern hybridization analyses. All solutions should be prepared using PCR-grade reagents and dispensed using aerosol-resistant pipet tips to minimize unwanted contamination of samples with other plasmid or genomic DNA.

1. To a 0.5-cm biopsy add 500 μL STES buffer and 10 μL proteinase K (from a freshly thawed 10 mg/mL stock). Mix well and incubate at 55°C overnight or until tissue is digested.
2. Add 500 μL phenol:chloroform solution to each sample. Mix well (do not vortex samples from this step onward, as this may sheer the genomic DNA) and spin for 5 min at 10,000*g* at room temperature in a microcentrifuge. Transfer the top aqueous layer containing the genomic DNA to a fresh microfuge tube. Discard the bottom layer as organic waste.
3. Add 500 μL chloroform:isoamyl alcohol (24:1). Mix well and spin for 5 min at 10,000*g* at room temperature in a microcentrifuge. Transfer the top aqueous layer containing the genomic DNA to a fresh microfuge tube. Discard the bottom layer as organic waste.

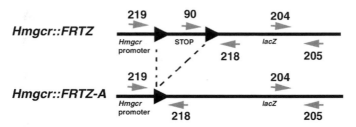

Fig. 1. Schematic example of PCR-genotyping primers capable of distinguishing the indicator transgene, *Hmgcr*::*FRTZ*, from its derivative, recombined product, *Hmgcr*::*FRTZ-A* ("A" denoting that this transgene configuration is now activated to express *lacZ*). Black line, transgene; black triangles, FRT sites; gray arrowheads, primers; dashed lines, the FLP-mediated excision. PCR amplification using primers 219 and 218 yield a 1.6-kb product for the *Hmgcr*::*FRTZ* transgene and a 0.45-kb product for the recombined *Hmgcr*::*FRTZ-A* transgene. Primers 90 and 218 amplify a 0.85-kb product for the *Hmgcr*::*FRTZ* transgene. Primers 204 and 205, directed against the shared *lacZ* sequence, amplify a 0.6-kb product for both *Hmgcr*::*FRTZ* and *Hmgcr*::*FRTZ-A*.

4. Add 40 µL 3 *M* NaOAc and 1 mL ice cold 100% ethanol. Mix well. Strands of genomic DNA will immediately precipitate. Scoop-out the DNA on the sealed and hooked-end of a glass capillary tube. Briefly air-dry 1 to 2 min.
5. Resuspend genomic DNA in 70 µL of TE (the final DNA concentration will be roughly 0.5 µg/µL). If the DNA does not readily release from the glass capillary, break off DNA-containing glass tip directly into the TE. Store genomic DNA at 4°C.
6. Use approx 0.5–1 µL of tail DNA per 25 µL PCR.

3.3.3. Identifying Recombinase and Indicator Transgenics by PCR-Genotyping DNA Isolated from Tail Biopsies

To distinguish the indicator transgene, *Hmgcr*::*FRTZ*, from the recombined derivative, *Hmgcr*::*FRTZ-A*, we typically analyze tail DNA using three distinct PCR amplifications (*see* **Fig. 1**):

Primer set 219/218 (*see* **Subheading 2.3.3.** for primer sequence) amplifies a 1.6-kb product for the *Hmgcr*::*FRTZ* transgene and a 0.45-kb product for the recombined *Hmgcr*::*FRTZ-A* transgene when used under the following cycle conditions.

1. Combine 12.5 µL sterile distilled water, 2.5 µL PCR buffer (10X), 2 µL MgCl$_2$ (25 m*M*), 3 µL primer 218 (20 µ*M* working stock), 3 µL primer 219 (20 µ*M* working stock), 0.5 µL dNTP stock (50×), 0.5 µL *Taq* DNA polymerase (5 U/µL), and 1 µL of tail DNA (0.5 µg/µL).
2. Amplify by cycling at 94°C for 5 min, 58°C for 1 min, and 72°C for 3 min for 1 cycle; 94°C for 1 min, 58°C for 1 min, and 72°C for 3 min. for 39 cycles; 72°C for 10 min (final extension).
3. Load entire reaction on a 1% agarose gel with 0.5 µm/mL ethidium bromide.

Primer set 90/218 (*see* **Subheading 2.3.3.** for primer sequence) amplifies a 0.85-kb product for the *Hmgcr*::*FRTZ* transgene when used under the following cycle conditions. These primers do not amplify the recombined *Hmgcr*::*FRTZ-A* transgene, as the 90 primer site has been removed by FLP-mediated excisional recombination.

1. Combine 12.5 µL sterile distilled water, 2.5 µL PCR buffer (10X), 2 µL MgCl₂ (25 m*M*), 3 µL primer 218 (20 µ*M* working stock), 3 µL primer 90 (20 µ*M* working stock), 0.5 µL dNTP stock (50X), 0.5 µL *Taq* DNA polymerase (5 U/µL), and 1 µL of tail DNA (0.5 µg/µL).
2. Amplify by cycling at 94°C for 5 min, 60°C for 1 min, and 72°C for 1 min. for 1 cycle; 90°C for 1 min, 60°C for 1 min, and 72°C for 1 min, for 39 cycles; 72°C for 10 min (final extension).
3. Load entire reaction on a 1 to 2% agarose gel with 0.5 µg/mL ethidium bromide.

Primer set 204/205 (*see* **Subheading 2.3.3.** for primer sequence) amplifies a 0.6-kb fragment from the *lacZ* sequence shared by both the *Hmgcr*::FRTZ and *Hmgcr*::FRTZ-A transgenes and serves to simply detect the presence of the indicator transgene regardless of configuration. This amplification provides a useful control for interpreting the above PCR results.

1. Combine 12.8 µL sterile distilled water, 2.5 µL PCR buffer (10X), 1.7 µL MgCl₂ (25 m*M*), 3 µL primer 204 (20 µ*M* working stock), 3 µL primer 205 (20 µ*M* working stock), 0.5 µL dNTP stock (50X), 0.5 µL *Taq* DNA polymerase (5 U/µL), and 1 µL of tail DNA (0.5 µg/µL).
2. Amplify by cycling at 94°C for 5 min, 60°C for 1 min, and 72°C for 1 min, for 1 cycle; 90°C for 1 min, 60°C for 1 min, and 72°C for 1 min, for 39 cycles; 72°C for 10 min (final extension).
3. Load entire reaction on a 1 to 2% agarose gel with 0.5 µg/mL ethidium bromide.

Primer set 222/223 (*see* **Subheading 2.3.3.** for primer sequence) amplifies a 0.7-kb product for the FLP recombinase transgene when used under the following cycle conditions.

1. Combine 12.5 µL sterile distilled water, 2.5 µL PCR buffer (10X), 2.0 µL MgCl₂ (25 m*M*), 3 µL primer 222 (20 µ*M* working stock), 3 µL primer 223 (20 µ*M* working stock), 0.5 µL dNTP stock (50X), 0.5 µL *Taq* DNA polymerase (5 U/µL), and 1 µL of tail DNA (0.5 µg/µL).
2. Amplify by cycling at 94°C for 5 min, 65°C for 1 min, and 72°C for 1 min. for 1 cycle; 90°C for 1 min, 65° for 1 min, and 72°C for 1 min, for 39 cycles; 72°C for 10 min (final extension).
3. Load entire reaction on a 1 to 2% agarose gel with 0.5 µg/mL ethidium bromide.

3.3.4. Determining the β-Gal Profile of an Activated Indicator Strain by X-Gal Histochemistry on Tissue Cryosections

Process staged embryos and adult tissues isolated from the activated (derivative) indicator strain as described in **Subheadings 3.1.1.** and **3.2.1.** The resultant β-gal profile determines the range of cell types that can be mapped reliably using the parental (unrecombined) indicator strain, establishing its overall utility.

3.4. Generation of a Recombinase-Based Fate Map

3.4.1. Identifying Recombinase–Indicator Double Transgenic Embryos by PCR-Genotyping DNA from Yolk Sac Membranes

Embryos can be genotyped by PCR amplification of transgene DNA directly from a yolk sac lysate. This is especially useful for early stage embyos (<9.5 dpc), in which the yolk sac is still quite small and the yield of DNA by conventional extraction methods is poor. To minimize contamination of samples with unwanted plasmid or genomic DNA, all solutions should be prepared using PCR-grade reagents, dispensed using aerosol-resistant tips, and stored or incubated in autoclaved microfuge tubes.

1. Isolate yolk sac membranes from each embryo individually, rinse in PBS, and place in microfuge tube on dry ice until all dissections are complete. Store at –20°C.
2. Without thawing the yolk sac tissue, add 100 µL of yolk sac lysis buffer and 1 µL proteinase K to each sample. Mix well and incubate at 55°C for 4 h to overnight or until completely digested.
3. Heat-inactivate the proteinase K by placing samples at 95°C for 15 min (use microfuge tube covers to prevent splash contamination). Store at –20°C
4. Use 1 to 2 µL/25 µL PCR genotyping reaction (*see* **Subheading 3.3.4.** above).

3.4.2. Detection of Progenitor Cells and Their Descendants by Staining for Heat-Resistant Alkaline Phosphatase Activity

Many different indicator mouse strains exist for use with Cre and FLP transgenics, each harboring different reporter molecules (e.g., transgenes encoding either β-gal *(18,39,41,42,48)*, PLAP *(43*, and Dymecki, unpublished data), or EGFP *(46,75)*. We present in **Subheading 3.2.1.** a protocol for β-gal detection, and here we present a method for PLAP detection. **Steps 2–7** are performed in glass staining dishes.

Prepare tissue sections as described in **Subheading 3.1.1.** with the following two modifications and then proceed with **step 1** below:

A. Replace PBS with PBS-M.
B. Fixation (in 4% fix solution) is typically reduced to 2–4 h at 4°C, although extended incubations may help reduce unwanted endogenous alkaline phosphatase activity.

1. Bring sealed box containing tissue sections from –80°C to room temperature prior to opening; this prevents unwanted water condensation on the tissue section.
2. Place slides in rack and submerge in 4% fix solution for 10 min on ice, after which discard fix solution as hazardous waste.
3. Transfer slides to ice-cold PBS-M for 10 min on ice to eliminate residual fixative. Repeat twice.
4. Transfer slides to prewarmed (65°C) PBS-M and incubate at 65°C for 30–90 min. This step will inactivate endogenous (heat-sensitive) alkaline phosphatase activity that would otherwise result in confounding background activity. The duration of this step will vary for different tissues depending on the amount of endogenous alkaline phosphatase activity (*see* **Note 12**).
5. Transfer slides to detection buffer for 30 min at room temperature.
6. Transfer slides to freshly prepared detection buffer containing 0.1 mg/mL BCIP, 1 mg/mL NBT, and 0.5 m*M* levamisole. Protect samples from light by wrapping the staining dish with aluminum foil. Incubate at room temperature until color has developed to the desired extent (typically 1–48 h). Replace staining solution after the first 8 h and then every 24 h thereafter (*see* **Note 8**) to prevent any unwanted precipitate from forming.
7. Transfer slides to EDTA wash solution for 10 min at room temperature. Repeat this step two additional times.
8. Coverslip slides as described in **Subheading 3.1.3.** using aqueous mounting media such as prewarmed (37°C) (*see* **Note 9**) gelvatol (*see* **Subheading 3.1.4.**) or Vectashield. EDTA (20 m*M*) may be included in the mounting media to inhibit further development of the alkaline phosphatase reaction.

9. Air-dry slides in a dark place until gelvatol is firm. Wipe slides clean of excess mounting media and examine by light microscopy. Slides may be stored at −80°C to prevent further development of the alkaline phosphatase reaction.

4. Notes

1. To inactivate any phosphatases present in the sheep serum that could cause nonspecific background, incubate at 70°C for 30 min, swirling every few minutes to prevent hardening. Aliquot and store at −80°C.
2. While most X-gal staining reactions use a reaction buffer comprised of 100 mM Na-phosphate, pH 7.4, 2 mM MgCl$_2$, 0.01% Na-deoxycholate, 0.02% NP40, we have found that increasing the concentration of Na-deoxycholate and NP40 10-fold (to 0.1 and 0.2%, respectively) can improve the β-gal detection without compromising tissue architecture.
3. To ensure that the aliquot of riboprobe is not degraded during the electrophoresis, precautions should be taken to avoid any contamination of the electrophoresis buffer and apparatus with RNases. Electrophoresis tanks suspected of contamination should be cleaned with a detergent solution, rinsed well in water, dried with ethanol, and then filled with a solution of 3% H$_2$O$_2$. After 10 min at room temperature, the electrophoresis tank should be rinsed thoroughly with DEPC-water *(76)*.
4. We have observed a significant reduction in nonspecific background staining when riboprobes are precipitated twice with LiCl and 100% ethanol.
5. Riboprobe penetration is increased by treating tissue sections with low concentrations of proteinase K (1–3 μg/mL). It is useful to vary both the concentration and duration of this treatment in order to determine the conditions that yield the best hybridization signal.
6. Prehybridization and hybridization solutions (containing probe) can be stored in slide mailers at −80°C and reused several times. With some probes, the degree of nonspecific background staining will decrease as the hybridization solution is reused.
7. While recommended, levamisole can be eliminated here and in all subsequent steps of Subheading **3.1.3.** without deletions increases in nonspecific alkaline phosphatase activity.
8. To prevent crystal deposits on tissue sections, change the alkaline phosphatase staining solution after the first 8 h and then every 24 h thereafter. To ensure that even small crystals are removed, dip slides into fresh NTMT-L and wash the slide mailer with NTMT-L between changes of the alkaline phosphatase staining solution. These steps will minimize nonspecific background staining.
9. Reducing the viscosity of gelvatol by prewarming to 37°C will facilitate mounting applications.
10. Some sediment will collect on tube bottoms after gelvatol centrifugation. Carefully aliquot the gelvatol into 50-mL polypropylene tubes, leaving sediment behind. Store indefinitely at 4°C.
11. While β-gal encoded by the bacterial *lacZ* gene is usually quite stable, enzyme activity can be diminished by over-fixation. Protocols show a range of fixatives including 0.2, 2.0, and 4.0% paraformaldehyde in PBS, as well as 0.2 and 2.0% glutaraldehyde. We have seen consistently strong X-gal precipitate with minimal to no endogenous background β-gal activity when tissue is fixed in 0.2% paraformaldehyde-PIPES buffer.
12. Background staining resulting from endogenous alkaline phosphatase activity can be reduced effectively by heat (the exogenous PLAP encoded by the transgene is heat-resistant). In addition to heating tissue sections, we recommend heating whole tissues to 65°C prior to incubating in the sucrose solution.

References

1. Galileo, D. S., Gray, G. E., Owens, G. C., Majors, J., and Sanes, J. R. (1990) Neurons and glia arise from a common progenitor in chicken optic tectum: demonstration with two retroviruses and cell type-specific antibodies. *Proc. Natl. Acad. Sci. USA* **87,** 458–462.

2. Cepko, C. L., Ryder, E. F., Austin, C. P., Walsh, C., and Fekete, D. M. (1993) Lineage analysis using retrovirus vectors. *Methods Enzymol.* **225,** 933–960.

3. Wetts, R. and Fraser, S. E. (1991) Microinjection of fluorescent tracers to study neural cell lineages. *Development* **Suppl. 2,** 1–8.

4. Le Douarin, N. (1982) *The Neural Crest.* Cambridge University Press, Cambridge.

5. Dymecki, S. and Tomasiewicz, H. (1998) Using Flp-recombinase to characterize expansion of *Wnt1*-expressing neural progenitors in the mouse. *Dev. Biol.* **201,** 57–65.

6. Zinyk, D., Mercer, E. H., Harris, E., Anderson, D. J., and Joyner, A. L. (1998) Fate mapping of the mouse midbrain-hindbrain constriction using a site-specific recombination system. *Curr. Biol.* **8,** 665–668.

7. Lakso, M., Sauer, B., Mosinger, B., Lee, E. J., Manning, R. W., Yu, S.-H., et al. (1992) Targeted oncogene activation by site-specific recombination in transgenic mice. *Proc. Natl. Acad. Sci. USA* **89,** 6232–6236.

8. Orban, P. C., Chui, D., and Marth, J. D. (1992) Tissue- and site-specific DNA recombination in transgenic mice. *Proc. Natl. Acad. Sci. USA* **89,** 6861–6865.

9. Gu, H., Marth, J. D., Orban, P. C., Mossmannand, H., and Rajewsky, K. (1994) Deletion of a DNA polymerase beta gene segment in T cells using cell type-specific gene targeting. *Science* **265,** 103–106.

10. Dymecki, S. M. (1996) Flp recombinase promotes site-specific DNA recombination in embryonic stem cells and transgenic mice. *Proc. Natl. Acad. Sci. USA* **93,** 6191–6196.

11. Sternberg, N. and Hamilton, D. (1981) Bacteriophage P1 site-specific recombination. I. Recombination between loxP sites. *J. Mol. Biol.* **150,** 467–486.

12. Jayaram, M. (1985) Two-micrometer circle site-specific recombination: the minimal substrate and the possible role of flanking sequences. *Proc. Natl. Acad. Sci. USA* **82,** 5875–5879.

13. Senecoff, J. F., Bruckner, R. C., and Cox, M. M. (1985) The FLP recombinase of the yeast 2-μm plasmid: characterization of its recombination site. *Proc. Natl. Acad. Sci. USA* **82,** 7270–7274.

14. McLeod, M., Craft, S., and Broach, J. R. (1986) Identification of the crossover site during FLP-mediated recombination in the Saccharomyces cerevisiae plasmid 2 microns circle. *Mol. Cell. Biol.* **6,** 3357–3367.

15. Argos, P., Landy, A., Abremski, K., Egan, J. B., Haggard-Ljungquist, E., Hoess, R. H., et al. (1986) The integrase family of site-specific recombinases: regional similarities and global diversity. *EMBO J.* **5,** 433–440.

16. Stark, W. M., Boocock, M. R., and Sherratt, D. J. (1992) Catalysis by site-specific recombinases. *Trends Genet.* **8,** 432–439.

17. Kilby, N. J., Snaith, M. R., and Murray, J. A. H. (1993) Site-specific recombinases: tools for genome engineering. *Trends Genet.* **9,** 413–421.

18. Rodriguez, C. I. and Dymecki, S. M. (2000) Origin of the precerebellar system. *Neuron* **27,** 475–486.

19. Lee, K. J., Dietrich, P., and Jessell, T. M. (2000) Genetic ablation reveals that the roof plate is essential for dorsal interneuron specification. *Nature* **403,** 734–740.

20. Sauer, B. (1993) *Manipulation of transgenes by site-specific recombination: use of Cre recombinase.* Academic Press, San Diego.

21. Torres, R. M. and Kuhn, R. (1997) *Laboratory protocols for conditional gene targeting.* Oxford University Press, Oxford.

22. O'Gorman, S., Fox, D. T., and Wahl, G. M. (1991) Recombinase-mediated gene activation and site-specific integration in mammalian cells. *Science* **251,** 1351–1355.
23. Dymecki, S. M. (1996) A modular set of *Flp, FRP*, and *lacZ* fusion vectors for manipulating genes by site-specific recombination. *Gene* **171,** 197–201.
24. Rodriguez, C. I., Buchholtz, F., Galloway, J., Sequerra, R., Kasper, J., Ayala, R., et al. (2000) High-efficiency deleter mice show that FLPe is an alternative to Cre-*loxP*. *Nat. Genet.* **25,** 139–140.
25. Kozak, M. (1987) An analysis of 5'-noncoding sequences from 699 vertebrate messenger RNAs. *Nucleic Acids Res.* **15,** 8125–8148.
26. Brinster, R. L., Allen, J. M., Behringer, R. R., Gelinas, R. E., and Palmiter, R. D. (1988) Introns increase transcriptional efficiency in transgenic mice. *Proc. Natl. Acad. Sci. USA* **85,** 836–840.
27. Chaffin, K. E., Beals, C. R., Wilkie, T. M., Forbush, K. A., Simon, M. I., and Perlmutter, R. M. (1990) Dissection of thymocyte signaling pathways by in vivo expression of pertussis toxin ADP-ribosyltransferase. *EMBO J.* **9,** 3821–3829.
28. Chung, J. H., Whiteley, M., and Felsenfeld, G. (1993) A 5' element of the chicken β-globin domain serves as an insulator in human erythroid cells and protects against position effect in *Drosophila. Cell* **74,** 505–514.
29. Pikaart, M. J., Recillas-Targa, F., and Felsenfeld, G. (1998) Loss of transcriptional activity of a transgene is accompanied by DNA methylation and historic deacetylation and is prevented by insulators. *Genes Dev.* **12,** 2852.
30. Hasty, P., Abuin, A., and Bradley, A. (2000) Gene targeting, principles, and practice in mammalian cells, in *Gene Targeting: A Practical Approach* (Joyner, A. L., ed.), Oxford University Press, Oxford, pp. 1–35.
31. Meyers, E. N., Lewandoski, M., and Martin, G. R. (1998) An *Fgf8* mutant allelic series generated by Cre- and Flp-mediated recombination. *Nat. Genet.* **18,** 136–141.
32. Lerner, A., D'Adamio, L., Diener, A. C., Clayton, L. K., and Reinherz, E. L. (1993) CD3 zeta/eta/theta locus is colinear with and transcribed antisense to the gene encoding the transcritpion factor Oct-1. *J. Immunol* **151,** 3152–3162.
33. Jacks, T., Shih, T. S., Schmitt, E. M., Bronson, R. T., Bernards, A., and Weinberg, R. A. (1994) Tumour predisposition in mice heterozygous for a targeted mutation in *Nf1. Nat. Genet.* **7,** 353–361.
34. Ohno, H., Goto, S., Taki, S., Shirasawa, T., Nakano, H., Miyatake, S., et al. (1994) Targeted disruption of the CD3 eta locus causes high lethality in mice: modulation of Oct-1 transcription on the opposite strand. *EMBO J.* **13,** 1157–1165.
35. Carmeliet, P., Ferreira, V., Breier, G., Pollefeyt, S., Kieckens, L., Gertsenstein, M., et al. (1996) Abnormal blood vessel development and lethality in embryos lacking a single *VEGF* allele. *Nature* **380,** 435–439.
36. Schwenk, F., Sauer, B., Kukoc, N., Hoess, R., Muller, W., Kocks, C., et al. (1997) Generation of Cre recombinase-specific monoclonal antibodies, able to characterize the pattern of Cre expression in cre-transgenic strains. *J. Immunol. Methods* **207,** 203–212.
37. Gagneten, S., Le, Y., Miller, J., and Sauer, B. (1997) Brief expression of a GFP cre fusion gene in embryonic stem cells allows rapid retrieval of site-specific genomic deletions. *Nucleic Acids Res.* **25,** 3326–3331.
38. Feng, G., Mellor, R. H., Bernstein, M., Keller-Peck, C., Nguyen, Q. T., Wallace, M., et al. (2000) Imaging neuronal subsets in transgenic mice expressing multiple spectral variants of GFP. *Neuron* **28,** 41–54.
39. Araki, K., Araki, M., Miyazaki, J.-I., and Vassalli, P. (1995) Site-specific recombination of a transgene in fertilized eggs by transient expression of Cre recombinase. *Proc. Natl. Acad. Sci. USA* **92,** 160–164.
40. Tsien, J. Z., Chen, D. F., Gerber, D., Tom, C., Mercer, E. H., Anderson, D. J., et al. (1996) Subregion- and cell type-restricted gene knockout in mouse brain. *Cell* **87,** 1317–1326.

41. Michael, S. K., Brennan, J., and Robertson, E. J. (1999) Efficient gene-specific expression of cre recombinase in the mouse embryo by targeted insertion of a novel IRES-Cre cassette into endogenous loci [published erratum appears in Mech Dev 1999 Aug;86(1–2):213]. *Mech. Dev.* **85,** 35–47.

42. Soriano, P. (1999) Generalized *lacz* expression with ROSA26 Cre reporter strain. *Nat. Genet.* **21,** 70–71.

43. Lobe, C. G., Koop, K. E., Kreppner, W., Lomeli, H., Gertsenstein, M., and Nagy, A. (1999) Z/AP, a double reporter for Cre-mediated recombination. *Dev. Biol.* **208,** 281–292.

44. Maxwell, I. H., Harrison, G. S., Wood, W. M., and Maxwell, F. (1989) A DNA cassette containing a trimerized SV40 polyadenylation signal which efficiently blocks spurious plasmid-initiated transcription. *BioTechniques* **7,** 276–280.

45. Niwa, H., Yamamura, K., and Miyazaki, J. (1991) Efficient selection for high-expression transfectants with a novel eukaryotic vector. *Gene* **108,** 193–200.

46. Novak, A., Guo, C., Yang, W., Nagy, A., and Lobe, C. G. (2000) Z/EG, a double reporter mouse line that expresses enhanced green fluorescent protein upon cre-mediated excision. *Genesis* **28,** 147–155.

47. Zambrowicz, B. P., Imamoto, A., Fiering, S., Herzenberg, L. A., Kerr, W. G., and Soriano, P. (1997) Disruption of overlapping transcripts in the ROSA βgeo 26 gene trap strain leads to widespread expression of β-galactosidase in mouse embryos and hematopoietic cells. *Proc. Natl. Acad. Sci. USA* **94,** 3789–3794.

48. Mao, X., Fujiwara, Y., and Orkin, S. H. (1999) Improved reporter strain for monitoring Cre recombinase-mediated DNA excisions in mice. *Proc. Natl. Acad. Sci. USA* **96,** 5037–5042.

49. Farley, F., Soriano, P., Steffen, L., and Dymecki, S. (2000) Widespread recombinase expression using FLPeR (flipper) mice. *Genesis* **28,** 106–110.

50. Fields-Berry, S. C., Halliday, A. L., and Cepko, C. L. (1992) A recombinant retrovirus encoding alkaline phosphatase confirms clonal boundary assignment in lineage analysis of murine retina. *Proc. Natl. Acad. Sci. USA* **89,** 693–697.

51. Golden, J. A., and Cepko, C. L. (1996) Clones in the chick diencephalon contain multiple cell types and siblings are widely dispersed. *Development* **122,** 65–78.

52. Chen, C. M., Smith, D. M., Peters, M. A., Samson, M. E., Zitz, J., Tabin, C. J., and Cepko, C. L. (1999) Production and design of more effective avian replication-incompetent retroviral vectors. *Dev. Biol.* **214,** 370–384.

53. Jiang, W. and Hunter, T. (1998) Analysis of cell-cycle profiles in transfected cells using a membrane- targeted GFP. *BioTechniques* **24,** 349–354.

54. Finley, K. D., Edeen, P. T., Foss, M., Gross, E., Ghbeish, N., Palmer, R. H., et al. (1998) Dissatisfaction encodes a tailless-like nuclear receptor expressed in a subset of CNS neurons controlling Drosophila sexual behavior. *Neuron* **21,** 1363–1374.

55. Buchholz, F., Angrand, P.-O., and Stewart, A. F. (1998) Improved properties of Flp recombinase evolved by cycling mutagenesis. *Nat. Biotechnol.* **16,** 657–662.

56. Buchholz, F., Ringrose, L., Angrand, P.-O., Rossi, F., and Stewart, A. F. (1996) Different thermostabilities of FLP and Cre recombinases: implications for applied site-specific recombination. *Nucleic Acids Res.* **24,** 4256–4262.

57. Groth, A. C., Olivares, E. C., Thyagarajan, B., and Calos, M. P. (2000) A phage integrase directs efficient site-specific integration in human cells. *Proc. Natl. Acad. Sci. USA* **97,** 5995–6000.

58. Shaikh, A. C. and Sadowski, P. D. (2000) Chimeras of the Flp and Cre recombinase: tests of the mode of cleavag by Flp and Cre. *J. Mol. Biol.* **302,** 27–48.

59. Kistner, A., Gossen, M. Z. F., Jerecic, J., Ullmer, C., Lybbert, H., and Bujard, H. (1996) Doxycycline-mediated quantitative and tissue-specific control of gene expression in transgenic mice. *Proc. Natl. Acad. Sci. USA* **93,** 10933–10938.

60. Holzenberger, M., Zaoui, R., Leneuve, P., Hamard, G., and Le Bouc, Y. (2000) Ubiquitous postnatal loxP recombination using a doxycycline auto-inducible Cre transgene (DAI-Cre). *Genesis* **26,** 157–159.

61. Logie, C. and Stewart, F. (1995) Ligand-regulated site-specific recombination. *Proc. Natl. Acad. Sci. USA* **92,** 5940–5944.

62. Metzger, D., Clifford, J., Chiba, H., and Chambon, P. (1995) Conditional site-specific recombination in mammalian cells using a ligand-dependent chimeric Cre recombinase. *Proc. Natl. Acad. Sci. USA* **92,** 6991–6995.

63. Feil, R., Brocard, J., Mascrez, B., LeMeur, M., Metzger, D., and Chambon, P. (1996) Ligand-activated site-specific recombination in mice. *Proc. Natl. Acad. Sci. USA* **93,** 10887–10890.

64. Kellendonk, C., Tronche, F., Monaghan, A.-P., Angrand, P.-O., Stewart, F., and Schutz, G. (1996) Regulation of Cre recombinase activity by the synthetic steroid RU 486. *Nucleics Acids Res.* **24,** 1404–1411.

65. Danielian, P. S., Muccino, D., Rowitch, D. H., Michael, S. K., and McMahon, A. P. (1998) Modification of gene activity in mouse embryos in utero by a tamoxifen-inducible form of Cre recombinase. *Curr. Biol.* **8,** 1323–1326.

66. Schwenk, F., Kuhn, R., Angrand, P. O., Rajewsky, K., and Stewart, A. F. (1998) Temporally and spatially regulated somatic mutagenesis in mice. *Nucleic Acids Res.* **26,** 1427–1432.

67. Schena, M., Shalon, D., Davis, R. W., and Brown, P. O. (1995) Quantitative monitoring of gene expression patterns with a complementary DNA microarray. *Science* **270,** 467–470.

68. Velculescu, V. E., Zhang, L., Vogelstein, B., and Kinzler, K. W. (1995) Serial analysis of gene expression. *Science* **270,** 484–487.

69. Angerer, L. M. and Angerer, R. C. (1992) In situ hybridization to cellular RNA with radiolabelled RNA probes, in *In Situ Hybridization: A Practical Approach* (Wilkinson, D. G., ed.), IRL Press, Oxford, pp. 15–32.

70. Wilkinson, D. G., (1992) Whole mount in situ hybridization of vertebrate embryos, in *In Situ Hybridization: A Practical Approach* (Wilkinson, D. G., ed.), IRL Press, Oxford, pp. 74–83.

71. Riddle, R. D., Johnson, R. L., Laufer, E., and Tabin, C. (1993) Sonic hedgehog mediates the polarizing activity of the ZPA. *Cell* **75,** 1401–1416.

72. Bao, Z. Z. and Cepko, C. L. (1997) The expression and function of Notch pathway genes in the developing rat eye. *J. Neurosci.* **17,** 1425–1434.

73. Rodriguez, J. and Deinhardt, F. (1960) Preparation of a semipermanent mounting medium for fluorescent antibody studies. *Virology* **12,** 316.

74. Nan, X. S., Tate, P., Li, E., and Bird, A. (1996) DNA methylation specifies chromosomal localization of MeCP2. *Mol. Cell. Biol.* **16,** 414–421.

75. Mao, X., Fujiwara, Y., Chapdelaine, A., Yang, H., and Orkin, S. H. (2001) Activation of EGFP expression by Cre-mediated excision in a new ROSA26 reporter mouse strain. *Blood* **97,** 324–326.

76. Sambrook, J., Fritsch, E. F., and Maniatis, T. (1989) Extraction, purification, and analysis of messenger RNA from eukaryotic cells, in *Molecular Cloning: A Laboratory Manual* vol. 1. CSH Laboratory Press, Cold Spring Harbor, N.Y., pp. 7.3.

25

From ES Cells to Mice: *The Gene Trap Approach*

Francesco Cecconi and Peter Gruss

1. Introduction

The gene trap approach, represented in **Fig. 1**, is based on the use of mouse embryonic stem (ES) cells and a class of vectors containing a splice-acceptor site upstream of the reporter and resistance genes, β-galactosidase (β-gal, *lac*Z) and neomycin resistance (*neo*), respectively. Integration of these vectors into a genomic locus, controlled by a functional promoter, results in the generation of a fusion transcript between the endogenous gene and the *lac*Z gene. Fusion mRNAs transcribed from the promoter of the tagged locus mimic endogenous gene expression, which can be monitored by visualizing *lac*Z activity *(1,2)*. The tagged genes can be identified by the use of anchored polymerase chain reaction (PCR) procedure *(3)*, performed on RNA extracted from the selected ES cell clones. In addition, the gene trap vectors could also act as insertional mutagens.

After genetic manipulation by electroporation or retroviral infection (for a review see **ref.** *4*), the cells remain pluripotent, and when reintroduced into morula by embryo-aggregation (or into blastocyst by microinjection), they retain the potential to contribute to all tissues of the mouse including the germ line *(5)*. Using this strategy, the expression patterns of the endogenous genes during embryogenesis and adulthood, as well as the phenotypic consequences of the gene trap mutations, may be analyzed.

One major flaw, of the gene trap approach we have used, is the necessity to generate a large number of mice from the corresponding ES cell clones to obtain few interesting genes. One way to circumvent this problem is offered by the possibility to preselect gene-trapped ES cell lines in vitro before generating the mice. This could be extremely useful when searching for specific classes of genes, for example, genes activated by specific signaling cascades. A simple and reproducible preselection protocol has been set up in our laboratory, in which gene-trapped ES cell lines are screened for the β-gal staining patterns in vitro and tested for their responsiveness to specific growth–differentiation factors such as follistatin, nerve growth factor, and retinoic acid *(6)*.

Huge effort is being carried out worldwide to create gene trap libraries, which could, in principle, saturate the mammalian genome. We believe that the gene trap approach will increase our ability to study gene activity in depth and to associate such genes to human diseases.

From: *Methods in Molecular Biology, vol. 185: Embryonic Stem Cells: Methods and Protocols*
Edited by: K. Turksen © Humana Press Inc., Totowa, NJ

1.Electroporation
of the gene trap cassette
into ES cells

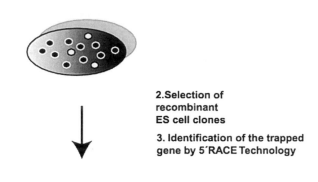

2.Selection of
recombinant
ES cell clones

3. Identification of the trapped
gene by 5´RACE Technology

4. Aggregation of the ES clone with
preimplantation mouse embryos

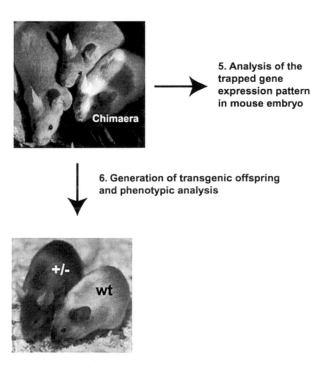

5. Analysis of the
trapped gene
expression pattern
in mouse embryo

6. Generation of transgenic offspring
and phenotypic analysis

Fig. 1. Strategy of the gene trap approach. The schematic represents the strategy used in our laboratory. The transgenic offspring at step 6 is derived by breeding the chimaeric with an inbred strain C57bl/6 or an outbred strain (NMRI). The progeny will be tested for germ line transmission. Hybrid crosses between F1 littermates will give rise to the homozygous specimen.

In this chapter, we describe the "routine" gene trap methodology which has allowed us in the last years to identify, clone, and mutate several genes important for vertebrate development and cell biology *(7–10)*. We will not discuss here the **morula** aggregation technique, extensively described elsewhere *(11)*, but we will focus on ES cells

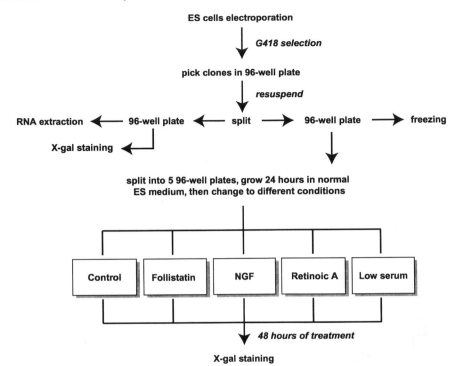

Fig. 2. ES clones in vitro preselection. Diagram of the screening used in our laboratory to preselect gene-trapped ES cell lines with soluble factors.

manipulation (*see* **Fig. 2**) and on the gene expression screening which follows the generation of a chimaeric mouse and precedes the phenotypic analysis of the mutant allele.

2. Materials

1. ES cell medium: Dulbecco's modified Eagle medium (DMEM) (Gibco, cat. no. 41965-039), 100 μM β-mercaptoethanol (Sigma, cat. no. M-7522), 2 mM glutamine (Gibco, cat. no. 25030-024), 1% stock solution of nonessential aminoacids (Gibco, cat. no. 11140-035), 1 mM Na-pyruvate (Gibco, cat. no. 11360-039), 15% (v/v) fetal calf serum (FCS) (Gibco, cat. no. 3703359A), 500 U/mL leukemia inhibitory factor (LIF) (Gibco, cat. no. 13275-029).

2. Phosphate-buffered saline (PBS) filter-sterilized. Composition for 1 L: add distilled water to 8.0 g NaCl, 0.2 g KCl, 1.15 g $Na_2HPO_4 \cdot 2H_2O$ and 0.2 g KH_2PO_4, adjust pH to 7.2, and add distilled water to 1 L.

3. Mitomycine C (Sigma, cat. no. M-0503). Stock concentration is 2 mg/vial diluted in 2 mL 1X PBS, filter-sterilized, and stored at 4°C in the dark. The inactivation medium is made of 10 mL DMEM, 10% FCS, 100 μL mytomicine C stock solution.

4. Gene Pulser II electroporation apparatus (Bio-Rad, cat. no. 165-2105/6).

5. 1X Trypsin-EDTA (Gibco, cat. no. 25300-069).

6. Geneticin (G418) (Gibco, cat. no. 11811-064): 200 μg/mL, dissolved in dH_2O.

7. ES cells fixing solution: 0.2% glutaraldehyde (Sigma, cat. no. G-6257), 2 mM $MgCl_2$, 100 mM K_2HPO_4 (pH 7.4), 5 mM EGTA (Sigma, cat. no. E-4378).

8. ES cells washing solution: 0.01% Na-deoxycholate (Sigma, cat. no. D-5670), 0.02% Igepal CA-630 (Sigma, cat. no. I-3021; chemically indistinguishable from Nonidet P-40), 2 mM $MgCl_2$, 0.1 M K_2HPO_4 (pH 7.4), 5 mM EGTA.

9. 5-Bromo-4-chloro-3-indolyl β-D-galactopyranoside (X-gal) (Sigma, cat. no. B-9146).

10. ES cells staining solution: 0.5 mg/mL X-gal, 10 mM K$_3$[Fe(CN)$_6$], 10 mM K$_4$[Fe(CN)$_6$], 0.01% Na-deoxycholate, 0.02% Igepal CA-630, 5 mM EGTA, 2 mM MgCl$_2$, 0.1 M K$_2$HPO$_4$ (pH 7.4).

11. 2X ES cells freezing medium: for 10 mL, 1 mL dimethyl sulfoxide (DMSO) (Sigma, cat. no. D-8779) cell culture grade, and 2 mL FCS to 7 mL ES cells medium.

12. Dissecting microscope (Leica Wild M3Z).

13. All *trans* retinoic acid (retinoic A) (Sigma, cat. no. R-2625).

14. Follistatin (Sigma, cat. no. F-1175).

15. Nerve growth factor-β (Sigma, cat. no. N-2393).

16. TRIZol (Gibco, cat. no. 15596-018).

17. 1:1:24 Phenol:Chloroform:Isoamylalcohol.

18. Chloroform.

19. Isopropylalcohol.

20. 70% Ethanol.

21. Lauryl sulfate sodium salt (SDS).

22. Ethanol absolute.

23. 5′ RACE System for Rapid Amplification of cDNA Ends kit, Version 2.0 (Gibco, cat. no. 18374-058).

24. 0.1-μm nylon filters (Millipore, cat. no. VCWP 025 00).

25. pGEM-T Vector System (Promega, cat. no. A3610).

26. Tris-EDTA buffer (TE): 10 mM Tris-HCl, pH 8.0, 1 mM EDTA, pH 8.0.

27. 8 mM Ammonium-acetate.

28. Electrocompetent cells ElectroMAX DH5α-E (Gibco, cat. no. 11319-019).

29. Transfer membranes Biodyne A (Pall-Gelman, cat. no. 60101).

30. N,N-Dimethylformamide (Sigma, cat. no. D-8654).

31. Formaldehyde (Sigma, cat. no. F-1635).

32. Glutaraldehyde (Sigma, cat. no. G-5882).

33. X-gal staining solution for tissues (embryos and adult brains):

	10 mL	25 mL	50 mL
1X PBS	9.05	22.625	45.25
X-gal (40 mg/mL in N,N-Dimethylformamide)	0.25	0.625	1.25
K$_3$[Fe(CN)$_6$] (200 mM)	0.25	0.625	1.25
K$_4$[Fe(CN)$_6$] (200 mM)	0.25	0.625	1.25
MgCl$_2$ (100 mM)	0.2	0.5	1.0

34. Fixative solution for tissues (embryos and adult brains):

Fixative A (FixA):	25 mL	50 mL	100 mL
37% Formaldehyde	0.675	1.35	2.7
25% Glutaraldehyde	0.2	0.4	0.8
10% Igepal CA-630	0.05	0.1	0.2
20X PBS	1.25	2.5	5.0
dH$_2$O	22.825	45.65	91.3

Fixative B (FixB):	25 mL	50 mL	100 mL
37% Formaldehyde	0.675	1.35	2.7
25% Glutaraldehyde	0.2	0.4	0.8
10% Igepal CA-630	0.5	1.0	2.0
1% Na-deoxycholate	2.5	5.0	10.0
20X PBS	1.25	2.5	5.0
dH$_2$O	19.875	39.75	79.5

35. Glycerol 15, 30, 50, 70, and 80% in PBS (v/v).

3. Methods

3.1. Electroporation of ES Cells

1. Wash the confluent ES cells with PBS.
2. Add 1X Trypsin-EDTA (2 mL for an 8.5-cm culture dish) and incubate for 5 min at 37°C.
3. Pipet the cells up and down in order to obtain a single-cell suspension.
4. Add an excess of ES cell medium (6 mL for an 8.5-cm culture dish) to stop trypsinization, and mix well.
5. Spin down the cells with a table-top centrifuge at 260g for 5 min.
6. Remove the supernatant and resuspend the cells in new ES cell medium.
7. Split the cells on newly inactivated mouse embryonic fibroblasts (MEF) feeder plates (*11*); *see* **Note 1**).
8. After 24 h, change ES cell medium.
9. Twenty-four hours after changing the medium, trypsinize the cells as described above (**steps 2–5**) and add ES cell medium to 10 mL, mix well, and spin down the cells as above (**step 5**).
10. Aspirate the supernatant and resuspend the cells in 30 mL 1X PBS. Determine the cell number. The optimal number is 1.5×10^7 cells.
11. Spin down the cells as in **step 5** above and resuspend them in 0.8 mL 1X PBS containing 25 μg/mL of linear DNA (*see* **Note 2**). Incubate the suspension at room temperature for 5 min.
12. Pipet the cells gently several times up and down and transfer 0.8 mL of the suspension to one electroporation cuvette, carefully avoiding air bubbles. Electroporate with one pulse of 500 μF and 250 V at room temperature. Incubate at room temperature for 5 min.
13. Transfer the cells to ES medium (to a final vol of 30 mL) and plate them on 7×8.5 cm newly inactivated MEF feeder plates. As a control plate, use cells at the same density electroporated without linear DNA.
14. After 24 h, change ES cell medium to a medium containing G418 for positive selection.
15. Change medium every day for 9 d.
16. Ten days after electroporation, check the control plate, which should not contain any ES cell clones. Large ES cell clones (>1000 cells) can be picked.

3.2. Picking of ES Cell Clones

1. Mark a circle under the clones to be picked.
2. Prepare a 96-well plate with 35 μL 1X trypsin-EDTA in each well.
3. Under a sterile hood and using a dissecting microscope pick, with a P20 Gilson pipet equipped with 10-μL PCR tips, every marked clone individually and transfer to the 96-well plate. Incubate at room temperature.
4. Check trypsinization under the microscope every 2 min, gently shaking the plate. When the clones are disaggregated into small cell clumps, add 65 μL ES medium and pipet several times up and down in order to obtain a single-cell suspension.
5. Transfer 50 μL of each suspension, using a multichannel pipet to two other 96-well plates with newly inactivated MEF feeders in 100 μL ES medium. One will be used for preselection experiments (*see* **Subheading 3.3.**) and for freezing the clones, the second for X-gal staining (*see* **Subheading 3.4.**) and RNA extraction (*see* **Subheading 3.5.1.**).
6. First plate. After 3 d, the clones will be ready for freezing and in vitro preselection. After trypsinizing the cells, from the first 96-well plate, with 35 μL 1X trypsin-EDTA for 5 min, add 2X freezing medium to 25 μL of the cell suspension, mix well, wrap the plate with paper, and store at –80°C in a styrofoam box. The cells will remain viable for several months (*see* **Note 3**).
7. Add 90 μL of ES medium to the remaining 10 μL of suspension.

8. Plate the suspension on a 96-well plate containing newly inactivated MEF feeders in 100 µL ES medium for pre-selection experiments.
9. After 3 d of incubation, expand the cells into five different 96-well dishes without MEF feeders, and culture with normal ES cell medium (*see* **Subheading 3.3.**).

3.3. Screening of Gene Trap ES Cell Lines Responsive to Soluble Factors

1. Let the 96-well plates from **Subheading 3.2., step 9** grow for 24 h. When small colonies are apparent, medium is removed, and the dishes are cultured under different conditions.
2. Add ES medium without LIF according to the following five different conditions, success-fully applied in our laboratory (*see* **Fig. 2** and **Note 4**) (*6*):
   a. ES medium with 20% FCS (control plate).
   b. ES medium with 20% FCS and 150 ng/mL follistatin (follistatin plate).
   c. ES medium with 1% FCS (low serum control plate).
   d. ES medium with 1% FCS and 100 ng/mL NGF (NGF plate).
   e. ES medium with 1% FCS and 0.2 m*M* retinoic A.
3. Culture cells for additional 48 h.
4. Process for X-gal staining according to **Subheading 3.4.**

3.4. X-Gal Staining of ES Cells Clones

1. Second Plate. After 3 d, the clones will be ready for X-gal staining. Trypsinize the cells from the second 96-well plate of **Subheading 3.2., step 5**, with 35 µL 1X Trypsin-EDTA for 5 min.
2. When the clones are disaggregated into small cell clumps, add 65 µL ES medium, and pipet several times up and down in order to obtain a single-cell suspension.
3. Transfer 50 µL of each suspension, using a multichannel pipet, to another two 96-well plates with newly inactivated feeders in 100 µL ES medium.
4. After three d, wash the cells from one plate in 1X PBS.
5. Fix the cells up to 3 min at room temperature in fixing solution.
6. Wash 3 to 4 times in washing solution at room temperature.
7. Incubate in staining solution at 37°C for up to 3 d. Stop the reaction by washing in 1X PBS.
8. Refix for 5 min in 2% formaldehyde in 1X PBS.
9. Wash 3 to 4 times in 1X PBS and store at 4°C.

3.5. Identification of the Tagged Gene by the 5′ RACE Procedure

3.5.1. RNA Extraction from ES Cells

1. After 3 d, trypsinize the cells from the second 96-well plate of **Subheading 3.4., step 3**, with 35 µL 1X trypsin/EDTA for 5 min.
2. When the clones are disaggregated into small cell clumps, add 65 µL ES medium and pipet several times up and down in order to obtain a single-cell suspension.
3. Transfer the suspension with a multichannel pipet to a 24-well plate with newly inactivated feeders in ES medium.
4. After an additional 3 d, expand the cells on a 6-well plate and let them grow to conflu-ence.
5. Wash cells twice with 1X PBS.
6. Add 0.75 mL of Trizol to a single well, swirl the plate for at about 30 sec, and collect the lysate in a 1.5-mL tube, passing it several times through a pipette (*see* **Note 5**).
7. Incubate the samples for 5 min at room temperature.

8. Add 0.14 mL of chloroform, shake the tube vigorously for 15 s, and incubate the sample for 2 to 3 min at room temperature. Centrifuge the sample at 12,000*g* for 15 min at 4°C. Following centrifugation, take the upper aqueous phase and transfer to a new tube.

9. Precipitate the RNA by mixing with isopropylalcohol (0.375 mL). Incubate the sample for 10 min at room temperature and centrifuge at 12,000*g* at 4°C for 10 min. The RNA precipitate will form a gel-like pellet.

10. Discard the supernatant and wash the pellet once with 0.75 mL of 70% ethanol.

11. At the end of the procedure, air-dry the pellet, keeping the lower half of the tube in dry ice. Do not use a vacuum. Dissolve the RNA in 20 μL RNase-free water, 0.5% SDS solution by repeated pipetting and incubating for 5 min at 55°C with gentle shaking.

3.5.2. The 5′ RACE Procedure

We have used the protocol provided from the Gibco Instruction manual *5′ Race System for rapid amplification of cDNA ends, Version 2.0*, derived from Frohman et al., *(3)* with some modifications:

1. Superscript II RT reaction. Problems with 5′ RACE due to the secondary structure of the target RNA can be easily avoided by increasing the temperature of the reverse transcription (RT) reaction (Superscript II RT; Gibco) (*see* **Note 6**). The gene trap approach is aimed to identify novel mRNAs whose structural features are unknown to the investigators. Therefore, during the first strand synthesis step, we shifted the primer/RNA mix directly from 70°C to 50°C and prewarmed the complete reaction mixture to 50°C, before adding it to the primer and RNA.

2. cDNAs size selection. To size-select cDNAs, in order to amplify only the most sequence-informative fragments, the reactions were spotted after dC-tailing on 0.1-μm nylon filters. Microdialysis took place against TE for 3 h, allowing the discharge of DNA fragments less than 200 bp. Size-selected cDNAs were then taken through two rounds of nested PCR amplification and microdialyzed after each round.

3. Isolation of 5′ RACE products. Extract the PCR samples twice with phenol: chloroform:isoamylalcohol, once with chloroform, and precipitate the RACE products on ice for 10 min by adding 50 μL TE, 33 μL of 8 *M* ammonium-acetate, and 330 μL of ethanol. Centrifuge for 10 min at room temperature at 12,000*g*, and wash the pellet with 70% ethanol. Dry the pellet and resuspend in 20 μL TE.

4. Cloning of 5′ RACE products (*see* **Note 7**). We have used the pGEM-T Vector system from Promega, based on the concept of T/A overhang *(12)*. Set up the ligation reaction as described below:

5. Use 0.5-mL tubes with low DNA-binding capacity.

T4 DNA ligase 10X buffer	1 μL
PGEM-T Vector	1 μL
PCR product	2 μL
T4 DNA Ligase (1 U/μL)	1 μL
dH$_2$O to a final volume of	10 μL

6. Incubate overnight at 15°C.

7. Remove a 2-μL aliquot for the transformation step.

8. Transform into high efficiency competent cells.

9. Screen colonies for the positive inserts by colony hybridization with a labeled internal oligonucleotide.

10. Sequence >20 clones and confirm the reliability of the clones by comparing with the most common databases and excluding clones with spurious sites of fusions.

3.6. Production of Mouse ES Cells Chimaeras by Morula Aggregation

The generation of ES cell-derived chimaeric mice by blastocyst injection or morula aggregation is not described in this chapter. For extensive and accurate descriptions, *see* **refs.** *11,13–15*. Upon transfer to a pseudopregnant recipient, the ES cells participate in normal development of the chimaeric embryo and contribute to all cell types, including the germ line. Once germ line chimaeras are generated, they represent sources for heterozygous mice carrying the inserted trapping vector. The activity of β-gal can be detected in whole embryos and in sectioned postnatal (pn) tissues by a chemical reaction that involves the cleavage of the X-gal substrate by the β-gal (*lac*Z product) encoded by the transgene. This reaction leads to a blue coloring of positive tissues, as described in the next section.

3.7. Screening for lacZ Transgene Expression during Embryogenesis

1. Preparation: embryos or tissues are dissected in cold 1X PBS (4°C).
2. Embryos are fixed according to **Table 1**.
3. Postnatal brains (7–10 pn) are fixed for 30 min in FixB, embedded in paraffin wax, cut in 100-μm sections, fixed for an additional 1 h in fresh FixB. Older brains (more than 3 mo) are prefixed for 45 min, and then sectioned and refixed in fresh FixB for additional 60 min. Fixation should occur always on ice with gentle shaking, and the fixative should be freshly made.
4. Washing and X-gal staining: The embryos or tissues are washed twice for 20 min in 1X PBS at room temperature.
5. Add the staining solution. The staining is performed overnight in the dark at 30°C.
6. Clearing: The staining is stopped with two washes of 1X PBS. The embryos are then cleared by subsequent washes in 15, 30, 50, 70, and 80% glycerol-1X PBS. Each washing step should be done for 1 d.
7. Staining patterns are then classified for both spatial and temporal differences between mouse lines, with particular attention to the domain of interest of the laboratory. Because this method utilizes an extremely simple enzymatic reaction, the full characterization of the expression pattern of the captured gene can be completed in an efficient manner (*see* **Note 8**).

4. Notes

1. To prepare the MEF feeder, proceed as follows:
   1.1. Remove about ten embryos at d 13 postcoitum (p.c.). Remove and discard the head, the liver, and the internal organs. Remove as much blood as possible by washing the bodies twice in 1X PBS.
   1.2. Cut the bodies into small pieces, and wash over a sieve with 1X PBS.
   1.3. Incubate in 50 mL 1X trypsin-EDTA solution in an erlenmeyer flask containing glass beads for 30 min at 37°C under gentle agitation on a magnetic stirrer.
   1.4. Add other 50 mL of 1X trypsin-EDTA solution and let shake for additional 30 min.
   1.5. Pour the MEF suspension over a sieve, collect in 50-mL Falcon tubes, and centrifuge for 10 min at 260*g*.
   1.6. Plate 5×10^6 cells/150-mm plate. After 24 h, change the medium. When semiconfluent, freeze the cells (1 plate/1 vial) as described in **Subheading 3.2., step 6**.
   1.7. For inactivation, revive 1 vial of MEFs in a 100-mm plate.
   1.8. After about 48 h, split to 5×100-mm plates.
   1.9. After about 48 h, split each 100-mm plate to a 150-mm plate.

Table 1
Fixation of Embryos from Different Developmental Stages

Embryonic age	Fixation time	Fixative solution
Day 9.5 p.c.	30 min	FixA
Day 10.5 p.c.	30 min	FixA
Day 11.5 p.c.	50 min	FixA
Day 12.5 p.c.	50 min	FixA
Day 13.5 p.c.	30 min	FixA; cut in the middle
	30 min	Fresh FixA
Day 15.5 p.c.	30 min	FixA; cut in the middle
	50 min	Fresh FixA

1.10. When confluent, treat with mitomycin C. Remove the medium and add 10 mL inactivation medium to each 150-mm plate. Incubate about 2.5 h, change the medium 3 times. After 24 h, inactivated MEFs should be trypsinized and plated as follows: 60-mm plates, 2×10^6 cells; 100-mm plates, 4.5×10^6 cells.

1.11. When the MEFs are adherent (after about 5 h), the ES cells could be plated over the feeder.

2. The rationale behind the use of reporter constructs is to tag and detect *cis*-regulatory sequences by locating the reporter gene within an endogenous gene. The reporter construct may also disrupt endogenous gene function and act as a mutagen. All trapping vectors devised so far consist of modifications on one basic structure. Endogenous genomic sequences are placed next to a reporter gene, which is then brought under their influence. **Fig. 3** shows the schemes of two gene trap vectors used successfully in our laboratory. In those vectors, the 5′ region contains a splice-acceptor site derived from the *engrailed2* (*eng2*) gene (*eng2* intron element [ei] and exon element [ee], respectively). The reporter gene is *lac*Z, and the selectable marker is neomycin phosphotransferase (*neo*) fused together into the *lacZ/neo* gene (βgeo). The presence of an internal ribosomal entry site (IRES) derived from the encephalomyocarditis virus, allows the reporter and resistance gene to be expressed without requiring an in-frame fusion with the coding region of the endogenous gene (**Fig. 3**, IRESβgeo) *(16)*. Salminen et al. *(17)* have constructed a new poly(A) trap vector. In IRESβgalNeo(-pA), the transcription of *neo* is under the control of the β-*actin* constitutive promoter (βa), while the β-gal expression is dependent on the activity of the trapped gene. A *Pax2* splice-donor element is placed at the 3′ of the bi-cistronic structure. When the ES cells are cultured under conditions that allow differentiation, a 3-fold increase in positive-stained clones is observed with IRESβgalNeo(-pA). As a consequence, this vector will be very useful in the search for genes responding to specific regulatory factors in vitro.

3. Cells should not be too confluent at time of freezing. Freeze gradually: 2 h at –20°C, then store at –80°C. To resuspend, thaw quickly in a waterbath at 37°C, add 100 μL of medium, and let the cells grow.

4. Staining was carefully compared for each cell line in the 5 different dishes to detect gene trap lines that responded to one (or more) factor(s). Selected cell lines were then thawed, expanded to 35-mm plates, and screened again with the above protocol to confirm their response to the factors.

5. In order to optimize the quality of the RNA template used for RT reaction, we performed 16 different RNA preparations using four different cell lines and four different RNA

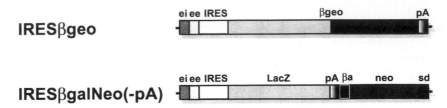

Fig. 3. Examples of gene trap vectors. For references, see text. ei, *eng2* intron; ee, *eng2* exon; βgeo, *lacZ-neo* fusion; pA, polyadenylation signal; IRES, internal ribosomal entry site; βa, human β-actin promoter; sd, mouse *Pax2* splice donor site.

preparation protocols (**Table 2**). The ES cell lines were divided onto four groups taking into account the different degrees of relative abundance of the trapped mRNAs, as revealed by X-gal staining of cell colonies; four lines were taken as samples, one from each group. Three commercial kits for RNA preparation were used. We report here that the best quality total RNA from ES cell lines grown to confluence in 6-well plates, measured by Northern blot and PCR, was obtained using 0.75 mL TRIzol. As reported by Townley et al. *(18)* an additional phenol extraction, prior to RNA precipitation, was found to improve the quality of the RNA.

6. First strand cDNA. The reaction was carried out with random primers and with vector-specific primers. Although PCR performed using the equally abundant hypoxanthine phosphoribosyltransferase (HPRT) mRNA specific primers has revealed that a certain level of unspecificity is generated also by the vector specific primers, this is the most useful approach.

7. Southern blot analysis of 5′ RACE products. In order to make a strategic choice between the direct sequencing of 5′ RACE products and their cloning prior to sequencing, we have tested the degree of purity of the amplification products obtained with this method. Five microliters from the second round RACE products were run on 2% agarose gel and transferred to transfer membranes. Following the transfer, the filter was probed with a labeled 42-mer primer corresponding to the *eng2* splice-acceptor site, present in all the analyzed vectors immediately 3′ of the integration site. Hybridizations were performed at 65°C. Filters were washed in stringent conditions and exposed to film for 2 h. Although positive hybridization signals were visible in all lanes of the electrophoresis gel, they either did not correspond to the more represented bands or there was more than one signal in one lane. If it can be argued that multiple bands represent different sized amplification products of the same cDNA, this is uncertain *a priori* and does not ensure a safe way to routinely proceed. In only a very few cases, we observed a sharp band that represented a reasonably good template for direct sequencing *(19)*.

8. It should be mentioned that the expression of the reporter gene in some cases does not mimic that of the endogenous trapped gene. This could be caused, for example, by the disruption of intronic regulative elements of the endogenous gene by the gene trap insertion. It is, therefore, essential to confirm the expression of the trapped gene, once it has been successfully cloned, by *in situ* hybridization *(20)*.

Acknowledgments

We thank K. Chowdhury, B. I. Meyer, and A. Stoykova for contributing to the scientific managing of the gene trap strategy in our laboratory and P. Bonaldo for setting up the ES cell in vitro preselection protocol. We gratefully acknowledge the

Table 2
RNA Preparation Test on Four Different ES Cell Clones Generated by Gene Trap in Our Laboratory

Sample	GT Vector	X-gal staining	Beads	Matrix	TRIzol[a]
EI-9	pKC405	++	+	++	++++
XIX-18	pKC405	++++	+	++	++++
XXII-5	pM26-22	+	+	++	+++
3a-91	pKC405	+++	+	++	++++

[a]Gibco-BRL.

excellent technical assistance of M. Daniel, S. Geisendorf, S. Hille, R. Libal, S. Mahsur, S. Schlott, R. Scholz, T. Schulz, and M. Walther. Work in our laboratory was supported by program grants from Amgen Inc. (Thousand Oak, CA) and The Max Planck Society. F. C. is presently in Rome as *Assistant Telethon Scientist* (Grant 38/cp). He was supported in Göttingen by an EMBO long-term fellowship, the Human Frontier Science Program, and the TMR Marie Curie program from EU.

References

1. Gossler, A., Joyner, A. L., Rossant, J., and Skarnes, W. C. (1989) Mouse embryonic stem cells and reporter constructs to detect developmentally regulated genes. *Science* **244,** 463–465.
2. Pascoe, W. S., KemLer, R., and Wood, S. A. (1992) Genes and functions: trapping and targeting in embryonic stem cells. *Biochim. Biophys. Acta* **1114,** 209–221.
3. Frohman, M. A., Dush, M. K., and Martin, G. R. (1988) Rapid production of full-length cDNAs from rare transcripts: amplification using a single gene-specific oligonucleotide primer. *Proc. Natl. Acad. Sci. USA* **85,** 8998–9002.
4. Friedrich, G. and Soriano, P. (1993) Insertional mutagenesis by retroviruses and promoter traps in embryonic stem cells. *Methods Enzymol.* **225,** 681–701.
5. Robertson, E., Bradley, A., Kuehn, M., and Evans, M. (1986) Germ-line transmission of genes introduced into cultured pluripotential cells by retroviral vector. *Nature* **323,** 445–448.
6. Bonaldo, P., Chowdhury, K., Stoykova, A., Torres, M., and Gruss, P. (1998) Efficient gene trap screening for novel developmental genes using IRESbetageo vector and in vitro preselection. *Exp. Cell Res.* **244,** 125–136.
7. Cecconi, F., Alvarez-Bolado, G., Meyer, B. I., Roth, K. A., and Gruss, P. (1998) Apaf1 (CED-4 homolog) regulates programmed cell death in mammalian development. *Cell* **94,** 727–737.
8. Salminen, M., Meyer, B. I., Bober, E., and Gruss, P. (2000) Netrin 1 is required for semicircular canal formation in the mouse inner ear. *Development* **127,** 13–22.
9. Voss, A. K., Thomas, T., and Gruss, P. (2000) Mice lacking HSP90beta fail to develop a placental labyrinth. *Development* **127,** 1–11.
10. Torres, M., Stoykova, A., Huber, O., Chowdhury, K., Bonaldo, P., Mansouri, A., et al. (1997) An alpha-E-catenin gene trap mutation defines its function in preimplantation development. *Proc. Natl. Acad. Sci. USA* **94,** 901–906.
11. Robertson, E. J. (ed.) (1987) Embryo-derived stem cell lines, in *Teratocarcinomas and embryonic stem cells, a practical approach.* IRL Press, Oxford, pp. 71–112.

12. Lail-Trecker, M. (1997) Cloning PCR products utilizing the T/A overhang and a kit, in *Methods in Molecular Biology, Vol. 67: PCR cloning protocols, From Molecular Cloning to Genetic Engineering* (White, B. A., ed.), Humana Press, Totowa.

13. Joyner, A. (ed.) (1993) *Gene targeting: a practical approach.* Oxford University Press, Oxford.

14. Wassarman, P. M. and DePamphilis, M. L. (eds.) (1993) *Methods in Enzymology, vol. 225: Guide to techniques in mouse development,* Academic Press, San Diego.

15. Hogan, B., Beddington, R., Costantini, F., and Lacy, E. (eds.) (1994) *Manipulating the Mouse Embryos. A Laboratory manual,* 2nd ed., CSH Laboratory Press, Cold Spring Harbor, N.Y.

16. Skarnes, W. C., Moss, J. E., Hurtley, S. M., and Beddington, R. S. (1995) Capturing genes encoding membrane and secreted proteins important for mouse development. *Proc. Natl. Acad. Sci. USA* **92,** 6592–6596.

17. Salminen, M., Meyer, B. I., and Gruss, P. (1998). Efficient poly(A) trap approach allows the capture of genes specifically active in differentiated embryonic stem cells and in mouse embryos. *Dev. Dyn.* **212,** 326–333.

18. Townley, D. J., Avery, B. J., Rosen, B., and Skarnes, W. C. (1997) Rapid sequence analysis of gene trap integrations to generate a resource of insertional mutations in mice. *Genome Res.* **7,** 293–298.

19. Cecconi, F. and Meyer, B. I. (2000) Gene trap: a way to identify novel genes and unravel their biological function. *FEBS Lett.* **480,** 63–71.

20. Deng, J. M. and Behringer, R. R. (1995) An insertional mutation in the BTF3 transcription factor gene leads to an early postimplantation lethality in mice. *Transgenic Res.* **4,** 264–269.

26

Functional Genomics by Gene-Trapping in Embryonic Stem Cells

Thomas Floss and Wolfgang Wurst

1. Introduction

The pace of sequencing whole genomes by far exceeds our increase of knowledge on gene functions. Twenty-two genomes have been already completed and both the human and mouse genome are close to be completed as well. The number of sequences in the National Center for Biotechnology Information (NCBI) gene bank doubles approximately every other 15 months *(1)*. The gain of information on expressed sequence tags (ESTs) from a multiplicity of organisms and cell types rose exponentially and is close to reaching saturation.

Almost two million human and one million mouse ESTs can be found inside the NCBI database by today (www.ncbi.nlm.gov/dbest_summary). In fact, over 63% of the 4.6 million different sequences in the Genbank are ESTs *(1)*.

In the post-sequencing area, large-scale and high-throughput technologies have to be developed and applied in order to understand the physiological rolc of any given gene–protein in a living organism. Computational analysis based on the identification of homologous genes in different species will be necessary, but not sufficient. In reported cases, homologous genes fulfill completely different functions in different species.

An impressive demonstration of this phenomenon has been given by the functional comparison of similar molecules in the two closely related nematodes *P. pacificus* and *Caenorhabditis elegans*, which revealed that homologous genes can be recruited to serve completely new functions in new regulatory linkages. While remaining in the original developmental program, genes were found to have changed their molecular specificity and at the same time, retained other functions. Genes have jumped in or moved out of regulatory circuits and are suddenly expressed in a spatially and temporarily different manner (*for a review see* **ref. 2**). Therefore, mere sequence or structural similarity does not necessarily allow conclusions on possible protein functions *(3)*.

It is clear that the actual function of each gene needs to be studied in the entire organism. The most straightforward way has been to analyze the phenotype of organisms carrying either gain- or loss-of-function mutations within the gene of interest. Traditionally, the phenotype-driven approach (forward genetics) has been most fruitful in organisms with comparably small genomes.

From: *Methods in Molecular Biology, vol. 185: Embryonic Stem Cells: Methods and Protocols*
Edited by: K. Turksen © Humana Press Inc., Totowa, NJ

The source of natural mouse mutants is relatively limited *(4)*. At present however, the resource of natural mutants experiences a remarkable increase by large-scale chemical mutagenesis screens with subsequent screening for dominant and recessive mutants *(5,6)*. Ethylnitrosourea (ENU)-screens are perfect tools in order to identify causes of monogenic disease. Redundant factors however, as well as complex traits, are falling through its net. The most important disadvantage in the analysis of naturally and chemically produced mutants is, however, that the identity of the responsible gene has to be subsequently determined. Although improvements have been made *(7)*, the identification of point mutations can still turn out to be a tedious venture and can take several years in individual cases *(8–10)*.

The targeted gene inactivation, which is now being performed, since more than 12 yr in the mouse *(11,12)* has led to an exponential growth of information on the function of a multiplicity of genes and has established the mouse as the favorite organism when considering the generation of models for human genetic disease. The number of the published mouse genes with a targeted mutation, so far, can only be estimated according to the current mouse locus catalogue around two thousand (Paul Szauter, personal communication; http://www.informatics.jax.org). The biomed knock-out database (http://reasearch.bmn.com/mkmd) shows 1818 entries for the term "knock-out."

A satisfactory answer to the question on gene function however, can often not be given from one single knock-out animal. In general, only the earliest phenotypes will be determined by conventional gene inactivation experiments and will, in many cases, lead to lethality. Depending upon construct design, different mutations have led to variable phenotypes *(13,14)* and it turns out that, increasingly, a promoter-driven resistance cassette is often disturbing to neighboring genes, which makes an interpretation of the mutant phenotypes even more difficult *(15–19)*. Therefore, it appears that a number of knock-outs that have been performed in the past need to be revisited in a complementary approach *(20)*.

It is important to point out, however, that despite world-wide activities in the past 10 yr, no more than 5% of the assumed mouse genes have been inactivated, which would add up to 200 yr or so in order to study each gene by applying conventional knock-out technology. The knock-out technology has experienced substantial improvements from different areas. Particularly, the construction of knock-out vectors could be shortened from several months initially, to only a few weeks at present *(21–23)*.

The possibility of a high through-put functional analysis, which is capable of complementing the current genome sequencing, can be accomplished by a large-scale insertional mutagenesis, namely the gene-trapping *(24–33)* (**Table 1**). Therefore, in the following chapter we want to give a general overview on the latest development of gene-trapping activities in mouse embryonic stem (ES) cells, highlighting in particular, the different approaches, advantages, and accomplishments, as well its inherent difficulties and pitfalls.

1.1. Gene-Trapping Strategies

1.1.1. Enhancer Trap

The trapping principle is based on random insertional mutagenesis of a vector ideally containing a promoterless reporter gene. The trap vector both serves as a sequence tag for the identification of the mutated locus and can provide a reporter–selection marker for monitoring endogenous gene activity. Originally, by utilizing bacteriophage

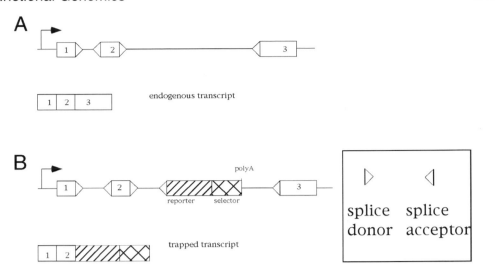

Fig. 1. The gene-trapping principle is based on random integration of a reporter–selector cassette into a genome to both mutate and identify the trapped locus. (**A**) Wild-type locus; (**B**) trap vector integration.

transposable elements, a reporter cassette provided with a minimal promoter has been introduced into the *Escherichia coli* genome *(34,35)*. This enhancer trapping approach has been demonstrated to be straightforward, for both the identification and mutation of bacterial genes. The original enhancer trap approach, which had been successfully applied to the bacterial genome, turned out to be less straightforward in more complex genomes. Although useful in order to identify tissue or cell type-specific regulatory elements, eucaryotic enhancers are often located far away from the regulated genes. In addition, enhancer trap vectors are in general not mutagenic in eukaryotes *(24)*.

1.1.2. Exon Trap

In principle, exon trap vectors are without splicing acceptors (SA). Consisting only of a selector or reporter–selector cassette, their activation requires either in-frame exon integration or integration in 5′ untranslated regions (UTRs) of actively expressed genes. Therefore, this vector type appears extremely inefficient in respect to identifying resistant colonies. Due to the limitations of exon trap vectors, mostly gene-trap vectors are being used for large-scale insertional mutagenesis screens in eucaryotic genomes.

1.1.3. Gene-Trap

Typical gene-trap vectors are promoterless and function by generating a fusion transcript with the endogenous gene (**Fig. 1**). The presence of splice donor (SD) or SA elements leads to the generation of a fusion protein, even in case of an intron integration.

Gene-trapping has been applied with great success in *Drosophila (36)*, *C. elegans (37)*, and murine cells *(24)*. In *Drosophila*, insertional mutagenesis has been performed using a nonretroviral transposon called P-element, which can be utilized as an enhancer detector and sequence tag in order to both identify and isolate affected loci *(38,39)*.

Table 1
Summary of the Advantages and Disadvantages of Mutagenesis Strategies to Determine Gene Function

Method	Advantages	Disadvantages
Knock-out	1. Defined mutation. 2. Possibility of conditional mutagenesis. 3. Laborious. 4. Secondary effects. 5. Knowledge of target sequence is a prerequisite.	1. Time-consuming. 2. Cost-effective.
ENU-mutagenisis	1. Fast and easy to apply. 2. Knowledge of phenotype. 3. Possibility of high-throughput analysis. 4. Detection of mutation can be tedious.	1. Large numbers of mice necessary. 2. Factors that are involved in complex traits get lost. 3. Recessive lethal mutations are hard to identify.
Gene-trap	1. No knowledge of target necessary. 2. Fast and easy to apply. 3. Easy to monitor. 4. Knowledge of genotype. 5. Possibility of high-throughput analysis. 6. Factors involved in complex traits are stored. 7. Possibility of conditional mutagenesis.	1. Necessity to generate external primers or probes to genotype F2 generation. 2. Splicing-around the vector can lead wild-type phenotypes as well as hypo- or hypermorphs.

The development of the mouse takes place *in utero* and is not accessible directly to genetic manipulation. Since the breakthrough of the ES cell system *(11,12)*, it suddenly became feasable to manipulate mouse embryonic cells ex vivo and generate animals carrying a single predetermined mutation. By combining both, insertional mutagenesis and the ES cell system, Gossler et al. *(24)* have pioneered the gene-trapping approach in mammalian cells. Friedrich and Soriano *(25)* subsequently generated twenty-four mouse strains carrying gene-trap mutations demonstrating the mutagenicity of the integrated vector in living organisms.

In higher organisms, gene-trapping turned out to be not only a tool for mutagenesis, but also as a means for gene expression studies and especially for gene discovery by capturing open reading frames **(Fig. 2.)**. Besides ES cells, the gene-trapping principle has also been successfully applied to other cell types *(40)*.

1.1.4. SA-Type Vectors

Most commonly used vectors contain a SA or both an SA and an SD. The typical SA type vector uses an ATG-less reporter–selector cassette, which is flanked by a 5′

SA sequence. In eucaryotic cells, nuclear pre-mRNA introns are excised by a large ribonucleoprotein complex known as the splicosome *(41)*, which recognizes sites at the 5′ and 3′ ends of the intron (the donor and acceptor splice sites, respectively). A minimal SA consists of a branch point sequence followed by a polypyrimidine tract and the actual splice site, which is recognized by the splicosome. Widely used SA in gene-trap vectors are the murine *engrailed-2* and the adenovirus late major transcript SA *(42)*. More recently, the human *BCL-2* gene intron2/exon3 SA has been introduced *(43)*. Proper integration and splicing will generate a fusion protein with the trapped locus (**Fig. 1**). The fusion transcript will ideally be prematurely terminated at the poly(A) site of the selection–reporter cassette.

In order to identify genes independent of reading frames, an internal ribosomal entry site (IRES) *(44–46)* from the encephalomyocarditis virus is utilized 3′ of the SA.

1.1.5. SD-Type Vectors: Poly(A) Trap

A limitation of the commonly used SA vector type is that mostly expressed genes can be identified. In contrast, the poly(A) trap vector type allows the identification of all intron-containing genes independent of their expression status. The simplest poly(A) trap vector contains a promoter-driven selector gene without a poly(A) sequence *(47)*. Thus, drug resistance is only obtained after successfully capturing a poly(A) signal in the genome. Furthermore, poly(A) trap vectors have been developed that contain an additional SD sequence *(32,47,48)*.

The latest poly(A) trap vectors contain, in addition, a reporter gene with a SA in front, followed by a selector gene driven by a strong promoter fused to a SD. SD vectors, thereby, allow to monitor the regulation of the trapped gene in vivo assessing reporter gene expression.

The advantage of SD-type vectors is their independence of actively expressed genes due to the ubiquitously active promoter of the selection unit. The only known genes that are not polyadenylated in metazoan organisms, are the major histone genes *(49)*. Therefore, poly(A) trap vectors will basically allow the entrapment of most genes.

A variety of SD sequences that are derived from the mouse *Pax-2* *(48)* and Hprt genes *(43)*, and a synthetic SD consensus sequence *(32)*, have been successfully utilized.

1.1.6. Reporter–Selector Cassettes

Several reporter assays have been described that are applicable for high-throughput screening procedures *(50,51)*.

A widely used reporter gene for high-throughput screening is the bacterial β-galactosidase (β-gal) (lacZ), which has the advantage to be well characterized, stable, nontoxic, and additionally, has inexpensive substrates. Highly sensitive fluorescent and chemiluminescent substrates are available for β-gal. A limitation of β-gal is some endogenous activity in the mouse, starting around embryonic d 11.5. In older embryos or newborns, this problem may be circumvented by utilizing commercially available lacZ antibodies.

An alternative to β-gal is the bacterial β-lactamase, for which no endogenous activity has been described in the mouse *(53)*. Human placental alkaline phosphatase (AP) has been utilized successfully in gene-trap vectors *(52)*.

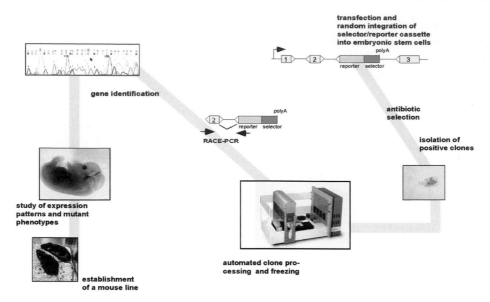

Fig. 2. The gene-trapping principle and its possibilities. A subsequent selection after random integration of the trap vector into the genome of ES cells (*see also* **Fig. 1**) allows the isolation of individual clones and the identification of the trapped gene by semiautomated RACE PCR and sequencing *(56)*. After picking of ES clones, further propagation and freezing of two copies from each clone is performed automatically (Plato 7; Qiagen). A prescreening of ES cells for a variety of inducing signals is possible by taking advantage of the lacZ reporter and additionally by microarraying RACE fragments *(30,57)*. Mouse lines can be established from potentially interesting clones. In these, the lacZ reporter usually reflects the endogenous expression of the trapped gene. After breeding to homozygosity, the phenotypical consequences of the mutation can be studied in vivo.

The jellyfish green fluorescent protein (GFP), which is available in multiple forms, has the advantage that no substrate is needed and no endogenous activity is observed in murine cells. In addition, antibodies for GFP detection postmortem are commercially available. Recently though, evidence of GFP toxicity in mouse cells has been reported when highly expressed *(54)*.

Resistance markers in gene-trap vectors are traditionally neomycin phosphotransferase or lacZ/neo fusions (βgeo), but other resistance markers like phleomycin have been used successfully *(42)*. LacZ/hygromycin cassettes have been utilized in order to trap genes in transfected cells and cells carrying a targeted mutation *(55)*. Poly(A) trap vectors are available with a neomycin *(43,48)* or puromycin resistence cassette *(32)*.

1.1.7. Bias of Vector Integration

Apparently, the integration of a trap vector is not entirely random. Often, integrations into the 5′ end of genes have been observed. 5′ integrations result in relatively small fusion proteins, whereas 3′ insertions can lead to large fusion complexes between the resistance marker and the trapped gene. Therefore, in case of 3′ integrations, because of inadaequate protein folding, the resistence marker may become inactive. Accordingly,

SA trap vector integrations could be more or less random, but 3' insertions might simply not be functional. Given, that this hypothesis is correct, one would expect 3' integrations to be functional for SD vectors. In contrast however, for SD-type vectors, it has been reported for 105 traps in total that 88% of integrations occured either 5' or in the middle of the encoded cDNA (www.lexgen.com/questions.php3#8).

Chowdhury et al. *(59)* did not observe a preferred 5' integration when using the SA-type vector types SAβgeo and IRESβgeo. Instead, integrations occured unbiased among 55 examined cDNAs. In the report however, the authors did not distinguish between vector integrations with and without an IRES sequence. Interestingly, we have found all 3' integrations described within the report of Chowdhury et al. to be derived from the IRES vector. Misfolding reasons, though, should not apply to IRES vectors, since translation of the selector gene takes place independently of the trapped locus from a bicistronic transcript.

Bronchain et al. *(60)* have addressed the misfolding problem by introducing a glycine stretch 5' of a GFP reporter cassette for a gene-trap approach in *Xenopus*. It has not been reported yet, whether the glycine-stretch has an impact on the 5'–3' integration ratio.

Only the analysis of larger numbers of gene-trap integrations will give a satisfactory answer to this issue. Assuming that there is a preference to identify 5' integrations, this will have the positive side-effect of generating preferably null mutations, since 3' integrations might only partially disrupt the function of the trapped gene.

1.2. Advantages and Disadvantages (Table 2) of Individual Vector Types

1.2.1. SA-Type Vectors

Each gene-trap vector has its own inherent advantages and disadvantages. A limitation of both the SA-type as well as the exon trap vectors is mainly the dependency on integration in active genes. It has been speculated that a large number of genes are expressed in ES cells probably due to an undermethylated status, and the analysis of 7000 gene-trap integrations in our database *(33)* revealed that we are still far from saturation.

Trap vectors have been used without SA but with an IRES sequence instead *(33,52; for an overview see* **ref.** *61)*. 5' rapid amplification of cDNA ends (RACE) sequences from gene-traps of this type usually yield an intronic sequence, which is under-represented in the public databases. IRES vectors without an SA which have integrated into intronic regions, have sometimes failed to be mutagenic due to splicing around the integration site or of unknown reasons *(52,62)*.

1.2.2. SD-Type Vectors

Very little data is available on mouse strains carrying a poly(A) trap mutation. In contrast to the SA-type vectors, it has been speculated, that the integration of ubiquitously active transcriptional units could interfere with neighboring genes and thereby complicate the interpretation of phenotypic observations *(17–19)*. Therefore, it is advised to finally remove the promoter element of the gene-trap vector by taking advantage of the Cre or Flp recombinase systems *(18,63–65)*.

From the view of functional genomics, one of the major issues concerning the poly(A) trap approach is whether the sequences that are being discovered are actually coding for proteins. The integration in the vicinity of canonical and noncanonical poly(A) sites, which are not related to genes, appears as a hypothetical disadvantage of the poly(A) trap approach. This integration might lead to a high number of false positive gene-trap clones. The consensus sequence for a canonical poly(A) site is AATAAA, but variations of GU or U-rich sequences are productive and may also lead to proper polyadenylation *(49,66)*. Recent studies of public EST databases have shown that the consensus AATAAA hexamer is absent from more than half of all 3′ UTRs *(67)*. Taking into account only the canonical site, the statistical occurence is 4^6, meaning that every other 4096 bp statistically, there is a poly(A) signal. This adds up to approximately 750,000 canonical poly(A) sites within the mouse genome. Given that two noncanonical sites are also capable of polyadenylating a transcript, this number adds up to more than two million sites. This number exceeds the estimated number of mouse genes by almost 20–30 times.

Naturally, there is more to proper polyadenylation than just a poly(A) signal. The poly(A) signal needs to be followed by a more complex signal that is not yet fully characterized *(68)*. In particular, the RNA polymerase needs to be released from the DNA template, and the transcript needs to be cleaved to be polyadenylated. Therefore, the number of false positives will be significantly smaller than 30 times. But in a high-throughput screen, this fact may lead to an unacceptable number of false positive clones, which could then only be identified after sequencing by the fact that no splicing has had occured.

This limitation of poly(A) trapping has been recently addressed by the use of mRNA instability elements 3′ of the SD, which lead to the degradation of polyadenylated transcripts, that have remained unspliced. It has been demonstrated that the use of instability elements diminishes the total number of clones by 50% *(43)*.

Additionally, as illustrated in **Fig. 3**, in the case of poly(A) trap vectors, single vector integrations are mandatory. Therefore, retroviral vectors should be used for vector delivery, since tandem integrations can lead to splicing into the neighboring reporter cassette and can, thereby, lead to the isolation of false positive clones.

1.2.3. Specialized Trapping Strategies

It has been estimated, that in humans, secretory molecules account for about one-tenth of the human proteome. Those include major factors of signaling pathways, blood coagulation, immune defense and carcinogenesis, as well as digestive enzymes and components of the extracellular matrix and blood plasma *(69)*. It has been speculated, that β-gal activity is lost when βgeo is fused to a signal peptide. Therefore, insertions in secreted molecules may not be detected *(70)*. In order to overcome this constraint, a gene-trap vector of the SA-type provided with a transmembrane domain has been developed *(70)*, which prevents the reporter–selector fusion protein from being secreted after integrating into a intercellular signaling molecule gene. Alternatively, the use of an IRES sequence will lead to an independent translation of βgeo from the bicistronic fusion transcript. Therefore, the reporter–selector protein should remain inside the cell *(59)*.

The application of exogenous retinoic acid is known to induce profound effects on patterning of various tissues *(71)*. In order to identify genes that are initially inactive in ES cells, but respond to challenges like retinoic acid, transcription factors, or signaling molecules in undifferentiated or differentiated ES cells, the induction trap strategy has

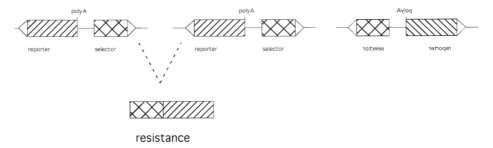

Fig. 3. The use of poly(A) trap vectors can put a selection pressure on tandem integration, since in this case, the reporter cassette is capable of providing a poly(A) signal to the selection marker. Therefore, when using poly(A) traps, retroviral delivery, which usually leads to a single integration per cell, is mandatory.

been developed *(30,57,72)*. In vectors used for an induction screen, the seletion marker is generally driven by a ubiquitously active promoter, whereas the reporter cassette is promoterless. As expected, only very few resistant clones (0.2–5%) are initially lacZ positive. Clones that turn "blue" in response to the challenge can be isolated and further characterized during this screen. In case a βgeo type of vector is used, only expression level difference can be monitored.

More recently, the induction strategy has been developed further to identify target genes of transcription factors. This approach is based on the internalization and nuclear addressing of exogenous homeodomain-containing proteins, using an *Engrailed* homeodomain (EnHD) as an example. An ES cell gene-trap library has been screened to identify targets of *Engrailed*. In this screen, 8 integrated gene-trap loci responded to EnHD. One of the genes identified was the bullous pemphigoid antigen 1 (BPAG1). By combining in vivo electroporation with organotypic cultures, it has been shown that a BPAG1 enhancer–promoter is differentially regulated by *Engrailed* in the embryonic spinal cord and mesencephalon. This strategy can, therefore, be used to identify and mutate homeoprotein targets. Because homeodomain third helices can internalize proteins, peptides, phosphopeptides, and antisense oligonucleotides, the strategy should also be applicable to other intracellular targets for characterizing genetic networks and dissecting different pathways involved in a large number of physiopathological states *(57)*.

1.3. Gene Identification

1.3.1. RACE Polymerase Chain Reaction

In both the SD and the SA trap vectors, the site of integration can be identified by 5′ (SA trap) or 3′ (SD trap) RACE and subsequent sequencing *(73)*. A major advantage of using RACE polymerase chain reaction (PCR) to identify the integration site, lies in the possibility to use automated RACE procedures. RACE products can be sequenced without subsequent cloning in case a tagged primer is being used for the last amplification *(56)*.

Trapping with poly(A) trap vectors is advantagous in respect to RACE procedures, since 3′ RACE is more robust than the 5′ RACE. This observation is probably based on the fact that tailing of the PCR product is not necessary because a poly(T) primer can be used in addition to a primer derived from the vector. A second disadvantage of performing 5′ RACE is the under-representation of 5′ sequences in EST databases. This limitation will be overcome, with the sequencing of the entire mouse genome. At present, we find approximately 30% of the obtained 5′ RACE sequences to be unknown *(33)*.

1.3.2. Plasmid Rescue

The utilization of plasmid rescue to identify the trapped locus is possible in case the bacterial origin of replication (ori) and ampicillin resistance are retained in the gene-trap insertion *(31)* (**Fig. 6**). Araki et al. *(74)* report for a total number of 109 clones that were generated by electroporation, the presence of the bacterial *amp* and *ori* in only 37% of the cases.

In contrast, presumably, resulting from homologous recombination between the long terminal repeats (LTRs), when using retroviral vectors, 70–80% of examined clones retain the bacterial sequences (Geoff Hicks, personal communication). Automation of the subsequent cloning step, which is involved in the protocol prior to sequencing, has not yet been achieved, therefore this method does not seem to be applicable in a high-throughput screen at present.

1.3.3. SupF Complementation

A retroviral entrapment vector of the SA-type has been described, which facilitates the isolation of trapped loci by taking advantage of the bacterial supF tRNA gene *(75,76)*. The protocol involves cloning of genomic DNA fragments into λ phages, plating on a nonpermissive bacterial strain, and screening with internal probes *(52)*.

1.3.4. Construction of Phage Libraries

In some reported cases, the rapid identification of trapped loci by simple PCR or cloning-based techniques did not turn out to be straightforward. In these cases, the construction of a phage library from mutant cells or animals, and subsequent screening with vector sequences, has yielded the desired information *(77)*.

1.4. Mutagenicity of Gene-Trap Vectors

Ideally, the integration of the gene-trap knocks the endogenous gene out or at least down. The phenotypes of mutant gene-trap mice can be phenocopies of null mutants, but can also be more variable, depending on the site of integration and the integrity of the vector cassette in the genome.

Two parts of the gene-trap vector mainly influence its mutagenicity in respect of interrupting the endogenous transcript: *(1)* the relative strength of the SA is critical; and *(2)* effective termination of the fusion transcript at the introduced poly(A) signal is required.

A large number of different SA sequences have been used in gene-trap vectors, which have been described elsewhere *(42)*. However, all vectors described show to some extent "splicing-around" the inserted vector. Thus, wild-type gene product can be made, and splicing-around leads to variable hypomorphic mutations as compared to null mutants *(78–82)*. Hypomorphic phenotypes of gene-trap mice are sometimes

Table 2
Summary of Advantages and Disadvantages of the Different Gene-Trap Methods

Gene-Trap	Method Characteristics	Advantage	Disadvantage
Exon trap	1. No SA. 2. ATG-less.	1. highly mutagenic.	1. small numbers of colonies/introduced vector.
Enhancer trap	1. ATG 2. SA-less.	1. Reporter lines.	1. Small numbers of colonies/introduced vector. 2. 5′ RACE yields noncoding sequence. 3. Enhancers are often located far away from the gene in eukaryotes. 4. Not necessarily mutagenic.
SA-type	1. SA. 2. No promoter. 3. lacZ with or without ATG.	1. Large numbers of colonies/introduced vector. 2. Induction trap possible. 3. lacZ functional in majority of the cases.	1. Only 50% mutagenicity. 2. 5′ RACE results are under-represented in database.
SD-type	1. SA. 2. SD. 3. Selection marker with its own promoter but without poly(A) signal. 4. lacZ with or without ATG.	1. 3′ RACE possible. 2. Induction gene-trap possible.	1. Introduced promoter might interfere with neighboring loci. 2. lacZ-activity/inactivity needs to be determined in vivo. 3. Use of retroviral vector necessary. 4. Polyadenylation at intronic canonical poly(A) sites possible.

informative as compared to complete knock-outs. As an example, Couldrey and coworkers *(83)* have reported a gene-trap insertion into the histone 3.3A gene, which resulted in partial neonatal lethality in only 50% of the examined cases, stunted growth, neuromuscular deficits, and male subfertility. The wild-type transcript was found to be reduced by 4–7 times in homozygotes, as compared to wild-type littermates *(83)*. However, there are also examples in which no phenotype has been obtained *(84)*.

Recently, a vector system has been developed, which allows to test the relative strength of SA sites in vitro *(43)*. The vector pSAT consists of a ubiquitously active cytomegalovirus (CMV) promoter and partial IL4 cDNA/IgE receptor α chain, followed by a SD and an intron sequence. 3′ to this gene cassette an ATG-less human placental AP with its endogenous SA but lacking the secretion signal is included and serves as a reporter. In the wild-type situation, splicing will lead to an active AP, which is easily detectable in vitro. The introduction of a SA vector in front of the AP cassette will suppress AP activity, if a strong SA has been used. Thus, the strength of splice sites can be pretested in vitro.

A recent survey of EST sequences in the public database has revealed alternative splicing to be more a rule than an exception *(85)*. Concerning the gene-trap screen, this means that there might not be a final solution to the splicing-around problem. Focusing on transcriptional termination sequences to prematurely terminate trapped transcripts may turn out to be the way to go. Transcriptional termination consensus sequences have been described *(86,87)* and tested in vitro. However, the 3′ UTR is suspected to interact with the splicing machinery itself *(88–90)*. As Barabino and Keller *(49)* put it: "why does 3′ end processing require a 1 MDa multicomponent machinery, if not to establish a network of weak cooperative interactions?" In case strong polyadenylation sites are capable of activating cryptic splice sites within the reporter–selector cassette, false negative expression results may be observed at least in specialized cell types.

1.5. Genotyping Mice with a Gene-Trap Mutation

1.5.1. Identification of Heterozygous Mutants

F1 generation mutants are easy to genotype by using probes or primers recognizing vector sequences. Dot blot analysis or a genomic PCR can be utilized to distinguish between positive and negative offspring and to give an estimate on copy number in the case of multiple integrations *(91)*. In case the trapped gene is expressed in the tail tip, β-gal staining of tail biopsies have been applied successfully *(92)*.

1.5.2. Identification of Homozygotes Using Quantitative Hybridization

Genotyping F2 gene-trap mice is a more laborious task. Internal probes or primers are not appropriate to distinguish between heterozygotes and homozygotes, unless an unrelated genomic probe is used as a loading control *(93)* **(Fig. 4)**.

1.5.3. Identification of Mutants Using Restriction Fragment-Length Polymorphisms

Often, cDNA, EST fragments, or PCR products are available, which can be used as external probes. In these cases, restriction fragment length polymorphisms (RFLPs) can generally be found by utilizing restriction enzymes that cut within the gene-trap vector **(Fig. 5)**.

1.5.4. Identification of Mutated Loci by Plasmid Rescue

Quantitative hybridization is subject to interpretation and, therefore, cannot be the method of choice. Some researchers take advantage of the bacterial ori and amp to use a plasmid rescue in order to isolate flanking probes, which are outside to the integrated gene-trap vector *(31,74)* **(Fig. 6)**. It is recommended to check for a retainment of bacterial sequences prior to plasmid rescue by PCR or dot blot methods. Araki et al. *(74)* have reported the loss of the ampicillin resistence cassette in more than 60% of examined cases. Probably due to homologous recombination between the LTRs when

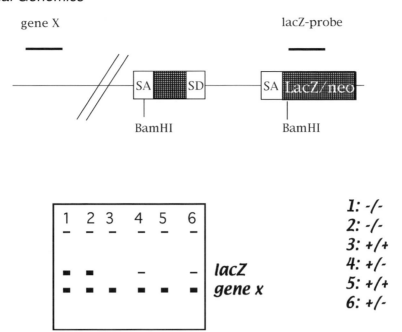

Fig. 4. In a quantitative hybridization experiment, a vector and a non-vector-related probe is used to estimate the copy number of the gene-trap vector integrations in the F2 generation. The copy number can be estimated both from a Southern hybridization or from a simple dot blot. By using restriction enzymes, which cut outside the vector, multiple integrations can be detected.

utilizing retroviral trap-vectors, only 20–30% of examined clones lose the bacterial sequences (Geoff Hicks, personal communication).

In order to perform plasmid rescue, enzymes need to be found, which will release genomic fragments of reasonable length. Genomic fragments over 20 kb have been cloned successfully *(31)*. For transformation, it is recommended to use a bacterial strain with a defective DNA recombination system in order to be able to propagate repetitive sequences that are frequent in intronic DNA. A highly recommended strain is the *stbl2* bacterial strain *(96)*, which is suitable for the cloning of unstable inserts such as retroviral sequences or direct repeats.

1.5.5. Identification of Mutants by Inverse PCR-Generated Probes

In our hands, the most successful method by far to generate external genomic probes for yet uncharacterized loci is inverse PCR (*overview in* **refs. *42,97,98***; Protocol: H.V. Melchner, personal communication *see* **Fig. 7**).

1.5.6. Identification of Trapped Loci by Adapter-Mediated PCR

Recently, an adapter-mediated PCR approach to generate an external probe has been proposed *(99)*. The protocol is straightforward in order to genotype mice with the scrambler mutation, but can be adapted on genotyping gene-trap mutations.

Briefly, the protocol involves the digestion of genomic DNA, anchoring with a blunt-end adapter, and subsequent PCR with both adapter-specific and gene-specific primers.

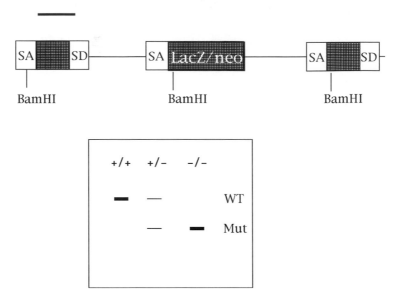

Fig. 5. RFLPs can be utilized in order to identify F2 gene-trap mutants. In some reported cases, the 5′ or 3′ RACE fragment has been successfully utilized as an external probe *(94,95)*. However, RACE fragments are usually small and derived from distant exons which could make it difficult to find RFLPs. Since there is usually no information on the exon–intron structures of trapped genes, the same problems apply to finding external primer pairs.

1.5.7. Identification of Homozygosity by Breeding

In case mutant mice are viable and fertile, a reliable way to verify homozygosity is achieved by breeding potential homozygotes to wild-type mice. In case of homozygosity, naturally each of the offspring will be heterozygous for the trap mutation.

1.6. Accomplishments of the Gene-Trap Technology in the Mouse

Large-scale insertional mutagensis is performed by a number of laboratories worldwide *(27,31–33,59)* and has already led to an attractive collection of trapped loci, which are being made available to the scientific community on both a profit and a nonprofit basis. Within the German gene-trap consortium, we have trapped 14,000 ES clones, of which by now more than 7000 have been identified by 5′ RACE already *(33)* (http://tikus.gsf.de). More than 1000 of the trapped loci are yet unknown, and more than 900 carry mutations in previously identified mouse ESTs. Overall, more than 2000 hits show homology to independent mouse or human unigene clusters, and more than 300 traps have links to the Online Mendelian Inheritance in Man (OMIM) database and are, therefore, human disease-related. Even mutagenesis with alkylating agents is never completely random. Chowdhury et al. *(59)* have demonstrated more or less stochastic integration for 55 integrations of both a IRESβgeo and a SAβgeo vector into the mouse genome. To show that there is no preferred locus of integration, we plotted gene-trap hits in our database on the corresponding human chromosomes. The vectors that have been used are the PT1βgeo and the retroviral U3βgeo *(33)*. Three hundred-eighty-two hits have links to the radiation hybrid mapping panel and can

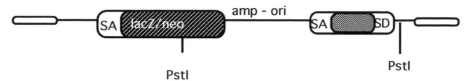

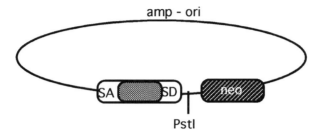

Fig. 6. Plasmid rescue as a means of isolating external probes from gene-trap mutations.

thereby be allocated to the corresponding genomic region (http://www.ncbi.nlm.nih.gov/genemap/). This analysis has revealed a more or less random integration of the different gene-trap vectors into the genome, taking into account the individual size of the chromosome (Ruiz P. et al, in preparation).

Nevertheless, "hot spots," with more than 15 integrations having an e-value less than 10^{10}, are being observed (Ruiz P. et al., in preparation). Examples are *"Bruce"* (17 hits) and the *"Jumonji"* locus (30 hits). Interestingly, *Bruce* has been trapped by Bill Skarnes' group (socrates.berkeley.edu/~skarnes/resource.html) and *Jumonji* has also been trapped independently by two other groups *(100,101)*. Another example is *"R-PTP-κ,"* which has been trapped once by Salminen et al. *(48)*, Skarnes et al. *(70)*, and twice by our gene-trap consortium (unpublished). Each group has used different, but more or less related, trap-vectors. Whether this phenomenon is based on the individual expression level, chromatin status, or vector homology is under investigation.

Since the publication of the first 60% of the human genome sequence, 47 previously "unknown" trapped loci show homology to the unfinished human genome sequence tags, emphasing the use of gene-trapping as an appropriate tool for gene discovery.

In summary, the gene-trap technology fullfills minimum requirements for a large-scale functional analysis of the mouse genome:

1. Gene-trap vectors randomly integrate into the genome.
2. The site of vector integration can be determined.
3. The reporter gene mimics endogenous gene expression.
4. The tagged gene is most likely mutated by gene-trap vector integration.
5. The mutation can be transmitted through the germ line using the ES cell technology.

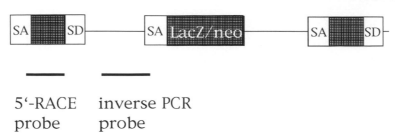

Fig. 7. External probes for genotyping can be generated by RACE or inverse PCR.

Since both the production of ES clones and the identification using RACE PCR methods have been already semiautomated within the German gene-trap consortium, we will be able to saturate the mouse genome with gene-trap vector integrations within the next 5 yr.

1.7. Some Future Aspects

1.7.1. Modifications of Trapped Loci

From many different areas, there is constant influence on the gene-trap technology. On the one hand, a new generation of vectors is on the rise, which will allow also somatic mutations of gene-trap integrations at any given time point or in a given tissue by using consensus or modified Cre/loxP or Flp/frt systems (*43,65,102*). The use of modified recombination sites not only allows to rescue gene-trap phenotypes and, thereby, demonstrate a causal relationship of phenotype and genotype, but also to irreversibly bring new genes under the control of the promoter of a trapped gene. Gene-trap vectors with modified loxP sequences are already in use (*74*). The use of loxP or FRT sites in gene-trap vectors will eventually saturate the mouse genome with these sites, which will allow the introduction of large chromosomal rearrangements after breeding two independent gene-trap lines (*103,104*).

1.7.2. Nuclear Transfer

We see a second major improvement of gene-trapping within the nuclear transfer of ES cells (*105*) and somatic cells (*106,107*). A major limitation of the gene-trap technology is the generation of chimeras and breeding for germ line transmission. Applying nuclear transfer technology to the trapped ES clones will speed up germ line transmission.

Recently, it has been shown that transgenic mice can be generated by replacing the nucleus of an unfertilized oocyte with the nuclei of wild-type and even targeted ES cells (*105*). Embryonic development is subsequently triggered by treatment in strontium chloride. Interestingly, it turns out that cells in G0/G1 are favored for nuclear transfer techniques.

Since 35–40% of ES cells, when maintained under standard conditions, are in the S-phase, a prerequisite of cloning ES cells is a cell cycle arrest. Low serum conditions are, therefore, facilitating the cloning of ES cells. Although the authors succeeded in cloning of cells carrying a targeted mutation, their data also prove the difficulties of cloning higher passages of ES cells. The reason for this fact might be the loss of epigenetic markers associated with imprinted genes during culture, which has been described by Dean et al. *(108)* for high-passage ES cells *(105)*.

In addition, the technique does not seem to be applicable on pure genetic 129 background cells: the authors report F1 outbred cells of $129 \times C57$ to be most efficient. Therefore, the nuclear transfer is unlikely to be applicable on existing gene-trap clones, which are mostly 129-derived. This aspect needs to be considered in the future. By using the cloning technique, a successful generation of transgenic mice will not be dependent anymore on the maintenance of an undifferentiated status of ES cells, since it has been demonstrated that differentiated cells are suitable for cloning as long as a normal karyotype is maintained (*for a review see* **ref. 107**). The generation of mice with targeted or gene-trap insertions by nuclear cloning and, thereby, circumventing the chimera stage is feasable and will save a considerable amount of both time and money in the future. It will make the gene-trap technology even more suitable for high-throughput mutagenesis screens.

1.7.3. DNA Chip Technology

The third major improvement currently arising is the DNA chip technology *(109)*. As a systematic approach to identify expression profiles of novel genes and to characterize biologically active substances, microarrays have already proven to be extremely valuable in complementing the genome sequencing effort *(110,111)*. The majority of genes are still unknown, and whole genome expression profiling will provide information on the expression levels and patterns of large numbers of ESTs and unknown genes at the same time. DNA microarrays, containing sequence information of trapped genes, will combine the expression profiling with the gene-trap technology, to be able to identify differentially expressed genes for which mutant ES cells exist and mouse models can be established. Joint efforts are already being undertaken to achieve this exciting goal.

A worldwide complementation of the sequencing effort is already underway in order to fill in the "phenotype gap" *(112)*. Large numbers of gene-trap mutants are currently under analysis and will contribute considerably to functional genomics in the near future. What is required is a worldwide coordination of the different activities to a joint effort to avoid redundancy in the future.

As a first step towards this goal, gene identification data by gene-trapping can be submitted to the NCBI Genbank (www.ncbi.nlm.nih.gov/Genbank/index.html). A number of laboratories applying the gene-trap technology use their own nomenclature, which is confusing to the scientific community (**Table 3**). A common nomenclature for gene-trap mutations is urgently needed.

Expression data of genes are made available through the mouse expression database (www.informatics.jax.org/menus/expression_menu.shtml), which can be linked to gene-trap libraries. Selected databases of gene-trap screens are already accessible in the World Wide Web and are summarized in **Table 3**.

In order to perform a large-scale phenotypical analysis of gene-trap mutants without overt phenotypes, phenotype centers should be established, which coordinate standardized quantitative analytical methods as well as noninvasive tests of function *(113)*. This includes the examination of clinically relevant parameters such as clinical–chemical and biochemical parameters, dysmorphology including pathological assessment,

Table 3
Selected Web Sites of Gene-Trap Mutations with Free and Restricted Access

Organization	Term for RACE Fragment	URL
Searchable Web Sites:		
Omnibank®	OST	www.lexgen.com
University of California (F)		socrates.berkeley.edu/~skarnes/resource.html
GSF (F)	GTST	tikus.gsf.de
Nonsearchable Web Sites:		
University of Manitoba (F)	PST	www.umanitoba.ca/institutes/manitoba_institute_cell_biology/
Max-Planck Goettingen (F)		www.mpibpc.gwdg.de/abteilungen/160/
Kumamoto University (F)		card.medic.kumamoto-u.ac.jp/card/english/index.html

F, free access; ®, restricted access; OST, Omnibank sequence tag; GTST, gene-trap sequence tag; Pst, promoter proximal sequence tag.

neurological analysis, and behavioral determination. A demonstration of handling large numbers of mice with a standardized phenotyping protocols is currently given by the ENU mutagenesis project and needs to be applied on the gene-trap screen as well *(6,114)*.

2. Materials

2.1. Recovery of Early Postimplantation Embryos (adapted from ref. 42)

1. 1 Pair of coarse forceps (FST, cat. no. 11373-22).
2. 1 Pair of fine scissors (FST, cat. no. 14071-12).
3. 2 Pairs of fine forceps (FST, cat. no. 11370-40).
4. 2 Pairs of watchmaker's forceps (Dumont, cat. no. 5).
5. Pasteur pipets.
6. Pasteur pipets with "wide opening."
7. Phosphate-buffered saline (PBS).
8. Petri dishes.
9. Dissecting microscope (Leica MZ8).

2.2. β-Gal Staining of Cultured Cells, Whole Embryos, and Tissues (42)

1. 10X PBS: 80 g NaCl, 2 g KCl, 14.4 g Na_2HPO_4, 2.4 g KH_2PO_4 to 800 mL with H_2O, HCl to pH 7.6, H_2O to 1 L.
2. Solution A: Kanolinite phenylphosphonate (KPP): 100 mM potassium phosphate buffer, pH 7.4, store at room temperature.
3. Solution B: 0.2% glutaraldehyde (GDA) in solution A containing 5 mM EGTA and 2 mM $MgCl_2$ (store at –20°C).
4. Solution C: 0.01% Na-desoxycholate and 0.02% Nonidet P-40 (NP40) in solution A containing 5 mM EGTA, and 2 mM $MgCl_2$ (store at room temperature).
5. Solution D: 0.5 mg/mL 5-bromo-4-chloro-3-indolyl-β-D-galactopyranoside (X-gal), 10 mM $K_3[Fe(CN)_6]$, and 10 mM $K_4[Fe(CN)_6]$ in solution C (store at –20°C in the dark).

2.3. In Vivo Staining of ES Cell Colonies (ref. *42*; adapted from ref. *52*)

1. Fluorodeoxyglucose (FDG) or Imagene.
2. Loading medium for FDG: dilute FDG stock solution (20 m*M* FDG in 10% dimethyl sulfoxide [DMSO]) 1:10 with sterile water; a 1:1 mixture of tissue culture medium and the 1:10 diluted FDG solution is used as loading medium. FDG stocks may vary depending on batch and supplier.
3. Fluorescence microscope.
4. ES cells grown on tissue culture dishes.

2.4. Paraffin Embedding of β-Gal-Stained Tissues (42)

2.4.1. Material

1. Paraffin for histology (e.g., Histowax; Reichert-Jung, FRG).
2. Isopropanol.
3. Screw cap tubes (e.g., 50 mL Falcon tubes).
4. Casting mould (e.g., Reichert-Jung, FRG; several sizes are available).
5. Small spatula.
6. Microtome and holder for fixing the paraffin block to the microtome.
7. Xylene.

2.5. Counter Staining Sections of lacZ-Stained Material (42)

1. Histological staining trays.
2. Hematoxilin and eosin solution (Sigma).
3. 60, 80, 96, and 100% ethanol.
4. Xylene.
5. Embedding medium.

2.6. ES-Cell Culture Media and Solutions*

1. PBS without Mg^{2+} and Ca^{2+} (Gibco).
2. Gelatinized plates (Nunc).
3. Geneticin (G418) (e.g., Gibco, cat. no. 11811-03) stock 200 mg/mL (4°C).
4. 0.05% Trypsin solution in Tris-saline-EDTA buffer (ICN Flow).
5. Gelatin solution: 0.1% in *aq* (Fresenius)
6. 2X concentrated freezing medium ES-Zellen: 50% fetal calf serum (FCS), 20% DMSO.
7. Nucleosides 100 ml 100X stock solution: 80 mg Adenosine (Sigma A-4036), 73 mg Cytidine (Sigma C-4654), 85 mg Guanosine (Sigma G-6264), 24 mg Thymidine (Sigma T-1895), 73 mg Uridine (Sigma U-3003), dissolve in 100 mL H_2O (pre-warm to 37°C and filter-sterilize. Aliquots (e.g. 5 ml in 15 ml Bluecap) can be stored at 4°C. Precipitate re-dissolves at 37°C.

2.6.1. tbv-2 ES Medium

Dulbeccos' Modified Eagles Medium (DMEM), high glucose, without Na-pyruvate containing: 15% FCS (heat-inactivated for 30 min at 56°C), 1 m*M* Na-pyruvate, 2 m*M* glutamine, $10^{-4}M$ β-mercaptoethanol (Gibco), use up within 2 wk. Add before use: 1000 U/mL leukemia inhibiting factor (LIF) (Gibco).

---

*ES-cell culture media and solutions kindly provided by S. Bourier and E. M. Fruechtbauer.

2.6.2. R1 ES-Medium **(58)**

DMEM, high glucose, without Na-pyruvate containing: 20% FCS, 1 mM Na-pyruvate, 2 mM glutamine, $10^{-4}M$ β-mercaptoethanol (Gibco), use up within 2 wk. Add before use: 1000 U/mL LIF (Gibco), 5.7 mL nonessential amino acids (from 100X stock; Gibco); optional: 50 U/mL Pen–Strep.

2.6.3. CJ7 ES-Medium

DMEM, high glucose, without Na-pyruvate containing 15% FCS, 2 mM glutamine, 5.7 mL nonessential amino acids (from 100X stock; Gibco), $10^{-4}M$ β-mercaptoethanol (Gibco), 5 mL nucleoside (100X), 1000 U/mL LIF (Gibco); optional: 50 U/mL Pen–Strep.

2.7. Electroporation of ES Cells* ("Gene-Trap Conditions")

1. ES cell medium.
2. PBS without Mg^{2+} and Ca^{2+}.
3. 10-cm gelatinized plates (Nunc).
4. G418 (e.g., Gibco, cat. no. 11811-03) stock (1000X) 200 mg/mL (4°C)
5. Electroporation apparatus (gene pulser; Bio-Rad).
6. Electroporation cuvets (0.4-cm gap; Bio-Rad, cat. no. 165-2088).

2.8. Infection of ES Cells with Retroviral Vectors (42)

1. Gelatinized tissue culture plates.
2. Medium-containing retrovirus.
3. Polybrene (Sigma, cat. no. H 9268).
4. ES cell medium containing LIF- or Buffalo rat liver (BRL)-conditioned medium.

2.9. Cloning Flanking Sequences by Inverse PCR (42)

1. Genomic DNA carrying the transgene insertion.
2. Oligonucleotide primers.
3. Restriction enzymes.
4. T4 Ligase (Boehringer, Mannheim).
5. *Taq* DNA polymerase, appropriate reagents and equipment for PCR.

2.10. Novel PCR-Based Technique of Genotyping Applied to Identification of Scrambler Mutation in Mice (adapted from ref. 99)

1. Microcentrifuge (Heraeus Instruments).
2. Minigel electrophoresis system (Desaga).
3. UV transilluminator (MWG-Biotech).
4. Thermal cycler (Biometra).
5. Agarose (Gibco BRL, cat. no. 540-5510UB).
6. Proteinase K (DNase-free) (Sigma, cat. no. P4914).
7. *Rsa*I and *Hae*III restriction endonucleases (NEBiolabs, cat. nos. 167S and 108S, respectively) in supplement with the buffers.

*Kindly provided by S. Bourier

8. T4 DNA ligase (Promega, cat. no. M180A) in supplement with the buffer. BioTherm DNA polymerase (GeneCraft) in supplement with the buffer DNA polymerization mixture (20 m*M* dNTPs) (Pharmacia Biotech, cat. no. 27-2094-01).
9. 100-bp DNA ladder (Gibco BRL, cat. no. 15628-019).

3. Methods

3.1. Recovery of Early Postimplantation Embryos (adapted from ref.42)

1. Sacrifice the pregnant female by cervical dislocation.
2. Lay mouse on its back, make a large V-shaped incision into the skin with the tip of the V just anterior to the vagina.
3. Fold the skin back, cut the abdominal wall similarly to the skin and fold back.
4. Find the uterus, hold the uterine horns with coarse forceps at the cervical end and cut at the uterine-cervical junction.
5. Pull uterine horns slightly, trim away the mesometrium (part of the broad ligament that is attached to one side of each uterine horn) along the uterine wall, and cut off the uterine horns at their anterior ends.
6. Put uterine horns into Petri dish containing PBS.
7. Cut the uterine horns between the decidual swellings.
8. Under the dissecting microscope, tear the uterine muscle on the opposite side to the attached mesometrium with fine watchmaker forceps.
9. Free the decidual swelling from the uterine wall by holding the muscle with one fine watchmaker forceps and sliding along between the torn muscle and deciduum with the second fine watchmaker forceps.
10. Transfer deciduas into fresh dish containing PBS.
11. Hold the deciduum on the mesometrial (broad) end with one fine watchmaker forceps. Insert the point of the closed second forceps in the midline above the redish streak (which is the embryo), through the deciduum, and open fine watchmaker forceps splitting the deciduum.
12. Grasp the split parts and pull apart. The embryo, surrounded in membranes, usually remains attached to one decidual half.
13. Gently push the embryo with the tip of the closed watchmaker forceps until the embryo and its membranes are entirely free.
14. Transfer the embryo with a Pasteur pipet (6.5 and 7.5 days postcoitum [dpc]) or a wide opening Pasteur pipet (8.5 and 9.5 dpc). Older specimens can be transferred by "scooping" them with a curved forceps.

3.1.1. Removal of Reichert's Membrane (6.5–8.5 dpc)

Grasp Reichert's membrane at the extraembryonic portion of the egg cylinder (away from the embryo) with both watchmaker forceps and gently tear it open. Most of the membrane can be torn off leaving behind only some remnants at the ectoplacental cone that do not impair staining and analysis of the embryo. It is advisable to leave the membranes attached to the embryo to discover extra-embryonic staining as well.

3.1.2. Removal of Extra-Embryonic Membranes (>d 8.5)

Grasp visceral yolk sac with both forceps. Tear until the embryo is freed from the yolk sac but still connected with it by the umbilical cord. Hold the cord with one forceps and tear off the yolk sac distally with second forceps. If the amnion, a very thin cellular membrane, is still surrounding the embryo remove that analogously to the yolk sac.

3.2. β-Gal Staining of Cultured Cells, Whole Embryos, and Tissues (42)

1. Wash in PBS: for cells: aspirate the medium and replace with PBS; repeat. For embryos: transfer into PBS, gently swirl around. For tissues: transfer into PBS, gently swirl around.
2. Fix in buffer B: for cells: add sufficient buffer B to the plate such that the cells are well covered and leave for 5 min at room temperature. For embryos: up to d 9.5, add 1 mL buffer B for 10–20 embryos and leave for 5 min at room temperature. For d 10.5 to 12.5 embryos, add 5–10 mL buffer B for 10 embryos and leave for 15 min at room temperature (*see* **Note 1**). For tissues: add about ten times the volume of the tissue of buffer B and leave 15–60 min (depending on size) at room temperature.
3. For all fixation steps: aspirate well wash buffer before adding buffer B to prevent dilution.
4. Wash 3 times with 10 mL buffer C at room temperature: for cells: 5 min each. For embryos up to day 9.5, 5 minutes each. For embryos up to d 10.5–12.5: 15 min each. For tissues: 15–60 min (depending on size).
5. Replace buffer C with buffer D and incubate at 37°C. Before adding buffer D, aspirate well buffer C: for cells: add sufficient buffer to the plate such that cells are well covered and that the solution will not evaporate (*see* **Notes 2, 3, 4** and **5**): for embryos up to d 9.5, add 1 mL buffer; for 10–20 embryos, for d 10.5–12.5, add 5–10 mL for 10 embryos. For tissues: add about ten times the volume of the tissue.
6. After staining, wash samples 3 times in 10 mL buffer C.
7. Samples can be stored for short term (a few days) in solution C at 4°C, but for prolonged storage, the specimens should be fixed again in 4% paraformaldehyde for 2 h at room temperature and kept in 70% ethanol at 4°C.

3.3. In Vivo Staining of ES Cell Colonies (ref. 42; adapted from ref. 52)

1. FDG: aspirate tissue culture medium and add sufficient loading medium to cover the cells (i.e., 1 mL/30-mm dish, 2 mL/60-mm dish, 3 mL/90-mm dish). Incubate for 1 min. Change back to regular tissue culture medium. Imagene: add dye directly to the medium at a final concentration of 33 μ*M*.
2. Incubate at 37°C for 1 h (FDG) or 2 h (Imagene).
3. Identify fluorescing colonies with a fluorescence microscope using 10× or 20× objectives and filters for fluorescein. Mark the positions of positive clones with a dot on the bottom of the dish (*see* **Note 5**).

3.4. Paraffin Embedding of β-Gal-Stained Tissues (42)

1. Place the dehydrated sample in 10 mL 100% isopropanol for 2 h with one change of the isopropanol at room temperature. Alternatively, samples can be dehydrated directly in isopropanol similarly as given below for ethanol. Using isopropanol, the steps are 50, 75, 90%, and 2 times 100%).
2. Preinfiltrate with paraffin : isopropanol (1 : 1) at 60°C.
3. Infiltrate with paraffin at 60°C. The times given in **Table 4**.
4. Place into prewarmed (60°C) mould and orientate the specimen with a needle or spatula and fill mould with paraffin.
5. Depending on the mould used, directly cast the holder for fixing the paraffin block to the microtome. After hardening, remove the paraffin block from the mould, and prepare 10 μ*M* sections.
6. Dewax sections 1 to 2 min in xylene. Sections can now be embedded or processed for counterstaining.

Table 4
Paraffin Incubation Times

	Embryos			Adult Tissue (e.g., brain)
	up to d 10.5	d 11.5–16.5	>d 16.5	
Incubation time	2 h	12 h	24 h	24 h

3.5. Counter Staining Sections of lacZ Stained Material (42)

1. Submerge sections for 1 min in hematoxilin solution in staining tray.
2. Rinse through staining tray for 5 min with tap water.
3. Submerge sections for 2 min in eosin solution.
4. Rinse 5 min with distilled water.
5. Dehydrate frozen sections and paraffin sections in ascending ethanol (60, 80, 96%, and 2X 100%, 2 min each).
6. Remove the ethanol by two incubations in xylene (1 min each), add Eukitt, and put coverslip on. Any other commercially available embedding solution can be used. Methacrylate sections can be dried and embedded directly.

3.6. Electroporation of ES Cells* ("Gene-Trap Conditions")

3.6.1. Expansion of ES Cells

1. Thaw one vial of early-passage ES cells, wash as usual, and plate the cells on a 60-mm gelatin-coated Petri dish with primary embryonic feeder cells (EMFI). EMFI cells should be confluent on the plate.
2. Change the ES cell medium once each day.
3. After 2 d, trypsinize and expand the cells on one to two 90-mm gelatin-coated Petri dishes (with EMFI cells) dilution 1:3 up to 1:8, depending on the cell density.
4. Wait another 2 d and transfer the cells to fresh feeder plates again dilution 1:3 or 1:4.
5. After 2 d expand the cells to at least 2×200 mm or 6×90 mm gelatinized Petri dishes (only a dilution of 1:2) and start the transfection 36 h later in one electroporation cuvette. The cell number on 2×200 mm subconfluent Petri dishes should be approx 1×10^8 cells. Change medium in the plates 6 h before the electroporation.

3.6.2. Electroporation and Selection

1. Trypsinize ES cells for 10 min in 3 mL trypsin/dish, and pipet them gently up and down to aqcuire a single-cell suspension, add 3 mL medium/dish, and transfer the cells to 2 Falcon tubes.
2. Centrifuge the cells for 3 min at 270*g*.
3. Resuspend the cells in 10 mL PBS.
4. Dilute an aliquot 1:10 and count the cells (keep the cells on ice).
5. Centrifuge 1×10^8 cells 5 min at 270*g* for one cuvette.
6. Add to the pellet 500 µL cold PBS and 100 µL of the DNA (120 µg), no more than 700–800 µL in total (Vector linearization: digest with adaequate enzyme 1 U/µg for at least 4 h, phenol-extract, and precipitate the DNA at 70°C for 10–15 min. Wash in 70% ETOH and air-dry under sterile conditions).

*Kindly provided by S. Bourier.

7. Transfer the suspension to the electroporation cuvette.

8. Set up the electroporation conditions in advance (0.8 kV, 3 µF for the Bio-Rad gene pulser).

9. Transfer the cuvette into the cuvette-holder with electrodes facing the output leads and deliver electric pulse.

10. Remove the cuvette from the cuvette-holder and leave it at room temperature or on ice for 10–20 min.

11. Transfer the cell suspension from one cuvette into 12 mL ES cell medium. Seed the electroporated cells at a density of $2.5–5 \times 10^6$ cells/90-mm dish on a gelatinized plate in medium containing LIF. The cell concentration per plate must be adjusted depending on the vector such that no more than 200–500 neo^R colonies are obtained on each plate.

12. Change the ES medium the next day.

13. Two days after electroporation, add the drugs for selection to the ES medium (e.g., G418: 200 µg/mL [active]; puromycin: 1 µg/mL).

14. Change the selection medium every day for the first 3 d, than every other 2 d.

15. About 6–8 d of selection, drug-resistant colonies should have appeared.

16. After 8 to 9 d of selection, colonies are picked and plated on 96-well feeder plates containing ES medium. Stop the selection-pressure and exchange the ES cell medium each day. After 1 d of growth, tryplate the colonies, after another 2 d, trypsinize and dilute the clones 1:3. After another 2 d, split each 96-well replica-plate 1:1 on 2 × 48 well feeder plates.

3.7. Infection of ES Cells with Retroviral Vectors (42)

1. Plate ES cells on gelatinized tissue culture dishes at a density of 3×10^6 cells/90-mm dish in ES cell medium supplemented with LIF- or BRL-conditioned medium.

2. After 24 h, aspirate medium, add 5 mL fresh medium containing the retroviral particles at a multiplicity of infection (moi) <1 and 5 µg/mL polybrene to obtain single integration per clone.

3. After overnight culture (14 h) remove virus-containing medium, add 10 mL fresh medium, and culture for an additional 24 h.

4. Change the medium to selection medium and change medium every other day. Drug-resistant colonies should become visible between 7–10 d.

3.8. Generation of ES Cell-Chimeric Embryos for lacZ Expression Analysis (42)

1. Inject 10–15 albino-derived blastocysts with ES cells derived from a pigmented mouse for each early embryonic stage to be analyzed.

2. Inject an additional 10–15 blastocysts for the d 12.5 control.

3. Transfer the embryos to foster females in groups of 10–15 (1 foster/stage to be analyzed, and 1 additional for the control).

4. Recover embryos at desired stages and stain for β-gal activity.

5. Recover control embryos at d 12.5, and monitor the embryos for pigmented cells in the eye (which can be easily seen by the naked eye or under a dissecting microscope), and stain for β-gal activity.

3.9. Cloning Flanking Sequences by Inverse PCR (ref. 42; H.v. Melchner, personal communication)

1. Digest genomic DNA to completion with a restriction enzyme that does not cut within vector DNA sequences between oligonucleotides 1 and 2, and that cuts at least 0.5–1 kb

Table 5
PCR Conditions

No. of Cycles	94°C (denaturation)	T_m (annealing)	72°C (extension)
1	10 min	2 min	2 min
30	1.5 min	2 min	2 min
1	1.5 min	2 min	10 min

outside the vector in the genomic flanking DNA. Heat-inactivate (15 min at 65°C) the restriction enzyme when digestion is complete.

2. Ligate the digested DNA with 8 Weiss U of T4 DNA ligase at 14°C for 16 h at a concentration of 1 µg DNA/mL in a total volume of 600 mL. This DNA concentration favors intramolecular as opposed to inter-molecular ligation. Heat-inactivate (15 min at 65°C) the ligation mixture.
3. Linearize DNA with appropriate restriction enzyme. Add the enzyme and its 10X incubation buffer (see manufacturer's data sheet) directly to the ligation reaction. After appropriate incubation, heat-inactivate (15 min at 65°C) the restriction enzyme.
4. Precipitate the DNA in high salt and isopropanol in the presence of 10 µg/mL tRNA or 1 µg/mL glycogen (Roche, cat. no. 901393).
5. Set up the amplification reaction by combining:
   a. 1–100 ng genomic DNA (cut, ligated, and recut).
   b. 5 m*M* dNTPs each.
   c. 1 to 2 U *Taq* DNA polymerase.
   d. 0.1–0.5 µ*M* primers each.
   e. 5 µL 10X *Taq* incubation buffer (manufacturer recommended buffer).
   f. H$_2$O to a total vol of 50 µL (**Table 5**).
6. 20 microliters of each test reaction is analyzed on a 1% agarose gel. A band corresponding in size to the fragment deduced from primer placement and genomic Southern blot data should be visible on an ethidium bromide-stained gels. Confirm that the correct piece of DNA has been amplified by Southern blot analysis of the gel by hybridization with vector sequences contained in the amplified fragment.
7. Digest the rest of the test reaction, or additional reactions, with restriction enzymes, which lie outside the first primer pair but within vector sequences. Alternatively, PCR primers that contain internal restriction sites can be designed. Analyze an aliquot of this digestion on an agarose gel. The amplified fragment should have shifted downward by the anticipated number of base pairs.
8. Purify the DNA fragment from an agarose gel (by electroelution, low melting point agarose, etc.).
9. Ligate the digested and purified PCR fragment to a plasmid vector cut with the same enzymes. If no convenient restriction enzymes (which do not cut within the genomic flank) exist, clone the purified PCR product blunt-ended into a plasmid vector, or use appropriate plasmids for the direct cloning of PCR products (e.g., TA cloning® kit [Invitrogen], AdvanTAge™ PCR cloning kit [Clontech]).
10. Transform competent bacteria with an aliquot of the ligation reaction and screen transformants for the presence of the desired clone.
11. The efficiency of the first steps of the protocol can be estimated on agarose gels. If multiple bands are visible after the PCR step, try less input DNA or raise the annealing temperature above the melting temperature (T_m).

3.10. Novel PCR-Based Technique of Genotyping Applied to Identification of Scrambler Mutation in Mice (adapted from ref. 99)

3.10.1. Genomic DNA Extraction

1. Place a cut tip of mouse tail into 0.5 mL of lysis buffer (100 mM Tris-HCl, pH 8.5, 5 mM EDTA, 0.2% SDS, 200 mM NaCl, 200 mg/mL proteinase K).
2. Shake overnight at 55°C.
3. Centrifuge for 20 min at 15,000g, at 4°C.
4. Collect supernatant and add 0.5 mL of isopropanol. Mix gently. Genomic DNA should become visible as a fibrous substance.
5. Centrifuge for 2 min at 15,000g, and discard the supernatant.
6. Add 1 mL of 80% ethanol, centrifuge for another 2 min, and discard the supernatant.
7. Repeat this washing with 80% ethanol.
8. Dry the pellet and dissolve it in 100 mL of Tris-HCl, pH 8.0. Normally, this requires 2 h of shaking at 55°C.

3.10.2. Endonuclease Digestion

1. Take 5 μL of genomic DNA solution from **Subheading 3.10.1.** Set the digestion in the total vol of 50 μL using 15 U of an enzyme that will generate blunt ends.
2. Incubate for 2 h at 37°C.
3. Extract digested DNA with phenol:chloroform and precipitate with sodium-acetate and ethanol.
4. Dissolve in 10 μL of 10 mM Tris-HCl, pH 8.0.
5. Run 1 μL of this solution on 1.5% agarose gel to estimate the quantity: each lane should contain 50–100 ng of DNA.

3.10.3. Adapter Ligation

1. The long strand of pseudo-double-stranded adapter was: 5′-AGCAGCGAACTCAGTACA ACAACTCTCCGACC-TCTCACCGAGT-3′.
2. The short strand was: 5′-ACTCG- GTGA-3′.
3. Perform the adapter ligation reaction in the final vol of 10 μL. The mixture should contain 200–400 ng of DNA, 2 μM of each adapter strand, and 1 U of T4 DNA ligase in the reaction buffer provided by the enzyme supplier.
4. The reaction is carried out for 3 h at room temperature or overnight at 12–16°C.

3.10.4. PCR Conditions

1. Primers used for PCR amplification of the scrambler-mutation were:
   a. Sc1: 5′-TTTTGTCCTTCTCTATAACT-3′.
   b. Sc2: 5′-CCTGGGA-TAATGGGGTAAG-3′.
   Instead, nested primers can be designed according to the gene-trap vector sequence. Annealing temperatures and extension times have to be determined individually.
2. The distal adapter primer (DAP): 5′-AGCAGCGAACTCAGTACAACA-3′ (corresponds to the 5′ part of the long strand of adapter).
3. Add to the ligation mixture: 4.5 μL of 10X PCR buffer, 10 nmol of each dNTP, 15 pmol of Sc1 primer, and water to the final vol of 45 μL, and cover the mixture with mineral oil.
4. To perform the "hot start," heat to 94°C in the PCR machine. Add 2.5 U of BioTherm DNA polymerase in 5 μL of 1X PCR buffer to the reaction mixture, while avoiding a cooling of the tube.
5. Carry out 22 cycles of PCR: 10 s at 94°C, 30 s at 55°C, and 60 s at 72°C.

6. When the amplification is complete, dilute the PCR mixture 40-fold with water.
7. Transfer 1 μL of this dilution into 25 μL of the second PCR mixture. The latter contains PCR buffer components, 5 nmol of dNTPs, 8 pmol of Sc2 primer, and 5 pmol of DAP, 2 U of BioTherm DNA polymerase (which should be added directly to the cold mixture).

Run 23 cycles of PCR: 10 s at 94°C, 30 s at 60°C, and 60 s at 72°C (please note that the annealing temperature and extension time is given for the scrambler mutation).

3.10.5. Analysis of PCR-Amplified Fragments

Separate PCR products by horizontal electrophoresis through a 1.5% agarose gel in parallel with the DNA marker for size estimation. To visualize the DNA, stain the gel with ethidium bromide.

3.11. In Vivo Staining of ES Cell Colonies (ref. 42; adapted from ref. 52)

Verify positive clones afterwards by X-gal staining. The fluorescence signal of a true positive clone can vary tremendously, and most of the signals localize to just a portion of the colony. Use a cell line known to express *lacZ* at clearly detectable levels as a positive control.

4. Notes

4.1. β-Gal Staining of Cultured Cells, Whole Embryos, and Tissues (42)

1. Always prepare fresh fixing solution before use or store at –20°C; other fixations can also be used (e.g., 2% glutaraldehyde or 4% paraformaldehyde [PFA] are possible).
2. When solution D is prepared freshly chill on ice for 10 min, spin down precipitate, and aliquot supernatant.
3. The staining solution D can be reused several times; filter after each use, and keep in dark at –20°C.
4. A 50 mg/mL X-gal stock solution (=100×) can be made both in DMSO or dimethylformamide (DMF). Keep at –20°C. DMF has the advantage to stay liquid at –20°C.
5. $K_3[Fe(CN)_6]$ and $K_4[Fe(CN)_6]$ should be kept as 0.5 M stock solutions (=50X) in dark bottles at (–20°C).

Acknowledgments

We are grateful to Jordi Guimera, Alessandro DeGrandi, and Karen Floss for critical comments, Geoffrey Hicks, Susanne Bourier, Paul Szauter, Harald von Melchner, Ernst-Martin Fuechtbauer, and Kamal Chowdhury for personal communication.

References

1. Benson, D. A., Karsch-Mizrachi, I., Lipman, D. J., Ostell, J., Rapp, B. A., and Wheeler, D. L. (2000) GenBank. *Nucleic Acids Res.* **28**, 15–18.
2. Eizinger, A., Jungblut, B., and Sommer, R. J. (1999) Evolutionary change in the functional specificity of genes. *Trends Genet.* **15**, 197–202.
3. Oliver, S. G. (1996) From DNA sequence to biological function. *Nature* **379**, 597–600.
4. Green, M. C. (1989) Catalogue OF mutant gene and polymorphic loci, in *Gene TIC Variants and Strains OF the laboratory mouse* (Lyons, M. F. and Searle, A. G., eds.), Oxford University, Oxford, pp. 12–403.
5. Balling, R., Brown, S., Hrabe de Angelis, M., Justice, M., Nadeau, J., and Peters, J. (2000) Great times for mouse genetics: getting ready for large-scale ENU-mutagenesis. *Mamm. Genome* **11**, 471.

6. Nolan, P. M., Peters, J., Vizor, L., Strivens, M., Washbourne, R., Hough, T., et al. (2000) Implementation of a large-scale ENU mutagenesis program: towards increasing the mouse mutant resource. *Mamm. Genome* **11,** 500–506.

7. Beier, D. R. (2000) Sequence-based analysis of mutagenized mice. *Mamm. Genome* **11,** 594–597.

8. Antoch, M. P., Song, E. J., Chang, A. M., Vitaterna, M. H., Zhao, Y., Wilsbacher, L. D., et al. (1997) Functional identification of the mouse circadian Clock gene by transgenic BAC rescue. *Cell* **89,** 655–667.

9. King, D. P., Zhao, Y., Sangoram, A. M., Wilsbacher, L. D., Tanaka, M., Antoch, M. P., et al. (1997) Positional cloning of the mouse circadian clock gene. *Cell* **89,** 641–653.

10. Li, Z., Otevrel, T., Gao, Y., Cheng, H. L., Seed, B., Stamato, T. D., et al. (1995) The XRCC4 gene encodes a novel protein involved in DNA double-strand break repair and V(D)J recombination. *Cell* **83,** 1079–1089.

11. Thomas, K. R. and Capecchi, M. R. (1987) Site-directed mutagenesis by gene targeting in mouse embryo-derived stem cells. *Cell* **51,** 503–512.

12. Doetschman, T., Maeda, N., et al. (1988) Targeted mutation of the Hprt gene in mouse embryonic stem cells. *Proc. Natl. Acad. Sci. U.S.A.* **85,** 8583–8587.

13. Olson, E. N., Arnold, H. H., et al. (1996) Know your neighbors: three phenotypes in null mutants of the myogenic bHLH gene MRF4. *Cell* **85,** 1–4.

14. Yoon, J. K., Olson, E. N., et al. (1997) Different MRF4 knockout alleles differentially disrupt Myf-5 expression: cis-regulatory interactions at the MRF4/Myf-5 locus. *Devel. Biol.* **188,** 349–362.

15. Pham, C. T., MacIvor, D. M., Hug, B. A., Heusel, J. W., and Ley, T. J. (1996) Long-range disruption of gene expression by a selectable marker cassette. *Proc. Natl. Acad. Sci. USA* **93,** 13,090–13,095.

16. Hug, B., Wesselschmidt, R. L., Fiering, S., Bender, M. A., Epner, E., Groudine, M., and Ley, T. (1996) Analysis of mice containing a targeted deletion of beta-Globin locus control region 5' hypersensitive site 3. *Mol. Cell. Biol.* **16,** 2906–2912.

17. Floss, T., Arnold, H. H., and Braun, T. (1996) Myf-5(m1)/Myf-6(m1) compound heterozygous mouse mutants down-regulate Myf-5 expression and exert rib defects: evidence for long-range cis effects on Myf-5 transcription. *Dev. Biol.* **174,** 140–147.

18. Walters, M. C., Magis, W., et al. (1996) Transcriptional enhancers act in cis to suppress position-effect variegation. *Genes Dev.* **10,** 185–195.

19. Leder, A., Daugherty, C., et al. (1997) Mouse zeta- and alpha-globin genes: embryonic survival, alpha-thalassemia, and genetic background effects. *Blood* **90,** 1275–1282.

20. Kaul, A., Köster, M., Neuhaus, H., and Braun, T. (2000) Myf-5 revisited: loss of early myotome formation does not lead to a rib phenotype in homozygous myf-5 mutant mice. *Cell* **102,** 17–19.

21. Akiyama, K., Watanabe, H., Tsukada, S., and Sasai, H. (2000) A novel method for constructing gene-targeting vectors. *Nucleic Acids Res.* **16,** I–VI.

22. Wattler, S., Kelly, M., et al. (1999) Construction of gene targeting vectors from lambda KOS genomic libraries. *BioTechniques* **26,** 1150–1160.

23. Westphal, C. H. and Leder, P. (1997) Transposon-generated "knock-out" and "knock-in" gene-targeting constructs for use in mice. *Curr. Biol.* **7,** 530–533.

24. Gossler, A., et al. (1989) Mouse embryonic stem cells and reporter constructs to detect developmentally regulated genes. *Science* **244,** 463–465.

25. Friedrich, G. and Soriano, P. (1991) Promoter traps in embryonic stem cells: a genetic screen to identify and mutate developmental genes in mice. *Genes Dev.* **5,** 1513–1523.

26. von Melchner, H., DeGregori, J. V., Rayburn, H., Reddy, S., Friedel, C., and Ruley, H. E. (1992) Selective disruption of genes expressed in totipotent embryonal stem cells. *Genes Dev.* **6,** 919–927.

27. Wurst, W., Rossant, J., Prideaux, V., Kownacka, M., Joyner, A., Hill, D. P., et al. (1995) A large-scale gene-trap screen for insertional mutations in developmentally regulated genes in mice. *Genetics* **139,** 889–899.

28. Skarnes, W. C., Auerbach, B. A., and Joyner, A. L. (1992) A gene trap approach in mouse embryonic stem cells: the lacZ reported is activated by splicing, reflects endogenous gene expression, and is mutagenic in mice. *Genes Dev.* **6,** 903–918.

29. Scherer, C. A., Chen, J., et al. (1996) Transcriptional specificity of the pluripotent embryonic stem cell. *Cell Growth Differ.* **7,** 1393–1401.

30. Forrester, L. M., Nagy, A., Sam, M., Watt, A., Stevenson, L., Bernstein, A., et al. (1996) An induction gene trap screen in embryonic stem cells: Identification of genes that respond to retinoic acid in vitro. *Proc. Natl. Acad. Sci. USA* **93,** 1677–1682.

31. Hicks, G. G., Shi, E. G., Li, X. M., Li, C. H., Pawlak, M., and Ruley, H. E. (1997) Functional genomics in mice by tagged sequence mutagenesis. *Nat. Genet.* **16,** 338–344.

32. Zambrowicz, B. P., et al. (1998) Disruption and sequence identification of 2,000 genes in mouse embryonic stem cells. *Nature* **392,** 608–611.

33. Wiles, M. V., Vauti, F., Otte, J., Fuchtbauer, E. M., Ruiz, P., Fuchtbauer, A., et al. (2000) Establishment of a gene-trap sequence tag library to generate mutant mice from embryonic stem cells. *Nat. Genet.* **24,** 13–14.

34. Casadaban, M.-J. and Cohen, S.-N. (1979) Lactose genes fused to exogenous promoters in one step using a Mu-lac bacteriophage: In vivo probe for transcriptional control sequences. *Proc. Natl. Acad. Sci. USA* **92,** 4530–4533.

35. Bellofatto, V., Shapiro, L., and Hodgson, D. A. (1984) Generation of a Tn5 promoter probe and its use in the study of gene expression in Caulobacter crescentus. *Proc. Natl. Acad. Sci. USA* **81,** 1035–1039.

36. O'Kane, C. J. and Gehring, W. (1987) Detection in situ of genomic regulatory elements in *Drosophila. Proc. Natl. Acad. Sci. USA* **84,** 9123–9127.

37. Hope, I. A. (1991) Promoter trapping in *Caenorhabditis elegans. Development* **113,** 399–408.

38. Rubin, G. M. and Spradling, A. C. (1982) Genetic transformation of *Drosophila* with transposable element vectors. *Science* **218,** 348–353.

39. Cooley, L., Kelley, R., and Spardling, A. (1988) Insertional mutagenesis of the *Drosophila* genome with single P-elements. *Science* **239,** 1121–1128.

40. Gogos, J. A., Thompson, R., Lowry, W., Sloane, B. F., Weintraub, H., and Horwitz, M. (1996) Gene trapping in differentiating cell lines: regulation of the lysosomal protease cathepsin B in skeletal myoblast growth and fusion. *J. Cell Biol.* **134,** 837–847.

41. Moore, M. J. and Sharp, P. A. (1993) Evidence for two active sites in the spliceosome provided by stereochemistry of pre-mRNA splicing. *Nature* **365,** 364–368.

42. Wurst, W. and Gossler, A. (2000) Gene trap strategies in ES cells, in *Gene targeting, A practical approach, 2nd ed.* (Joyner, A., ed.) Oxford University Press, New York.

43. Ishida, Y. and Leder, P. (1999) RET: a poly A-trap retrovirus vector for reversible disruption and expression monitoring of genes in living cells. *Nucleic Acids Res.* **27,** e35.

44. Ghattas, I. R., Sanes, J. R., and Majors, J. E. (1991) The encephalomyocarditis virus internal entry site allows efficient coexpression of two genes from a recombinant provirus in cultured cells and in embryos. *MCB* **11,** 5848–5859.

45. Jang, S. K. and Wimmer, E. (1990) Cap-independent translation of encephalomyocarditis virus RNA: structural elements of the internal ribosomal entry site and involvement of a 57-kD RNA-binding protein. *Genes Dev.* **4,** 1560–1572.

46. Kang, H. M., Kang, N. G., Kim, D. G., and Shin, H. S. (1997) Dicistronic tagging of genes active in embryonic stem cells. *Mol. Cell* **7,** 502–508.

47. Yoshida, M., Yagi, T., Furuta, Y., Takayanagi, K., Kominami, R., Takeda, N., et al. (1995) A new strategy of gene trapping in ES cells using 3′RACE. *Transgenic Res.* **4,** 277–287.

48. Salminen, M., Meyer, B. I., and Gruss, P. (1998) Efficient poly A trap approach allows the capture of genes specifically active in differentiated embryonic stem cells and in mouse embryos. *Dev Dyn.* **212,** 326–333.
49. Barabino, S. M. and Keller, W. (1999) Last but not least: regulated poly(A) tail formation. *Cell* **99,** 9–11.
50. Dhundale, A. and Goddard, C. (1996) Reporter assays in the high-throughput screening laboratory: A rapid and robust first look? *J. Biomol. Screen* **1,** 115–118.
51. Suto, C. M. and Ignar, D. M. (1997) Selection of an optimal reporter gene for cell-based high throughput screening assays. *J. Biomol. Screen* **2,** 7–9.
52. Xiong, J. W., Battaglino, R., Leahy, A., and Stuhlmann, H. (1998) Large-scale screening for developmental genes in embryonic stem cells and embryoid bodies using retroviral entrapment vectors. *Dev. Dyn.* **212,** 181–197.
53. Silverman, L., Campbell, R., and Broach, J. R. (1998) New assay technologies for high-throughput screening. *Curr. Opin. Chem. Biol.* **2,** 397–403.
54. Liu, H. S., Jan, M. S., Chou, C. K., Chen, P. H., and Ke, N. J.(1999) Is green fluorescent protein toxic to the living cells? *Biochem. Biophys. Res. Commun.* **260,** 712–717.
55. Natarajan, D. and Boulter, C. A. (1995) A lacZ-hygromycin fusion gene and its use in a gene trap vector for marking embryonic stem cells. *Nucleic Acids Res.* **19,** 4003–4004.
56. Townley, D. J., et al. (1997) Rapid sequence analysis of gene trap integrations to generate a resource of insertional mutations in mice. *Genome Res.* **7,** 293–298.
57. Mainguy, G., Luz Montesinos, M., Lesaffre, B., Zevnik, B., Karasawa, M., Kothary, R., et al. (2000) An induction gene trap for identifying a homeoprotein-regulated locus. *Nat. Biotechnol.* **18,** 746–749.
58. Nagy, A., Rossant, J., Nagy, R., Abramow-Newerly, W., and Roder, J. C. (1993) Derivation of completely cell culture-derived mice from early-passage embryonic stem cells. *Proc. Natl. Acad. Sci. USA* **90,** 8424–8428.
59. Chowdhury, K., et al. (1997) Evidence for the stochastic integration of gene trap vectors into the mouse germline. *Nucleic Acids Res.* **25,** 1531–1536.
60. Bronchain, O. J., Hartley, K. O., and Amaya, E. (1999) A gene trap approach in *Xenopus*. *Curr. Biol.* **9,** 1195–1198.
61. Martinez-Salas, E. (1999) Internal ribosome entry site biology and its use in expression vectors. *Curr. Opin. Biotechnol.* **10,** 458–464.
62. Xiong, J. W., Leahy, A., Lee, H. H., and Stuhlmann, H. (1999) Vezf1: A Zn finger transcription factor restricted to endothelial cells and their precursors. *Dev. Biol.* **206,** 123–141.
63. Sternberg, N. and Hamilton, D. (1981) Bacteriophage P1 site-specific recombination. I. recombinattori between lox P sites. *J. Mol. Biol.* **150,** 467–486.
64. Betz, U. A., Vosshenrich, C. A., Rajewsky, K., and Muller, W. (1996) Bypass of lethality with mosaic mice generated by Cre-loxP-mediated recombination. *Curr. Biol.* **6,** 1307–1316.
65. Seibler, J., Schubeler, D., et al. (1998) DNA cassette exchange in ES cells mediated by Flp recombinase: an efficient strategy for repeated modification of tagged loci by marker-free constructs. *Biochemistry* **37,** 6229–6234.
66. Lafon, I., Carballes, F., Brewer, G., Poiret, M., and Morello, D. (1998) Developmental expression of AUF1 and HuR, two c-myc mRNA binding proteins. *Oncogene* **16,** 3413–3421.
67. Claverie, J.-M. (1997) Computational methods for the identification of genes in vertebrate genomic sequences. *Hum. Mol. Genet.* **6,** 1735–1744.
68. Kozak, M. (1996) Interpreting cDNA sequences: some insights from studies on translation. *Mamm. Genome* **7,** 563–574.
69. Ladunga, I. (2000) Large-scale predictions of secretory proteins from mammalian genomic and EST sequences. *Curr. Opin. Biotechnol.* **11,** 13–18.

70. Skarnes, W. C., Moss, J. E., Hurtley, S. M., and Beddington, R. S. (1995) Capturing genes encoding membrane and secreted proteins important for mouse development. *Proc. Natl. Acad. Sci. USA* **92,** 6592–6596.

71. Linney, E. (1992) Retinoic acid receptors: transcription factors modulating gene regulation, development, and differentiation. *Curr. Top. Dev. Biol.* **27,** 309–350.

72. Hill, D. P. and Wurst, W. (1993) Gene and enhancer trapping: mutagenic strategies for developmental studies. *Curr. Top. Dev. Biol.* **28,** 181–206.

73. Frohman, M. A., Dush, M. K., and Martin, G. R. (1988) Rapid production of full-length cDNAs from rare transcripts: amplification using a single gene-specific oligonucleotide primer. *Proc. Natl. Acad. Sci. USA* **85,** 8998–9002.

74. Araki, K., et al. (1999) Exchangeable gene trap using the Cre/mutated lox system. *Cell Mol. Biol.* **45,** 737–750.

75. Reik, W., Weiher, H., and Jaenisch, R. (1985) Replication-competent Moloney murine leukemia virus carrying a bacterial suppressor tRNA gene: selective cloning of proviral and flanking host sequences. *Proc. Natl. Acad. Sci. USA* **82,** 1141–1145.

76. Soriano, P., Gridley, T., and Jaenisch, R. (1987) Retroviruses and insertional mutagenesis in mice: proviral integration at the Mov 34 locus leads to early embryonic death. *Genes Dev.* **1,** 366–375.

77. Friedrich, G. A., Hildebrand, J. D., and Soriano, P. (1997) The secretory protein Sec8 is required for paraxial mesoderm formation in the mouse. *Dev. Biol.* **192,** 364–374.

78. Shawlot, W., Deng, J. M., Fohn, L. E., and Behringer, R. R. (1998) Restricted beta-galactosidase expression of a hygromycin-lacZ gene targeted to the beta-actin locus and embryonic lethality of beta-actin mutant mice. *Transgenic Res.* **7,** 95–103.

79. Faisst, A. M. and Gruss, P. (1998) Bodenin: a novel murine gene expressed in restricted areas of the brain. *Dev. Dyn.* **212,** 293–303.

80. McClive, P., Pall, G., Newton, K., Lee, M., Mullins, J., and Forrester, L. (1998) Gene trap integrations expressed in the developing heart: insertion site affects splicing of the PT1-ATG vector. *Dev. Dyn.* **212,** 267–276.

81. Voss, A. K., Thomas, T., and Gruss, P. (1998) Compensation for a gene trap mutation in the murine microtubule-associated protein 4 locus by alternative polyadenylation and alternative splicing. *Dev. Dyn.* **212,** 258–266.

82. Sam, M., Wurst, W., Kluppel, M., Jin, O., Heng, H., and Bernstein, A. (1998) Aquarius, a novel gene isolated by gene trapping with an RNA-dependent RNA polymerase motif. *Dev. Dyn.* **212,** 304–317.

83. Couldrey, C., et al. (1999) A retroviral gene trap insertion into the histone 3.3A gene causes partial neonatal lethality, stunted growth, neuromuscular deficits and male sub-fertility in transgenic mice. *Hum. Mol. Genet.* **8,** 2489–2495.

84. Gasca, S., Hill, D. P., Klingensmith, J., and Rossant, J. (1995) Characterization of a gene trap insertion into a novel gene, cordon-bleu, expressed in axial structures of the gastrulating mouse embryo. *Dev Genet.* **17,** 141–154.

85. Hanke, J., Brett, D., Zastrow, I., Aydin, A., Delbruck, S., Lehmann, G., et al. (1999) Alternative splicing of human genes: more the rule than the exception? *Trends Genet.* **15,** 389–390.

86. Ashfield, R., Patel, A. J., Bossone, S. A., Brown, H., Campbell, R. D., Marcu, K. B., and Proudfoot, N. J. (1994) MAZ-dependent termination between closely spaced human complement genes. *EMBO J.* **13,** 5656–5667.

87. Yonaha, M. and Proudfoot, N. J. (1999) Specific transcriptional pausing activates polyadenylation in a coupled in vitro system. *Mol. Cell.* **3,** 593–600.

88. Niwa, M. and Berget, S. M. (1991) Mutation of the AAUAAA polyadenylation signal depresses in vitro splicing of proximal but not distal introns. *Genes Dev.* **5,** 2086–2095.

89. Niwa, M., MacDonald, C. C., and Berget, S. M. (1992) Are vertebrate exons scanned during splice-site selection? *Nature* **360,** 277–280.

90. Cooke, C., Hans, H., and Alwine, J. C. (1999) Utilization of splicing elements and polyadenylation signal elements in the coupling of polyadenylation and last-intron removal. *Mol. Cell Biol.* **19,** 4971–4979.

91. Gashler, A. L., Swaminathan, S., and Sukhatme, V. P. (1993) A novel repression module, an extensive activation domain, and a bipartite nuclear localization signal defined in the immediate-early transcription factor Egr-1. *Mol. Cell. Biol.* **13,** 4556–4571.

92. Bullock, S. L., Fletcher, J. M., Beddington, R. S. P., and Wilson, V. A. (1998) Renal agenesis in mice homozygous for a gene trap mutation in the gene encoding heparan sulfate 2-sulfotransferase. *Genes Devel.* **12,** 1894–1906.

93. Stoykova, A., Chowdhury, K., Bonaldo, P., Torres, M., and Gruss, P. (1998) Gene trap expression and mutational analysis for genes involved in the development of the mammalian nervous system. *Dev. Dyn.* **212,** 198–213.

94. Cecconi, F., Alvarez-Bolado, G., Meyer, B. I., Roth, K. A., and Gruss, P. (1998) Apaf1 (CED-4 homolog) regulates programmed cell death in mammalian development. *Cell* **94,** 727–737.

95. Hildebrand, J. D. and Soriano, P. (1999) Shroom, a PDZ domain-containing actin-binding protein, is required for neural tube morphogenesis in mice. *Cell* **99,** 485–497.

96. Pierson, V. L. and Barcak, G. J. (1999) Development of *E. coli* host strains tolerating unstable DNA sequences on ColE1 vectors. *Focus* **21,** 1.

97. Hartl, D. L. and Ochman, H. (1994) Inverse polymerase chain reaction. *Methods Mol. Biol.* **31,** 187–196.

98. Triglia, T. (2000) Inverse PCR (IPCR) for obtaining promoter sequence. *Methods Mol. Biol.* **130,** 79–83.

99. Usman, N., Tarabykin, V., and Gruss, P. (2000) The novel PCR-based technique of genotyping applied to identification of scrambler mutation in mice. *Brain Res. Brain Res. Protoc.* **5,** 243–247.

100. Baker, R. K., Haendel, M. A., Swanson, B. J., Shambaugh, J. C., Micales, B. K., and Lyons, G. E. (1997) In vitro preselection of gene-trapped embryonic stem cell clones for characterizing novel developmentally regulated genes in the mouse. *Dev. Biol.* **185,** 201–214.

101. Takeuchi, T., Yamazaki, Y., Katoh-Fukui, Y., Tsuchiya, R., Kondo, S., Motoyama, J., and Higashinakagawa, T. (1995) Gene trap capture of a novel mouse gene, jumonji, required for neural tube formation. *Genes Dev.* **9,** 1211–1222.

102. Araki, K., Araki, M., and Yamamura, K. (1997) Targeted integration of DNA using mutant lox sites in embryonic stem cells. *Nucleic Acids Res.* **25,** 868–872.

103. Zheng, B., Sage, M., Sheppeard, E. A., Jurecic, V., and Bradley, A. (2000) Engineering mouse chromosomes with Cre-loxP: range, efficiency, and somatic applications. *Mol. Cell Biol.* **20,** 648–655.

104. Herault, Y., Beckers, J., Gerard, M., and Duboule, D. (1999) Hox gene expression in limbs: colinearity by opposite regulatory controls. *Dev. Biol.* **208,** 157–165.

105. Rideout, III, W., Wakayama, T., Wutz, A., Eggan, K., Jackson-Grusby, L., Dausman, J., et al. (2000) Generation of mice from wild-type and targeted ES cells by nuclear cloning. *Nat. Genet.* **24,** 109–110.

106. Wakayama, T., Tateno, H., Mombaerts, P., and Yanagimachi, R. (2000) Nuclear transfer into mouse zygotes. *Nat. Genet.* **24,** 108–109.

107. Wilmut, I., Young, L., and Campbell, K. H. (1998) Embryonic and somatic cell cloning. *Reprod. Fertil. Dev.* **10,** 639–643.

108. Dean, W., Bowden, L., Aitchison, A., Klose, J., Moore, T., Meneses, J. J., et al. (1998) Altered imprinted gene methylation and expression in completely ES cell-derived mouse fetuses: association with aberrant phenotypes. *Development* **125,** 2273–2782.

109. Young, R. A. (2000) Biomedical discovery with DNA arrays. *Cell* **102,** 9–15.
110. Perou, C. M., Jeffrey, S. S., van de Rijn, M., Rees, C. A., Eisen, M. B., Ross, D. T., et al. (1999) Distinctive gene expression patterns in human mammary epithelial cells and breast cancers. *Proc. Natl. Acad. Sci. USA* **96,** 9212–9217.
111. Alizadeh, A. A., Eisen, M. B., Davis, R. E., Ma, C., Lossos, I. S., Rosenwald, A., et al. (2000) Distinct types of diffuse large B-cell lymphoma identified by gene expression profiling. *Nature* **403,** 503–511.
112. Brown, S. D. and Peters, J. (1996) Combining mutagenesis and genomics in the mouse—closing the phenotype gap. *Trends Genet.* **12,** 433–435.
113. Martin, J. E. and Fisher, E. M. (1997) Phenotypic analysis—making the most of your mouse. *Trends Genet.* **13,** 254–256.
114. Hrabe de Angelis, M. H., Flaswinkel, H., Fuchs, H., Rathkolb, B., Soewarto, D., Marschall, S., et al. (2000) Genome-wide, large-scale production of mutant mice by ENU mutagenesis. *Nat. Genet.* **25,** 444–447.

27

Phage-Displayed Antibodies to Detect Cell Markers

Jun Lu and Steven R. Sloan

1. Introduction

Immunologic reagents such as antibodies can be valuable for identifying and isolating cells with particular characteristics. While antibodies derived from immunized animals are the most common immunologic reagents used for this purpose, "antibodies" selected by phage display have also proven to be useful. The advantage of this technique is that one can relatively rapidly clone specific antibodies that will be available in unlimited supply. Additionally, phage display may allow for the cloning of antibodies to antigens to which an animal's immune system would not normally respond due to tolerance. Finally, strategies have been developed to select for some antibodies and deselect for other antibodies using phage display approaches. This can enable investigators to select for antibodies that, for instance, bind to antigens on the surface of one cell type but do not bind to antigens on the surface of another cell type.

In antibody phage display, a portion of an antibody is expressed on a filamentous bacteriophage such as M13 or f1. A fusion gene encoding an antibody fused to the pIII bacteriophage surface protein is cloned in a phagemid (*see* **Note 1**). Bacteria harboring the phagemid are infected with helper phage. The bacteria produce phage particles that display the antibody attached to the pIII protein on their surface and contain the phagemid DNA. Phage particles displaying antibodies that bind to a target antigen can be selected, and the antibody gene can be isolated from the phagemid in the phage particles.

Antibody phage display is still a developing technique, and several problems may be encountered. Most of these problems are related to the fact that the antibodies isolated from phage display libraries are not usually complete antibody molecules, but instead are Fabs or single-chain fusion proteins of variable chains (scFvs). Unlike full-length immunoglobulin proteins that must be glycosylated, Fabs and scFvs can usually be expressed in bacteria. However, Fabs, and especially scFvs, may not always fold in the desired conformation and may form aggregates that may precipitate in some solutions. Furthermore, enzyme and fluorochrome-conjugated anti-IgG antibodies do not recognize scFvs and Fabs. Thus, if a scFv or Fab is to be visualized (for staining cells, for example), it must be tagged with an epitope that is recognized by a secondary antibody.

Prior to cloning an antibody, one first has to make or obtain a library. One can use a naive library constructed from germ line variable genes, or one can use a library made from the immune system of a previously immunized animal or person.

From: *Methods in Molecular Biology, vol. 185: Embryonic Stem Cells: Methods and Protocols*
Edited by: K. Turksen © Humana Press Inc., Totowa, NJ

There are two potential advantages with using a library made from a previously immunized animal. First, the antibody of interest will be enriched. This advantage is minor, since in vitro selection methods are quite efficient, and this advantage might be nonexistent if the animal fails to respond to the antigen. Probably more important, however, is the fact that the in vivo immune response involves selection and mutation (affinity maturation) of antibodies. Hence, some antibodies that would not be present in a library made from the germ line genes might be present in a library made from DNA (or cDNA) from lymphocytes of immunized animals.

There are also advantages with using a naive library. Good naive libraries contain antibodies that would normally be deleted from an animal's immune system by tolerance mechanisms. Furthermore, some good naive libraries have already been constructed by investigators who are usually willing to share the resource for noncommercial purposes. This allows one to bypass substantial work.

Given these considerations, we recommend initially cloning antibodies from a naive library. Two groups have made phage display libraries of high complexity that have been widely used (*1,2*). The methods we describe here focus on using the library from Winter's group known as the Griffin.1 library (*1*), but most of these methods should apply to using other antibody phage display libraries. An outline of the entire process is depicted in **Fig. 1**.

Once the library is obtained, one selects for phage that bind to the antigen of interest and amplify those phage in bacteria. As a control, one can select for phage that are selected in the absence of the antigen of interest. The process is repeated approximately 3–6 times. After each selection, one can calculate the ratio of the number of phage selected in the presence of antigen to the number of phage selected in the absence of antigen. The higher this ratio, the higher the proportion of selected phage that display antibodies that bind to the target antigen. In addition, one can also determine whether the process is succeeding by using an enzyme-linked immunosorbent assay (ELISA) or similar assay. This is not usually worthwhile after the first selection and amplification, because the clone(s) of interest are likely to be very dilute at this stage.

Investigators may desire to make antibodies to cloned purified antigens, moderately pure antigens, or whole cells. In this chapter, we describe a cell selection technique, a protein selection technique, and a peptide selection technique. Undesirable antibodies that bind to contaminating antigens may be selected with any technique. The purer the antigen preparation, the lower the risk of selecting for undesirable antibodies. However, no selection technique is perfect, and selection for undesirable antibodies often occurs. Hence, we recommend using two or more different selection techniques if possible (*3*). With this approach, one selects phage using one technique, amplifies phage, and then selects phage again using a different technique in the next round of selection.

Finally, when there is evidence that the library has been enriched for phage expressing the antibody of interest, individual clones are identified. Once identified, the antibodies can be expressed in bacteria and purified. If the quality of an antibody expressed in bacteria is poor, we suggest expressing the antibody in insect cells and, therefore, have included a technique for insect cell expression.

2. Materials (*see* Note 2)

1. Antibody phage display library: the Griffin.1 library (*1*), a human synthetic VH plus VL scFv phage display library in a phagemid vector pHEN2 was constructed by the Winter

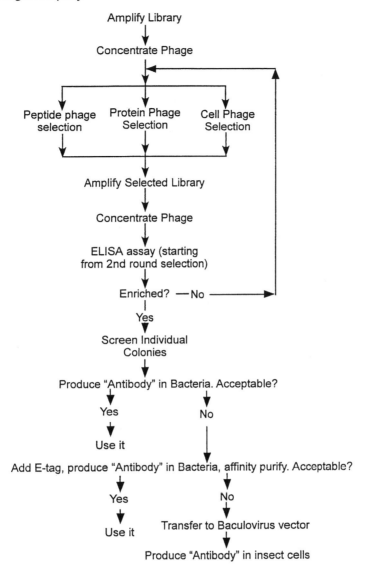

Fig. 1. Overview of Antibody Phage Selection Strategy. As indicated by the branched line, the selection step uses any one of three techniques. Some steps need to be performed multiple times as indicated the arrows that point back to earlier steps.

Group (The Medical Research Council, Centre for Protein Engineering, Cambridge, England). Other antibody phage display libraries are also available *(2)*.

2. Biotinylated peptide, purified protein, or cultured cells to serve as an immunogen for selection of phage displaying antibody.

3. *Escherichia coli* TG1Tr *(1)*, and *E. coli* HB2151 *(4)*. (Other F′ strains of *E. coli* can also work, but the Griffin.1 library requires a supE+ and a supE– strain.)

4. Phosphate-buffered saline (PBS): 5.84 g/L NaCl, 4.72 g/L Na$_2$HPO$_4$, 2.64 g/L NaH$_2$PO$_4$•2H$_2$O

5. 2X TY medium: 16 g tryptone, 10 g yeast extract, 5 g NaCl per liter.

6. Helper phage M13KO7 (1×10^{12} plaque forming unit [pfu]/mL) from Amersham Pharmacia Biotech. Other M13-based helper phage may also work, but kanamycin-resistant phage is especially useful for the Griffin.1 library.

7. PEG solution: 20% polyethylene glycol 6000, 2.5 M NaCl.
8. Ampicillin (Fisher, cat. no. ICN19014605) 100 mg/mL in H_2O; kanamycin (Sigma, cat. no. K4000) 50 mg/mL H_2O; isopropyl-β-D-thiogalactopyramoside (IPTG) (ISC Bioexpress, cat. no. C-5539-5) 100 mM in H_2O. Each should be filter-sterilized and stored at −20°C.
9. Anti-M13 Horseradish peroxidase (HRP) conjugate (Amersham Pharmacia Biotech, cat. no. 27-8411-01), anti-myc (Santa Cruz, cat. no. SC40, clone 9E10), anti-mouse HRP conjugate (Promega, cat. no. W4021).
10. ABTS: 2,2′-Azino-bis(3-ethylbenzothiazoline-6-sulfonic acid) diammonium salt (Fisher, cat. no. NC9596795): 2.2 mg in 10 mL sterile 0.05 M citric acid (pH 4.0), store at 4°C. The solution is stable for at least 2 wk.
11. H_2O_2 30%.
12. HRP-conjugated anti-E tag, which recognizes the synthetic E-tag amino acid sequence GAPVPYPDPLEPR (Amersham-Pharmacia, cat. no. 27-9413-01).
13. Streptavidin magnetic beads (Vector Laboratories, Burlingame, CA, cat. no. N100-15).
14. Triethylamine (Fisher, cat. no. 04884-100).
15. BaculoGold Starter Package (Pharmingen, cat. no. 21001K).
16. RPAS Purification Module (Amersham-Pharmacia, cat. no. 17136201).
17. Treated microtiter plates (Fisher, cat. no. 12-565-224).
18. Magnetic particle concentrator (Dynal Corporation, cat. no. 120.20).
19. TYE: 15 g bacto-agar, 8 g NaCl, 10 g tryptone, 5 g yeast extract per liter.

3. Methods

3.1. Phage Display Library Amplification

This protocol was developed for the Griffin.1 library *(5,6)*, but should work for most libraries cloned into phagemids.

1. Grow *E. coli* bacteria harboring the antibody phage display library in 500 mL 2X TY media with 100 µg/mL ampicillin, 1% glucose, shaking at 37°C (*see* **Notes 3,4**).
2. When the OD_{600} = 0.5, add M13 KO7 helper phage to an aliquot of 25 mL of bacteria at a ratio of 1:20 (bacteria:helper phage) and incubate without shaking at 37°C for 30 min to infect the bacteria (*see* **Note 5**).
3. Shake the remaining 475 mL bacteria at 37°C for another 2 h, then centrifuge at 3300*g* for 10 min at room temperature, and resuspend in 10 mL 2X TY containing 15% glycerol. Store at −80°C.
4. Centrifuge the helper phage-infected bacteria in **step 3** at 3300*g* for 10 min at room temperature, resuspend in 500 mL of prewarmed 2X TY containing 100 µg/mL ampicillin and 25 µg/mL kanamycin, and shake overnight at 30°C.
5. Centrifuge the culture at 3300*g* for 10 min, and transfer the supernatant to a clean tube. Centrifuge the supernatant again at 10,800*g* for 10 min, and transfer the supernatant containing phage to a clean container. The solution is now ready for phage concentration. Proceed to **Subheading 3.2.**

3.2. Concentration of Unselected Phage

1. Mix supernatant (from **Subheading 3.1., step 5**) containing phage with 1/5 volume PEG solution, chill at 4°C for at least 1 h.
2. Centrifuge at 10,800*g* for 30 min. Discard the supernatant.
3. Resuspend the phage pellet in 40 mL PBS. Centrifuge at 10,800g for 5 min.

4. Transfer the supernatant to a new tube, mix with 1/5 volume PEG, and incubate at 4°C for at least 20 min.
5. Centrifuge at 10,800*g* for 10 min at 4°C. Discard the supernatant.
6. Resuspend the phage pellet in 5 mL of PBS to form a concentrated phage solution. Start selection in no more than 2 h following resuspension. Proceed to **Subheading 3.3., 3.4.**, or **3.5.**

3.3. Peptide Bead Phage Selection

In this technique, phage-expressing antibodies that bind to a specific peptide are selected. This technique requires biotinylated antigen. We have used biotinylated peptide, which can be synthesized by one of a number of peptide synthesis services. Theoretically, a biotinylated protein could also be used, but there would be no significant advantage over coated protein selection. Peptide phage selection is a modified version of a technique described elsewhere *(7,8)*. One disadvantage of this technique is that antibodies to streptavidin are also selected and phage binding to streptavidin can out-compete almost everything else in the library. This problem can be minimized by preclearing the library (*see* **step 2** below) and/or alternating this selection technique with other technique(s) in different selection cycles.

An advantage of this technique is that one can simultaneously deselect for some antibodies and select for others by adding a competing nonbiotinylated antigen in excess (*see* **step 5**). For example, one could select for antibodies that bind to CD34 and deselect those that bind to c-kit.

1. Separately block 1 mL concentrated phage and streptavidin magnetic beads in PBS-2% fat free milk powder at room temperature for 30 min.
2. Add 100 µL of streptavidin magnetic beads to the concentrated phage and rotate for 0.5 h on a rotator. This step removes phage that bind to irrelevant antigens such as streptavidin or other epitopes on the magnetic beads.
3. Recover streptavidin magnetic beads from the solution by putting the tube on a magnetic rack for 2 min. Transfer the supernatant to a new tube, and discard the magnetic beads.
4. Repeat **steps 2** and **3** once. Transfer the supernatant to a new tube.
5. Add biotinylated peptide to the phage solution to a final concentration of 20 n*M* (first round) or 2 n*M* (subsequent rounds), and incubate the mixture for 1–2 h. Optionally, add 100 to 1000-fold excess of a competing nonbiotinylated protein or peptide for simultaneous negative selection.
6. Add 50 µL of the milk powder-blocked streptavidin magnetic beads (for the first round) or 5 µL (for the subsequent rounds) to the phage–peptide solution, and rotate the solution for 15 min.
7. Recover streptavidin magnetic beads from the solution by putting the tube on a magnetic rack for 2 min.
8. Wash the beads by resuspending the beads in PBS, 0.1% Tween-20. Recover the beads using a magnetic rack.
9. Repeat washing the beads 2 more times with PBS, 0.1% Tween-20.
10. Wash the beads 3 times as in **step 8** with PBS, and resuspend the beads in 100 µL PBS.
11. Infect 1 mL of exponentially growing TG1 *E. coli.* bacteria with selected phage by adding 1/2 of the magnetic beads in PBS and incubate at 37°C without shaking for 30 min (*see* **Note 6**).
12. Spread the bacteria on 2× TY plates containing 1% glucose and 100 µg/mL ampicillin, and grow at 30°C for 12–16 h (*see* **Notes 3,4,7**). Proceed to **Subheading 3.6.** or **3.8.**

3.4. Coated Protein Phage Selection

In this technique, phage-expressing antibody that bind to a specific protein are selected. This technique was described elsewhere previously *(5,6)*. In the original descriptions, phage were selected on protein-coated immunotubes. We have found this method to work well in the same microtiter tray wells normally used for ELISAs.

1. Coat 3 microtiter plate wells with purified protein in PBS (10 µg/mL) for 12–18 h at room temperature.
2. Separately block microtiter wells and phage with PBS-2% fat free milk powder for at least 2 h at room temperature.
3. Incubate concentrated phage solution (concentration approx 10^{12}–10^{13} pfu/mL) in microtiter wells for at least 1 h at room temperature.
4. Discard the phage solution, and wash the microtiter wells ten times with PBS, 0.1% Tween-20 (*see* **Note 8**).
5. Wash the tube ten times with PBS.
6. Elute selected phage by adding freshly diluted 100 m*M* triethylamine to the immunotubes, and incubate for 10 minutes at room temperature (*see* **Note 9**).
7. Transfer the elution solution to a clean tube and neutralize the elution solution with 1/2 vol 1 *M* Tris-HCl, pH 7.4, and store at 4°C.
8. Use 1/5 to 1/2 of the neutralized elution solution to infect exponentially growing TG1 *E. coli* bacteria by mixing the bacteria and the elution solution for 30 min at 37°C without shaking. Then spread the mixture onto 2× TY, 1% glucose, 100 µg/mL ampicillin plates and incubate at 30°C overnight (*see* **Notes 3,4,7**). Proceed to **Subheading 3.6.** or **3.8.**

3.5. Cell Phage Selection

In this technique, which has been previously described *(9)*, phage-expressing antibodies, which bind to specific cultured cells, are selected. Other techniques using fixed cells have also been described, but we prefer this technique, in which live cells with native antigens are exposed.

1. Grow cells in a 24-cm^2 tissue culture flask until almost confluent.
2. Change media three times.
3. Incubate for 1 h at 37°C.
4. Remove media. Add 10^{11}–10^{13} pfu phage in 2 mL media with supplements such as serum.
5. Shake flask gently at room temperature for 2 h.
6. Remove media and rinse rapidly ten times with PBS at room temperature.
7. Add 2 mL elution buffer (0.1 *M* glycine, pH 2.2, 0.1% bovine serum albumin [BSA]) for 10 min at room temperature.
8. Neutralize by adding 0.375 mL 1 *M* Tris-HCl, pH 9.1. Phage are in neutralized buffer. Remove buffer containing phage from flask and use to infect bacteria (as in **Subheading 3.4., step 8**). Proceed to **Subheading 3.6.** or **3.8.**

3.6. Growth of Selected Libraries

After selecting library and infecting bacteria, the infected bacteria harboring the selected library are stored. An aliquot is grown, and phage are produced by the bacteria. These phage are concentrated and can be assayed by ELISA and/or can be subjected to further rounds of selection.

1. Add 2× TY, 15% glycerol to a plate of bacterial colonies from **Subheading 3.1., 3.2., 3.3.**. Add about 0.5–1 mL to a 100-mm Petri dish or 5 to 6 mL to a bioassay dish.

2. Spread 2X TY, 15% glycerol with glass spreader to loosen colonies.
3. Inoculate 50–100 μL of bacteria into 100 mL of in 2X TY, 1% glucose, and 100 μg/mL ampicillin. Freeze the rest of the bacteria at −70°C for storage (*see* **Notes 3,4**).
4. Grow with shaking at 37°C for about 2 h until $OD_{600} \approx 0.5$.
5. Infect 10 mL of culture with M13K07 helper phage by adding phage at a ratio of 20:1 (helper phage:bacteria) (*see* **Notes 5, 10**). Incubate at 37°C without shaking for 30 min.
6. Centrifuge at 3300*g* for 10 min, and resuspend the pellet in 50 mL 2X TY, 100 μg/mL ampicillin, and 25 μg/mL kanamycin. Shake at 30°C overnight.
7. Centrifuge 40 mL at 10,800*g* for 30 min and add 8 mL PEG-NaCl to the supernatant. Mix and leave at 4°C for at least 1 h for the phage to precipitate.
8. Centrifuge at 10,800*g* for 10 min or at 3300*g* for 30 min to remove the supernatant, respin briefly, and aspirate any remaining PEG-NaCl.
9. Resuspend pellet in 1 to 2 mL PBS, then spin at 11,600*g* for 10 min in a microcentrifuge.
10. Transfer supernatant to fresh tubes and store phage at 4°C. Proceed to **Subheading 3.3.**, **3.4.**, **3.5.**, or **3.7.**

3.7. ELISA Using Phagemid Particles or Single-Chain Antibody

ELISAs are performed at three stages in the process. First, an ELISA can be performed using the polyclonal selected library to help determine whether the selection is succeeding. Second, an ELISA can be performed on individual phage clones to identify the phage of interest. Third, an ELISA can be performed to test the quality of a fresh preparation of a previously cloned antibody. ELISAs are performed essentially as previously described *(10)*.

1. Phage supernatant, concentrated phage and bacteria culture supernatant containing single-chain antibody or purified single-chain antibody can be used in ELISA.
2. Coat microtiter plate wells with 200 μL of the immunogen protein (1–10 μg/mL PBS) for 2–18 h. Alternatively, coat wells with cell extracts or other cell preparations that contain the target antigen.
3. Block the wells with 2% dry milk in PBS (blocking buffer). Block phage or single-chain antibody with 1% dry milk in PBS.
4. Incubate single-clone phage supernatant or concentrated selected phage library or single-chain antibody (1–10 μg/mL) in microtiter wells for 1 h, then wash wells six times with PBS, 0.05% Tween-20.
5. Add 200 μL HRP conjugated:
   a. anti-M13 antisera (for phage ELISA) diluted 5000:1 in blocking buffer; OR
   b. anti-epitope tag antibody such as anti-myc (clone 9E10) for the Griffin.1 library, or anti-E-tag antibody diluted 1000:1 in blocking buffer.
6. Incubate for at least 1 h.
7. Wash the wells six times with PBS, 0.05% Tween-20.
8. Mix ABTS substrate solution with H_2O_2 (21 mL ABTS with 36 μL H_2O_2). Then add 200 μL of the ABTS-H_2O_2 mixture to the wells for color development at room temperature or at 37°C. Measure optical density (415 nm) in a microplate reader 30 min after substrate is added (*see* **Note 11**).

3.8. Screening Individual Clones

The number of clones that should be picked depends on the degree of enrichment of the library, which is determined by the ratio of the clones selected in the presence of an antigen to the clones selected in the absence of an antigen. From poorly enriched libraries, one may want to test 96 clones; from highly enriched clones, one could test far fewer clones.

1. Pick individual colonies of TG1 *E. coli* harboring phagemids into 100 μL of 2X TY media with 1% glucose and 100 μg/mL ampicillin in a 96-well microtiter plate, and grow overnight at 37°C with vigorous shaking.
2. Transfer 2 μL of these cultures into a second microtiter plate containing 200 μL fresh 2X TY media with 1% glucose and 100 μg/mL ampicillin per well, and shake vigorously for 1 h.
3. Make glycerol stocks of the original 96-well plate by adding glycerol to a final concentration of 15%, and then store the plate at –80°C.
4. To each well of the second plate, add 25 μL 2X TY containing 100 μg/mL ampicillin and 10^9 pfu M13-K07 helper phage (*see* **Note 5**).
5. Stand for 30 min at 37°C, then shake vigorously for 1 h at 37°C.
6. Spin the plate at 1800*g* for 10 min at room temperature, and then aspirate off the supernatant.
7. Add 200 μL 2X TY containing 100 μg/mL ampicillin and 50 μg/mL kanamycin to each well to resuspend the bacteria pellet. Shake overnight at 30°C.
8. Spin at 1800*g* for 10 min, and use the supernatant from these cultures in phage ELISA assays (**Subheading 3.7.**).

3.9. Expressing Single-Chain Antibody in E. coli Strain HB2151 (see Note 12)

The Griffin.1 library vector contains an amber stop codon between the scFv gene and the region encoding the pIII phage protein. This amber stop codon is not recognized as a stop codon in TG1 bacteria (supE strain), resulting in expression of the fusion protein. In HB2151, the codon is recognized as a stop codon, and scFvs are produced. For other libraries, it may be necessary to subclone the antibody gene into a bacterial expression vector to express the antibody not fused to the pIII protein.

When expressed in bacteria, the cloned antibody may be secreted into the culture media or may accumulate in the bacterial periplasmic space. The location of antibody expression of each clone must be determined empirically as follows:

1. Infect 200 μL of exponentially growing HB2151 bacteria by adding 10 μL of supernatant from **Subheading 3.7., step 8**, and incubating 37°C for 30 min.
2. Plate about 1 μL of infected bacteria diluted in 100 μL of 2X TY on a TYE plate, 100 μg/mL ampicillin, 1% glucose, and incubate overnight at 37°C (*see* **Notes 3,4**).
3. Pick colonies and grow in 2X TY with 100 μg/mL ampicillin, 1% glucose overnight.
4. For harvesting antibodies from supernatant
   a. Transfer 100 μL of the overnight culture into 100 mL 2X TY with 1% glucose, 100 μg/mL ampicillin, and shake vigorously at 37°C until OD_{600} reaches 0.5–0.9.
   b. Centrifuge at 3300*g* for 10 min at room temperature and discard the supernatant.
   c. Resuspend the bacteria pellet in an equal vol of 2X TY containing 100 μg/mL ampicillin and 1 m*M* IPTG. Shake the culture at 30°C for 12–16 h.
   d. Centrifuge at 3300*g* for 10 min. Transfer the supernatant to a clean tube and centrifuge again at 10,000*g* for 10 min at room temperature.
   e. Transfer the supernatant to a clean container and neutralize with HCl (final pH 7.0–8.0).

     f. If desired, test supernatant in an ELISA (*see* **Subheading 3.7.**).

     g. Purify antibody (**Subheading 3.10.**).

5. For harvesting antibodies from periplasmic space:

     a. Innoculate 50 mL 2X TY media with 2% glucose, 100 µg/mL ampicillin.

     b. When the OD_{600} reaches 0.7–1.0, centrifuge culture at 3000*g* for 10 min at room temperature. Resuspend in 50 mL 2X TY with 100 µg/mL ampicillin, and 1 m*M* IPTG.

     c. Grow at 30°C for 3–6 h. Centrifuge at 3000*g* at 4°C for 15 min. Resuspend in cold PBS, 1 *M* NaCl, 1 m*M* EDTA. Keep on ice for 15 min.

     d. Centrifuge at 3000*g* at 4°C for 15 min. Transfer supernatant containing the periplasmic fraction to a new tube. Add $MgCl_2$ to final concentration of 1 to 2 m*M*.

     e. If desired, test supernatant in an ELISA.

     f. Purify antibody as described in **Subheading 3.11.**

3.10. Expressing Single-Chain Antibody in Baculovirus System

In our experience, scFvs expressed and secreted from insect cells are usually in the correct conformation and not significantly degraded *(11)*. To express an antibody in insect cells, the antibody gene must be subcloned into a baculovirus expression vector, which encodes a signal sequence for protein secretion that will be fused to the antibody. The antibody can be recovered from the culture media. One can start with the E-tagged construct made in **Subheading 3.11.** or with the original cloned antibody.

1. Using custom designed primers based on the sequence of the antibody, amplify the antibody gene, keeping the *Nco*I site at the 5′ end of the gene, and adding an *Eco*RI site to the 3′ end of the gene (*see* **Note 13**). Digest the polymerase chain reaction (PCR)-amplified DNA with *Nco*I and *Eco*RI, and purify the digested DNA using standard molecular biology techniques *(12,13)*. Ligate this DNA fragment into the *Nco*I and *Eco*RI sites of Baculovirus vector pAcGP67B (BaculoGold Starter Package; BD Pharmingen) resulting in a pAcGP67B-scFv construct.

2. Transfect the pAcGP67B-SvAb DNA into insect cell line Sf9 (*see* **Note 14**), according to the BaculoGold Starter Package's instruction manual, and harvest the culture supernatant containing virus particles containing the antibody cDNA by centrifuging the culture supernatant at 3300*g* for 10 min, and store the supernatant at 4°C (*see* **Note 15**). The virus in the supernatant will be stable for 6 mo.

3. Amplify the virus following the BaculoGold Starter Package's instruction manual, so that the titer of the virus particles in the supernatant reaches 2×10^8.

4. Infect Sf9 cells (*see* **Note 14**) with amplified baculovirus containing single-chain antibody cDNA at a ratio of 1 : 3–10 (Sf9 cells : λvirus).

5. After 3 d at 27°C, Sf9 culture supernatant should contain a sufficient concentration of antibody for most purposes.

6. Harvest Sf9 culture supernatant expressing antibody by centrifugation at 3300*g* for 10 min at room temperature.

7. Transfer the supernatant to a clean container and neutralize with 1/10 volume of 1 *M* pH 7.4 Tris-HCl.

8. Proceed to **Subheading 3.11., step 1** (*see* **Note 16**).

3.11. Purifying Antibodies

For some purposes, satisfactory purification can be achieved using a nickle column to purify a histidine-tagged antibody such as those cloned from the Griffin.1 library (*see* **Note 17**). In addition, some antibodies bind to protein A and/or protein L, and can be purified using protein A or protein L affinity columns. We have had success with immunoaffinity purification, which results in a purer preparation of functional antibody than nickle columns. To this end, we suggest adding an "E-tag" and purifying antibodies using the RPAS module from Amersham-Pharmacia Biotech. Using this procedure, the concentration of the purified antibody in the most concentrated fraction is usually between 200–400 µg/mL. The purified antibody can be used in a range of 1–10 µg/mL for ELISA and Western blot assays. Higher concentrations may be needed for staining cells.

1. Introduce an artificial E-tag to the 3′ end of the antibody cDNA by PCR using selected positive single-chain antibody single-clone phagemid DNA as the template. The nucleotide sequence of E-tag is: 5′-GGT GCG CCG GTG CCG TAT CCG GAT CCG CTG GAA CCG CGT-3′. For details on using PCR to add nucleotides to sequences see *(14)*.
2. After ligating the PCR product to the plasmid DNA, transform competent *E. coli* cells (*see* **Note 18,19**). Spread the bacteria onto a 2X TY 1% glucose 100 µg/mL ampicillin plate, and incubate at 30°C overnight (*see* **Notes 3,4**). Pick a single colony in 3 mL 2X TY, 1% glucose, 100 µg/mL ampicillin, and grow overnight at 37°C.
3. Grow and harvest supernatant or periplasmic fraction following **Subheading 3.9., step 4** or **5**, or express the antibody in insect cells as described in **Subheading 3.10.**
4. Filter the supernatant through a 0.45-µm syringe filter.
5. Use the anti-E-tag column provided in RPAS module (Amersham-Pharmacia Biotech) to purify the antibody.
   a. Wash the new column with 15 mL of elution buffer (provided with the RPAS module) by pushing through the elution buffer with a syringe at a rate of 5 mL/min.
   b. Equilibrate by applying 25 mL of binding buffer (provided with the RPAS module).
   c. Bind antibody by applying neutralized filtered extract or supernatant to the column.
   d. Wash column with 25 mL of binding buffer.
   e. Elute with 15 mL of elution buffer. Discard the first 4.5 mL. Collect the next five 1-mL fractions into tubes containing 100 µL of neutralizing buffer (provided with the RPAS module). The neutralizing buffer should be 1/10 the volume of the eluted antibody.
   f. Re-equilibrate the column with 25 mL of binding buffer.
6. Store the single-chain antibody at 4°C (*see* **Notes 20,21**).

4. Notes

1. A phagemid is a plasmid that contains a phage origin of replication and will be incorporated into phage particles. Although most antibodies displayed on pIII are fused to pIII, theoretically, other phage surface proteins such as pVIII could also be used.
2. Precautions need to be taken throughout the protocol to avoid any carryover of phage from flasks and centrifuge bottles, etc., as autoclaving alone is not sufficient enough to remove all phage contamination. All nondisposable plasticware can be soaked for 1 h in 2% (v/v) hypochlorite (>1000 ppm free chlorine), followed by extensive washing, and then autoclaving. Glassware should be baked at 200°C for at least 4 h. The use of polypropylene tubes is recommended as phage may absorb nonspecifically to other plastics.
3. Glucose is added when bacteria are grown to amplify the library. Glucose represses transcription of the Griffin.1 library or any other library whose transcription is driven by the lac promoter. This improves growth of the bacteria and removes selective advantages that some clones have over others during growth.
4. The Griffin.1 library confers ampicillin resistance to bacteria. If another phagemid library carries a different antibiotic resistance gene, then the corresponding antibiotic should be substituted for ampicillin.

5. Other helper phage such as VCSM13 (Stratagene) also work. It is advantageous to use a helper phage that carries a kanamycin resistance gene, so that the helper phage and the phagemid can be simultaneously selected with two different antibiotics.

6. In contrast to the other selection techniques, the phage do not need to be eluted from the solid support (beads). Phage can be eluted from the beads using acid, but in our experience this does not improve the results.

7. After the first round, one needs to plate many colonies. We recommend using a Nunc Bioassay dish (Fisher, cat. no. 12-565-224) or a plate with similar dimensions or multiple smaller plates. With further rounds of selection, fewer colonies are needed, and 100-mm round Petri dishes are convenient.

8. To empty the wash solution, the microtiter plate is usually inverted and drained on a paper towel. Although the plate can be tapped onto the paper towel, tapping with too much force can disrupt binding of desirable phage.

9. By using more stringent conditions to elute phage in coated protein phage selection, higher affinity single-chain antibody can be obtained. (*see* de Bruin, R., Spelt, K., Mol, J., Koes, R., and Quattrocchio, F., 1999, Selection of high-affinity phage antibodies from phage display libraries, *Nat. Biotechnol.* **17**, 397–399).

10. At an $OD_{600} = 0.5$, the bacterial concentration is about 4×10^8 bacteria/mL.

11. For individual clones and for moderately enriched polyclonal phage libraries, the reaction is easily identified by the blue-green color. The microplate reader provides a quantitative measurement that may be useful for choosing individual clones.

12. Expression of single-chain antibody in the HB2151 *E. coli* strain may not be consistent. Sometimes a significant proportion of antibody may be degraded or incorrectly folded. Rather than spending significant time troubleshooting, we recommend expressing the antibody in the baculovirus system if the bacterially-expressed antibody is of poor quality.

13. The antibody must be fused in frame with the signal sequence for protein secretion. Antibodies cloned in the *Nco*I site of pHEN2 will be in the correct reading frame using the strategy described. If not using an antibody cloned into the *Nco*I site of pHEN2, then consider using pAcGP67A or pAcGP67C (also included in BaculoGold Starter Package; Pharmingen, cat. no. 21001K), which are in different reading frames.

14. Instead of the Sf9 cell line, another insect cell line such as Sf21 may be used.

15. Since yeastolate, which is supplemented in some of the TNM-FH media (such as provided in the BaculoGold Starter Package), can cause severe precipitation of scFvs when neutralizing the media with 1*M* Tris-HCl, pH 7.4 (**step 7**), Hink's TNM-FH Medium without yeastolate should be used. The antibody will be expressed equally well in this media.

16. BAC purity note: baculovirus culture media has fewer interfering substances than bacterial culture media, and antibody in the supernatant may not need to be purified.

17. For nickle column purification, please see protocols from Qiagen or Novagen.

18. The goal is to be able to pick individual colonies. The amount of infected bacteria can be adjusted.

19. The bacteria should be supE-, such as strain HB2151, if an amber stop codon needs to be recognized as a stop codon. This is true for antibodies in the Griffin.1 library.

20. BSA (15 mg/mL) can be added to stabilize the single chain antibody. Thymerosal 0.01% can be added as a preservative.

21. Antibodies and specifically scFvs will degrade over time. Each specific clone has a different shelf life, which usually varies between a week to a few months. The quality of the preparation also affects the shelf life, with preparations from older affinity columns producing preparations with shorter shelf lives. We recommend using the purified antibody shortly after purification.

References

1. Griffiths, A. D., Williams, S. C., Hartley, O., Tomlinson, I. M., Waterhouse, P., Crosby, W. L., et al. (1994) Isolation of high affinity human antibodies directly from large synthetic repertoires. *EMBO J.* **13,** 3245–3260.
2. Barbas, C. F. d., Bain, J. D., Hoekstra, D. M., and Lerner, R. A. (1992) Semisynthetic combinatorial antibody libraries: a chemical solution to the diversity problem. *Proc. Natl. Acad. Sci. USA* **89,** 4457–4461.
3. Lu, J. and Sloan, S. R. (1999) An alternating selection strategy for cloning phage display antibodies. *J. Immunol. Methods* **228,** 109–119.
4. Carter, P., Bedouelle, H., and Winter, G. (1985) Improved oligonucleotide site-directed mutagenesis using M13 vectors. *Nucleic Acids Res.* **13,** 4431–4443.
5. Marks, J. D., Hoogenboom, H. R., Bonnert, T. P., McCafferty, J., Griffiths, A. D., and Winter, G. (1991) Bypassing immunization: human antibodies from V-gene libraries displayed on phage. *J. Mol. Biol.* **222,** 581–598.
6. Griffiths, A. D., Malmqvist, M., Marks, J. D., Bye, J. M., Embleton, M. J., McCafferty, J., et al. (1993) Human anti-self antibodies with high specificity from phage display libraries. *EMBO J.* **12,** 725–734.
7. McCafferty, J. and Johnson, K. S. (1996) Construction and screening of antibody display libraries, in *Phage display of peptides and proteins: a laboratory manual* (Kay, B. K. J. W., and McCafferty, J., eds.), Academic Press, San Diego, pp. 79–111.
8. Hawkins, R. E., Russell, S. J., and Winter, G. (1992) Selection of phage antibodies by binding affinity: mimicking affinity maturation. *J. Mol. Biol.* **226,** 889–896.
9. Cai, X. and Garen, A. (1995) Anti-melanoma antibodies from melanoma patients immunized with genetically modified autologous tumor cells: selection of specific antibodies from single-chain Fv fusion phage libraries. *Proc. Natl. Acad. Sci. USA* **92,** 6537–6541.
10. Coligan, J. E. (1996) *Current protocols in immunology.* J. Wiley & Sons, New York.
11. Kretzschmar, T., Aoustin, L., Zingel, O., Marangi, M., Vonach, B., Towbin, H., and Geiser, M. (1996) High-level expression in insect cells and purification of secreted monomeric single-chain Fv antibodies. *J. Immunol. Methods* **195,** 93–101.
12. Sambrook, J., Maniatis, T., and Fritsch, E. F. (1989) *Molecular cloning: a laboratory manual.* CSH Laboratory Press, Cold Spring Harbor, N.Y.
13. Ausubel, F. M. (1987) *Current protocols in molecular biology.* Published by Greene Pub. Associates and Wiley-Interscience: J. Wiley & Sons, New York.
14. White, B. A. (1997) *PCR cloning protocols: from molecular cloning to genetic engineering.* Humana Press, Totowa.

28

Gene Transfer Using Targeted
Filamentous Bacteriophage

**David Larocca, Kristen Jensen-Pergakes, Michael A. Burg,
and Andrew Baird**

1. Introduction
1.1. Conferred Tropism

Phage-mediated gene transfer offers an alternative method of introducing genes into specific cell types, including cell lines *(1–4)* and primary cell cultures (*see* **Fig. 1**). Recent studies demonstrate that filamentous bacteriophages can be engineered to transfer genes to mammalian cells by attaching a targeting ligand to the phage surface either noncovalently *(1)* or genetically *(2–4)* and, thus, directing phage particles to specific cell surface receptors (*see* **Fig. 1**). Successful gene transfer and subsequent protein expression is measured using a reporter gene such as green fluorescent protein (GFP), neomycin phosphotransferase, or β-galactosidase. In fact, any gene with an appropriate mammalian transcriptional promoter and polyadenylation signal can be incorporated into a ligand-targeted phage vector (*see* **Note 1**). It is this combination of ligand retargeting and insertion of a mammalian expression cassette that confers mammalian tropism to bacteriophage. These modified phage act like nonproductive animal viral vectors but can be propagated and manipulated genetically with all the conveniences of a phage vector.

Like viral gene transfer, phage gene transfer is time- and dose-dependent and specific for cell surface receptors. Transduced cells begin to appear at about 48 h after the addition of phage, and the percentage of cells expressing GFP increases with time. Most importantly, the mechanism of phage internalization is through the interaction of the targeting ligand with its cognate receptors on the cell surface. Accordingly, ligand-targeted phage transduction is inhibited by competition with the free ligand or with a neutralizing anti-receptor antibody *(2,3)*. Little or no transduction occurs in the absence of a targeting ligand, because phage particles have no native tropism for mammalian cells. In addition, transduction occurs with concentrations of phage as low as approximately 100 phage/cell and continues to increase up to the highest concentration tested (about 1×10^6 phage/cell). At these higher doses, internalization of the ligand-targeted phage is highly efficient, and phage protein is detectable in almost all cells. Transduction efficiencies of up to 4% have been described, thus far, using

From: *Methods in Molecular Biology, vol. 185: Embryonic Stem Cells: Methods and Protocols*
Edited by: K. Turksen © Humana Press Inc., Totowa, NJ

A

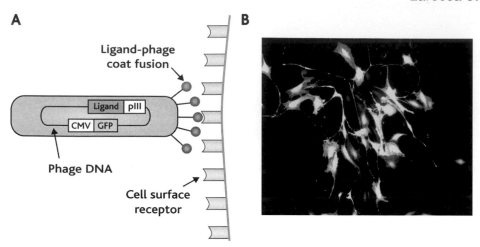

Ligand-phage
coat fusion

Phage DNA

Cell surface
receptor

B

Fig. 1. Transduction of mammalian cells by ligand-targeted phage. (**A**) Phage vectors are genetically modified by insertion of a mammalian promoter-regulated reporter gene (GFP) and fusion of a ligand to a surface coat protein (pIII). The resulting phage particles deliver the reporter gene to targeted cells that express the appropriate receptor. (**B**) Auto-fluorescent GFP-positive cells, resulting from transduction of primary rat olfactory bulb cell culture by epidermal growth factor (EGF)-targeted phagemid particles, observed 96 h after phage addition. Explanted cells were grown for 8 d on polylysine–laminin-coated plates before transfection by phage. Original magnification: ×200. (*See* color plate 11, following p. 254).

ligand- *(1–3)* and antibody-targeted phage *(4)*. Stable transformants can be isolated from phage-transduced cells using G418 drug selection *(1)* or by repeated selection of GFP-positive cells by fluorescence-activated cell sorting (FACS) (unpublished observation). We have found that genotoxic treatments improve the transduction efficiency of single-stranded phage (see below); yet, further improvements in transduction efficiency are possible by incorporating peptides into the phage coat protein, which facilitates its trafficking in the cell, and by applying directed molecular evolution to genetically select improved phage from combinatorial libraries *(2)*.

Phage vectors are simple and convenient to produce in bacteria, can be specifically targeted to cells, and have the potential to be evolved genetically for specific applications. In addition, filamentous phage have an inherent capacity to package large DNA inserts, because they are not limited in size by a preformed capsid, but instead form their protein coat as they are extruded from bacteria. We have successfully transduced cells with phage vectors approaching 10 kb in length including both the targeting ligand sequence and the mammalian expression cassette. For larger gene inserts, which tend to be unstable in phage, we have recently engineered phagemid vectors that are much simpler and smaller (approx 6 kb) (unpublished observation). Phagemid vectors contain no phage sequences except the origin of replication and are, therefore, prepared by rescue with helper phage *(5,6)*. The simple genetics of filamentous phage vectors make them particularly adaptable for a wide variety of targeted gene transfer applications.

1.2. Identification of Targeting Ligands

The flexible structure of the pIII coat protein is well suited for displaying of a variety of biologically active peptide and protein sequences while retaining the structural integrity of the phage particle *(5,7)*. For example, biologically active hormones, cytokines, and growth factors have been displayed on phage *(8–14)*. To date, phage-mediated gene delivery has been performed with targeting ligands on pIII, but it is conceivable that fusion to pVIII would provide a similarly targeted phage. In fact, recent studies show that phage displaying multiple copies of a peptide on the major coat protein are rapidly internalized into mammalian cells *(15)*. While there are many examples of active pIII fusion proteins, not all ligands are equally displayed. Differences in the ability of the fusion proteins to be secreted into the bacterial periplasmic space for packaging into the phage particle can significantly affect surface display. Thus, while many ligands are functional when displayed on phage, insufficient display is a limitation that should be considered when using phage display to identify new ligands that can target phage for gene delivery to cells (*see* **Note 2**). Alternatively, noncovalent display of the targeting ligand *(1)* can be used when genetic display is not applicable (*see* **Note 3**).

We have used naturally occurring ligands for phage targeting; however, many types of targeting ligands are possible including those identified from phage libraries. For example, Poul and Marks have targeted M13 phage to HER2 expressing breast carcinoma cell lines using an anti-Her2 single-chain antibody *(4)*. These results open up the possibility of targeting cells through a variety of receptors against which internalizing antibodies can be raised. We have also demonstrated that phage displaying selected peptides are capable of targeted gene delivery to cells (unpublished observation). Accordingly, it is possible to select new targeting ligands from phage libraries of peptides, antibodies, or cDNAs against a given cell type without prior knowledge of the targeted receptor. In previous studies, ligand selection was performed by panning phage on cells and selecting for phage that bind or both bind and internalize *(16–18)*. Recently, we have developed a novel selection strategy called ligand identification via expression (LIVE) that directly selects those ligands that both bind and internalize and target phage to cells for gene delivery *(2)*. Repeated rounds of phage transfection and recovery of targeted phage from the GFP expressing cell population are performed to select phage displaying gene targeting ligands. We have demonstrated, in this system, that targeted phage displaying a functional ligand are enriched 1 million-fold after 3 to 4 rounds of LIVE selection *(2)*.

1.3. Strand Conversion

Phage-mediated gene transfer, like adeno-associated virus (AAV), involves the introduction of single-stranded DNA that must be converted to double-stranded DNA for transgene expression. For this reason, transduction by AAV vectors is enhanced by treatments that induce endogeneous DNA repair *(19)*. Similarly, we have found that phage-mediated transduction is increased by the same genotoxic treatments that enhance AAV transduction efficiency. We have used camptothecin, hydroxyurea, heat-shock, and UV irradiation and have found that these treatments increase transduction efficiency as much as 10-fold in certain cell lines (unpublished observations). The degree of enhancement in transduction efficiency varies among cell lines, presumably due to individual differences in response to genotoxic stress (*see* **Note 4**).

2. Materials

2.1. Genetically Targeted Phage Vector

Recombinant M13 phage vectors are adapted for gene transfer to mammalian cells by inserting a mammalian expression cassette into the intergenic region of the phage genome *(3)*. The modified phage vector, MG3, contains the GFP gene expression cassette from pEGFP-N1 (Clontech, Palo Alto, CA, cat. no. 6085-1), which encodes a mutagenized GFP *(20)* that is optimized for visualization by fluorescent microscopy or FACS. It also contains the simian virus 40 (SV40) origin of replication from pEGFP-N1. Any phage vector that can be engineered for phage display (i.e., fUSE5 *[21]*, fAFF1 *[22]*, M13East *[23]*) can be adapted for gene delivery in this manner, including phagemid vectors. When phagemids are rescued with helper phage, both wild-type pIII and the ligand-pIII fusion protein are incorporated into the phagemid particle *(6)*, resulting in monovalent display of the targeting ligand. Monovalent display, however, is sometimes less optimal, because the number of ligands on the phage surface can significantly effect binding and internalization *(24)*. In this case, the system can be adapted for multivalent display by rescuing the phagemid with a gene III deleted helper phage, as described by Rakonjac et al. *(25)*.

2.2. Preparation of Targeted Phage Particles

1. Host F' bacterial strain (XLI-Blue Competent Cells; Stratagene, cat. no. 200249).
2. Replicative form phage DNA.
3. SOC Media (Gibco BRL, cat. no. 15544-034).
4. Luria Bertani medium (LB) plates: 1% tryptone, 0.5% yeast extract, 0.5% NaCl, pH 7.0, 2% agar, with and without antibiotics (60 μg/mL ampicillin).
5. 2X Yeast tryptone medium (YT) broth: 1.6% tryptone, 1% yeast extract, 0.5% NaCl, pH 7.0, plus 60 μg/mL ampicillin.
6. 1.5 M NaCl 30% polyethylene glycol (PEG) 8000.
7. Phosphate-buffered saline (PBS) plus 0.2 mM (4-(2-aminoethyl)-benzene sulfonyl fluoride (AEBSF) (Roche, cat. no. 1585916).
8. 1 M MgCl$_2$.
9. DNaseI (Sigma, cat. no. D4513).
10. 0.5 M EDTA.
11. 10% Triton X-114.
12. Glycerol.
13. Sterile labware: Falcon 2059 tubes, centrifuge bottles, pipets, 0.45-μm syringe filters, syringes, cryovials.
14. Top agar (LB with 0.8% agar)

2.3. Transfection of Cultured Cells

1. PC-3 cell line (ATCC, cat. no. CRL-1435).
2. PC-3 culture medium: RPMI 1640 plus 10% fetal bovine serum (FBS), 0.1 mM nonessential amino acids, 1 mM sodium pyruvate, 2 mM L-glutamine, 50 μg/mL gentamicin.
3. 0.25% trypsin (Gibco BRL, cat. no. 25200-056).
4. Targeted phage particles containing reporter gene (GFP).

5. Fixative buffer: 0.925% formaldehyde, 0.02% sodium azide, 2% glucose in PBS, pH 7.4.
6. Sterile labware: 12-well tissue culture dishes, pipets, Falcon 2054 tubes.

2.4. Genotoxic Treatments

1. Camptothecin: stock solution (10 mM) of camptothecin (Sigma, cat. no. C9911) are prepared in dimethyl sulfoxide (DMSO) and stored at –20°C.
2. Hydroxyurea: stock solution (1.0 M) of hydroxyurea (Sigma, cat. no. H8627) are prepared in PBS and stored at –20°C.
3. Heat shock: tissue culture incubator set at 42.5°C.
4. UV irradiation: Stratalinker UV crosslinker (Stratagene, La Jolla, CA).

3. Method
3.1. Genetically Targeted Phage Vector

3.1.1. Transformation of Host Bacteria with Replicative Form (RF) of the Recombinant Phage Vector Containing a Mammalian Expression Cassette

1. Thaw 100 μL XL1-Blue competent cells on ice.
2. Add 1.7 μL of 1.42M β-mercaptoethanol to cells and incubate on ice for 10 min.
3. Mix 50 ng replicative form phage DNA with cells and incubate on ice 30 min.
4. Heat shock cells for 45 s at 42°C.
5. Place cells on ice for 2 min.
6. Add 900 μL SOC medium to cells and incubate 1 h with shaking at 250 rpm at 37°C.
7. Spread cells on LB plate containing 60 μg/mL ampicillin and incubate overnight at 37°C.

3.2. Preparation of Targeted Phage Particles

3.2.1. Concentration of Phage Particles from Bacterial Culture Medium (see **Note 5**).

1. Innoculate approx 20 bacterial colonies (*see* **Subheading 3.1.1., step 7**) per liter of medium (2X YT plus 60 μg/mL ampicillin) and grow overnight at 37°C with shaking at 300 rpm.
2. Centrifuge bacterial culture at 6000g for 10 min at 4°C.
3. Save the supernatant and add 1/5 volume cold 1.5 M NaCl, 30% PEG. Mix well and incubate on ice for 2 h to precipitate phage.
4. Centrifuge at 15,000g for 30 min at 4°C to pellet phage.
5. Remove supernatant and all residual liquid.
6. Resuspend the phage pellet in PBS containing 0.2 mM AEBSF and incubate at 37°C for 10 min followed by 4°C for 30 min.
7. Centrifuge at 20,000g for 20–30 min to remove debris.
8. Repeat (**steps 3–6**) if further concentration of phage is necessary.
9. DNaseI treat phage by adding MgCl$_2$ to 10 μM and DNaseI at 125 U/mL of phage solution. Incubate 30 min at room temperature and stop the reaction by adding 10 μL 0.5 M EDTA/mL phage solution (*see* **Note 6**).
10. Immediately add 1/5 volume 1.5 M NaCl, 30% PEG. Mix well and incubate on ice for 2 h to precipitate phage.
11. Centrifuge at 15,000g for 30 min at 4°C to pellet the phage.
12. Remove supernatant and all residual liquid.
13. Resuspend the phage pellet in PBS containing 0.2 mM AEBSF and incubate 5–15 min at 37°C.
14. Incubate at 4°C for 30 min.
15. Centrifuge at 20,000g for 20–30 min to remove debris.
16. Filter phage through a 0.45-μm filter, freeze in 20% glycerol, and store at –70°C.

3.2.2. Endotoxin Removal by Triton X-114 Phase Partitioning

1. Add 100 µL 10% Triton X-114/mL sample and incubate on ice for 30 min with occasional vortex mixing.
2. Incubate at 37°C for 10 min.
3. Centrifuge at 16,000g in a microfuge (Eppendorf) for 10 min at room temperature and save the aqueous (upper) phase.
4. Repeat phase partitioning (**steps 1–3**) twice (*see* **Note 7**).

3.2.3. Phage Titering—Plaque Forming Units

1. Prewarm LB plates at 37°C, melt top agar (LB), and place in a 55°C water bath.
2. Grow F′ bacteria (XLI-Blue) to OD_{600} = 0.5 and aliquot 300 µL of cells to Falcon 2059 tubes for each dilution to be tested.
3. Set up serial dilutions in PBS by starting with a 100-fold dilution (5 µL diluted in 500 µL PBS) and repeating several times for desired dilution series.
4. Add 100 µL from each dilution to bacterial cells.
5. Add 3 mL of top agar to each tube, briefly vortex mix, and pour on top of prewarmed LB plates.
6. Allow top agar to harden and invert at 37°C overnight.
7. Count plaques and determine titer in plaque forming units per milliliter (pfu/mL) by multiplying the number of plaques times the dilution and dividing by the volume (*see* **Note 8**).

3.3. Transfection of Cultured Cells

1. Seed cells in 1 mL culture media on 12-well tissue culture dishes. Incubate at 37°C with 5% CO_2 overnight. The seeding density is determined by growth rate. After an overnight incubation, the cells should be 25% confluent in the 12-well dishes are plated PC-3 cells at 2×10^4 cells/well.
2. Culture medium is removed from the cells and replaced with culture media containing phage. A typical dose of targeted phage for highest transduction efficiency is 10^{11} pfu/mL. Cells are incubated at 37°C with 5% CO_2 for 72 h.
3. Cells are harvested for reporter gene analysis by removing the phage-containing culture medium and washing the cells with PBS. The cells are removed from the culture dishes by adding 150 µL 0.25% trypsin and incubating at 37°C for 2 to 3 min. Once the cells have detached from the plate, 350 µL of fixative buffer is added. The cells are now ready for analysis by flow cytometry (fluorescein isothiocyanate [FITC] filter set).

3.4. Genotoxic Treatments

3.4.1. Phage Are Added to Cell Cultures

As described in **Subheading 3.3.**, the cells are subjected to genotoxic treatments 40 h later. Treatments described here are for human carcinoma cell lines and should be optimized for each target cell line. Treatments are performed as follows:

1. Camptothecin: camptothecin is directly diluted into cell culture media containing 10% FBS at a final working concentration of between 1 and 100 µM. Medium containing phage is removed from the plate and replaced with camptothecin-containing medium, and the cells are incubated for an additional 7 h at 37°C.
2. Hydroxyurea: hydroxyurea is directly diluted into cell culture media containing 10% FBS at a final working concentration of between 10 and 100 mM. Medium containing phage is removed from the plate and replaced with hydroxyurea-containing medium, and the cells

are incubated for an additional 7 h at 37°C.

3. Heat-shock: phage-containing medium is removed from cells and replaced with fresh tissue culture medium containing 10% FBS. Cells are incubated for an additional 7 h at 42.5°C.

4. UV irradiation: medium is removed from the plate and the cells are immediately irradiated at doses ranging between 10–100 J/m². Fresh tissue culture media containing 10% FBS is added to cells.

5. Following the genotoxic treatment, cells are washed 3× in PBS, fresh cell culture media containing 10% FBS is added, and cells are incubated for an additional 24–48 h at 37°C. Cells are washed 3 times with PBS and prepared for FACS analysis as described in **Subheading 3.3.3.** Direct observation of GFP-expressing fluorescent cells under a fluorescent microscope (FITC filter set) can also be used to monitor the degree of transduction efficiency and assess the effect of genotoxic treatments.

4. Notes

1. The orientation of the reporter gene relative to the phage structural genes can affect vector transduction efficiency. Transduction efficiency of the MG4 phage vector *(2)*, in which the GFP cassette is in the antisense orientation relative to the phage sense strand, is about 3-fold higher than the same vector containing the GFP cassette in the opposite orientation (MG3) *(3)*.

2. The choice of targeting ligand will determine the specificity of targeted phage transduction. While there have been a wide variety of proteins expressed as fusion proteins to phage coat proteins, genetic targeting is limited to those targeting proteins that can be efficiently expressed and biologically active following secretion into the periplasmic space of the bacteria and subsequent incorporation into the phage particle. For example, we have found that while it is possible to target phage with a basic fibroblast growth factor (FGF2)-pIII fusion protein *(3)*, the efficiency of FGF2 display is relatively low presumably because its high isoelectric point (9.6) prevents efficient secretion into the periplasmic space. The capacity of phage to display a chosen targeting ligand needs to be determined empirically and optimized. Alternatively, the targeting ligand can be selected after display in a phage library. In this case, only those ligands that are efficiently displayed will survive the selection.

3. We have also targeted phage particles for gene delivery, nongenetically, using an avidin-biotin linkage *(1)*. This noncovalent attachment of the targeting ligand to the phage is advantageous for selecting ligands without concern for their ability to be displayed genetically. In this system, the phage–ligand complex is assembled directly on cell monolayer at 0°C, and then the cells are returned to 37°C to allow internalization.

4. The exact timing, duration, and doses of genotoxic treatments must be optimized for each mammalian cell line used. We obtained maximum enhancement of phage-mediated transduction (with minimal toxicity) on COS-1 and PC-3 cells using 5–10 μM camptothecin, 40 mM hydroxyurea, 50 J/m² UV irradiation, or 7-h heat-shock at 42.5°C.

5. Targeted phage will not produce infective phage particles after transfection of mammalian cells, because bacterial promoters regulate all of the phage structural genes. Even if the phage proteins were expressed, the mechanism for phage packaging and the differences in the intracellular environment of mammalian cells versus bacteria make the probability of a productive infection negligible. Nevertheless, we recommend following the same biohazard safety precautions for targeted phage as those used for working with nonreplication competent adenoviral vectors *(26)*. Overall biosafety level 2 (BSL-2) precautions are followed, including the use of laminar flow biosafety cabinets during all phage manipulations where there is the potential for aerosols, centrifuging phage in O-ring sealed tubes, and decontaminating disposable plasticware with bleach prior to sterilization by autoclaving.

6. DNase treatment is important to prevent nonspecific transfection of cells by any contaminating RF of the phage. Contamination by RF phage DNA should be monitored before and after DNase treatment by evaluating the phage on an agarose gel and staining with ethidium bromide. No double-stranded DNA should be detectable.

7. The phage particles themselves are generally not toxic to the mammalian cell lines that we have tested. However, because some cell lines are more sensitive to endotoxin contamination than others, it is important that endotoxin be removed. The number of extractions needed to reduce the endotoxin levels should be determined by testing samples of phage following endotoxin removal.

8. For example:

$$\text{number of plaques} \times \text{dilution factor} \div \text{volume (mL)} = \text{titer}$$
$$100 \text{ pfu} \times 10^8 \div 0.1 \text{ mL} = 1 \times 10^{11} \text{ pfu/mL}$$

Acknowledgments

This work was supported in part by funding from the National Institutes of Health (SBIR grant no. 1R43 CA80515-01).

References

1. Larocca, D., Witte, A., Johnson, W., Pierce, G. F., and Baird, A. (1998) Targeting bacteriophage to mammalian cell surface receptors for gene delivery. *Hum. Gene Ther.* **9,** 2393–2399.

2. Kassner, P. D., Burg, M. A., Baird, A., and Larocca, D. (1999) Genetic selection of phage engineered for receptor-mediated gene transfer to mammalian cells. *Biochem. Biophys. Res. Commun.* **264,** 921–928.

3. Larocca, D., Kassner, P., Witte, A., Ladner, R., Pierce, G. F., and Baird, A. (1999) Gene transfer to mammalian cells using genetically targeted filamentous bacteriophage. *FASEB J.* **13,** 727–734.

4. Poul, M. and Marks, J. D. (1999) Targeted gene delivery to mammalian cells by filamentous bacteriophage. *J. Mol. Biol.* **288,** 203–211.

5. Smith, G. P. and Scott, J. K. (1993). Libraries of peptides and proteins displayed on filamentous phage. *Methods Enzymol.* **217,** 228–257.

6. Kay, B. K., Winter, J., and McCafferty, J. (1998) *Phage display of peptides and proteins: a laboratory manual* (Kay, B. K., Winter, J., and McCafferty, J., eds.), Academic Press, San Diego.

7. Smith, G. P. (1985) Filamentous fusion phage: novel expression vectors that display cloned antigens on the virion surface. *Science* **228,** 1315–1317.

8. Bass, S., Greene, R., and Wells, J. A. (1990) Hormone phage: an enrichment method for variant proteins with altered binding properties. *Proteins* **8,** 309–314.

9. Saggio, I., Gloaguen, I., and Laufer, R. (1995) Functional phage display of ciliary neurotrophic factor. *Gene* **152,** 35–39.

10. Buchli, P. J., Wu, Z., and Ciardelli, T. L. (1997) The functional display of interleuken-2 on filamentous phage. *Arch. Biochem. Biophys.* **339,** 79–84.

11. Gram, H., Strittmayer, U., Lorenz, M., Glück, D., and Zenke, G. (1993) Phage display as a rapid gene expression system: production of bioactive cytokine-phage and generation of neutralizing monoclonal antibodies. *J. Immunol. Methods* **161,** 169–176.

12. Souriau, C., Fort, P., Roux, P., Hartley, O., Lefranc, M. P., and Weill, M. (1997) A simple luciferase assay for signal transduction activity detection of epidermal growth factor displayed on phage. *Nucleic Acids Res.* **25,** 1585–1590.

13. Vispo, N. S., Callejo, M., Ojalvo, A. G., Santos, A., Chinea, G., Gavilondo, S., and Arana, M. J. (1997) Displaying human interleukin-2 on the surface of bacteriophage. *Immunotechnology* **3,** 185–193.

14. Merlin, S., Rowold, E., Abegg, A., Berglund, C., Klover, J., Staten, N., et al. (1997) Phage presentation and affinity selection of a deletion mutant of human interleukin-3. *Appl. Biochem. Biotechnol.* **67,** 199–214.

15. Ivanenkov, V. V., Felici, F., and Menon, A. G. (1999) Targeted delivery of multivalent phage display vectors into mammalian cells. *Biochim. Biophys. Acta* **1448,** 463–472.

16. Barry, M. A., Dower, W. J., and Johnston, S. A. (1996) Toward cell-targeting gene therapy vectors: selection of cell-binding peptides from random peptide-presenting phage libraries. *Nat. Med.* **2,** 299–305.

17. Pereira, S., Maruyama, H., Siegel, D., Van Belle, P., Elder, D., Curtis, P., and Herlyn, D. (1997) A model system for detection and isolation of a tumor cell surface antigen using antibody phage display. *J. Immunol. Methods* **203,** 11–24.

18. Watters, J. M., Telleman, P., and Junghans, R. P. (1997) An optimized method for cell-based phage display panning. *Immunotechnology* **3,** 21–29.

19. Yakinoglu, A. O., Heilbronn, R., Burkle, A., Schlehofer, J. R., and zur Hausen, H. (1988) DNA amplification of adeno-associated virus as a response to cellular genotoxic stress. *Cancer Res.* **48,** 3123–3129.

20. Cormack, B. P., Valdivia, R. H., and Falkow, S. (1996) FACS-optimized mutants of the green fluorescent protein (GFP). *Gene* **173,** 33–38.

21. Scott, J. K. and Smith, G. P. (1990) Searching for peptide ligands with an epitope library. *Science* **249,** 386–390.

22. Cwirla, S. E., Peters, E. A., Barrett, R. W., and Dower, W. J. (1990) Peptides on phage: a vast library of peptides for identifying ligands. *Proc. Natl. Acad. Sci. USA* **87,** 6378–6382.

23. Giebel, L. B., Cass, R. T., Milligan, D. L., Young, D. C., Arze, R., and Johnson, C. R. (1995) Screening of cyclic peptide phage libraries identifies ligands that bind streptavidin with high affinities. *Biochemistry* **34,** 15,430–15,435.

24. Becerril, B., Poul, M. A., and Marks, J. D. (1999) Toward selection of internalizing antibodies from phage libraries. *Biochem. Biophys. Res. Commun.* **255,** 386–393.

25. Rakonjac, J., Jovanovic, G., and Model, P. (1997) Filamentous phage infection-mediated gene expression: construction and propagation of the *gIII* deletion mutant helper phage R408d3. *Gene* **198,** 99–103.

26. CDC/NIH and U.S. Department of Health and Human Services. (1993) *Biosafety in microbiological and biomedical laboratories.* HHS Publication No. (CDC) 93-8395. 3rd ed.

29

Single-Cell PCR Methods for Studying Stem Cells and Progenitors

Jane E. Aubin, Fina Liu, and G. Antonio Candeliere

1. Introduction

Knowledge of the molecular and cellular events characterizing osteoblast development is growing as new markers, including important classes of regulatory molecules such as transcription factors (e.g., Cbfa-1 *[1]*), are elucidated. Nevertheless, a paucity of definitive and specific markers, especially for the more primitive progenitors and stem cells, slows advancement in the field in comparison to other lineages, such as the hemopoietic lineages *(2)*. One useful model, however, has been culture of mixed populations of freshly isolated cells derived from a variety of bones (e.g., 21-d fetal rat calvaria [RC]) or bone marrow stroma under conditions that favor osteoblast development *(2)*. For example, when such heterogeneous primary cultures are grown long-term (approx 3 wk) in medium supplemented with ascorbic acid and β-glycerophosphate, a low frequency (about 0.00001–1% of unfractionated freshly isolated populations) of osteoprogenitor cells present divide and differentiate to form 3-dimensional mineralized bone nodules *(3,4)*. These infrequent cells comprise the colony forming units or colony forming cells-osteoprogenitor (CFU-Os or CFC-Os, respectively) in populations from the whole tissue and appear analogous to the nonstem cell CFU/CFCs in lineages such as the hemopoietic. Notably, the frequency of such cells can be determined by limiting dilution, and they appear to have limited capacity for self-renewal *(3,4)*. On the other hand, morphological, immunohistochemical, and molecular analyses have confirmed that differentiation of CFU-Os and formation of bone nodules reproducibly recapitulates a proliferation–differentiation sequence from an early precursor cell to a mature osteoblast *(2)*. However, because osteoprogenitors comprise such a low fraction of cells in these populations, and the differentiation process is not synchronous, it can be difficult to analyze the expressed gene repertoires in osteoprogenitors, particularly during very early events in the maturational sequence. It was for these reasons that we explored use of single-cell and single-colony methods that would allow individual osteoprogenitors and their progeny to be studied.

This chapter does not directly cover studies on embryonic stem (ES) cells, but does summarize a variety of methods and protocols that we have used successfully on osteoblast lineage cells and that are easily adaptable to the ES situation. In particular,

From: *Methods in Molecular Biology, vol. 185: Embryonic Stem Cells: Methods and Protocols*
Edited by: K. Turksen © Humana Press Inc., Totowa, NJ

we cover methods useful when cell number or sample size is limited, such as when a particular cell type is present in only limited number in a much larger heterogeneous population, e.g., when an ES population is used to generate cells of a particular lineage, but the conversion frequency is relatively low. A comparable situation is often, if not normally, the case in our field, which has been to investigate the developmental program underlying the osteoblast lineage and bone formation. Our methodology developed for use on this lineage will serve as a useful paradigm for any lineage of interest, including many of those discussed elsewhere in this volume.

We have not outlined the osteoblast cell isolation and culturing procedures, since the single-cell methods we outline should be adapted for the cells and cultures to be used for each lineage of interest.

2. Materials

2.1. Cell Isolation

1. α-Modified Eagle medium (α-MEM) with antibiotics and fetal bovine serum (FBS).
2. 37°C CO_2 incubator.
3. Chondroitin sulfate (Fluka, cat. no. 27043).
4. Citrate saline.
5. Collagenase (Sigma, cat. no. C 0130).
6. DNase (Sigma, cat. no. D 4513).
7. D-Sorbital (Fisher, cat. no. S 459).
8. Forceps.
9. Fungizone (Sigma, cat. no. 10041).
10. Gentamycin (Sigma, cat. no. G1264).
11. Glass beads (Pyrex 4 mm; VWR, cat. no. 13782-554) (*see* **Note 1**).
12. Glass cloning rings (*see* **Note 2**).
13. Marker.
14. Metal spatula.
15. Methylcellulose (Stemcell Technologies, cat. no. GF H4434).
16. Microcaps VL/PK100 1 λ glass micropipet (VWR, cat. no. 53440-001) and suction bulb assembly (VWR, cat. no. 53507-268) (*see* **Note 3**).
17. Micromanipulator—optional.
18. Minivortex mixer (VWR, cat. no. 58816-121).
19. P2 and P200 pipet and filtered tips.
20. Phosphate-buffered saline (PBS).
21. Penicillin G (Sigma, cat. no. 13752).
22. Phase contrast microscope with adaptable stage for 35-mm dishes.
23. Polyester cloth (B & SH Thompson, Scarborough, ON, cat. no. HD7-1).
24. Silicone grease (VWR, cat. no. KT743206-0000).
25. Terizaki plates (Nunc, cat. no. 476546).
26. Tissue culture dishes (35-mm Falcon; Becton Dickinson, cat. no. 353001).
27. Trypsin (Gibco, cat. no. 15090-046).

2.1.1. Solutions to Prepare for Cell Isolation

2.1.1.1. 10X Antibiotic Solution; Penicillin–Gentamycin–Fungizone (*see* **Note 4**)

1. Prepare a 50X stock penicillin G solution by dissolving 8 g of penicillin G sodium salt in 200 mL α-MEM, then filter-sterilize; aliquot into 10-mL aliquots **and** store at –20°C.
2. Prepare a 100X stock solution of gentamycin by dissolving 500 mg **of gentamycin** sulfate in 100 mL of α-MEM, then filter-sterilize; aliquot into 5-mL aliquots **and** store at –20°C.

3. Prepare a 100X stock solution of fungizone by dissolving 3 mg of fungizone in 100 mL of α-MEM, then filter-sterilize; aliquot into 5-mL aliquots and store at –20°C.
4. To prepare the 10X antibiotic solution stock, mix 10 mL of stock penicillin G, 5 mL of stock gentamycin, and 5 mL of stock fungizone into 480 mL of α-MEM.
5. Aliquot the 10X antibiotics solution into 10-mL aliquots and store at –20°C. After thawing, 10X antibiotics solution can be stored at 4°C for up to 7 d.

2.1.1.2. α-MEM WITH 15% FBS (FOR 100 ML)

1. Sterilely aliquot 75 mL of α-MEM into a clean autoclaved glass bottle.
2. Add 10 mL of 10X antibiotic solution.
3. Add 15 mL of freshly thawed and heat-inactivated FBS and swirl gently.
4. Store unused supplemented medium at 4°C for up to 1 wk.

2.1.1.3. 1X PBS

We buy our sterile 1X PBS (pH 7.2) from a Tissue Culture Core Service, but comparable solutions are available from Gibco (cat. no. 20012027).

2.1.1.4. CITRATE SALINE

We buy our sterile citrate saline from a Tissue Culture Core Service, but comparable solutions are available from Sigma (cat. no. S0902).

2.1.1.5. 0.01% TRYPSIN

1. Prepare a 1% stock trypsin solution by dissolving 1 g of trypsin in 100 mL of citrate saline.
2. Filter-sterilize, aliquot into 10-mL aliquots and store at –20°C.
3. To prepare a 0.01% solution, mix 5 mL of the 1% stock solution with 495 mL of sterile citrate saline.
4. The 0.01% working solution can be re-aliquotted and stored in conveniently-sized aliquots at –20°C; we store 10-mL aliquots. Thawed aliquots can be stored for up to 5–7 d at 4°C.

2.1.1.6. KREB'S II A BUFFER WITH Zn++ (FOR 1 L)

NaCl	111.2 mM	6.4965 g
Tris Buffer (Base)	21.3 mM	2.5802 g
Glucose	13.0 mM	2.3421 g
KCl	5.4 mM	0.4026 g
$MgCl_2$	1.3 mM	0.2643 g
$ZnCl_2$	0.5 mM	0.0682 g

1. Combine the first 5 chemicals in 900 mL of distilled water, adjust pH to 7.4.
2. Add in $ZnCl_2$ and make up the volume to 1 L in a volumetric flask.
3. Filter-sterilize and store at 4°C.

2.1.1.7. COLLAGENASE DIGESTION MIXTURE

Kreb's II A Buffer	200 mL
Collagenase	0.6 g
D-Sorbital	3.644 g
Chondroitin sulfate	1.2 g
DNase	0.8 mL

1. Add the sorbitol and the chondroitin sulfate to the required buffer first, and warm gently.
2. Dissolve collagenase separately in a small quantity of the warm buffer.
3. Cool the rest of the buffer, and then add to the collagenase.
4. Lastly, add DNase.
5. Filter-sterilize and aliquot. Store at –20°C.

2.1.1.8. 0.1% METHYLCELLULOSE-PBS

1. In an Erlenmeyer flask, heat 100 mL of PBS just to boiling, and add 0.1 g of methylcellulose.
2. Cover the flask loosely with aluminum foil, and thoroughly resuspend powder by vigorous swirling of the flask until no clumps remain.
3. Heat the flask again, just to boiling, and then immediately remove from heat source.
4. Cool suspension to 40–50°C under running water.
5. Add distilled water for a final volume of 100 mL and swirl the flask to mix contents.
6. Immerse flask in ice for 2 h, then sterilely distribute into 4-mL aliquots and store at –20°C.
7. For use, tubes should be thawed at 4°C for at least 2 d, and they can be stored at 4°C for 4–6 wk.

2.2. Poly(A) Polymerase Chain Reaction

1. 1 M Tris Buffer (Base), pH 8.3
2. 1 M KCl.
3. 2 M KCl.
4. 1 M MgCl$_2$.
5. (dT)$_{24}$ Oligo nucleotide (synthesized commercially).
6. Avian myeloblastosis virus (AMV) RTase (Roche, cat. no. 1 495 062).
7. Bovine serum albumin (BSA) (Roche, cat. no. 711 454).
8. dATP, 100 mM solution (Amersham Pharmacia, cat. no. 27-2050-01).
9. dNTP set, 100 mM solutions (Amersham Pharmacia, cat. no. 27-2035-02).
10. Microcentrifuge.
11. Maloney murine leukemia virus (MMLV) RTase (Life, cat. no. 28025-013).
12. Nonidet P-40 (NP-40) (Sigma, cat. no. 56741).
13. Prime RNase Inhibitor (5′→3′, cat. no. 9-901109).
14. RNAguard (Amersham Pharmacia, cat. no. 27-0815-01).
15. *Taq* DNA polymerase (Qiagen, cat. no. 201203).
16. Terminal deoxynucleotidyl transferase (TdT) (Life, cat. no. 10533-016; Roche, cat. no. 220 582).
17. Thermal cycler.
18. Triton X-100 (Sigma, cat. no. T9284).
19. X-(dT)$_{24}$ Oligo nucleotide (synthesized commercially) (*see* **Note 5**).

2.2.1. Solutions to Prepare for Poly(A) Polymerase Chain Reaction

1. *25 mM dNTPs:* combine equal volumes of each dNTP; aliquot and store at –70°C.
2. *RNase inhibitor mixture:* prepare a 1 : λ1 mixture of RNAguard and Prime RNase Inhibitor; aliquot and store at –20°C.
3. *cDNA primer mixture:* prepare a mixture of 12.5 mM dNTPs and 6.125 OD$_{260}$/mL of oligo(dT)$_{24}$; store in 5-μL aliquots at –70°C.
4. *10X RTase buffer:* prepare a solution containing 500 mM Tris (pH 8.3), 750 mM KCl, and 30 mM MgCl$_2$; aliquot and store at –20°C.

5. *First lysis buffer:* prepare a mixture of 1X RTase buffer and 0.5% NP40; aliquot and store at –20°C.

6. *cDNA/lysis buffer*
   a. Prepare this mixture when you are ready to pick cells and keep on ice.
   b. Add 120 µL H_2O to a tube of cDNA primer mixture, then take 2 µL of this diluted mixture and add to 96 µL of First lysis buffer and 2 µL of RNase inhibitor mixture.

7. *RTase mixture:* prepare by combining both MMLV and AMV RTases; aliquot and store at –20°C.

8. *2X Tailing buffer:* prepare a mixture of 2X TdT buffer (use the 5X buffer that comes with the Life brand of TdT) and 1.5 mM dATP; aliquot and store at –70°C.

9. *TdT mixture:* prepare by combining both brands of TdT for a concentration of approximately 10 U/µL; aliquot and store at –20°C.

10. *Tailing mixture:* per sample, prepare by combining 4 µL of 2X Tailing buffer and 1 µL of TdT mixture.

11. *10X Taq buffer:* prepare a solution containing 100 mM Tris (pH 8.3), 500 mM KCl, 15–50 mM $MgCl_2$ (*see* **Note 6**), 1 mg/mL BSA, and 0.5% Triton X-100; aliquot and store at –20°C.

12. *Polymerase chain reaction (PCR) mix:* prepare a solution containing 1X *Taq* buffer, 1–10 OD_{260}/mL X-$(dT)_{24}$oligonucleotide (*see* **Note 7**), and 1 mM dNTPs.

2.3. cDNA Fingerprinting

1. AmpliTaq DNA polymerase and 1X AmpliTaq PCR buffer (Perkin Elmer, cat. no. N808-0160).
2. 4 mM $MgCl_2$.
3. Abritrary primer (20-mer).
4. Acrylamide.
5. Each of the four dNTPs (*see* **Subheading 2.2.**).
6. 50 mM NaOH.
7. ^{32}P-dCTP.
8. 1 M Tris, pH 8.0.
9. TOPO TA Cloning Kit (Invitrogen, cat. no. 45-0641).

2.3.1. Solutions to Prepare

2.3.1.1 Acrylamide Solution

Acrylamide	7.125 g
bis-acrylamide	0.375 g
Urea	75 g
10X Tris-borate EDTA (TBE) buffer	15 mL

1. Add all the ingredients to 75 mL H_2O and bring final vol to 150 mL.
2. Place solution on hot plate with a stirrer to completely dissolve. Never allow acrylamide solution to exceed temperatures of 50°C.

3. Methods

The poly(A) PCR procedure, described originally by Brady and Iscove *(5,6)*, is a method that can be used to generate microgram amounts of cDNA representative of the entire spectrum of polyadenylated mRNAs in samples as small as a single cell and up to 100 cells, while maintaining relative abundance relationships. This technique offers a unique advantage, because it makes it possible to identify patterns of coexpression of several genes within the same cells, whereas other techniques such as in situ hybridization and immunocytochemistry limit the number of genes that can be detected concomitantly within a single cell. Further advantages of the poly(A) PCR protocol are that the reactions are performed in a single tube, it reduces sample loss and contamination by eliminating extraction and precipitation steps, and it provides an inexhaustible source of cDNA through reamplification.

We have employed poly(A) PCR to investigate the simultaneous expression of multiple bone-related macromolecules in single cells from individual discrete osteoblastic colonies undergoing differentiation in vitro *(7,8)*. Moreover, we have adapted the poly(A) PCR procedure, described initially for hemopoietic cells *(5)*, for general use on cells that grow adherent to a substrate, and indeed for cells that require adherence for differentiation. Thus, individual cells analyzed by poly(A) PCR can be used in lieu of mass populations to extend investigation of stages in the progression of differentiation. The methodology has been used to study several other cell types, including hemopoietic cells *(9–12)* and Hodgkin and Reed-Sternberg cells *(13)*. There are already a few examples of poly(A) PCR use on ES-derived cells *(14,15)*. The poly(A) PCR approach generates cDNA libraries or expressed sequence tag (EST) pools from single cells or small colonies of cells at different developmental stages, and these are appropriate for use for gene screening protocols such as the cDNA fingerprinting strategy we describe further below.

3.1. Cell Isolation for Poly(A) PCR

The basis for our procedure rests in being able to isolate single cells or colonies or small homogeneous groups of cells using any one of a number of specific approaches. For cultured cells, one must first establish conditions so as ensure growth of colonies derived from single cells, i.e., colonies should be clonal in origin. We had previously determined from limiting dilution studies the conditions under which there was a statistically significant probability that osteoprogenitor colonies were clonally-derived (see, e.g., **refs.** *3,4*), and we used the same conditions for these studies. Similar strategies may be used for other cell types, including ES cells; in some cases recloning steps may be possible, which will lessen, if not obviate, the need for clonality in the initial culturing steps. There are, in addition, other approaches that may be used to obtain relevant cells from tissue culture or from tissue sections, including direct micromanipulation of cells out of the culture dish or tissues, and such techniques as laser capture microdissection *(16)*.

The protocols that follow have been designed for isolating single cells for the poly(A) PCR procedure. A simpler procedure can be used for cells that grow nonadherently or as colonies in semisolid media, such as methyl cellulose, since, in these latter cases, single cells or colonies can be plucked directly into micropipets without the trypsinization step. Variations of the procedure can be used to isolate whole colonies of cells for either poly(A) PCR or reverse transcription PCR (RT-PCR) as noted.

3.1.1. Isolation of Single Cells and Colonies

1. Under the microscope, choose individual colonies with morphologies of interest and that are well-separated from other colonies. Using a marker, circle the underside of the culture dish where the colony is located.

2. Rinse the dishes with PBS.

3. While holding a cloning ring with forceps, apply silicone grease with the spatula to one end of the cloning ring. Place the cloning ring over the marked colony and form a seal by lightly pushing the top of the ring against the dish using your thumb (*see* **Note 8**).

4. Add 50 µL of 0.01% trypsin (or a 1:1 mixture of 0.01% trypsin and collagenase-containing mixture when there is a heavy matrix) into the cloning ring, and incubate the dish at 37°C. Verify occasionally under the microscope that the cells are detaching from the dish.

5. Neutralize after cell release by adding 50 µL of α-MEM containing 15% FBS

6. Gently pipet up and down to form a single-cell suspension and transfer into a tube with 0.1% methylcellulose-PBS.

7. Vortex mix the tube to mix the contents and pour into a 35-mm tissue culture dish. Swirl the dish to distribute the mixture evenly.

8. Place Terizaki plate on ice with 4 µL/well of cDNA lysis buffer.

9. Under the microscope, move the dish around and focus to locate a single cell.

10. With a finely drawn out glass micopipet (with or without the aid of a micromanipulator) and gentle applied suction, draw the cell into the micropipet in a volume of not more than 1 µL (*see* **Note 1**).

11. With the aid of the microscope, transfer the cell into a Terizaki well. Then pipet the mixture into a PCR tube and keep on ice. Under the microscope, verify that the cell does not remain in the well. Continue with this step until the required number of cells has been collected.

12. The collected cells can now be stored at –70°C or continue to the poly(A) PCR steps.

3.1.2. Replica Plating to Isolate Single Cells and Colonies

As already mentioned, definitive biochemical or molecular markers to allow identification and/or isolation of primitive osteoprogenitors are not currently available. Our solution to the problem of identifying definitively the low frequency progenitors for further molecular characterization was to use the technique of replica plating on dishes plated at low density and sampled early in the development of colonies. Replica plating has been found to allow screening of large numbers of individual mammalian cell clones for the phenotype of interest, while still maintaining a master copy of the colonies; use of polyester cloth, in particular, has been found to provide high-fidelity copies for a variety of cell types (*17,18*).

The replica technique allowed us to identify the less than 1% osteoprogenitors present on the master plates, which then served as the basis of samples for molecular analysis. Clearly, since no recloning step was possible in this procedure, we used our optimized low density plating techniques so as to have high confidence that each colony was indeed clonal, i.e., a single-cell-derived colony.

1. Cultures at limiting dilution for single colony growth are established as usual.

2. Mark a sterile disc of 1-µm pore size polyester cloth (*see* **Note 9**) by notching; the notch is lined up with a mark on the culture dish.

3. Float the disc above the cells and weigh it down by gently pouring on a monolayer of 4-mm glass beads that have been washed several times in a standard tissue culture grade detergent, washed extensively, and sterilized by autoclaving.

4. Replica cloths are placed on cells at d 1 or d 4 and are removed on d 5 or d 11, respectively (*see* **Note 10**).

5. Transfer each cloth into a new dish.

6. Rinse the master dish and polyester cloth with PBS and then refeed both with appropriate cell culture medium.
7. Incubate the master dish at a low enough temperature so as to stall or minimize growth, e.g., 25°C (*see* **Note 11**).
8. Incubate the replica cloth at 37°C for times chosen as in **step 2**; maintain normal culture conditions, changing medium and adding supplements as usual for the cell type of interest.
9. Confirm the differentiation status of colonies on the replica cloth with methods suitable for the cell type of interest, e.g., histochemistry and/or immunohistochemistry for identification markers.
10. When colony types have been established for colonies on the replica cloth, the master dish is transferred to 37°C for 5–9 h (*see* **Note 12**).
11. Match the replica cloth to the master dish to localize colonies of interest; mark the colonies by indelible marker on the underside of the culture dish.
12. Rinse dishes with PBS and place a cloning ring around each colony of interest; collect cells as in **Subheading 3.1.1.** above or collect total RNA as in **step 13**.
13. Total RNA is extracted using either Trizol reagent or a miniguanidine thiocyanate method (*see* **Note 13**).

3.2. Poly(A) PCR (see Note 14)

1. Place the tubes into the thermal cycler for the following profile: 65°C for 1 min, 22°C for 3 min, and 4°C soak.
2. Add 0.5 µL of RTase mixture to each tube. Continue in the thermal cycler for the following profile: 37°C for 15 min, 65°C for 10 min, and 4°C soak.
3. Add 5 µL of Tailing mixture to each tube and place in the thermal cycler for the following profile: 37°C for 15 min, 65°C for 10 min, and 4°C soak.
4. Add 4 µL of PCR mixture to each tube and place in the thermal cycler for the following profile: 94°C for 5 min; 25 cycles consisting of 94°C for 1 min, 42°C for 2 min, 72°C for 6 min; then linked to another 25 cycles consisting of 94°C for 1 min, 42°C for 1 min, 72°C for 2 min; 72°C for 10 min; and 4°C soak.
5. The samples may be stored at –70°C; but due to deterioration with prolonged storage, it is preferable to expand the supply through reamplification or cloning (for more details, see **ref. 6**) (*see* **Note 15**).
6. Globally amplified cDNA pools are analyzed as usual by Southern blotting, taking care to ensure that probes are appropriately prepared so as to encompass approx 600 bp upstream of the poly(A) sequence.

3.3. cDNA Fingerprinting

cDNA fingerprinting is a powerful approach to identify markers of interest from biologically and minimally molecularly characterized samples obtained from sources in which the quantity of mRNA is extremely limited. The approach provides a functional genomics strategy which targets any physiological or even pathological differentiation stage of interest for any cell population in tissue culture or in tissue sections, in which a differentiation or developmental sequence can be identified. As already mentioned, there are multiple approaches that may be used to obtain relevant cells from tissue culture or from tissue sections *(19)* for cDNA fingerprinting; these include micromanipulation of cells and colonies, as we have used for osteoblastic cells *(7)*, and such techniques as laser capture microdissection *(16)*.

The specific example we describe comprises cDNA libraries prepared from replica-plated differentiating osteoblast colonies; the replica plating allowed unambiguous retrospective identification of low frequency osteoprogenitors, and colonies were staged as more or less differentiated based on the Southern hybridization signals obtained with a variety of osteoblast markers. Other arbitrarily primed PCR approaches have been described (summarized in **ref. 20**), but the cDNA fingerprinting approach has two advantages. One advantage is that the poly(A) PCR technique globally amplifies mRNA, while maintaining relative abundance and provides cDNA libraries that can be reamplified indefinitely *(5,9)*. Cell-type specific and differentiation-stage specific gene searches have and are being done on poly(A) PCR libraries by subtractive hybridization approaches *(6)*, but the subtraction strategy relies on comparison of only two populations or stages of interest at any one time. cDNA arrays, which allow simultaneous comparison of several hundred thousand genes at a time, is also theoretically possible and will undoubtedly also be used in the future, but it is likely that only a few samples will be studied at one time. The cDNA fingerprinting strategy, on the other hand, allows rapid comparison of numerous samples simultaneously; for example, we have compared over 20 colonies simultaneously. There are, however, also limitations. One of these relates to the possibility of isolating multiple ESTs of one size in one band, although strategies exist for overcoming this problem *(21)*. Second, "false positive" signals in secondary screens on the cDNA Southern and Northern blots have been seen with some small ESTs (<300 bp). Finally, cDNA fingerprinting is not appropriate for the selective amplification of particular family members, since the global amplification strategy amplifies relatively short gene tags (approx 0.6–0.7 kb) at the 3' end of mRNAs, frequently outside the coding region of genes of interest. Nevertheless, the strategy described here is applicable to many physiological or pathological developmental systems. By virtue of simultaneous analysis of expression profiles over multiple stages of interest, it also identifies and predicts potential temporally relevant interrelationships and interacting molecules in particular differentiation cascades.

3.3.1. Arbitrarily Primed PCR

The PCR conditions described have produced a usable fingerprint (detectable complex banding pattern) for all primers tested to date, although varying film exposures are required for different primers (*see* **Note 16**).

1. The final concentrations of reagents in a 2.5-µL reaction volume are 1X AmpliTaq PCR buffer, 4 mM MgCl$_2$, 250 mM each of the four dNTP's, 0.1 µCi/mL of P32-dCTP, 0.1 U/µL Perkin Elmer AmpliTaq DNA polymerase, 10 mM of an arbitrary primer (*see* **Note 17**), and a 1 : λ100 dilution of the first poly(A) PCR amplification.
2. An initial priming step consisting of two cycles of 94°C for 1 min, 35°C for 5min, and 72°C for 5 min, is followed by 25 cycles (*see* **Note 18**) of 94°C for 1 min, 50°C for 1 min, 72°C for 1 min.
3. The entire 2.5-µL reaction is then separated on a 5% denaturing gel *(22)* (*see* **Note 19**).

3.3.2. EST Identification

Interesting patterns of expression are identified on films exposed to dried-down gels and bands containing DNA of interest are isolated for reamplification by a method described previously *(22)*.

1. Align the autoradiograph with the dried gel.
2. To mark the bands, use a clean needle to poke holes through the film into the dried gel underneath. Cut the bands of interest out and try to minimize cross-contamination with similar sized bands not of interest (*see* **Note 20**).
3. The piece of dried gel containing the DNA fragment of interest is placed in 500 µL of 50 m*M* NaOH and boiled to elute DNA (*see* **Note 21**).
4. After 30 min, the solution is neutralized with 50 µL of 1 *M* Tris, pH 8.0.
5. An aliquot (1/10 volume of PCR) is then used for reamplification with standard PCR conditions *(22)* for up to 40 cycles at 94°C for 1 min, 50°C for 1 min, 72°C for 1 min (*see* **Note 22**).
6. PCR fragments are then subcloned into the the pCR 2.1-TOPO cloning vector for sequencing and for use as templates for making DNA probes.

4. Notes

1. We prepare all materials to ensure sterility, whether they are being used for prolonged culturing or for cell isolation as described here. Glass beads, glass cloning rings, filter tips, etc., are all autoclaved for 30 min; transfer silicone grease into an autoclavable glass jar prior to autoclaving.
2. We make glass cloning rings from pieces of glass tubing. Use tubing of 0.5–1.0 cm diameter, depending on your colony size of interest; our rings are 0.8 mm inside diameter. Cut tubing into lengths of 0.8–1 cm. Fire the cut edges to give smooth surfaces and autoclave the rings. Rings can be reused many times, after removing grease, washing extensively, and re-autoclaving.
3. There are several ways to pick up single cells for this reaction. Some laboratories may have access to micromanipulators and micropipets used for such applications as electrophysiology. In our experiments, we found micromanipulators to be relatively slow and cumbersome, and found that with time, we could isolate cells more quickly with manual gentle suction applied to glass micropipets. Glass micropipets of suitable size can be prepared easily in one of several ways. A glass λ micropipet can be drawn out over a low flame and used with its usual bulb assembly. However, a regular glass pasteur pipet can also be drawn out to very small size over a low flame and used with a regular rubber bulb suction device. The most important issue is to be able to control suction and volume so as to pick up single cells in a very small volume (1 µL) of medium.
4. This is a more complex antibiotic solution than some other available solutions. It was formulated for use with primary cell cultures from rodents. A simpler solution will work well with many cells and cell lines, including, e.g., penicillin–streptomycin (100X; Gibco, cat. no. 15140-122).
5. The X component can be any convenient sequence that can be used downstream for cloning and/or reamplification. It is recommended that it be between 15–36 bases (<15 bp tends to lead to reduced yield of PCR product). Examples of X component sequences are: *(i)* 5′-ATGTCGTCCAGGCCGCTCTGGACAAAATATGAATTC-3′ (contains an *Eco*RI restriction site); and *(ii)* 5′-CATCTCGAGCGGCCGC-3′ (contains a *Not*I site).
6. The MgCl$_2$ concentration must be optimized by doing a titration for each oligonucleotide when doing poly(A) PCR (for more details, see **ref. *6***).
7. The oligonucleotide concentration must be optimized by doing a titration for each oligonucleotide when doing poly(A) PCR *(6)*.
8. You can use whole colonies identified and isolated in this way to extract total RNA for RT-PCR as well. For example, we use Trizol reagent (Gibco BRL, cat. no. 15596-026) and add 200 µL directly into the cloning ring after briefly rinsing the colony with PBS. In single-colony RT-PCR, the entire total RNA sample (approx 100–400 ng/colony)

obtained from each single colony is used to synthesize cDNA. Other details for use of the Trizol reagent are according to the manufacturer's directions.

9. You should do pilot experiments to choose a pore size optimal for your specific cell type; one wants good nutrient exchange, but does not want cells crawling through to the noncontact side of the filter. Polyester cloth comes in large rolls from which you can cut discs to fit the size of culture dish you wish to use. Sterilize the discs by dipping briefly in 100% ethanol, air-drying, and autoclaving for 30 min; discs will stay flatter during autoclaving if each polyester disc is placed between Whatman (no. 50) filter discs.

10. These times were selected based on pilot experiments that established replication fidelity of >95% of osteoprogenitor colonies, i.e., that >95% of randomly chosen colonies faithfully differentiated on both master dishes and replica cloths. Pilot experiments should be done on each cell type of interest to optimize the day for placing and recovering filters for replication fidelity and the biological stage to be assayed. For example, one may want to minimize growth of the cells on the master dish prior to replication. Filters should also be recovered at times chosen so as to maximize faithful replication of colonies of interest.

11. The most effective temperatures for stalling growth are either freezing at –80°C or cooling cells to 4°C. However, low density cultures of many cell types of interest do not recover well from either of these temperatures or procedures. In our case, viable osteoprogenitor cells that had maintained differentiation capacity could not be recovered from either temperature. We, therefore, incubated master plates at either 25°C or 30°C, which was found to stall the proliferative and differentiation activities of progenitor cells. Other temperatures may prove to be more suitable for particular cell types of interest.

12. Pilot experiments should be performed to establish the optimal time for return of normal metabolic activity and growth–proliferation of cells that have been subjected to stalled growth and differentiation.

13. We now routinely achieve very high yields with Trizol reagent. However, total RNA can also be recovered with a mini-guanidine thiocyanate method that yields high-quality RNA for poly(A) PCR. Lyse the cells with vortex mixing in 25 µL of a solution of 5 M guanidine thiocyanate, 0.5% sarkosyl, 25 mM sodium citrate, pH 7.0, and 20 mM dithioerythritol. Precipitate RNA overnight with 0.5X vol of 7.5 M ammonium acetate and 800 µg/mL glycogen (Boehringer Mannheim Canada, Laval, PQ) and 3X volume ethanol. Wash the pellet twice with 75% ethanol and resuspend in 15 µL of a buffer containing 0.5% NP40 and 1 U Inhibit Ace (5′→3′, West Chester, PA). One microliter of this resuspension can be placed directly in 4 µL of prechilled lysis/1st strand buffer *(5,6)* for poly(A) PCR as outlined.

14. Brail, Iscove and colleagues have assessed, in detail, sources of variation that may affect the results of single-cell global amplification; it appears that the PCR step contributes most to this *(23)*.

15. Reamplification can be done multiple times to obtain a virtually limitless supply of cDNA from single cells or colonies.

16. Conditions for PCR were chosen so as to minimize specificity while preserving reproducibility, thus AmpliTaq at a high concentration was used. With this material, we have not observed that using two primers instead of one in the arbitrary amplification or that performing a second arbitrary amplification with a different primer has any advantage. The objective of this second arbitrary amplification is to reveal EST subpopulations not detected in the first round of arbitrary amplification.

17. The sequence of the arbitrary primer used to establish these conditions (and used to generate Fig. 3 in **ref. *20***) was TGTAGGAGCCAGAGGTGGTG. However, all 20-mers we have tried to date have worked.

18. Twenty-five cycles were chosen because that number was shown to maintain relative mRNA abundance in the original poly(A) PCR protocol (used for amplification steps *[5,6]*).

19. Since resolution capacity of the gel is important, we like to prepare acrylamide solution fresh from powder, because of the risk of possible degradation with time.

20. You can re-expose the dried gel to ensure no cross-contamination. Even a small amount of carryover contamination can cause problems during reamplification and subsequent analysis.

21. This is how we performed our original experiments. However, an alternative strategy that appears to work better is as follows. Add 40 µL of Tricine-EDTA buffer (10 m*M* Tricine, pH 9.5, 0.2 m*M* EDTA) to each tube (Clonetech manual, PT1173-1 version PR73484). This is a buffered reaction that also allows one to minimize the reaction volume. Overlay with mineral oil and heat to 100°C for 5 min. A 1:7 dilution is sufficient for PCR reamplification.

22. It is important to dilute the template enough so that any contaminants from the gel will not interfere with the PCR reamplification steps. However, if template is too dilute, it may be difficult to reamplify the bands.

Acknowledgments

This work was supported by a Canadian Institute of Health Research (CIHR) grant (MT-12390) to J.E.A. The authors also thank Norman Iscove for helpful discussions when we were setting up poly(A) PCR in our laboratory.

References

1. Ducy, P. (2000) CBFA1: a molecular switch in osteoblast biology. *Dev. Dyn.* **219,** 461–471.

2. Aubin, J. E. (1998) Bone stem cells. 25th Anniversary Issue: new directions and dimensions in cellular biochemistry. Invited chapter. *J. Cell Biochem.* **Suppl. 30/31,** 73–82.

3. Bellows, C. G. and Aubin, J. E. (1989) Determination of numbers of osteoprogenitors present in isolated fetal rat calvaria cells in vitro. *Dev. Biol.* **133,** 8–13.

4. Aubin, J. E. (1999) Osteoprogenitor cell frequency in rat bone marrow stromal cell populations: role for heterotypic cell-cell interactions in osteoblast differentiation. *J. Cell Biochem.* **72,** 396–410.

5. Brady, G., Barbara, M., and Iscove, N. N. (1990) Representative in vitro cDNA amplification from individual hemopoietic cells and colonies. *Methods Mol. Cell Biol.* **2,** 17–25.

6. Brady, G. and Iscove, N. N. (1993) Construction of cDNA libraries from single cells. *Methods Enzymol.* **225,** 611–623.

7. Liu, F., Malaval, L., Gupta, A., and Aubin, J. E. (1994) Simultaneous detection of multiple bone-related mRNAs and protein expression during osteoblast differentiation: polymerase chain reaction and immunocytochemical studies at the single cell level. *Dev. Biol.* **166,** 220–234.

8. Liu, F., Malaval, L., and Aubin, J. E. (1997) The mature osteoblast phenotype is characterized by extensive palsticity. *Exp. Cell Res.* **232,** 97–105.

9. Brady, G., Billia, F., Knox, J., Hoang, T., Kirsch, I. R., Voura, E., et al. (1995) Analysis of gene expression in a complex differentiation hierarchy by global amplification of cDNA from single cells. *Curr. Biol.* **5,** 909–922.

10. Cheng, T., Shen, H., Giokas, D., Gere, J., Tenen, D. G., and Scadden, D. T. (1996) Temporal mapping of gene expression levels during the differentiation of individual primary hematopoietic cells. *Proc. Nat. Acad. Sci. USA* **93,** 13,158–13,163.

11. Cumano, A., Paige, C. J., Iscove, N. N., and Brady, G. (1992) Bipotential precursors of B cells and macrophages in murine fetal liver. *Nature* **356,** 612–615.

12. Billia, F., Barbara, M., McEwen, J., Trevisan, M., and Iscove, N. N. (2001) Resolution of pluripotential intermediates in murine hematopoietic differentiation by global complementary DNA amplification from single cells: confirmation of assignments by expression profiling of cytokine receptor transcripts. *Blood* **97,** 2257–2268.

13. Trumper, L. H., Brady, G., Bagg, A., Gray, D., Loke, S. L., Griesser, H., et al. (1993) Single-cell analysis of Hodgkin and Reed-Sternberg cells: molecular heterogeneity of gene expression and p53 mutations. *Blood* **81,** 3097–3115.

14. Kennedy, M., Firpo, M., Choi, K., Wall, C., Robertson, S., Kabrun, N., and Keller, G. (1997) A common precursor for primitive erythropoiesis and definitive haematopoiesis. *Nature* **386,** 488–493.

15. Robertson, S. M., Kennedy, M., Shannon, J. M., and Keller, G. (2000) A transitional stage in the commitment of mesoderm to hematopoiesis requiring the transcription factor SCL/tal-1. *Development* **127,** 2447–2459.

16. Emmert-Buck, M. R., Bonner, R. F., Smith, P. D., Chuaqui, R. F., Zhuang, Z., Goldstein, S. R., et al. (1996) Laser capture microdissection. *Science* **274,** 998–1001.

17. Raetz, C. R., Wermuth, M. M., McIntyre, T. M., Esko, J. D., and Wing, D. C. (1982) Somatic cell cloning in polyester stacks. *Proc. Natl. Acad. Sci. USA* **79,** 3223–3227.

18. Esko, J. D. (1989) Replica plating of animal cells. *Methods Cell Biol.* **32,** 387–422.

19. Jensen, R. A., Page, D. L., and Holt, J. T. (1997) RAP-PCR using RNA from tissue microdissection. *Methods Mol. Biol.* **85,** 277–283.

20. Candeliere, G. A., Rao, Y., Floh, A., Sandler, S. D., and Aubin, J. E. (1999) cDNA fingerprinting of osteoprogenitor cells to isolate differentiation stage-specific genes. *Nucleic Acids Res.* **27,** 1079–1083.

21. McClelland, M., Arensdorf, H., Cheng, R. and Welsh, J. (1994) Arbitrarily primed PCR fingerprints resolved on SSCP gels. *Nucleic Acids Res.* **22,** 1770–1771.

22. Liang, P. and Pardee, A. B. (1998) Differential display. A general protocol. *Mol. Biotechnol.* **10,** 261–267.

23. Brail, L. H., Jang, A., Billia, F., Iscove, N. N., Klamut, H. J., and Hill, R. P. (1999) Gene expression in individual cells: analysis using global single cell reverse transcription polymerase chain reaction (GSC RT-PCR). *Mutat. Res.* **406,** 45–54.

30

Nonradioactive Labeling and Detection of mRNAs Hybridized onto Nucleic Acid cDNA Arrays

Thorsten Hoevel and Manfred Kubbies

1. Introduction

Cellular gene expression changes during ontogenetic development of cell physiological activation–inhibtion and differentiation. Classical molecular assays like Southern, Northern, or Western blotting display only a few genes at once. The analysis of complex alterations of gene expression patterns, therefore, requires large quantities of biological materials, has significant experimental inter- and intravariability, and is quite time-consuming. Some of these problems became negligible with the advent of microarray techniques *(1,2)*. cDNAs or oligonucleotids are immobilized on glass or membrane surfaces, cDNA or cRNA transcribed from cellular mRNA is hybridized, and signal detection is performed by radioactive or fluorescent techniques. The expression of up to several tens of thousands of genes can be made visible on small arrays, enabling investigators a rapid quantitative and qualitative analysis of pro- or eukaryotic gene expression patterns.

The biochemical conditions for the reverse transcription (RT) procedure of single genes (i.e., for detection in Northern blots) is optimized for each individual gene according to the primers used and the gene sequence. However, the optimal parameters vary between different genes, and it can be suggested that the RT procedure used for the generation of cDNA out of total mRNA might generate a quantitative bias of the mRNA mirror image of RT coupled with polymerase chain reaction (PCR)-generated cDNA used for hybridization onto high density gene arrays. This problem has been addressed recently comparing the gene expression patterns using cDNA and mRNA of human fibroblasts hybridized onto cDNA membrane arrays *(3)*. Applying a novel nonradioactive mRNA labeling technique we have shown that a significant number of genes display a decreased expression rate using cDNA in comparison to hybridized mRNA, indicative of a quantitative bias of the RT step. A typical example of the Atlas Human cDNA Expression Array (receptors, cell surface antigens, and cell adhesion; Clontech) comparing nonradioactive mRNA hybridization using the digoxigenin (DIG) ChemLink labeling technique and the classical radioactive ^{32}P technique is shown in **Fig. 1**. The overall dot images applying both techniques are comparable. However, overexposure of dots and signal crosstalks between dots is a common problem with the

From: *Methods in Molecular Biology, vol. 185: Embryonic Stem Cells: Methods and Protocols*
Edited by: K. Turksen © Humana Press Inc., Totowa, NJ

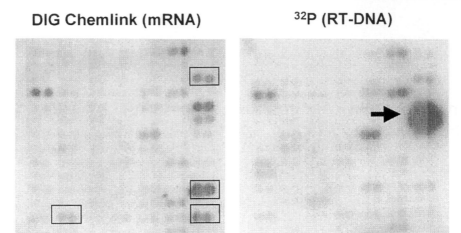

Fig. 1. Comparison of the nonradioactive mRNA hybridization and radioactive 32P technique on an Atlas human cDNA expression array.

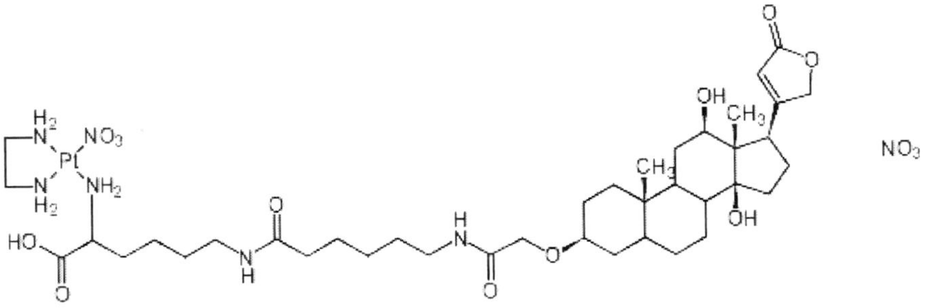

Fig. 2. Cisplatin derivative used for the nohnradioactive mRNA labeling procedure.

^{32}P technique (i.e., fibronectin receptor β-subunit; white arrow), which is not observed using the nonradioactive hybridization technique. On contrary, genes like IL2 receptor α chain, integrin α3, integrin-αL and β-catenin display a significant increased signal using the mRNA hybridization technique (**Fig. 1**, white boxes). We assume that this artifical bias is due to artefacts of the RT-step necessary for the generation of ^{32}P-labeled cDNA for array hybridization *(3)*.

The novel nonradioactive mRNA labeling procedure described below is based on the noncovalent binding of a cisplatin derivative to mRNA *(4)*. The cisplatin derivative is covalently modified either with biotin or DIG epitops, which are detected by streptavidin or monoclonal antibody binding (**Fig. 2**). Recording of the cisplatin-labeled hybridized mRNA is performed either by luminescence or fluorescence labeling techniques. The advantages using this ChemLink labeling technique are several fold: hybridization of the primary biological reporter molecule mRNA, nonradioactive

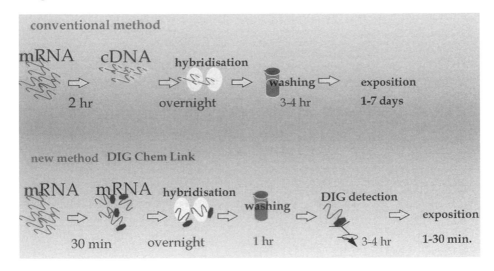

Fig. 3. Comparison of the mRNA Chem Link and cnventional cDNA labeling technique.

labeling and significant time-savings. A comparison of the conventional cDNA vs mRNA ChemLink labeling techniques is shown in **Fig. 3**. This article describes the generation of ChemLink-modified mRNA and its hybridization onto cDNA membrane arrays. mRNAs of various sources have proved to be useful for this novel nonradioactive mRNA labeling and array hybridization technique. Although this technique is developed with fibroblastic cells, it will be very useful in lineage studies using embryonic stem cell cultures.

2. Materials

1. Phosphate-buffered saline (PBS) (Roche, cat. no. 295868).
2. Dulbecco's modified Eagle medium (DMEM) Medium (LifeTech, cat. no. 11963-022).
3. Fetal calf serum (FCS), myoclone (LifeTech, cat. no. 10082147).
4. 0.25% (w/v) Trypsin (Roche, cat. no. 210234).
5. RNeasy maxi kit (Qiagen, cat. no. 75162).
6. mRNA isolation kit (Roche, 1741985).
7. Phenol, pH 4.3 (Sigma, cat. no. P-4682).
8. Phenol:chloroform (Sigma, cat. no. P-1944).
9. Phenol:chloroform:isoamylalcohol (25:λ24:λ1) (Sigma, cat. no. P-3803).
10. 8 M Lithium chloride (LiCl) (Sigma, cat. no. L-7026).
11. DIGChemlink Kit (Roche, cat. no. 1836463).
12. DIG wash and block buffer set (Roche, cat. no. 1585762).
13. DIG easy hyb solution (Roche, cat. no. 1603558).
14. Atlas membrane arrays (Clontech, cat. no. several fold).
15. CDP-Star ready to use substrate solution (Roche, cat. no. 2041677).
16. DNA, MB-grade (Roche, cat. no. 1467140).
17. Cot1 DNA (Roche, cat. no. 1581074).
18. Stripping buffer (50 mM NaOH, 0.1% sodium dodecyl sulfate [SDS]).
19. Stripping wash solutions 1: 2X standard saline citrate (SSC), 1% SDS.
20. Stripping wash solution 2: 0.1X SSC, 1% SDS.
21. Optional: Lumi-Imager (Roche, cat. no. 2012847).

22. Optional: DIG control teststrips (Roche, cat. no. 1669966) plus DIG quantification teststrips (Roche, cat. no. 1669958).

3. Methods

3.1. Human Fibroblast Cell Culture

1. Inoculate human fibroblasts (*see* **Note 1**) in DMEM, 10% FCS in cell culture flasks and expand cell culture until subconfluency. A total of 5×10^7 cells are sufficient to extract mRNA for several experiments.
2. Wash the cells twice in culture flasks with PBS (37°C).
3. Add a sufficient volume of trypsin to detach cells from the cell culture flask for 5–10 min at 37°C.
4. Wash the cells twice with PBS (37°C). Discard supernatant to reduce the protein content of the medium. This decreases the risk of clogging the column.
5. Continue with RNA isolation as quickly as possible, as a delay may induce apoptotic or necrotic processes of cells.

3.2. Extraction of Total RNA

The total RNA isolation described below is performed using the RNeasy Maxi Kit (*see* **Note 2**).

1. Resuspend up to 5×10^8 cells in 6 mL lysis buffer and shear it 5–6 times through a 18–20-gauge needle fitted to an RNase-free syringe. Do not apply extensive force, as this would result in air bubbles in the lysate.
2. Load the sample onto the column (*see* **Note 3**).
3. Wash columns with wash buffers (once with buffer RW1 and twice with buffer RPE) according to the protocol. Spin it down subsequently to drain it (*see* **Note 4**).
4. Elute RNA from the column twice in a suitable volume of RNase-free water into two separate elution tubes.
5. Measure the quantity and quality of total RNA using a spectrophotometer at wavelengths 260/280, avoiding absorption coefficients below 0.1 and above 0.6.

3.3. Extraction of mRNA

1. Dissolve total RNA in the kits' denaturation buffer, but not more than 2-fold.
2. Add biotin labeled oligo(dT)$_{20}$ primers. Allow mRNA oligonucleotide beads to attach onto streptavidin-coupled magnetic beads for 5 min at 37°C (*see* **Note 5**).
3. Wash beads twice in wash buffer.
4. Elute mRNA from beads in at least 100 µL RNase-free water twice for 2 min at 65°C.
5. Measure the quantity and quality of mRNA (**Subheading 3.2.**) (*see* **Note 6**).
6. Optional: check quality of mRNA on Tris-borote EDTA (TBE) agarose gel. Quality can be checked inspecting residues of 23S and 18S RNA. Band intensities should be greatly reduced using an aliquot of total RNA as a standard (*see* **Note 7**).

3.4. Increase Purity of mRNA

The quality of the mRNA is improved by additional phenol, phenol:chloroform, and phenol:chloroform:isoamylalcohol extraction steps (*see Current Protocols in Molecular Biology, Unit 4.1* **ref. 5**).

1. Add 1 vol of phenol to isolated mRNA (**Subheading 3.3.**). Shake vigorously for 30 s and spin for 5 min in a microcentrifuge at maximum speed.

2. Remove the aqueous (upper) phase and transfer it to a clean tube, carefully avoiding contamination with the lower phase (organic).

3. Repeat steps 1 and 2 with phenol:chloroform and phenol:chloroform:isoamylalcohol.

4. Add 1/10 vol of 8 *M* LiCl to aqueous phase and mix thoroughly.

5. Add 2.5 vol 100% ethanol and mix carefully.

6. Incubate for 30 min at –70°C or overnight at –20°C (formation of precipitates).

7. Recover RNA by centrifugation for 30 min at maximum speed in table-top centrifuge at room temperature.

8. Carefully rinse mRNA pellet in 1 mL 70% ethanol (pellet should be brownish for good quality of RNA, if pellet is white due to high salt precipitation, rinse again in 70% ethanol).

9. Allow pellet to air-dry (do not use a vacuum device, as pellets might get lost) and dissolve in an appropriate amount of RNase-free water to obtain a mRNA concentration of at least 100 ng/µL.

10. Determine the quantity of mRNA (**Subheading 3.2.**).

11. Store aliquots of mRNA at –70°C as frequent freeze-thaw steps decrease the quality of mRNA. Do not store at temperatures above –70°C to decrease hydrolytic activity of water.

3.5. Labeling of mRNA with DIGChem Link

This is performed using a DIGChemLink kit (components DIGChemLink, stop solution, and anti-digoxingen-AP, Fab fragments) (*see* **Note 8**).

1. Set up the labeling reaction. Mix 0.5 up to 2 µg mRNA, 0.5 up to 2 µg DIGChemLink, and 20 µL diethyl pyrocarbonate (DEPC) H$_2$O. For the fibroblasts described here, an amount of 0.5 to 1 µg mRNA was optimal.

2. Vortex mix and spin down for 4 s at 1000*g* force at room temperature.

3. Incubate for 30 min at 85°C (this results in the best labeling efficacy and slightly fragmented mRNAs, which tend to have a better hybridization performance).

4. Spin down for 4 s at 1000*g* at room temperature.

5. Stop reaction by adding 5 µL of stop solution. Short-term storage at 4°C or –20°C (long-time storage not recommended).

6. Check labeling efficiency with DIG control teststrips plus DIG quantification teststrips (labeling efficiency should be at least 0.3 pg/µL).

3.6. cDNA Array Hybridization

1. Dilute 5 mg DNA MB-grade (heat-denatured for 5 min at 95°C) in 50 mL of DIG easy hyb and heat up to 50°C (prehybridization mixture) (*see* **Note 10**).

2. Place cDNA array in a roller bottle without overlapping the membrane parts. Pour prehybridization mixture into roller bottle (*see* **Note 9**).

3. Incubate for 1 h at 50°C. Discard supernatant.

4. Dissolve labeled mRNA to a final concentration of 250–500 ng/mL in DIG easy hyb (8 mL in a roller bottle of 150 mm height). Incubate 5 µg of human Cot1 DNA in 100 µL RNase-free H$_2$O for 5 min at 95°C, mix well, spin down for 5 s at 1000*g* at room temperature, and add to DIG easy hyb.

5. Transfer hybridization solution to the array.

6. Hybridize overnight at 50°C.

7. Heat wash solutions up to 50°C. (stripping wash solution 1: 2X SSC, 1% SDS; and stripping wash solution 2: 0.1X SSC, 1% SDS).

8. Wash membrane with wash buffer 1 twice for 5 min at 50°C, and, subsequently, 2 times with wash buffer 2 for 15 min at 50°C, and proceed with DIG detection (**Subheading 3.7.**)

3.7. DIG Detection and Luminescence Recording

1. Wash membrane array in DIG wash buffer (supplied with DIG wash and block buffer set) for 5 min at room temperature.
2. Incubate for 30 min in DIG blocking buffer (supplied with DIG wash and block buffer set) and discard supernatant.
3. Incubate for 30 min in anti-DIG antibody solution (supplied in DIGChemLink kit, dilution 1:10000 in DIG blocking buffer) and discard supernatant.
4. Wash membrane array twice in DIG wash buffer for 15 min each at room temperature.
5. Incubate in DIG detection buffer (supplied with DIG wash and block buffer set) for 5 min at room temperature and discard supernatant.
6. The detection of DIG-labeled RNA is performed by luminescence technique using a Lumi-Imager instrument for signal recording. Use CDP-Star as a substrate for peroxidase enzyme coupled to the DIG antibody. For this purpose, put membrane on a solid plastic sheet (do not use saran wrap) and drop 2 mL of the substrate close to the membrane (do not drop it on the membrane as this might result in an irregular signal intensity on the membrane). Take a second plastic sheet and allow it to glide onto the membrane preparing an even fluidic film, avoiding air bubbles on the membrane.
7. Incubate for 5 min at 37°C in dark.
8. Put the cDNA membrane in a Lumi-Imager Device using suitable software, including nearest neighbor background subtraction algorithms.

3.8. Reuse of cDNA Array Membranes

Nucleic acid arrays can be reused limited times performing a stripping and reprobing procedure (see array manufacturers' protocol). The following protocol is designed for usage on mRNA-cDNA hybrids, which display a higher affinity compared to cDNA-cDNA hybrids.

1. Incubate cDNA membranes in RNase-free water for 1 min to wash off substrate.
2. Gently agitate membrane in stripping buffer for 5 min at 37°C to remove mRNA from the membranes and discard supernatant. Repeat procedure a second time.
3. Rinse in 2X SSC, 1% SDS for 5 min at room temperature and subsequently in 0.1X SSC, 1% SDS.
4. Array membrane can be stored for a short time (up to some days) semi-dry at 4°C. For long-term storage, membrane must be stored at –20°C.

4. Notes

1. Although in **Subheading 3.1.**, culture of human fibroblast cell culture is described, any cell type in addition to the human fibroblasts will be applicable to the DIGChemLink mRNA labeling technique.
2. The quality of RNA used for the procedure is of utmost importance, and, for this reason, it is of utmost importance to follow the extraction protocol of the RNeasy maxi kit as strictly as possible (as described in **Subheadings 3.2.–3.4.**). The general guidelines working with RNA must be exercised *(5)*.
3. For optimal extraction of total RNA, do not exceed the capacity of the column (as described by the manufacturer), for this would result in a lower performance and extraction quality.

4. Residual alcohol in the column reduces the quantity and quality of total RNA. Extraction of mRNA is performed in a 2-step procedure, as this increases stability and purity of mRNA.

5. The mRNA isolation kit is applied to extract mRNA out of total RNA. To extract mRNA post-translational polyadenylation is used, which is a common feature of the biogenesis of most eukaryotic mRNAs. The poly(A) tail allows mRNA extraction from total RNA using biotin-labeled oligo(dT)$_{20}$ primers bound to streptavidin-coupled magnetic beads. Do not allow the magnetic particles used for the mRNA extraction to dry out, as this may result in a nonreversible binding of mRNA to the beads.

6. The yield of mRNA extracted from total RNA should result in a yield of smaller than 5% of total RNA depending on the cells used.

7. Enrichment of mRNA from total RNA can be checked with a TBE gel comparing residual bands of ribosomal RNA in mRNA compared to total RNA.

8. For direct labeling of mRNA (**Subheading 3.5.**), the ChemLink molecule is utilized in the DIGChemLink kit. It consists of a digoxigenin-modified cisplatin, which binds noncovalently to mRNA. The pool of labeled mRNAs is hybridized onto cDNA membranes, and the binding of DIGChemLink to mRNA is identified by reporter molecule-labeled anti-DIG antibodies.

9. For prehybridization–hybridization, Cot 1 DNA and MB-grade DNA serve as competitor nucleic acids to reduce probe affinity to repetitive elements.

10. For hybridization (**Subheading 3.6.**), keep membranes wet and never let them dry out. Dry areas result in strong background signals. Use of roller bottles gives lower background heterogeneity on membranes compared to hybridization bags in a water bath.

References

1. Brown, P. O. and Botstein, D. (1999) Exploring the new world of the genome with DNA microarrays. *Nat. Genet.* **21,** 33–37.

2. Chee, M., Yang, R., Hubbell, F., Berno, A., Huang, X. C., Stern, D., et al. (1996) Accessing genetic information with high-density DNA arrays. *Science* **274,** 610–614.

3. Hoevel, T., Holz, H., and Kubbies, M. (1999) Cisplatin-digoxigenin mRNA labeling for nonradioactive detection of mRNA hybridized onto nucleic acid cDNA arrays. *Biotechniques* **27,** 1064–1067.

4. van Belkum, A., Linkels, E., Jelsma, T., van den Berg, F. M., and Quint, W. (1994) Non-isotopic labeling of DNA by newly developed hapten-containing platinum compounds [published erratum appears in Biotechniques 1995 Apr;18(4):636]. *Biotechniques* **16,** 148–153.

5. Ausubel, F. M. (1999) *Short protocols in molecular biology: a compendium of methods from current protocols in molecular biology.* Ausubel, F. M., et al., eds., Wiley & Sons, New York.

31

Expression Profiling Using Quantitative Hybridization on Macroarrays

Geneviève Piétu and Charles Decraene

1. Introduction

Expression studies performed on a genome scale have become very popular and provide an important link between the sequence and the function of a specific gene, thereby constituting the first step towards the elucidation of the function of specific genes. By correlating the modulation of gene expression with specific changes in physiology, it is possible to gain insights into the temporal modifications that occur in induced cells as well as the molecular differences between various cells or tissues samples.

With the advance of cDNA arrays, gene expression levels can be monitored by measuring the hybridization of mRNA to thousands of DNA fragments immobilized on a solid support. The experiment provides measurement of the relative abundance for each gene represented on the array, as well as the expression levels of the corresponding genes in the original sample. The method allows the collection of a large amount of data simultaneously.

According to the nomenclature recommended by Nature Genetics *(1)*, a probe will describe the molecules that are fixed on the support, whereas a target will describe the messenger RNAs or cDNAs in solution whose abundance is being detected.

The arrays are generated by polymerase chain reaction (PCR) amplification of cDNA library sets (using primers complementary to the vector portion or specific primers) or oligonucleotides, which are robotically printed onto the surface of Nylon filters *(2–4)* or glass slides *(5,6)* at defined locations. The target is produced by reverse transcription of RNA extracted from a given cell line or tissue with radioactive or fluorescent labeling. After hybridization, the image is acquired with an appropriated reader, scanner, or radioisotope detector. The scanned image is then analyzed, and the hybridization signal intensity is quantitated. Since the amount of probe attached to the solid support is in excess, the observed signal at any given position is a good estimate of the abundance of the corresponding species of the target mRNA.

cDNA arrays have been used for a variety of purposes, including changes in gene expression during treatment, disease states, and cancer *(7–13)*.

From: *Methods in Molecular Biology, vol. 185: Embryonic Stem Cells: Methods and Protocols*
Edited by: K. Turksen © Humana Press Inc., Totowa, NJ

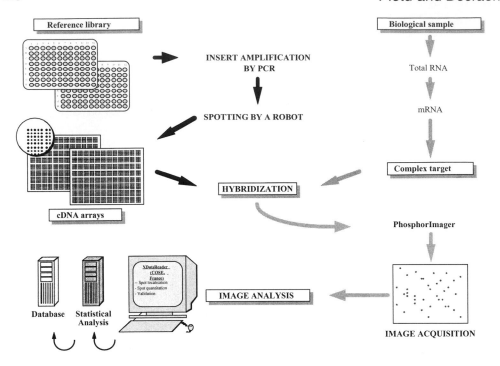

Fig. 1. General diagram of the hybridization signal collection on high-density filters.

These techniques are presently available in three formats; *(1)* high-density membranes also called macroarrays; *(2)* microarrays of DNA spots (a miniaturized version of the former technique); and *(3)* oligonucleotide chips.

The goal of this chapter is to describe the procedure of macroarray technology, in which PCR products are regularly arranged on a Nylon membrane and hybridized with radioactive complex targets. This method is well developed with a number of suitable arraying robots and detection systems commercially available making this technology adapted to any academic laboratory. Ready-to-use membranes carrying a few hundred cDNAs are sold by a number of manufacturers.

This technology has been developed in our laboratory *(4)*, applied to the study of the brain *(14)* and the muscle *(15)* transcriptomes, and is broadly applicable to the study of the embryonic stem (ES) cells transcriptome.

A general diagram of the hybridization signal collection on high density filters is presented in **Fig. 1**.

2. Materials

2.1. RNA Isolation

1. TRI Reagent (Euromedex, FR, cat. no. TR118).
2. Dynabeads mRNA DIRECT KIT (Dynal, Norway, cat. no. 610.11).

2.2. Probes and Macroarrays Preparation

1. Robots: Flexys from Perkin-Elmer, USA, or Biogrid TAS (BioRobotics, UK).
2. Nylon filters (Hybond N+) (Amersham Pharmacia Biotech, UK, cat. no. PRN119B).

3. Eurobiotaq® DNA polymerase, DNA, PCR buffer, and $MgCl_2$ (Eurobio, FR, cat. no. 018001).
4. d(N)TP (Amersham Pharmacia Biotech, UK, cat. no. 27-2035-01).
5. Primers for PCR were: M13 reverse primer (5′-AACAgCTATgACCAg-3′) or T3 primer (5′-TAACCCTCACTAAAgggA-3′) and M13 forward primer-20 (5′-gTAAAACgACggCCAgT-3′) or T7 primer (5′-AATACgACTCACTATAg-3′).
6. Thermal cycler Gene Amp PCR System 9600 (Perkin-Elmer, Norwalk, CT).
7. MicroAmp 96-well tray–retainer set (Perkin-Elmer, Norwalk, CT).
8. UV-Stratalinker 2400 (Stratagene, CA, USA).

2.3. Preparation of the Target and Labeling

1. Kit Superscript™ Preamplification System (Life Technologies, MD, USA, cat. no. 18089011).
2. d(T,C,G)TP (Amersham Pharmacia Biotech, UK, cat. non. 27-2035-01).
3. [α-^{33}P]dATP, 3000 Ci/mmol (Amersham Pharmacia Biotech, UK, cat. no. AH 9904).
4. ddTTP (Life Technologies, MD, USA, cat. no. 18246017).
5. Quick Spin™ Linker 6 column (Sephadex CL6B) (Boehringer Mannheim, Germany, cat. no. 1273973).

2.4. Hybridization

1. Hybridization oven (Appligene, FR).
2. Hybridization tubes (Appligene, FR).
3. ExpressHyb™ hybridization solution (Clontech, CA, USA, cat. no. 8015-3).
4. Sodium dodecyl sulfate (SDS) 20% (Amersham Pharmacia Biotech, UK, cat. no. 45-75832).
5. Standard saline citrate (SSC) 20X (Appligene, FR, cat. no. SSC20x01).
6. Washing solutions: solution 1 = 2X SSC, 0.1% SDS; solution 2 = 0.1X SSC, 0.1% SDS.
7. Striping solutions: solution 1 = 0.4 M NaOH, 0.1% SDS; solution 2 = 0.1 M NaOH, 0.1% SDS; solution 3 = 0.2 M Tris-HCl pH 8, 0.1X SSC, 0.1% SDS; solution 4 = 0.2 M Tris-HCl pH 8, 1X SSC, 0.1% SDS.

2.5. Hybridization Signal Analysis

1. PhosphorImager (Molecular Dynamics, CA, USA).
2. Storage Phosphor Screens (Molecular Dynamics, CA, USA).
3. XdotsReader software (Cose, France) (*see* **Note 1**).

3. Methods
3.1. RNA Isolation

1. Total RNA is isolated using the TRI Reagent (*see* **Note 2**) as recommended by the manufacturer.
2. Poly(A)⁺ mRNA ou mRNA is extracted from total RNA using the Dynabeads mRNA purification system kits according to the manufacturer's instructions. The main criterion is the production of high-quality RNA.

3.2. Probes and Macroarrays Preparation

3.2.1. cDNA Clones

Individual cDNA clones corresponding to gene transcripts are used as source of gene-specific probe in the arrays. In the recent years, the use of arrayed cDNA clone libraries has become an established tool *(16)*. Many high-quality cDNA libraries are available through commercial companies or distributors of IMAGE consortium resources (http://image.llnl.gov). Each clone that is analyzed exists as a permanent reference in form of frozen stock, with an address in a microtiter plate.

3.2.2. Inserts Amplification

Inserts from cDNA clones organized in 96-well plates are directly amplified from the master bacterial culture by PCR using primers that are complementary to vector sequences flanking both sides of the cDNA insert.

Amplifications are carried out with the following procedures:

1. Perform amplifications in 96-well plates in a 50 μL final reaction volume containing 5 μL of PCR buffer (10X), 1 μL of a d(N)TP mixture (10 mM each), 2 μL of MgCl$_2$ (50 mM), 0.3 μL of *Taq* DNA polymerase (5U/μL), 1 μL of each primer (M13 reverse primer, or T3 primer and M13 forward primer-20, or T7 primer) (25 μM), 37.7 μL of water, and 2 μL of the stock of bacterial culture.
2. Set the following cycle profile using a thermal cycler through thirty cycles: first denaturation for 10 min, 94°C; denaturation for 30 s at 94°C; annealing for 30 s at 59°C; elongation for 1 min at 72°C; and a last elongation of 10 min at 72°C.
3. Visualize one-tenth of the reaction on 1% agarose gel to confirm the success of amplification and the quantity and purity of the PCR products (*see* **Notes 3** and **4**). PCR products are stored at −20°C.

3.2.3. Spotting

1. Spot 10–20 ng of the PCR products from 96-well plates onto 8 × 12 cm Nylon membranes using a robot at a density of 25 × 96-well plates/filter (2400 clones) in a 5 × 5 format (25 PCR products/cm^2).
2. Denature membranes by incubating in 0.5 M NaOH, 1.5 M NaCl for 5 min, then neutralize in 1.5 M Tris-HCl, pH 8.0, NaCl and briefly rinse in 2X SSC.
3. Cross-link DNA to the membrane by UV radiation (1200 J/2 × 1 min). Membranes are prepared in batches, dried, and stored at 4°C before use.

3.3. Preparation of the Target and Labeling

All reagents should be sterilized and treated with diethyl pyrocarbonate (DEPC).

3.3.1. Preparation of the Target

The reverse transcription is performed using the SuperScript™ Preamplification System to generate the labeled complex target as follows:

1. Mix 250 ng poly(A)$^+$ RNA with 10 μL of random oligonucleotide primers (hexamers) (500 ng) in a 25 μL final reaction volume in DEPC-treated water, and incubate in a water bath at 70°C for 10 min.
2. Place on ice.
3. Add 5 μL of PCR buffer (10X), 5 μL of MgCl$_2$ (25 mM), 5 μL of dithiothreitol (DTT) (0.1 M), 2.5 μL of ddTTP (1 mM), and 5 μL of [α-^{33}P]dATP (50 μCi), 2.5 μL of mixture of each dCTP, dTTP, and dGTP (10 mM) (*see* **Notes 5** and **6**).
4. Incubate in a water bath at 25°C for 5 min.
5. Add 1 μL of Superscript II reverse transcriptase (200 U/μL).
6. Incubate at 25°C for 10 min, at 42°C for 50 min, and at 70°C for 15 min.

7. Add 1 µL of RNAse H (2 U/µL).
8. Incubate at 37°C for 20 min.
9. Quickly place on ice.

3.3.2. Purification of the Labeled Target

Remove unincorporated radioactive nucleotides from the labeled target by gel filtration on a Sepharose CL6B column according to the instruction manual.

To obtain interpretable results, the total target must have an activity around 75×10^6 counts per minute (cpm).

The probes are stored at –20°C before use, up to a maximum of 2 d.

3.4. Hybridization Conditions

For each cDNA target, two duplicate membranes are incubated in a roller bottle in a hybridization oven.

A general procedure for the hybridization is as follows:

1. Perform prehybridization at 68°C for 30 min in a vol of 10 mL ExpressHyb™ hybridization solution (*see* **Notes 7** and **8**).
2. Remove the solution and perform hybridization in 10 mL ExpressHyb™ hybridization solution, with the totality of the radiolabeled complex cDNA target at 68°C for 2 h.
3. Wash with agitation successively, twice in solution 1 at room temperature for 30 min, and twice in solution 2 at 55°C for 30–40 min.
4. After washing, wrap membranes with Saran Wrap, and expose to phosphor screens for 2 d.
5. Strip hybridized membranes with two successive immersions in solution 1 at 65°C for 30 min, 2×15 min in a solution 2, and 2×10 min in solution 3 at room temperature, then rinse the membranes in solution 4 for 10 min at room temperature. Membranes are used a maximum of 5 times.

3.5. Hybridization Signal Analysis

1. Image acquisition is performed by scanning the phosphor screens with the PhosphorImager imaging plate system. An image from a typical experiment is presented in **Fig. 2**.
2. The scanned 16-bit images are imported on a Sun workstation and image analysis is performed using the XdotsReader software specifically designed for this application. A grid is applied to the image of the membrane to enclose each individual signal.
3. The expression level of each gene is estimated by quantitating the hybridization signals intensity using the XdotsReader software. This software calculates the mean of all pixels for each spot after background subtraction in a local correction mode (*see* **Note 9**). Local mode is selected, since the background signals are not evenly distributed.
4. In our experiments, wholesale differences in mRNA levels are not expected to occur, thus, we can use the average overall hybridization for all clones on the filter to calculate a normalized signal. In these conditions, the intensity values of each individual hybridization signal are normalized by dividing for each spot the intensity obtained with a complex target with the average of the intensity from the set of the 2400 values collected for each filter (*see* **Note 10**).
5. Each experiment is conducted in duplicate with two complex targets representing four hybridization values for each spot. The intensity values for each spot are validated when up to 3 of the 4 hybridization intensity values are similar (standard deviation less than 25% of the average). Finally, each validated average is assigned to the corresponding cDNA clone.

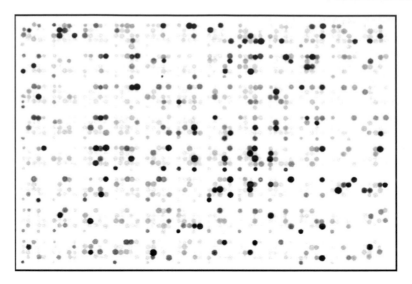

Fig. 2. Image acquisition by the PhosphorImager.

3.6. Exploitation of the Data

Data reports corresponding to the normalized and validated values are exported to a PC computer and analyzed using an Excel software (Microsoft).

Once searching for differential expression, data obtained from two or more experimental conditions are used to determine the ratio of hybridization intensity values of two biological samples. Selection of genes with modulated expression is performed for clones with a ratio of variation higher than 2.

4. Notes

1. The XDotsReader Software is commercially available from COSE (contact person Mr Brice Achddou, e-mail: brice.achddou@cose.fr). It runs on SUN work stations under UNIX (SunOs or Solaris) and on PC under LINUX. No particular hardware is required, and the software is compatible with any other software.
2. A number of artifacts [nonspecific hybridization due to repeat sequences or to poly(A) tracts in the probe hybridizing with poly(T) tracts] in the targets must be carefully eliminated, and their absence monitored to ensure the collection of meaningful data.
   Quality controls of the experiments are crucial. They include:
   a. Purification of PCR products: clones that give multiple PCR products are discarded.
   b. Resequencing of inserts: considerable effort is being put into providing sequence-verified cDNA arrays to ensure close identity between sequencing and array construction.
   c. Test for presence of repetitive sequences: filters are hybridized with Cot-1 DNA to identify cDNA clones that contain repeated sequences. Positive clones are scored, and only clones that fail to hybridize with the Cot-1 probe are analyzed further.
   d. Inclusion of control genes that are: human genomic DNA, several human highly abundant genes including GAPDH, β-actin, as well as useful negative controls, such as DNA sequences derived from *Arabidopsis thaliana* genes, oligo(dA) 80-mer, oligo(dT) 25-mer, pBluescript vector DNA, and water.

3. It is not necessary to quantitate the amount of PCR product. It is assumed that even with inefficient reaction, the amount of product will exceed the amount of target.

4. Adding dideoxyribonucleotides into the labeling reaction will increase the sensitivity and reproducibility of the hybridization procedure *(17)*.

5. ^{33}P isotope gives better signal-to-noise ratio than ^{32}P.

6. The accuracy for the low abundant genes may not be reliable due to detection limitations. In addition, the values of the transcripts of low abundance are more subject to influence by background correction. In our system, the sensibility is estimated at 1/10,000, based on the measurement of the abundance of known genes *(4)*.

7. ExpressHyb™ buffer hybridization solution could be changed for a buffer containing at a final concentration: SSC (4X), Dextran sulfate (8%), Denhart's (10X), EDTA (1 m*M*), formamide (25%), SDS (0.1%), and denatured salmon sperm (100 µg/mL). Prehybridization would be 4 h at 42°C and hybridization overnight at 42°C.

8. Normalization alternative: the hybridization intensity from appropriate controls, whose signals are not expected to vary, may be used to normalize the intensity.

9. For the first hybridization, the membranes are prehybridized overnight in the hybridization buffer.

10. RNeasy Kits (Qiagen, FR) or Fast Track mRNA solution Kit (Invitrogen, CA, USA) have also been successfully used for RNA isolation.

References

1. Brown, P. O. and Botstein, D. (1999) Exploring the new world of the genome with DNA microarrays. *Nat. Genet.* **21,** 33–37.

2. Zhao, N., Hashida, N., Takaashi, N., Misumi, Y., and Sakaki, Y. (1995) High density cDNA filter analysis: a novel approach for large scale quantitative analysis of gene expression. *Gene* **156,** 207–213.

3. Nguyen, C., Rocha, D., Granjean, S., Baldit, M., Bernard, K., Naquet, P., and Jordan, B. (1995) Differential gene expression in the murine thymus assayed by quantitative hybridization of arrayed cDNA clones. *Genomics* **29,** 207–216.

4. Piétu, G., Alibert, O., Guichard, V., Lamy, B., Bois, B., Leroy, E., ct. al. (1996) Novel gene transcripts preferentially expressed in human muscles revealed by quantitative hybridization of a high density cDNA array. *Genome Res.* **6,** 492–503.

5. Schena, M., Shalon, D., Davis, R. W., and Brown, P. O. (1995) Quantitative monitoring of gene expression patterns with a complementary DNA microarray. *Science* **270,** 467–470.

6. Lockhart, D. J., Dong, H., Byrne, M. C., Folletie, M. T., Gallo, M. V., Chee, M. S., et al. (1996) Expression monitoring by hybridization to high-density oligonucleotide arrays. *Nat. Biotechnol.* **14,** 1675–1680.

7. Wodicka, L., Dong, H., Mittmann, M., Ho, M., and Lockhart, D. J. (1997) Genome-wide expression monitoring in *Saccharomyces cerevisia. Nat. Biotechnol.* **15,** 1359–1367.

8. Lashkari, D. V., DeRisi, J. L., McCusker, J. H., Namath, A. F., Gentile, C., Hwang, S. Y., et al. (1997) Yeast microarrays for genome wide parallel genetic and gene expression analysis. *Proc. Natl. Acad. Sci. USA* **94,** 13,057–13,062.

9. Schena, M., Shalon, D., Heller, R., Chai, A., Brown, P., and Davis, R. D. (1996) Parallel human genome analysis: microarray-based expression monitoring of 1000 genes. *Proc. Natl. Acad. Sci. USA* **93,** 10,614–10,619.

10. Heller, R. A., Schena, M. Chai, A., Shalon, D., Bedilion, T., Gilmore, J., et al. (1997) Discovery and analysis of inflammatory disease-related genes using cDNA microarrays. *Proc. Natl. Acad. Sci. USA* **94,** 2150–2155.

11. Iyer, V. R., Eisen, M. B., Ross, D. T., Schuler, G., Moore, T., Lee, J. C. F., et al. (1999) The transcriptional program in the response of human fibroblasts to serum. *Science* **283,** 83–87.

12. DeRisi, J., Penland, L., Brown, P. O., Bittner, M. L., Meltzer, P. S., Ray, S., et al. (1996) Use of a cDNA microarray to analyse gene expression patterns in human cancer. *Nat. Genet.* **14,** 457–460.

13. Ross, D. T., Scherf, U., Eisen, M. B., Perou, C. M., Rees, C., Spellman, P., et al. (2000) Systematic variation in gene expression patterns in human cancer cell lines. *Nat. Genet.* **24,** 227–235.

14. Piétu, G., Mariage-Samson, R., Fayein, N. A., Matingou, C., Eveno, E., Houlgatte, R., et al. (1999) The Genexpress IMAGE knowledge base of the human brain transcriptome: a prototype integrated resource for functional and computational genomics. *Genome Res.* **9,** 195–209.

15. Piétu, G., Eveno, E., Soury-Segurens, S., Fayein, N. A., Mariage-Samson, R., Matingou, C., et al. (1999) The Genexpress IMAGE knowledge base of the human muscle transcriptome: a resource of structural, functional, and positional candidate genes for muscle physiology and pathologies. *Genome Res.* **9,** 1313–1320.

16. Lennon, G., Auffray, C., Polymeropoulos, M., and Soares, M. B. (1996) The I.M.A.G.E consortium: an integrated molecular analysis of genomes and their expression. *Genomics* **33,** 151–152.

17. Decraene, C., Reguigne-Arnould, I., Auffray, C., and Piétu, G. (1999) Reverse transcription in the presence of dideoxynucleotides to increase the sensitivity of expression monitoring with cDNA arrays. *BioTechniques* **27,** 962–966.

32

Isolation of Antigen-Specific Intracellular Antibody Fragments as Single Chain Fv for Use in Mammalian Cells

Eric Tse, Grace Chung, and Terence H. Rabbitts

1. Introduction

The control of gene function by the modulation of mRNA translation, stability, or protein activity has important implications for the treatment of disease, as well as in research programs designed to study the biological role of proteins. For instance, preventing protein function in specific cell types in development can produce a phenotypic knock-out due to the effective protein loss. Antibodies are ideally suited for this purpose, as they have evolved to bind macromolecules with high affinity and can neutralize their function as a result. Ablation of protein function in vivo with antibodies has been achieved by microinjecting whole antigen-specific immunoglobulin molecules into the cell cytoplasm (*1*). However, complete antibody comprises four chains (two heavy and two light chains) held by interchain disulfide bonds. This is not suitable for expression from DNA vectors, as the individual chains would not assemble in the cytoplasm. Therefore, antibody fragments have been employed in the single chain Fv (scFv) format (*2*), which are single polypeptide chains comprising a heavy chain variable region (VH) and a light chain variable region (VL) held by a short linker sequence. ScFv folds into an antibody combining site (which binds antigen) but no effector function is present. Many uses of scFv expressed inside cells (herein called intracellular antibody or ICAbs) have been described in cell systems (*3,4*) and in whole organisms. For example, in vivo expression of a scFv directed against the coat protein of artichoke mottle crinkle virus has conferred resistance to viral infection in plant cells (*5*).

The potential for isolation of scFv, or single VH (or perhaps VL) domains (*6*), which specifically bind antigen, should make it possible to derive a spectra of scFv for any in vivo antigenic target. Consequently, expressing scFv inside a given cell to ablate the target antigen is possible. The initial source of cDNA for scFv expression can be from phage display libraries (*7*) or from immunized mouse spleen (*8*). The antigen-specific scFv could subsequently be expressed from a vector (e.g., retrovirus) and any effects on the cell could be assessed. If the scFv are used in embryonic stem (ES) cells with tissue-specific transgene–targeted genes, the effect of scFv-protein binding during development could be determined. A further potentially major area of use will be in functional genomics. The human and mouse genome sequences are almost complete,

From: *Methods in Molecular Biology, vol. 185: Embryonic Stem Cells: Methods and Protocols*
Edited by: K. Turksen © Humana Press Inc., Totowa, NJ

and the full set of gene sequences will soon be known. Many of these derived mRNAs will have no specified function other than those guessed from sequence homologies. Using intracellular scFv to abolish specific protein function will be a powerful tool to obtain first generation data on the role of specific proteins in cells and, later, could be used for phenotypic ablation in mice by transgenic expression of antigen-specific scFv. Finally, intracellular scFv have great potential in human disease treatment by serving as antigen-specific reagents against disease-related proteins inside cells.

Despite this potential of ICAbs, there are few antibodies that can function effectively in an intracellular environment. When scFv are expressed in the cell cytoplasm, folding and stability problems often occur, which results in a nonfunctional low level of expression and a short half-life. These problems are caused by the reducing environment of cell cytoplasm, which hinders the formation of the intrachain disulfide bonds in the immunoglobulin VH and VL domains *(9)*. Although some scFv can tolerate the absence of the disulfide bond *(10)*, there is no general rule that predicts such characteristics. Two main approaches can be adopted to circumvent these problems. One would be to derive scFv scaffold(s), which effectively fold in vivo and are expressed well, and then to use these as backbones onto which antigen-specific complementarity-determining regions (CDRs) could be introduced. A second more simple and generic approach is to use selection methods that allow isolation of those scFv, which will bind in vivo from a repertoire of different scFv. Taking the latter approach, we have recently developed an in vivo antibody–antigen interaction screening method to identify functional intracellular binders *(11)*. Similar selection approaches have subsequently been described *(12,13)*.

Our selection method *(11)*, intracellular antibody capture (IAC) technology, takes advantage of the ability of protein interactions to be detected in cellular environments, such as shown for the yeast two-hybrid system *(14)*. It was based on the fact that some antibody scFv fragments can fold adequately in vivo to bind antigen in a VH-VL dependent way (i.e., via the antibody combining site) and, thus, using a library of diverse antibody specificities should facilitate their identification, if sufficient scFv could be screened. In addition, ideally, one would like to isolate a small repertoire of intracellular scFv, which would bind to different epitopes on the target antigen. Our screening system comprised of yeast cell expression of a "bait" antigen fused to the LexA DNA binding domain and a library of "prey" scFv fused to the VP16 transcription activation domain. Interaction between the antigen bait and any specific antibody scFv fragment in the yeast intracellular environment results in the formation of a protein complex, in which the DNA binding domain and the activation domain are in close proximity. This results in the activation of yeast chromosomal reporter genes, such as *HIS3* and *LacZ*, facilitating the identification and thus isolation of the yeast carrying the DNA vectors encoding the scFv, which in turn can be isolated to yield the DNA sequence of the antigen-specific scFv. The main limitation of this approach is the number of scFv-VP16 fusion preys that can be screened in yeast antibody–antigen interaction system (conveniently up to 2–5×10^6). This figure is well below the size of scFv repertoires. Thus, it is necessary to limit the numbers of scFv to be screened in vivo in yeast. We currently use one round of in vitro phage scFv library screening (panning) using, for instance, bacterially produced antigen coated on a surface, prior to the in vivo yeast antibody–antigen interaction screening. With this combined approach, we have successfully isolated intracellular scFv specific to the breakpoint cluster region (BCR) antigen *(15)*.

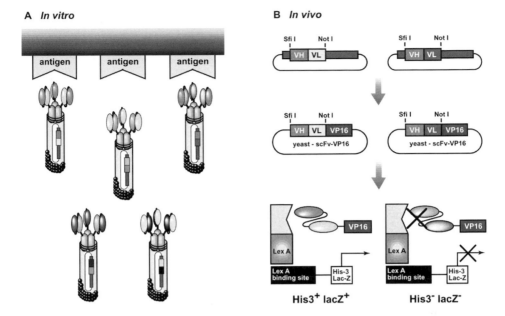

Fig. 1. Diagram illustrating the strategy for the selection of specific intracellular antibodies: intracellular antibody capture (IACT) technology. (**A**) A phage library displaying a repertoire of scFv is screened in vitro with an antigen coated onto a surface. Only phages displaying the specific scFv bind to the antigen. (**B**) The phagemid DNA from the bound phage in panel **A** is prepared, and the scFv DNA inserts are subcloned into the yeast VP16 expression vector (shown in **Fig. 2**) to give the yeast scFv-VP16 library. The library is screened in yeast with the antigen bait. Only those scFv that retain the specific binding ability in vivo can activate the reporter genes, His3 and LacZ. (*See* color plate 12, following p. 254).

In this chapter, we describe a protocol to isolate functional intracellular scFv binders to specific antigens, Intracellular Antibody Capture (IAC). In summary, the overall strategy (**Fig. 1**) involves:

1. (**A**) In vitro screening of a phage scFv display library (which should have a diversity >10^9) with the target antigen (e.g., made by bacterial or bacullovirus expression) to produce a sublibrary enriched for antigen-specific scFv.
2. (**B**) Subcloning the first round scFv DNA from the phage vector into the yeast vector to make a library of the enriched scFv fused with VP16 transactivation domain.
3. Selection of intracellular scFv binders by performing the antibody–antigen interaction screening in yeast using the scFv-VP16 prey library together with the antigen–DNA binding domain bait.
4. When intracellular scFv binders have been isolated, these are validated by binding of the isolated scFv to the antigen by Western blotting (in vitro). Other functional or in vivo assays may then be tested.

2. Materials

2.1. In Vitro Phage Antibody Library Screening with Antigen

1. Nunc Maxisorp Immunotubes.
2. Phosphate-buffered saline (PBS): 0.1 M NaCl, 17 mM Na$_2$HPO$_4$, 8 mM NaH$_2$PO$_4$•2 H$_2$O, adjust pH to 7.0–7.2 with HCl.

3. 100 m*M* Triethylamine solution (freshly prepared).
4. Dried milk powder.
5. 1 *M* Tris-HCl, pH 7.4, at room temperature.
6. TG1 *Escherichia coli* bacteria.
7. TYE agar plate (per liter): 15 g agar, 140 m*M* NaCl, 10 g Bacto-tryptone (Haarlem, England, cat. no. T1332), 5 g Bacto-yeast extract (Oxoid, England, cat. no. X589B).
8. 2X TY medium (per liter): 16 g Bacto-tryptone, 10 g Bacto-yeast extract, 85 m*M* NaCl, pH adjusted to 7.4 with HCl.

2.2. Preparation of the scFv Sublibrary, Enriched for the Antigen-Specific scFv, from the First Round Phage Infected TG1

1. Glycerol.
2. QIAGEN Plasmid Midi Kit (Qiagen [http://www.qiagen.com], cat. no. 12143).
3. QIAEX II Gel Extraction Kit (Qiagen, cat. no. 20021).
4. *Not*I and *Sfi*I restriction enzymes, NEB buffer 2 (New England Biolabs [http://www.uk.neb.com], cat. no. RO189S and RO123S).
5. Elution buffer: 10 m*M* Tris-HCl, pH 8.5, at room temperature.

2.3. Construction of Yeast scFv-VP16 In Vivo Expression Library for Antibody–Antigen Yeast Interaction Selection (11)

1. Yeast pVP16 vector (**Fig. 2**) (gift from Prof. A. Cattaneo).
2. T4 DNA ligase and ligase buffer (New England Biolabs, cat. no. MO202S).
3. DH5α *E. coli*.

2.4. Hanahan Method of Bacterial Transformation (16)

1. TFB solution: 10 m*M* 2-(N-morpholino)ethanesulfonic acid (adjusted to pH 6.2 with potassium hydroxide), 100 m*M* KCl, 45 m*M* manganese chloride, 10 m*M* CaCl$_2$, 3 m*M* hexamine cobalt trichloride.
2. Dimethylformamide (DMF) (Fisher Scientific, England, cat. no. D/3841/17).
3. 2.25 *M* Dithiothreitol (DTT) (Melford Laboratories Ltd., England, cat. no. MB1015) in 40 m*M* potassium acetate, pH 6.0.

2.5. Yeast Antibody–Antigen Interaction Screening of scFv-VP16 Library with the Antigen "Bait"

1. pBTM116 yeast expression vector (**Fig. 2**) (gift from Prof. A. Cattaneo).
2. L40 yeast: *Mata* his3 Δ200 trp1-901 leu2-3, 112 ade2 LYS:: (lexAop)$_4$-HIS3 URA3:: (lexAop)$_8$-LacZ GAL4 (gift from Prof. A. Cattaneo).
3. YPD medium (per liter): 10 g yeast extract, 20 g Bacto-peptone, 2% glucose.
4. YC medium (per liter): 1.2 g yeast nitrogen base without amino acid and ammonium sulfate, 5 g ammonium sulfate, 10 g succinic acid, 6 g NaOH, 0.75 g amino acid mixture (1 g each of adenine sulfate, arginine, cysteine, threonine, and 0.5 g each of aspartic acid, isoleucine, methionine, phenylalanine, proline, serine, and tyrosine), 2% glucose, 0.1 g of tryptophan (W), 0.1 g of leucine (L), 0.05 g of histidine (H), 0.1 g of uracil (U), and 0.1 g of lysine (K). YC–WLHUK is YC medium lacking tryptophan (W), leucine (L), histidine (H), uracil (U), and lysine (K). YC-WL is YC medium lacking tryptophan (W), leucine (L). YC-L is YC medium lacking leucine (L).
5. 10X TE buffer: 0.1 *M* Tris-HCl, 10 m*M* EDTA, pH 7.5, at room temperature.
6. 10X LiAc: 1 *M* lithium acetate, adjusted to pH 7.5 with acetic acid.
7. PEG-LiAc: 40% polyethylene glycol (PEG) 4000, 1× TE buffer, and 1× LiAc.

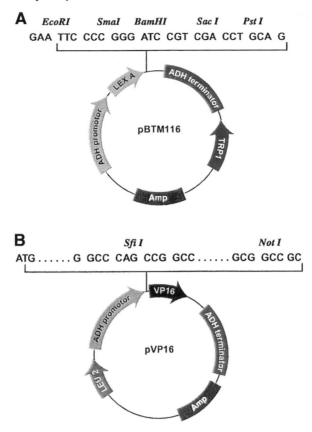

A   *EcoRI*   *SmaI*   *BamHI*   *Sac I*   *Pst I*

GAA TTC CCC GGG ATC CGT CGA CCT GCA G

B   *Sfi I*   *Not I*

ATG . . . . . . G GCC CAG CCG GCC . . . . . . GCG GCC GC

Fig. 2. Diagram showing the restriction maps and polylinker sequences of the yeast expression vectors, **(A)** pBTM116 and **(B)** pVP16. **(A)** Antigen baits are cloned into pBTM116 to create an in-frame fusion with the lexA DNA binding domain. The codons are indicated as triplets. **(B)** ScFv are cloned into pVP16 as in-frame fusion with the ATG start codon in the vector. Cloning is typically between the *Sfi*I and *Not*I sites, and the codons are indicated as triplets. (*See* color plate 13, following p. 254).

8. 100% Dimethyl sulfoxide (DMSO) (Fisher Scientific, cat. no. D/4121/PB08).
9. 20 mg/mL Salmon sperm DNA (Sigma, cat. no. D1626): sheared with fine bore needles (gauge 25) and boiled for 5 min before use. Denatured salmon sperm DNA can be stored at –20°C.

2.6. Confirmation of Antibody–Antigen Interaction in Yeast Clones by β-Galactosidase Filter Assay

1. Z buffer (per liter): 16.1 g $Na_2HPO4 \cdot 7 \ H_2O$, 5.5 g $NaH_2PO_4 \cdot H_2O$, 0.75 g KCl, 0.246 g $MgSO_4 \cdot 7 \ H_2O$, adjusted to pH 7.0 with HCl. Can store at room temperature for at least 1 yr.
2. X-gal stock solution: 20 mg of 5-bromo-4-chloro-3-indolyl-β-D-galactopyranoside (X-gal) (Melford Laboratories Ltd., England., cat. no. MB1001) in DMF. Store at –20°C and can keep for 1 yr.
3. Z buffer/X-gal: 100 mL Z buffer, 0.27 mL β-mercaptoethanol (BDH Laboratories Supplies, England, cat. no. 441433A), 1.67 mL X-gal stock solution. Prepare fresh.
4. Liquid nitrogen.

2.7. Plasmid DNA Extraction from Individual Selected Yeast Colonies

1. Yeast lysis solution: 2% Triton X-100, 15% sodium dodecyl sulfate (SDS), 100 m*M* NaCl, 10 m*M* Tris-HCl, pH 8.0, and 1 m*M* EDTA.
2. Phenol.
3. Chloroform.
4. Acid washed glass beads (425–600 μm) (Sigma [http://www.sigma.sial.com], cat. no. G8772).
5. 3 *M* sodium acetate, pH 5.2 (adjusted with glacial acetic acid).
6. 70% and 100% Ethanol.
7. Dry ice.
8. QIAprep Spin Miniprep Kit (Qiagen, cat. no. 27104).

2.8. Retesting the scFv-VP16 Plasmid Using Yeast Antibody–Antigen Interaction Assay

Materials are the same as in **Subheading 2.5.**

2.9. Characterization of the Antigen-Specific Intracellular scFv

1. Sequencing primers
   Forward primer: 5′-TTG TTT CTT TTT CTG CAC AAT-3′
   Back primer: 5′-CAA CAT GTC CAG ATC GAA-3′
2. AmpliTaq DNA polymerase, 10X polymerase chain reaction (PCR) buffer, 25 m*M* MgCl$_2$ and 20 m*M* dNTP (Perkin Elmer, cat. no. N801-0060).
3. *Bst*N I restriction enzyme (New England Biolabs, cat. no. RO168S).
4. NuSieve 3:1 agarose (Flowgen [http://www.philipharris.co.uk/flowgen], cat. no. 50090).
5. Thermal cycler (PTC-225 DNA Engine Tetrad; MJ Research, USA).

2.10. Binding of scFv to Antigen by Immunodetection Assay (Western Blot)

1. pHEN2 vector (gift from Dr. Heather Griffin) (http://www.mrc-cpe.cam.ac.uk).
2. HB2151 *E. coli* (gift from Dr. H. Griffin).
3. 1X TES buffer: 0.2 *M* Tris-HCl, 0.5 m*M* EDTA, 0.5 *M* sucrose.
4. 9E10 Anti-myc monoclonal antibody (Santa Cruz Biotechnology [http://www.scbt.com], cat. no. sc-40) and horseradish peroxidase (HRP)-conjugated anti-mouse antibody (Amersham, cat. no. NA931).

3. Methods

3.1. In Vitro Phage Antibody Library Screening with Antigen (see Note 1)

The method for in vitro phage screening below is adapted from Dr. Heather Griffin (http://www.mrc-cpe.cam.ac.uk and click on "phage display").

1. Coat an immunotube with 4 ml of the antigen (concentration 50–100 μg/mL in PBS) and leave it standing at 4°C overnight. Also, inoculate a single colony of TG1 into 5 mL of 2X TY and grow at 37°C overnight with shaking.
2. Pour off the antigen the next day and wash the immunotube three times with PBS.
3. Fill the tube with 5 mL of PBS containing 2% (w/v) dried milk powder (M-PBS) for blocking to avoid nonspecific binding of phage particles to the surface of the immunotube. Incubate at 37°C for 2 h and wash the tube three times with PBS.

4. Dilute the phage scFv aliquot containing 10^{12} to 10^{13} phage in 4 mL of 2% M-PBS and add it to the immunotube. Seal the tube with Parafilm, rotate it on a turntable wheel at room temperature for 30 min, and stand for a further 90 min at room temperature.

5. Pour off the unbound phage and wash the immunotube ten times with PBS containing 0.1% Tween-20 and ten times with PBS (all at room temperature).

6. Add 1 mL of freshly made 100 m*M* triethylamine to elute the bound phage. Seal the tube with Parafilm and rotate on a turntable wheel for 10 min at room temperature.

7. Pipet the eluted phage into a tube containing 0.5 mL of 1 *M* Tris-HCl, pH 7.4 for neutralization. Store it at 4°C (*see* **Note 2**).

8. Dilute 0.5 mL of the overnight TG1 bacterial culture into 50 mL of fresh 2X TY and grow at 37°C with shaking until OD_{600} is between 0.5–0.6. This usually takes 1.5–2 h.

9. Add the eluted phage to 8.5 mL of the exponentially growing culture of TG1. Incubate at 37°C for 30 min without shaking, during which phage infection of the TG1 bacteria occurs.

10. Titer bacteria by taking 100 μL of the infected TG1 bacteria and make 10-fold serial dilutions. Plate these dilutions on TYE agar plates containing 100 μg/mL ampicillin and 1% glucose. Grow overnight at 30°C and count the number of colonies for each dilution.

11. Pellet the remaining infected TG1 culture by centrifuging at 2000*g* for 10 min at 4°C. Resuspend cells in 1 mL of 2X TY and plate on a large square plate (23 × 23 cm) with TYE agar containing 100 μg/mL ampicillin and 1% glucose. Grow overnight at 30°C (*see* **Note 3**).

12. Add 5 mL of 2X TY with 100 μg/mL ampicillin and 15% glycerol to the large square plate containing the infected TG1 colonies and scrape off the colonies with a glass spreader (*see* **Note 4**). Proceed to isolate phagemids carrying scFv as in **Subheading 3.2.**

3.2. Preparation of the scFv Sublibrary, Enriched for Antigen-Specific scFv, from the First Round Phage Infected TG1

The in vitro selection of scFv displayed on phage in **Subheading 3.1.** reduces the initial complexity of the phage to an enriched population, some of which bind antigen (if a phage library of 5×10^9 is used, about 10^5 phage will be recovered). It is then necessary to isolate the phagemid clones encoding these scFv (**Subheading 3.2.**) and transfer these to the yeast vector (using *Sfi*I/*Not*I restriction sites, *see* **Note 5**) for the in vivo selection of intracellular scFv, which bind antigen.

1. Take 100–200 μL from the 5 mL bacteria recovered by scraping (**Subheading 3.1., step 12.**) and dilute into 10 mL of 2X TY containing 100 μg/mL ampicillin. This is plated onto ten large square plates of TYE agar containing 100 μg/mL ampicillin and 1% glucose. Grow at 30°C overnight (*see* **Notes 4** and **6**).

2. Store the remaining bacteria at –70°C as glycerol stock in aliquots. This is the sublibrary stock enriched for phage expressing antigen-specific scFv, this will include those scFv, which do or do not bind the target antigen in vivo.

3. Add 5 mL of 2X TY to each plate prepared in **step 1** and scrape off all the bacteria colonies. Phagemid DNA can be extracted from the recovered bacteria using QIAGEN Plasmid Midi Kit (Qiagen) according to the manufacturer's instructions. Quantitate the amount of phagemid DNA obtained by measuring the UV spectrum between 220 and 300 nm. The peak at 258 nm is used to calculate the amount of DNA. DNA concentration (ng/μL) = $OD_{258} \times 50$.

4. Digest 6 μg of phagemid DNA with 20 U of *Not*I enzyme in a 50-μL reaction in a 1.5-mL Eppendorf tube with 1X NEB buffer 2. Incubate at 37°C for 3 h.
5. Add 40 U of *Sfi*I enzyme to the *Not*I-digested DNA and mix. Cover with a drop of mineral oil and incubate overnight at 50°C.
6. Run the digested sample in a 1.5% agarose gel at 50 V for about 45 min and cut out the gel band containing the scFv DNA fragment mixture (800–900 bp). Extract the DNA using QIAEX II Gel Extraction Kit according to manufacturer's instructions. Elute DNA with 20 μL of elution buffer. This is the purified scFv (*Sfi*1/*Not*1) DNA fragment, which is used to construct the yeast scFv-VP16 expression library.

3.3. Construction of Yeast scFv-VP16 In Vivo Expression Library for Antibody–Antigen Yeast Interaction Screening (11)

When the extraction of phagemid DNA encoding the sublibrary of scFv has been accomplished, it is subcloned into the yeast VP16 antibody–antigen interaction assay vector (*see* **Note 7**). The size of the yeast scFv-VP16 library necessary will depend on the number of phage that was obtained in the first round in vitro screening (**Subheading 3.1.–3.2.**). Ideally, the yeast library should represent three to ten times the number of phage obtained in **Subheading 3.1.**, to include as many potential binders in the yeast library as possible.

1. Digest 1.5 μg of yeast pVP16 vector DNA with 20 U of *Not*I enzyme in a 50-μL reaction with 1X NEB buffer 2. Incubate at 37°C for 3 h.
2. Add 40 U of *Sfi*I enzyme to the *Not*I-digested vector DNA and mix. Cover with a drop of mineral oil and incubate overnight at 50°C.
3. Run the digested sample in a 1% agarose gel at 50 V for about 45 min. Cut out the gel band containing the yeast pVP16 DNA of 7.5 kb. Extract the DNA using the QIAEX II Gel Extraction Kit according to manufacturer's instructions. Elute DNA with 28 μL of elution buffer. This is the purified yeast pVP16 vector (*Sfi*I/*Not*I).
4. Set up a 10-μL ligation reaction containing 1X T4 DNA ligase buffer, 1 μL of purified yeast pVP16 vector (Sfi/Not) DNA, 0.5 μL of purified scFv (Sfi/Not) DNA, and 400 U of T4 DNA ligase. Also set up another 10-μL ligation reaction without scFv DNA to check for the presence of incompletely cut vector background. Incubate overnight at 15°C.
5. Transform 5 μL of ligation reaction into 100 μL DH5α bacteria by the method of Hanahan *(16)* (*see* **Subheading 3.4.**). Plate the transformed bacteria onto large square plate (23 × 23 cm) of TYE agar containing 100 μg/mL ampicillin. Incubate overnight at 37°C and count the number of colonies obtained, typically around 10,000.
6. Set up more ligations and perform transformation to achieve the required number of colonies. Generally, 40–50 ligations are required (i.e., 40–50 × 10^4 colonies).
7. After overnight growth on the large plates, scrape off all the bacteria by adding 5 mL of 2X TY with 100 μg/ml ampicillin and 15% glycerol, and by using a sterile glass spreader. Dilute 100–200 mL of the recovered bacteria into 10 mL of 2X TY and plate on 10 large square plates of TYE agar containing 100 μg/mL ampicillin (*see* **Notes 4** and **6**). Incubate overnight at 37°C. Store the remaining bacteria at −70°C in aliquots. This is the antigen-specific scFv sublibrary.
8. Recover the bacteria by scraping into 5 mL 2X TY, and extract the plasmid DNA from it using QIAGEN Plasmid Midi Kit. Quantify the DNA by the UV spectrum from 220 to 300 nm. This DNA is the yeast scFv-VP16 library DNA to be used for the yeast antibody–antigen two-hybrid screening *(11)*. The presence of the scFv insert can be tested by digesting 0.5 μg of DNA with *Sfi*I and *Not*I. This should yield a vector band of approx 4 kb and an insert band (a mixture of scFv fragments) of approx 800 kb.

1.7. Hanahan Method of Bacterial Transformation

The bacterial transfection method described by Hanahan (16) allows for a higher efficiency than the standard $CaCl_2$ approaches. Other methods such as electroporation may also be used to achieve high efficiency cloning.

1. Inoculate a single colony of DH5α into 5 mL 2X TY and grow overnight at 37°C.
2. Dilute 0.5 mL of overnight culture into 50 mL of 2X TY and grow for about 2 h with shaking until an OD_{600} of 0.4–0.6 is obtained. Pellet the bacteria by centrifuging at $1000g$ for 8 min at 4°C, and resuspend in 20 mL of TFB solution. Incubate on ice for 20 min.
3. Pellet the bacteria again by centrifuging at $1000g$ for 8 min at 4°C, and gently resuspend in 4 mL of TFB solution. Add 140 µL of DMF and mix gently. Incubate on ice for 5 min.
4. Add 140 µL of 2.25 M DTT in 40 mM potassium acetate, pH 6.0, and mix gently. Incubate on ice for 10 min.
5. Add 140 µL of DMF and mix gently. The bacteria are now ready for transformation. Only use on the day of preparation.
6. Mix 5 µL of ligation reaction with 200 µL of competent bacteria. Incubate on ice for 1 h.
7. Heat shock for 2 min at 42°C and incubate on ice for 1 min.
8. Add 200 µL of 2X TY and incubate for 20 min at 37°C.
9. Plate all bacteria onto a TYE agar plate containing 100 µg/mL ampicillin. Incubate overnight at 37°C.

3.5. Yeast Antibody–Antigen Interaction Screening of the scFv-VP16 Library with the Antigen Bait

The yeast antibody–antigen in vivo interaction screen (11) followed the strategy of Fields and Song (14) in having a bait, in this case antigen linked to a DNA binding element, such as LexA DNA binding domain (for instance in the fusion vector, pBTM116). To detect those antibody fragments (scFv), which can bind antigen in vivo, the scFv is cloned as a fusion with the VP16 transcriptional activation domain. Yeast containing the plasmid that expresses scFv specific to the antigen and with which it interacts, can grow on medium without histidine, due to transcriptional activation of the *HIS3* gene, and express β-galactosidase, which can be assayed histochemically (*see* **Subheading 3.6.**).

1. Inoculate about 5 colonies of L40 yeast into 150 mL of YPD medium, and vortex mix to ensure that there are no clumps of yeast. Incubate overnight with shaking at 30°C.
2. Add 50–100 mL of this overnight yeast culture into 1 L of prewarmed YPD medium (in a 2-L Erhlenmeyer flask) to produce an OD_{600} between 0.2 to 0.3. Incubate at 30°C for about 3 h with shaking until OD_{600} reaches 0.4 to 0.6.
3. Pellet the yeast by centrifuging at $1000g$ for 5 min at 21°C, and resuspend in 500 mL of sterile distilled water.
4. Centrifuge the yeast cells at $1000g$ for 5 min at 21°C, and resuspend in 8 mL of freshly made sterile 1X TE-LiAc. This is the competent yeast ready for transformation. Keep at 4°C until use (preferably not more than 1 h).
5. Mix 1 mg of bait DNA (LexA-DBD-antigen), 500 µg of yeast scFv-VP16 library DNA, and 20 mg of salmon sperm carrier DNA (*see* **Note 7**). Add the mixture of DNA to 8 mL of competent yeast and mix well. (*See* **Note 8**).
6. Add 60 mL of freshly prepared sterile PEG-LiAc solution and mix vigorously. Incubate at 30°C for 30 min with shaking.
7. Add 7 mL of DMSO and mix gently by inversion. Do not vortex mix.
8. Heat shock at 42°C for 15 min and swirl to mix every 2 to 3 min.

9. Transfer to ice for 2 min and pellet the cells by centrifuging at 1000*g* for 5 min. Resuspend in 500 mL of YC medium (-WLHUK) and incubate in a 1-L flask for 3 h at 30°C with shaking.

10. Pellet the cells by centrifuging at 2000*g* for 10 min and resuspend in 10 mL of 1X TE. Plate 10 µL onto a 100-mm dish of YC(-WL) agar and incubate at 30°C for 3 to 4 d. Count the number of yeast colonies and multiply this number to 1000 to give the number of yeast screened.

11. Plate the remainder of the transformed yeast onto 10 large square plates of YC(-WLHUK) agar. Incubate at 30°C for 4 to 5 d. Pick those yeast colonies that have grown to about 1 to 2 mm in diameter and separately streak onto a new YC(-WLHUK) plate (*see* **Note 9**).

3.6. Confirmation of Antibody–Antigen Interaction in Yeast Clones by β-Galactosidase Filter Assay

A β-galactosidase filter assay is performed to confirm the interaction between the antigen and the scFv intracellularly in those colonies that were able to grow in the absence of histidine.

1. Streak the histidine-independent yeast colonies onto a sterile Whatman no. 1 filter paper.
2. The yeast are lysed by immersing in liquid nitrogen. Use a pair of forceps, hold the filter paper and submerge it into a pool of liquid nitrogen for 15 s. Allow to thaw at room temperature for 1 min.
3. Place the filter, colony side up, onto another filter paper presoaked with Z buffer/X-gal solution. Incubate the filters at 30°C and check periodically for the appearance of blue colonies, usually within 1 h. Yeast colonies that are both histidine independent and β-galactosidase positive are further analyzed.

3.7. Plasmid DNA Extraction from Individual Selected Yeast Colonies

After the identification of yeast clones, which can grow in the absence of histidine and activate β-galactosidase, the yeast scFv-VP16 expression plasmid is extracted and retested in fresh transfections. This eliminates false positives due to spontaneous mutation in yeast, giving rise to histidine independence and β-galactosidase activation, independent of the interaction between the antigen bait and the antibody prey.

1. Inoculate the positive yeast colony into 2 mL of YC(-L) medium and incubate overnight at 30°C with shaking in a 20-mL tube.
2. Pellet the cells by spinning in a benchtop microfuge at 11,600*g* for 1 min at room temperature, and resuspend in 200 µL of yeast lysis solution.
3. Add 200 µL phenol:chloroform (1:1 v:v) and 0.3 g of acid-washed glass beads. Vortex mix hard for 2 min, and spin at 11,600*g* for 5 min.
4. Carefully pipet out the supernatant (plasmid-containing) without disturbing the interphase. Add 20 µL of 3 *M* sodium acetate, pH 5.2, and 500 µL of 100% ethanol to the supernatant to precipitate the plasmid DNA.
5. Incubate on dry ice for 1 h, and spin again at 11,600*g* for 5 min.
6. Wash the DNA pellet once with 500 µL 70% ethanol, respin, and dry at room temperature for 10–15 min. Do not over-dry using a vacuum.
7. Resuspend the pellet with 20 µL of sterile water. Transform the 20 µL into DH5α cells using the Hanahan method (**Subheading 3.4.**) and plate on TYE with 100 µg/mL ampicillin. Incubate the plate overnight at 37°C.
8. Pick several bacterial colonies and inoculate individually into 2 mL of 2X TY containing 100 µg/mL ampicillin. Grow overnight at 37°C with shaking.

9. Extract plasmid DNA from each bacterial overnight culture using QIAprep Spin Miniprep Kit according to manufacturer's instruction. Elute DNA with 50 µL of elution buffer.

10. To check for the identity of yeast scFv-VP16 plasmid, digest 5 µL of DNA with *Sfi*I and *Not*I enzyme and fractionate on a 1.5% agarose gel. The 800- to 900-bp scFv fragment and the 7.5-kb vector band should be seen after digestion.

11. A quick comparison of the various scFv-VP16 clones can be achieved by a gel-DNA fingerprinting method of the product on agarose gels (**Subheading 3.9.**). Definitive comparison is obtained from DNA sequence comparison of the scFv (**Subheading 3.9.**).

3.8. Retesting the scFv-VP16 Plasmid Using the Yeast Antibody–Antigen Interaction Assay

Final verification of the intracellular binding of scFv with antigen requires retransfection of individual scFv-VP16 clones with the original antigen bait and assaying histidine-independent growth and β-galactosidase activation. Thus, retesting is similar to the initial yeast in vivo screen, except that it is done on a smaller scale and with individual isolated yeast scFv-VP16 plasmid instead of a library mixture.

1. Prepare competent yeast strain L40, as described in **Subheading 3.5.1., steps 1–4**.
2. Mix 200 ng of bait plasmid DNA, (this must be a fish bait DNA in this re-test) 100 ng of the isolated yeast scFv-VP16 plasmid DNA, and 100 ng of salmon sperm carrier DNA (final volume should not exceed 10 µL). Add the DNA mixture into 100 µL of competent yeast and mix well.
3. Add 600 µL of freshly made PEG-LiAc solution and vortex mix. Incubate at 30°C for 30 min with shaking.
4. Add 70 µL of DMSO and mix gently by inversion. Do not vortex mix.
5. Heat shock at 42°C for 15 min. Place on ice for 2 min.
6. Pellet the cells by spinning in a benchtop microfuge at 11,600*g* for 10 s. Resuspend cells in 1X TE buffer.
7. Plate the yeast cells onto YC(-WLHUK) agar and incubate at 30°C for 3 to 4 d. Histidine-independent colonies should grow to a size of 1 to 2 mm in diameter. These should be tested for β-galactosidase activity using the β-galactosidase filter assay (**Subheading 3.6.**).

3.9. Characterization of the Antigen-Specific Intracellular scFv

The various scFv-VP16 clones can be compared by using DNA fingerprinting after digestion with a frequent cutting restriction enzyme such as *Bst*NI (CCA/TGG). This comprises PCR amplification of insert scFv DNA followed by *Bst*NI digestion and analysis of the product on agarose gel. The isolated intracellular scFv can also be characterized by DNA sequencing of the yeast scFv-VP16 clone.

The primers for PCR and sequence analysis of yeast scFv-VP16 are:

Forward primer: 5′-TTG TTT CTT TTT CTG CAC AAT-3′

Back primer: 5′-CAA CAT GTC CAG ATC GAA-3′

1. Set up a 20-µL PCR with 2 µL of 10X PCR buffer, 2 µL of 25 m*M* MgCl$_2$, 2 µL of 2 m*M* dNTP, 1 µL of each primer (100 ng/mL), 1 mL of 50 ng/mL yeast scFv-VP16 plasmid, 10.8 µL of H$_2$O, and 1 µL of AmpliTaq polymerase. Overlay the reaction mixture with a drop of mineral oil.
2. Put onto a thermal cycler and heat up to 95°C for 10 min. Denature at 95°C for 1 min, anneal at 55°C for 1 min, and extend at 72°C for 1.5 min, for a total of 25 cycles.

3. Take 5 μL of the PCR and analyze on a 1.4% agarose gel. An approx 900-bp band should be seen.
4. Add to the remaining PCR product 17.5 μL of H_2O, 2 μL of NEB buffer 2, and 0.5 μL of *Bst*NI restriction enzyme. Mix well and incubate at 60°C for 3 h.
5. Fractionate 20 μL of the digested PCR product on a 4% NuSieve 3:1 agarose gel. Run at 75 V for about 1 h. Different clones should give rise to different patterns depending on the occurrence of the *Bst*NI sites.

3.10. Binding of scFv to Antigen by Immunodetection Assay (Western Blot)

The methods described above facilitate the isolation of scFv antibody fragments that can effectively bind to antigen in vivo for potential applications involving modification of function *(4)* and antigen-specific cell killing *(17)*. The soluble scFv fragment can be prepared for use in immunoblots (Western blots) by recloning in phage display vectors such as pHEN2 and production of periplasmic protein in *E. coli*.

1. Set up a digestion of 5 μg yeast scFv-VP16 plasmid DNA with 20 U of *Not*I enzyme in a 50-μL of reaction with 1X NEB buffer 2. Incubate at 37°C for 3 h.
2. Add 40 U of *Sfi*I enzyme to the *Not*I-digested DNA and mix. Cover with a drop of mineral oil and incubate overnight at 50°C. Run the digested sample on a 1.5% agarose gel at 50 V for about 45 min, and cut out the gel containing the scFv DNA fragment of around 800–900 bp. Extract the DNA using QIAEX II Gel Extraction Kit according to manufacturer's instructions. Elute DNA with 20 μL of elution buffer. Digest 1 μg of pHEN2 vector with *Not*I and *Sfi*I enzymes and purify the linearized DNA.
3. Set up a 10-μL ligation reaction with 1 μL of the purified *Sfi*I/*Not*I linear pHEN2 vector, 1 μL of the purified scFv DNA *Sfi*I/*Not*I fragment. 1 μL of T4 DNA ligase, and 1 μL of 10X ligase buffer. Incubate overnight at 15°C.
4. Transform DH5α bacteria with the ligation reaction using the Hanahan method (**Subheading 3.4.**). Identify bacteria containing the ligated plasmid (i.e., pHEN2-scFv) by screening with the scFv fragment (**ref. *18***) or making DNA miniprep and restriction enzyme digestion.
5. For periplasmic protein production, transform HB2151 bacteria with pHEN2-scFv using the Hanahan method. Select colonies on TYE plates with 100 μg/mL ampicillin.
6. Inoculate a single colony of HB2151 containing pHEN2-scFv into 5 mL of 2X TY with 100 μg/mL ampicillin and 1% glucose. Incubate overnight at 30°C with shaking.
7. Put 0.5 mL of the overnight culture into 50 mL of 2X TY with 100 μg/mL ampicillin. Incubate at 30°C with shaking for about 2 to 3 h until OD_{600} is 0.5–0.6.
8. Add 100 μL of 500 m*M* isopropyl-β-Δ-thiogalactopyranoside (IPTG) to the culture to induce scFv protein, and continue incubation at 30°C with shaking for another 3.5 h.
9. Pellet the cells by centrifuging at 2000*g* for 20 min and resuspend in 400 μL of ice-cold 1X TES buffer. Transfer to a 2-mL Eppendorf tube.
10. Add 600 μL of 1:5 TES buffer (ice-cold) and mix gently by inversion. Place on ice for 30 min.
11. Spin at 4°C in a benchtop microfuge at 11,600*g* for 15 min. Keep the supernatant, which contains the periplasmic protein containing the scFv.
12. Use the periplasmic protein fresh for Western blots *(19)*. Dilute periplasmic protein to 1:50 for immunodetection for Western blot. Incubate with the antigen immobilized on a nitrocellulose membrane overnight at 4°C with shaking.
13. Use 9E10 anti-myc tag mouse monoclonal antibody as the secondary antibody, because the scFv in pHEN2 is fused in-frame to a myc-tag.

4. Notes

1. Phage antibody libraries (7), displaying scFv, consisting of greater than 5×10^9 members should be used. In order to select the highest diversity of antigen-specific scFv binders, one round of in vitro panning of the library using antigen-coated immunotubes is used. This selection is carried out under low stringency (high concentration of antigen and short washes) to retain rare binders, resulting in a mixture of specific and nonspecific binding scFv-phage. Typically, one would expect 10^4–10^5 in vitro binders from a library of about 5×10^9 initial complexity.

2. The eluted phage can be kept for about 1 wk at 4°C, but the titer will decrease with time.

3. The number of colonies obtained can be estimated by multiplying the number of colonies counted from (*see* **Subheading 3.5.1., step 10**) to the dilution factor. Typically, 10^4 to 10^5 colonies should be obtained. Between 10^5 to 5×10^6 colonies can usually be plated onto one large square plate.

4. When plating the bacteria on agar plates, the surface of the agar should be flat to avoid uneven spreading of bacteria. The surface should also be dried, by leaving the agar plate with the lid off inside a sterile hood for 30 min.

5. The *Sfi*I restriction site (GGCCNNNNNGGCC) is variable. The *Sfi*I site (GGCCCAGCCG GCC) of the yeast VP16 vector described in this protocol is compatible with that found in phage libraries cloned in pHEN/pHEN2.

6. Amplification of the first round phage library and of the yeast scFv-VP16 library should preferably be done by plating on TYE agar plates containing ampicillin. Growing in liquid culture medium may result in loss or preferential enrichment of some clones.

7. For the salmon sperm carrier DNA used in yeast transformation, boiling for 5 min and then cooling to room temperature just before use can increase the transformation efficiency substantially.

8. If using other yeast expression vectors for the scFv-VP16 fusion, make sure that the scFv is cloned in-frame with the VP16. Also, we found that cloning scFv DNA at the 3′ end of the VP16 sequence (i.e., fusing scFv at the carboxy terminus of the VP16 molecule instead of the amino terminus) can affect the interaction of the scFv with the antigen to a certain extent. Also a greater efficiency can be achieved by using a yeast bait strain stably expressing the hex A-DBD-Ag to transfect with the scFv-vp16 sub library.

9. The yeast colonies, which grow in the absence of histidine, do so, in theory, due to the interaction between the antigen and the scFv antibody fragment. However, in practice, we find a number of false-positive colonies on HIS-plating. Therefore it is prudent to reassess the colonies by testing their ability to activate the *LacZ* gene and to isolate the scFv-VP16 clone from a selected colony and retransform this either alone or with the original bait plasmid preparation (the latter is essential, as the bait itself can mutate and give a false positive).

Acknowledgment

Eric Tse was supported by the Croucher Foundation.

References

1. Morgan, D. O. and Roth, R. A. (1988) Analysis of intracellular protein function by antibody injection. *Immunol. Today* **9,** 84–88.
2. Bird, R. E., Hardman, K. D., Jacobson, J. W., Johnson, S., Kaufman, B. M., Lee, S.-M., et al. (1988) Single-chain antigen-binding proteins. *Science* **242,** 423–426.
3. Wright, M., Grim, J., Deshane, J., Kim, M., Strong, T. V., Siegal, G. P., and Curiel, D. T. (1997) An intracellular anti-erbB-2 single-chain antibody is specifically cytotoxic to human breast carcinoma cells overexpressing erbB-2. *Gene Ther.* **4,** 317–322.

4. Cochet, O., Kenigsberg, M., Delumeau, I., Virone-Oddos, A., Multon, M. C., Fridman, W. H., et al. (1998) Intracellular expression of an antibody fragment-neutralizing p21 ras promotes tumor regression. *Cancer Res.* **58,** 1170–1176.
5. Tavladoraki, P., Benvenuto, E., Trinca, S., De Martinis, D., Cattaneo, A., and Galeffi, P. (1993) Transgenic plants expressing a functional single-chain Fv antibody are specifically protected from virus attack. *Nature* **366,** 469–472.
6. McCafferty, J., Griffiths, A. D., Winter, G., and Chiswell, D. J. (1990) Phage antibodies: filamentous phage displaying antibody variable domains. *Nature* **348,** 552–554.
7. Sheets, M. D., Amersdorfer, P., Finnern, R., Sargent, P., Lindqvist, E., Schier, R., et al. (1998) Efficient construction of a large nonimmune phage antibody library: the production of high-affinity human single-chain antibodies to protein antigens. *Proc. Natl. Acad. Sci. USA* **95,** 6157–6162.
8. Clackson, T., Hoogenboom, H. R., Griffiths, A. D., and Winter, G. (1991) Making antibody fragments using phage display libraries. *Nature* **352,** 54–56.
9. Biocca, S., Ruberti, F., Tafani, M., Pierandrei-Amaldi, P., and Cattaneo, A. (1995) Redox state of single chain Fv fragments targeted to the endoplasmic reticulum, cytosol and mitochondria. *Biotechnology (N.Y.)* **13,** 1110–1115.
10. Worn, A. and Pluckthun, A. (1998) An intrinsically stable antibody ScFv fragment can tolerate the loss of both disulfide bonds and fold correctly. *FEBS Lett.* **427,** 357–361.
11. Visintin, M., Tse, E., Axelson, H., Rabbitts, T. H., and Cattaneo, A. (1999) Selection of antibodies for intracellular function using a two-hybrid *in vivo* system. *Proc. Natl. Acad. Sci. USA* **96,** 11,723–11,728.
12. De Jeager, G., Fiers, E., Eeckhout, D., and Depicker, A. (2000) Analysis of the interaction between single-chain variable fragments and their antigen in a reducing intracellular environment using the two-hybrid-system. *FEBS Lett.* **467,** 316–320.
13. Portner-Taliana, A., Russell, M., Froning, K. J., Budworth, P. R., Comiskey, J. D., and Hoeffler, J. P. (2000) In vivo selection of single-chain antibodies using a yeast two-hybrid system. *J. Immunol. Meth.* **238,** 161–172.
14. Fields, S. and Song, O. (1989) A novel genetic system to detect protein-protein interactions. *Nature* **340,** 245–246.
15. Tse, E., Lobato-Caballero, M. N., Forster, A., Tanaka, T., Chung, G. T. Y., and Rabbitts, T. H. (2002) Intracellular antibody capture technology: application to selection of single chain Fv recognizing the BCR-ABL on cogenic protein. *Submitted.*
16. Hanahan, D. (1983) Studies on transformation of *Escherichia coli* with plasmids. *J. Mol. Biol.* **166,** 557–580.
17. Tse, E. and Rabbitts, T. H. (2000) Intracellular antibody-caspase mediated cell killing: a novel approach for application in cancer therapy. *Submitted.*
18. Buluwela, L., Forster, A., Boehm, T., and Rabbitts, T. H. (1989) A rapid method for colony screening using nylon filters. *Nucl. Acids Res.* **17,** 452.
19. Harlow, E. and Lane, D. (1998) *Antibodies: a laboratory manual.* CHS Laboratory Press, Cold Spring Harbor, N.Y.

33

Detection and Visualization of Protein Interactions with Protein Fragment Complementation Assays

Ingrid Remy, André Galarneau, and Stephen W. Michnick

1. Introduction

A first step in defining the function of a novel gene is to determine its interactions with other gene products in an appropriate context; that is, because proteins make specific interactions with other proteins as part of functional assemblies, an appropriate way to examine the function of the product of a novel gene is to determine its physical relationships with the products of other genes. This is the basis of the highly successful yeast two-hybrid system *(1–6)*. The central problem with two-hybrid screening is that detection of protein-protein interactions occurs in a fixed context, the nucleus of *S. cerevisiae*, and the results of a screening must be validated as biologically relevant using other assays in appropriate cell, tissue or organism models. While this would be true for any screening strategy, it would be advantageous if one could combine library screening with tests for biological relevance into a single strategy, thus tentatively validating a detected protein as biologically relevant and eliminating false-positive interactions immediately. It was with these challenges in mind that our laboratory developed the protein-fragment complementation assay (PCA) strategy. In this strategy, the gene for an enzyme is rationally dissected into two pieces. Fusion proteins are constructed with two proteins that are thought to bind to each other, fused to either of the two probe fragments. Folding of the probe protein from its fragments is catalyzed by the binding of the test proteins to each other and is detected as reconstitution of enzyme activity. We have already demonstrated that the PCA strategy has the following capabilities: *(1)* Allows for the detection of protein–protein interactions in vivo and in vitro in any cell type; *(2)* allows for the detection of protein–protein interactions in appropriate subcellular compartments or organelles; *(3)* allows for the detection of induced versus constitutive protein–protein interactions that occur in response to developmental, nutritional, environmental, or hormone-induced signals; *(4)* allows for the detection of the kinetic and equilibrium aspects of protein assembly in these cells; and *(5)* allows for screening of cDNA libraries for protein–protein interactions in any cell type.

In addition to the specific capabilities of PCA described above, are special features of this approach that make it appropriate for screening of molecular interactions,

From: *Methods in Molecular Biology, vol. 185: Embryonic Stem Cells: Methods and Protocols*
Edited by: K. Turksen © Humana Press Inc., Totowa, NJ

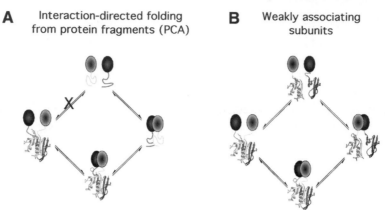

Fig. 1. Two alternative strategies to achieve complementation. **(A)** The PCA strategy requires that unnatural peptide fragments be chosen that are unfolded prior to association of fused interacting proteins. This prevents spontaneous association of the fragments (pathway X) that can lead to a false signal. **(B)** Naturally occurring subunits that are already capable of folding can be mutated to interact with lower affinity. However, to some extent, this will always occur, requiring the selection of cells that express protein partner fusions at low enough levels that background is not detected. (*See* color plate 14, following p. 254).

including: *(1)* PCAs are not a single assay but a series of assays; an assay can be chosen because it works in a specific cell type appropriate for studying interactions of some class of proteins; *(2)* PCAs are inexpensive, requiring no specialized reagents beyond those that are necessary for a particular assay and off the shelf materials and technology; *(3)* PCAs can be automated and high-throughput screening could be done; *(4)* PCAs are designed at the level of the atomic structure of the enzymes used; because of this, there is additional flexibility in designing the probe fragments to control the sensitivity and stringencies of the assays; and *(5)* PCAs can be based on enzymes for which the detection of protein–protein interactions can be determined differently, including by dominant selection or production of a fluorescent or colored product.

The selection of enzymes and design of PCAs have been discussed in detail *(7)*, and here, we will review only the most basic ideas. Polypeptides have evolved to code for all of the chemical information necessary to spontaneously fold into a stable, unique 3-dimensional structure *(8–10)*. It logically follows that the folding reaction can be driven by the interaction of two peptides that together contain the entire sequence, and in the correct order, that a single peptide will fold. This was demonstrated in the classic experiments of Richards *(11)* and Taniuchi and Anfinsen *(12)*. In practice, this is not easily done. Protein fragments will tend to aggregate rather than fold together into the three dimensional structure of the complete protein. However, if one fuses soluble binding proteins to the fragments, which by interacting with each other increase the effective concentration of the fragments, correct folding could be favored over any other nonproductive process *(13–15)*. If the protein that folds from its constitutive fragments is an enzyme, whose activity could be detected in vivo, then the reconstitution of its activity can be used as a measure of the interaction of the fused binding proteins (**Fig. 1A**). Further, this binary all or none folding event provides for a very specific measure of protein interactions dependent on not mere proximity, but the absolute requirement that the peptides must be organized precisely in space to allow for folding of the enzyme from the polypeptide chain. We select to dissect proteins into fragments that are not capable of spontaneously folding along with their complementary fragment, into a functional and complete protein. These facts distinguish the PCA strategy from complementation of naturally occurring and weakly associating subunits of enzymes *(16)*, in which some spontaneous assembly occurs, as illustrated in **Fig. 1B**.

All applications of PCA, thus far, have been performed in model cell lines as described below. However, it is not difficult to imagine how the basic approach could be combined with functional genomics approaches using embryonic stem (ES) cells. The most obvious applications would be in gene-trap strategies, in which a reporter PCA fragment is inserted randomly or directionally into specific genes in the same way that an individual reporter gene such as lacZ or green fluorescent protein (GFP) might be incorporated *(17–21)*. The PCA reporter fragments would be inserted into the genome of ES cells, and the cells induced to differentiate in vitro. The goal, then, would be to identify differentiated cells in which two marked gene products interact and then to identify the interacting partners. We have already developed 5 PCAs based on dominant-selection, colorimetric, or fluorescent outputs *(7)*. Specific features and applications of PCAs are given in **Table 1**. All of these assays have potential applications to gene trapping in ES cells. Here, we will discuss the most well developed PCA, based on the enzymes murine dihydrofolate reductase (mDHFR) and tiethylene-melamine (TEM) β-lactamase.

The DHFR PCA can be used in a variety of applications to perform simple survival-selection as readout, but simultaneously, as a fluorescent assay allowing quantitative detection of protein interactions as well as the determination of the cellular location of protein interactions can be performed *(22,24)*. The β-lactamase assay can be used as a very sensitive in vivo or in vitro quantitative detector of protein interactions as, unlike DHFR, one measures the continuous conversion of substrate to colored or fluorescent product. However, it should be noted that the generation of a product by an enzyme does not guarantee that signal-to-background would be superior to that of fixed fluorophore reporters like GFP and fluorescein-conjugated methotrexate (fMTX) bound to DHFR. Observable signal-to-background depends, for example, on the quantum yield of the fluorophore, retention of fluorophore by a cell, the optical properties of the cells used, and the extent to which fluorophores are retained in individual cellular compartments. For instance, in spite of no enzymatic amplification, the DHFR fluorescence assay only requires between 1000 to 3000 molecules of reconstituted DHFR to clearly distinguish a positive response from background.

Reconstitution of DHFR activity can be monitored in vivo by cell survival in DHFR-negative cells (CHO-DUKX-B11, for example) grown in the absence of nucleotides. The principle of the DHFR PCA survival assay is that cells, simultaneously expressing complementary fragments of DHFR fused to interacting proteins or peptides, will survive in media depleted of nucleotides. Alternative, recessive selection can be achieved in DHFR-positive cells by using DHFR PCA fragments containing one or more of several mutations that render the refolded DHFR resistant to the anti-folate drug methotrexate. The cells are then grown in the absence of nucleotides with selection for methotrexate resistance. The survival DHFR PCA is an extraordinarily sensitive

Table 1.
Existing Protein-fragment Complementation Assays (PCA)
and Their Applications

Enzyme	Molecular Weight	Function	Assays	Demonstrated Applications	Organism Restrictions
DHFR	21 kD/monomeric	Reduces dihydrofolate to tetrahydrofolate	-Fluorescence -Survival selection	-Survival selection -Localization of protein interactions in living cells -Quantitation of induced associations in vivo -Translocation	None (Universal)
ß-lactamase	27 kD/monomeric	Hydrolyzes ß-lactam antibiotics (e.g. cephalosporin, ampicillin)	-Survival selection in bacteria -Fluorescence -Colorimetric	-In vivo fluorescence (e.g. with CCf2/AM) -In vitro colorimetric assay (nitrocefin)	None (Universal)
GFP	28 kD/monomeric	Spontaneously fluorescent protein	Intrinsic fluorescence	-Localization of protein interactions in living cells -Quantitation of induced associations in vivo -Translocation	None (Universal)
GAR Transformylase	30 kD/monomeric	Transfer of formyl group to glycineamide ribonucleotide	Survival selection in bacteria	Survival selection	Restricted to bacterial auxotrophs lacking GARTase
Aminoglycoside phosphotransferase	35 kD/monomeric	Phosphorylation of aminoglycosides (e.g. neomycin/G418) antibiotic	Survival selection in many cell types (bacteria, yeast mammalian, etc…)	Survival selection	None (Universal)
Hygromycin B phosphotransferase	35 kD/monomeric	Phosphorylation of hygromycin B	Survival selection in many cell types (bacteria, yeast mammalian, etc…)	Survival selection	None (Universal)

assay. In mammalian cells, survival is dependent only on the number of molecules of DHFR reassembled, and we have determined that this number is approx 25–100 molecules of DHFR per cell *(22)*. The second approach is a fluorescence assay based on the detection of fMTX binding to reconstituted DHFR. The basis of the DHFR PCA fluorescence assay is that complementary fragments of DHFR, when expressed and reassembled in cells, will bind with high affinity (K_d = 540 pM) to fMTX in a 1:1 complex. fMTX is retained in cells by this complex, while the unbound fMTX is actively and rapidly transported out of the cells *(25,26)*. In addition, binding of fMTX to DHFR results in a 4.5-fold increase in quantum yield. Bound fMTX, and by inference reconstituted DHFR, can then be monitored by fluorescence microscopy, fluorescence-activated cell sorting (FACS), or spectroscopy *(22–24)*. It is important to note that, though fMTX binds to DHFR with high affinity, it does not induce DHFR folding from the fragments in the PCA. This is because the folding of DHFR from its fragments is obligatory; if binding of the oligomerization domains does not induce folding, no binding sites for fMTX are created. Therefore, the number of complexes observed, as measured by number of fMTX molecules retained in the cell, is a direct measure of the equilibrium number of complexes of oligomerization domain complexes formed, independent of binding of fMTX *(22,24,26)*. The other obvious application of the DHFR PCA fluorescence assay is in determining the location in the cell of interactions, as illustrated in a number of cell types (**Fig. 2**).

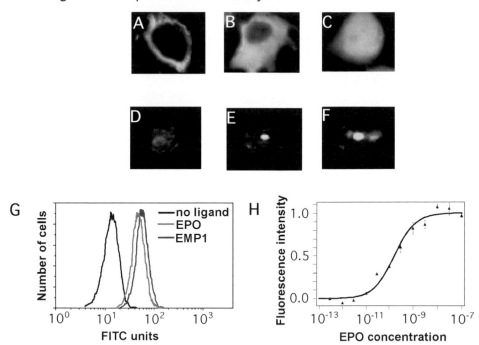

Fig. 2. Applications of the DHFR PCA to detecting the localization of protein complexes and quantitating protein interactions. **(A–C)** Different protein pairs showing **(A)** plasma membrane, **(B)** cytosol, and **(C)** whole cell localization in transiently transfected COS cells. **(D–F)** cytosolic and nuclear localization of interacting proteins in potato protoplasts. (D) cytosolic, **(E)** nuclear localization, **(F)** DAPI co-staining of **(E)**, **(G)** FACS results of DHFR PCA. CHO cells expressing the erythropoietin (EPO) receptor fused to complementary DHFR fragments. Receptor activation (conformation change) induced by EPO or a peptide agonist (EMP1) lead to an increase in fluorescence. **(H)** Dose-response curve for EPO-induced fluorescence as detected by FACS results in **(G)**. (*See* color plate 15, following p. 254).

β-Lactamase is strictly a bacterial enzyme and has been genetically deleted from many standard *Escherichia coli* strains. It is not present at all in eukaryotes. Thus, the β-lactamase PCA can be used universally in eukaryotic cells and many prokaryotes, without any intrinsic background. Also, assays are based on catalytic turnover of substrates with rapid accumulation of product. This enzymatic amplification should allow for relatively weak molecular interactions to be observed. The assay can be performed simultaneously or serially in a number of modes, such as the in vitro colorimetric assay, the in vivo fluorescence assay, or the survival assay in bacteria. Assays can be performed independent of the measurement platform and can easily be adapted to high-throughput formats requiring only one pipetting step.

2. Materials

2.1. DHFR PCA Survival Assay

1. 12-Well plates, tissue culture treated (Corning Costar, cat. no. 3513); 6-well plates, tissue culture treated (Corning Costar, cat. no. 3516).

2. Minimum essential medium: α-medium without ribonucleosides and deoxyribonucleosides (α-MEM) (Life Technologies, cat. no. 12000022).
3. Dialyzed fetal bovine serum (FBS) (Hyclone, cat. no. SH30079-03).
4. Adenosine (Sigma, cat. no. A-4036); desoxyadenosine (Sigma, cat. no. D-8668); thymidine (Sigma, cat. no. T-1895).
5. Lipofectamine Plus reagent (Life Technologies, cat. no. 10964013).
6. Trypsin-EDTA (Life Technologies, cat. no. 253100062).
7. Cloning cylinders (Scienceware, cat. no. 37847-0000).

2.2. DHFR PCA Fluorescence Assay

1. 12-Well plates, tissue culture treated (Corning Costar, cat. no. 3513).
2. Dulbecco's modified Eagle medium (DMEM) (Life Technologies, cat. no. 12100046); α-MEM (Life Technologies, cat. no. 12000022).
3. Cosmic calf serum (Hyclone, cat. no. SH3008703); dialyzed FBS (Hyclone, cat. no. SH30079-03).
4. Lipofectamine Plus reagent (Life Technologies, cat. no. 10964013).
5. fMTX (Molecular Probes, cat. no. M-1198).
6. Dulbecco's phosphate-buffered saline (PBS) (Life Technologies, cat. no. 21600069).
7. Geltol aqueous mounting medium (Immunon, cat. no. 484950).
8. Trypsin-EDTA (Life Technologies, cat. no. 253100062).
9. Microcover glasses, 18-mm circles, no 2 (VWR Scientific, cat. no. 48382041).
10. Microscope slides, glass, $25 \times 75 \times 1.0$ mm (any supplier).
11. 96-Well black microtiter plates (Dynex no 7805; VWR Scientific, cat. no. 62402-983).
12. Bio-Rad protein assay (Bio-Rad, Cat. No.: 500-0112).

2.3. β-Lactamase PCA Colorimetric Assay

1. 12-Well plates, tissue culture treated (Corning Costar, cat. no. 3513).
2. DMEM (Life Technologies, cat. no. 12100046).
3. Cosmic calf serum (Hyclone, cat. no. SH3008703).
4. Fugene 6 transfection reagent (Roche Diagnostics, cat. no. 1814443).
5. Trypsin-EDTA (Life Technologies, cat. no. 253100062).
6. Dulbecco's PBS (Life Technologies, cat. no. 21600069).
7. Nitrocefin (Becton Dickinson Microbiology Systems, cat. no. 89-7065-0).
8. 96-Well plate, (Corning Costar, cat. no. 3595).

2.4. β-Lactamase PCA Fluorometric Assay

1. 12-Well plates, tissue culture treated (Corning Costar, cat. no. 3513).
2. DMEM (Life Technologies, cat. no. 12100046).
3. Cosmic calf serum (Hyclone, cat. no. SH3008703).
4. Fugene 6 transfection reagent (Roche Diagnostics, cat. no. 1814443).
5. Trypsin-EDTA (Life Technologies, cat. no. 253100062).
6. Dulbecco's PBS (Life Technologies, cat. no. 21600069).
7. CCF2-AM (kindly provided by Roger Tsien UCJD).
8. 96-Well white microtiter plates (Dynex no 7905; VWR Scientific, cat. no. 62402-980).
9. Normal saline: 140 mM, NaCl, 5 mM KCl, 2 mM CaCl$_2$, 10 mM Hepes, 6 mM Sucrose, 10 mM Glucose, pH 7.35.
10. Physiological saline solution: 10 mM HEPES, 6 mM sucrose, 10 mM glucose, 140 mM NaCl, 5 mM KCl, 2 mM MgCl$_2$, 2 mM CaCl$_2$, pH 7.35.
11. 15-mm Glass coverslip (Ted Pella, cat. no: 26021).

3. Method

3.1. DHFR PCA Survival Assay

1. Split CHO DUKX-B11 (DHFR-negative; could also be done in other cells lines) (*see* **Note 1**) cells 24 h before transfection at 1×10^5 in 12-well plates in α-MEM medium enriched with 10% dialyzed FBS and supplemented with 10 µg/mL of adenosine, desoxyadenosine, and thymidine.

2. Co-transfect cells with the PCA fusion partners (*see* **Note 2**) using Lipofectamine Plus reagent according to the manufacturer's instructions.

3. Forty-eight hours after the beginning of the transfection, split cells at approx 5×10^4 in 6-well plates in selective medium consisting of α-MEM enriched with dialyzed FBS, but without addition of nucleotides (*see* **Note 3** and **4**).

4. Change medium every 3 d. The appearance of distinct colonies usually occurs after 4 to 10 d of incubation in selective medium. Colonies are observed only for clones that simultaneously express both interacting proteins fused to one or the other complementary DHFR fragments. Only interacting proteins will be able to achieve normal cell division and colony formation.

For further analysis of the interacting protein pair:

5. Isolate 3 to 5 colonies per interacting partners by trypsinization (trypsin-EDTA) using cloning cylinders and grow them separately.

6. Select the best expressing clone by immunoblot (Western blot) or using the DHFR PCA fluorescence assay (*see* **Subheading 3.2.**). Amplification of the expressed gene using methotrexate resistance can be done afterwards if desired, to obtain clones with increased expression *(27)*.

7. Carry out functional analysis of the clone stably expressing your interacting proteins pair fused to the complementary DHFR fragments by using the DHFR PCA fluorescence assay.

3.2. DHFR PCA Fluorescence Assay

3.2.1. Fluorescence Microscopy

1. Split COS cells (this assay can be used with any other cell line) (*see* **Note 5**) 24 h before transfection at 1×10^5 on 18-mm circles glass coverslips in 12-well plates in DMEM medium enriched with 10% Cosmic calf serum.

2. Transiently co-transfect cells with the PCA fusion partners (*see* **Note 2**) using Lipofectamine Plus reagent according to the manufacturer's instructions.

3. The next day, change medium and add fMTX to the cells at a final concentration of 10 µ*M* (*see* **Note 6**).
   For stable cell lines:
   For CHO DUKX-B11 cells (or other cell line) stably expressing PCA fusion partners, seed cells to approximately 2×10^5 on 18-mm glass coverslips in 12-well plates in α-MEM medium enriched with 10% dialyzed FBS. The next day, fMTX is added to the cells at a final concentration of 10 µ*M*.

4. After an incubation with fMTX of 22 h at 37°C, remove the medium, and wash the cells with PBS 1X, and re-incubate for 15–20 min at 37°C in the culture medium to allow for efflux of unbound fMTX (*see* **Note 7**). Remove the medium and wash the cells four times with cold PBS 1X on ice, and finally mount the coverslips on microscope glass slides with an aqueous mounting medium.

5. Fluorescence microscopy is performed on live cells (*see* **Note 8**). These experiments must be performed within 30 min of the wash procedure. If the negative control (untransfected cells treated with fMTX) is too fluorescent, the wash procedure must be modified (*see* **Note 9**).

3.2.2. Flow Cytometry Analysis

Preparation of cells for FACS analysis is the same as described for fluorescence microscopy, except that following the PBS 1X wash (just two times in this case), cells are gently trypsinized (trypsin-EDTA), suspended in 500 μL of cold PBS 1X, and kept on ice prior to flow cytometric analysis within 30 min. Data are collected on a FACS analyzer with stimulation with an argon laser tuned to 488 nm with emission recorded through a 525 nm band width filter.

3.2.3. Fluorometric Analysis

Preparation of cells for fluorometric analysis is the same as described for fluorescence microscopy, except that following the PBS 1X wash (just 2 times in this case), cells are gently trypsinized (trypsin-EDTA). Plates are put on ice, and 100 μL of cold PBS is added to the cells. The total cell suspensions are transferred to 96-well black microtiter plates, and keep on ice prior to fluorometric analysis. The assay can be performed on any microtiter plate reader; we use a Perkin-Elmer HTS 7000 Series Bio Assay Reader in the fluorescence mode. The excitation and emission wavelengths for the fMTX are 497 nm and 516 nm, respectively. Afterward, the data are normalized to total protein concentration in cell lysates (Bio-Rad protein assay).

3.3. In Vitro β-Lactamase PCA Colorimetric Assay

1. Split COS or HEK 293 T cells (this assay can be use with any other cell line) 24 h before transfection at 1×10^5 in 12-well plates in DMEM medium enriched with 10% Cosmic calf serum.
2. Transiently co-transfect cells with the PCA fusion partners (*see* **Note 10**) using Fugene 6 Transfection reagent according to the manufacturer's instructions.
3. Forty-eight hours after transfection, cells are washed three times with cold PBS, resuspended in 300 μL of cold PBS, and kept on ice. Cells are then centrifuged at 4°C for 30 s, the supernatant discarded, and cells resuspended in 100 μL of cold phosphate buffer 100 mM, pH 7.4 (β-lactamase reaction buffer).
4. Freezing in dry ice-ethanol for 10 min and thawing in a water bath at 37°C for 10 min then lyse cells with three cycles of freeze and thaw. Cell membrane and debris are removed by centrifugation at 4°C for 5 min (10,000g). The supernatant whole cell lysate is then collected and stored at –20°C until assays are performed.
5. Assays are performed in 96-well microtiter plates. For testing β-lactamase activity, 100 μL of phosphate buffer 100 mM, pH 7.4, is allocated into each well. To this, add 78 μL of H$_2$O and 2 μL of 10 mM Nitrocefin (final concentration of 100 μM). Finally, add 20 μL of unfrozen cell lysate (final buffer concentration of 60 μM).
6. The assays can be performed on any microtiter plate reader; we use a Perkin-Elmer HTS 7000 Series Bio Assay Reader in the absorption mode with a 492-nm measurement filter.

3.4. In Vivo Enzymatic Assay and Fluorescent Microscopy with CCF2/AM

1. Split COS or HEK293 T cells 24 h before transfection at 1×10^5 in 12-well plates in DMEM enriched with 10% Cosmic calf serum.

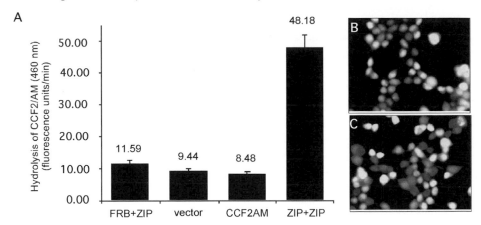

Fig. 3. β-Lactamase PCA using the fluorescent substrate CCF2/AM. (**A**) ZIP (GCN4 leucine zipper-forming sequences) are tested in HEK 293 cells as described in the text. FRB (rapamycin-FKBP binding domain of FRAP) is used as a negative control. pMT3 is the expression vector alone, and ZIP plus ZIP is the positive control. Data recorded in white microtiter plates on a Perkin Elmer HTS 7000 plate reader. (**B, C**) Fluorescent micrographs of cells expressing β-lactamase PCA showing negative (**B**) (FRB + ZIP) or positive (**C**) (ZIP plus ZIP) response. (*See* color plate 16, following p. 254).

2. Transiently co-transfect cells with the PCA fusion partners (*see* **Note 10**) using Fugene 6 Transfection Reagent according to the manufacturer's instructions.
3. Twenty-four hours after transfection, cells are split again to assure 50% confluency the following day (1.5×10^5) (*see* **Note 11**). They are split either onto 12-well plates for suspension enzymatic assay or onto 15-mm glass coverslips for fluorescent microscopy.
3. Forty-eight hours after transfection, cells are washed three times with PBS to remove all traces of serum (*see* **Note 12**).
4. Cells are then loaded with the following: 1 μ*M* of CCF2/AM diluted into a physiologic saline solution for 1 h.

   For in vivo enzymatic assay:

5. Cells are then washed twice with the physiologic saline. The cells are then resuspended into the same solution, and 1×10^6 cells are aliquoted into a 96-well fluorescence white plate and are read for blue fluorescence with a Perkin Elmer HTS 7000 Series Bio Assay Reader in the fluorescence Top reading mode with a 409-nm excitation filter and a 465-emission filter.
   For fluorescence microscopy:
6. Cells are washed twice with the physiologic saline as in **step 5**, prior to examination under the microscope (*see* **Note 13**).

We used two substrates to study the β-lactamase PCA. The first one is the cephalosporin called nitrocefin *(28)*. This substrate is used in the in vitro colorimetric assay. β-lactamase has a *kcat*/km of 1.7×10^4 m*M*$^{-1}$*s^{-1}. Substrate conversion can be easily observed by eye; the substrate is yellow in solution, while the product is a distinct ruby red color. The rate of hydrolysis can be monitored quantitatively with any spectrophotometer by measuring the appearance of red at 492 nm. Signal-to-background, depending on the mode of measurement, can be greater than 30 to 1.

We have also developed an in vivo fluorometric assay using the substrate CCF2/AM *(28)*. While not as good a substrate as nitrocefin (kcat/km of 1260 mM^{-1}*s^{-1}) CCF2/AM has unique features that make it a useful reagent for in vivo PCA. First, CCF2/AM contains butyryl, acetyl, and acetoxymethyl esters, allowing diffusion across the plasma membrane, in which cytoplasmic esterases catalyze the hydrolysis of its ester functionality releasing the polyanionic (4 anions) β-lactamase substrate CCF2. Because of the negative charge of CCF2, the substrate becomes trapped in the cell. In the intact substrate, fluorescence resonance energy transfer (FRET) can occur between a coumarin donor and fluorescein acceptor pair covalently linked to the cephalosporin core. The coumarin donor can be excited at 409 nm with emission at 447 nm, which is within the excitation envelope of the fluorescence acceptor (maximum around 485 nm), leading to remission of green fluorescence at 535 nm. When β-lactamase catalyzes hydrolysis of the substrate, the fluorescein moiety is eliminated as a free thiol. Excitation of the coumarin donor at 409 nm then emits blue fluorescence at 447 nm, whereas the acceptor (fluorescein) is quenched by the free thiol.

4. Notes

1. Alternatively, recessive selection can be achieved in eukaryotic cells by using DHFR fragments containing one or more of several mutations (for example F31S mutation, see below) that reduce the affinity of refolded DHFR to the anti-folate drug methotrexate and growing cells in the absence of nucleotides with selection for methotrexate resistance. This would obviously be necessary in working with mouse ES cells as, with all eukaryotes, DHFR activity is present.

2. The best orientations of the fusions for the DHFR PCA are: protein A-DHFR(1,2) plus protein B-DHFR(3) or DHFR(1,2)-protein A plus protein B-DHFR(3), where proteins A and B are the proteins to test out for interaction. We typically insert a 10 amino acid flexible polypeptide linker consisting of (Gly.Gly.Gly.Gly.Ser)$_2$ between the protein of interest and the DHFR fragment (for both fusions). DHFR(1,2) corresponds to amino acids 1–105, and DHFR(3) corresponds to amino acids 106–186 of murine DHFR. The DHFR(1,2) fragment that we use also contains a phenylalanine to serine mutation at position 31 (F31S), rendering the reconstituted DHFR resistant to MTX treatment.

3. It is crucial that cell density is kept to a minimum and that cells are well separated when split, to avoid cells "harvesting" nutrients from adjacent cells on dense plates or colonies that might appear to be forming from clumps of cells that were not sufficiently separated during the splitting procedure.

4. The choice of dialyzed FBS manufacturer is crucial. Cells need very little nucleotide in the medium to propagate, and this will result in false positives. The Hyclone dialyzed FBS has proven a particularly reliable source.

5. The fluorescence DHFR PCA assay is universal and, in theory, can be used in any cell type or organism. This assay has already been shown to work in several mammalian cell lines as well as in plant cells and insect cells.

6. A stock solution of 1 mM fMTX should be prepared as follows: dissolve 1 mg of fMTX in 1 mL of dimethyl formamide (DMF). To facilitate the dissolution, incubate 15 min at 37°C and vortex mix every 5 min. Protect the tube from light. Keep at –20°C.

7. Complementary fragments of DHFR fused to interacting protein partners, when expressed and reassembled in cells, will bind with high affinity (K_d = 540 pM) to fMTX in a 1:1 complex. fMTX is retained in cells by this complex, while the unbound fMTX is actively and rapidly transported out of the cells.

8. All of the work reported to date has been performed in live cells. While cells can be fixed, there is a significant reduction in observable fluorescence.

9. Particular attention must be given to optimizing the fMTX load and "wash" procedures. Important variables include the time of loading, temperatures at which each wash step is performed, the number and length of wash steps and the time between washing and visualization. Too little washing will mean that background cannot be distinguished from a positive result. One should scrutinize the relevant parameters in the same sense as one would for say, a Western blot. Results may also vary with the way the cells are plated and the types of cells used. Generally, as in other fluorescent microscopy procedures, the shape of cells and the localization of the fluorophore will result in better or worse results. For stable cell lines, the intensity of fluorescence will also depend on the levels of expression of the fusion proteins. The loading times and concentrations of fMTX (22 h and 10 μM) used may result in a nonspecific and punctate fluorescence that is observed with any filter set. We do not know the source of this background, but it should not be mistaken for the real fluorescence signal produced by the PCA, which should be observed strictly with a filter that is optimal for observation of fluorescein. We have observed that loading fMTX for between 2 and 5 h at lower (5 μM) concentrations prevents this nonspecific signal, although fewer cells are labeled. Loading times and concentrations must be optimized for specific cell types.

10. The best orientations of the fusions for the β-lactamase PCA are: protein A-BLF[1] plus protein B-BLF(2) or BLF(1)-protein A plus protein B-BLF[2], where proteins A and B are the proteins to test out for interaction. We typically insert a 15 amino acid flexible polypeptide linker consisting of (Gly.Gly.Gly.Gly.Ser)$_3$ between the protein of interest and the β-lactamase fragment (for both fusions). BLF(1) corresponds to amino acids 26–196 (Ambler numbering), and BLF(2) corresponds to amino acids 198–290 of TEM-1 β-lactamase.

11. The maximum loading efficiency of CCF2-AM is observed at 50% confluence.

12. Serum may contain esterases that can destroy the substrate.

13. We perform fluorescence microscopy on live HEK 293 or COS cells with an inverse Nikon Eclipse TE-200 (objective plan fluor 40X dry, numerically open at 0.75) Images were taken with a digital change-coupled device (CCD) cooled (–50°C) camera, model Orca-II (Hamamatsu Photonics (exposure for 1 s, binning of 2 × 2 and digitalization 14 bits at 1.25 MHz). Source of light is a Xenon lamp Model DG4 (Sutter Instruments). Emission filters are changed by an emission filter switcher (model Quantoscope) (Stranford Photonics). Images are visualized with ISee software (Inovision Corporation) on an O2 Silicon Graphics computer. The following selected filters are used: filter set no. 31016 (Chroma Technologies); excitation filter: 405 nm (passing band of 20 nm); dichroic mirror: 425 nm DCLP; emission filter no. 1: 460 nm (passing band of 50 nm); emission filter no. 2: 515 nm (passing band of 20 nm).

References

1. Drees, B. L. (1999) Progress and variations in two-hybrid and three-hybrid technologies. *Curr. Opin. Chem. Biol.* **3**, 64–70.
2. Evangelista, C., Lockshon, D., and Fields, S. (1996) The yeast two-hybrid system—prospects for protein linkage maps. *Trends Cell Biol.* **6**, 196–199.
3. Fields, S. and Song, O. (1989) A novel genetic system to detect protein-protein interactions. *Nature* **340**, 245–246.
4. Vidal, M. and Legrain, P. (1999) Yeast forward and reverse 'n'-hybrid systems. *Nucleic Acids Res.* **27**, 919–929.

5. Walhout, A. J., Sordella, R., Lu, X., Hartley, J. L., Temple, G. F., Brasch, M. A., et al. (2000) Protein interaction mapping in *C. elegans* using proteins involved in vulval development. *Science* **287,** 116–122.

6. Uetz, P., Giot, L., Cagney, G., Mansfield, T. A., Judson, R. S., Knight, J. R., et al. (2000). A comprehensive analysis of protein-protein interactions in *Saccharomyces cerevisiae*. *Nature* **403,** 623–627.

7. Michnick, S. W., Remy, I., C.-Valois, F.-X., V.-Belisle, A, and Pelletier, J. N. (2000) Detection of protein-protein interactions by protein fragment complementation strategies, in *Methods in enzymology*, vol. 328 (Abelson, J. N., Emr, S. D., and Thorner, J., eds.), Academic Press, New York, pp. 208–230.

8. Anfinsen, C. B., Haber, E., Sela, M., and White, F. H., Jr. (1961) The kinetics of formation of native ribonuclease during oxidation of the reduced polypeptide chain. *Proc. Natl. Acad. Sci. USA* **47,** 1309–1314.

9. Anfinsen, C. B. (1973) Principles that govern the folding of protein chains. *Science* **181,** 223–230.

10. Gutte, B. and Merrifield, R. B. (1971) The synthesis of ribonuclease A. *J. Biol. Chem.* **246,** 1922–1941.

11. Richards, F. M. (1958) On the enzymatic activity of subtilisin-modified ribonuclease. *Proc. Natl. Acad. Sci. USA* **44,** 162–166.

12. Taniuchi, H. and Anfinsen, C. B. (1971) Simultanious formation of two alternative enzymically active structures by complementation of two overlapping fragments of staphylococcal nuclease. J. Biol. Chem. **216,** 2291–2301.

13. Pelletier, J. N., Campbell-Valois, F., and Michnick, S. W. (1998) Oligomerization domain-directed reassembly of active dihydrofolate reductase from rationally designed fragments. *Proc. Natl. Acad. Sci. USA* **95,** 12,141–12,146.

14. Pelletier, J. N. and Michnick, S. W. (1997) A protein complementation assay for detection of protein-protein interactions in vivo. *Prot. Eng.* **10,** 89.

15. Johnsson, N. and Varshavsky, A. (1994) Split ubiquitin as a sensor of protein interactions in vivo. *Proc. Natl. Acad. Sci. USA* **91,** 10,340–10,344.

16. Rossi, F., Charlton, C. A.. and Blau, H. M. (1997) Monitoring protein-protein interactions in intact eukaryotic cells by beta-galactosidase complementation. *Proc. Natl. Acad. Sci. USA* **94,** 8405–8410.

17. Hill, D. P. and Wurst, W. (1993) Gene and enhancer trapping: mutagenic strategies for developmental studies. *Curr. Top. Dev. Biol.* **28,** 181–206.

18. Nakano, T., Kodama, H., and Honjo, T. (1994) Generation of lymphohematopoietic cells from embryonic stem cells in culture. *Science* **265,** 1098–1101.

19. Gossler, A., Joyner, A. L., Rossant, J., and Skarnes, W. C. (1989) Mouse embryonic stem cells and reporter constructs to detect developmentally regulated genes. *Science* **244,** 463–465.

20. Wang, R., Clark, R., and Bautch, V. L. (1992) Embryonic stem cell-derived cystic embryoid bodies form vascular channels: an in vitro model of blood vessel development. *Development* **114,** 303–316.

21. Thomas, K. R. and Capecchi, M. R. (1987) Site-directed mutagenesis by gene targeting in mouse embryo-derived stem cells. *Cell* **51,** 503–512.

22. Remy, I. and Michnick, S. W. (1999) Clonal selection and in vivo quantitation of protein interactions with protein fragment complementation assays. *Proc. Natl. Acad. Sci. USA* **96,** 5394–5399.

23. Remy, I., Wilson, I. A., and Michnick, S. W. (1999) Erythropoietin receptor activation by a ligand-induced conformation change. *Science* **283,** 990–993.

24. Israel, D. I. and Kaufman, R. J. (1993) Dexamethasone negatively regulates the activity of a chimeric dihydrofolate reductase/glucocorticoid receptor protein. *Proc. Natl. Acad. Sci. USA* **90,** 4290–4294.

25. Kaufman, R. J., Bertino, J. R., and Schimke, R. T. (1978) Quantitation of dihydrofolate reductase in individual parental and methotrexate-resistant murine cells. Use of a fluorescence activated cell sorter. *J. Biol. Chem.* **253,** 5852–5860.

26. Kaufman, R. J. (1990) Selection and coamplification of heterologous genes in mammalian cells. *Methods Enzymol.* **185,** 537–566.

27. O'Callaghan, C. and Morris, A. (1972) Inhibition of beta-lactamases by beta-lactam antibiotics. *Antimicrob. Agents Chemother.* **2,** 442–448.

28. Zlokarnik, G., Negulescu, P. A., Knapp, T. E., Mere, L., Burres, N., Feng, L., et al. (1998) Quantitation of transcription and clonal selection of single living cells with beta-lactamase as reporter. *Science* **279,** 84–88.

34

Direct Selection of cDNAs by Phage Display

Reto Crameri, Gernot Achatz, Michael Weichel, and Claudio Rhyner

1. Introduction

Dozens of genomes will be partly or completely sequenced over the next few years. In the post genomic area, we are now facing the challenge of functionally characterizing thousands of genes generated by the genome projects. Selective enrichment of clones encoding a desired gene product and rapid handling of large numbers of individual clones resulting from screening of molecular libraries bear the potential to facilitate progress in this field. Phage surface display technology, first described by Smith *(1)*, enables the construction of large combinatorial peptide and antibody libraries. The basic concept of phage display links the phenotype, expressed as fusion together with a phage surface coat protein, to its genetic information integrated into the phage genome. This procedure allows to survey large libraries for the presence of specific clones using the discriminative power of affinity selection *(2–4)*. The selection of cognate phage (biopanning) is achieved by interaction between a solid phase-coated ligand and the phage library applied in fluid phase during multiple rounds of phage growth and selection. The field of phage display technology has developed rapidly, and antibodies, enzymes, enzyme inhibitors, hormones, DNA binding molecules, and allergens have been successfully selected from molecular libraries *(5–11)*. These examples clearly demonstrate the general applicability of linking genotype and phenotype to phage coat proteins. Molecular libraries allow the rapid identification of peptide–ligand interactions. In combination with high-throughput screening technology, they will play an important role in a rational approach for the identification of gene products *(12,13)*. However, one of the limitations of filamentous phages as display vectors for cDNA libraries is a direct consequence of the capsid structure. The integrity of the carboxy terminus of both pIII and pVIII, the phage coat proteins mostly used in phage display *(14)*, is essential for phage assembly and hampers insertion of foreign peptides. Therefore, insertion of heterologous genes into pIII or pVIII to generate fusion proteins can only be tolerated at the amino terminus. Translational stop codons present at the 3′ end of nontranslated regions of eukaryotic mRNA prevent, however, construction of fusion proteins N terminal to the pIII and pVIII phage coat proteins *(15,16)*. In contrast, the minor coat protein pVI has been shown to tolerate insertions at its C terminus and has been proposed as cDNA display system to overcome this limitation *(17)*.

From: *Methods in Molecular Biology, vol. 185: Embryonic Stem Cells: Methods and Protocols*
Edited by: K. Turksen © Humana Press Inc., Totowa, NJ

However, the filamentous phage vector most used for construction and screening of cDNA libraries is based on an indirect fusion strategy, in which cDNA inserts fused to the 3′ end of the Fos leucine gene are co-expressed with a gene encoding a Jun leucine zipper fused N terminal to pIII *(13,15,16,18)*. The high affinity between the Jun and Fos leucine zippers *(19)* efficiently links the cDNA gene products outside of the phage capsid. Thus, the fusion products become solvent exposed, and phage displaying any given gene product can be affinity purified from nonbinding phage by interactive panning cycles against immobilized ligands *(16,18)*. Although most of the cDNA libraries constructed and screened using the pJuFo vector were devoted to the isolation of IgE binding molecules from complex allergenic sources *(13)*, many other applications have been reported *(20–23)*, demonstrating the versatile applicability of the cloning system. In particular, it has been demonstrated that pJuFo cloning strategy can be used to select ligands for receptors expressed in differentiated U937 cells *(23,24)*. Therefore, the technology might be potentially applicable to select gene products able to interact with whole cells during different time points of differentiation. In this chapter, we describe a procedure for the direct generation of cDNA phage surface display libraries starting from isolated mRNA.

2. Materials

1. pJuFo II Vector: free, available from nonprofit research organizations.
2. cDNA synthesis kit for construction of directional libraries (Stratagene, La Jolla, CA, cat. no. 200401).
3. DNA restriction and modifying enzymes (Roche Diagnostics, Mannheim, Germany): *EcoRI* (cat. no. 703737), *XhoI* (cat. no. 899194), calf intestinal phosphatase (CIP) (cat. no. 713023), T4 DNA-ligase (cat. no. 481220).
4. *Escherichia coli* XL1-Blue strain (cat. no. 200301), VCSM13 helper phage (cat. no. 200251), and premade λ-ZAP libraries (Stratagene).
5. Media:
   a. Super broth (SB): 30 g tryptone, 20 g yeast extract, 10 g MOPS per liter, pH adjusted to 7.0 with HCl. Autoclave at 121°C.
   b. Luria-Bertani medium (LB): 10 g tryptone, 5 g yeast extract, 10 g NaCl, pH adjusted to 7.5 with NaOH. Autoclave at 121°C.
   c. LB agar: make up liquid LB medium and, just before autoclaving, add 15 g agar per liter.
6. RNA and plasmid purification kits Qiagen (Hilden, Germany): Oligotex direct mRNA kit (cat. no. 72041), Qiagen plasmid mini kit (cat. no. 12123), Qiagen plasmid maxi kit (cat. no. 12162).
7. Solutions:
   a. Tris-buffer (1 *M*): dissolve 121.1 g Tris (Fluka, Buochs, CH, cat. no. 93362) in 800 mL of H_2O. Adjust the pH to 7.4 by adding concentrated HCl (Fluka, cat. no. 84411), add H_2O to 1 L, and autoclave at 121°C.
   b. EDTA (0.5 *M*): add 185.1 g of disodium ethylene diamine tetraacetate (Fluka, cat. no. 03609) to 800 mL H_2O. Stir vigorously on a magnetic stirrer. Adjust the pH to 8.0 with NaOH pellets (Fluka, cat. no. 71687), add H_2O to 1 L, and autoclave at 121°C.
   c. Tris-EDTA (TE): mix 10 m*M* Tris (pH 7.4) and 1 m*M* EDTA (pH 8.0).
   d. Tris-acetate EDTA (TAE): prepare a 50X concentrated stock solution composed of 242 g Tris, 57.1 mL glacial acetic acid (Fluka, cat. no. 45732), and 100 mL 0.5 *M* EDTA (pH 8.0). Add H_2O to 1 L.

e. Phosphate-buffered saline (PBS): 50 mM sodium phosphate, pH 7.4, (Fluka, cat. no. 04278), 150 mM NaCl. Autoclave at 121°C.

f. Tris-buffered saline (TBS): 50 mM Tris, pH 7.5, 150 mM NaCl, sterilze by filtration.

g. NaOAc (3 M): Dissolve 408.1 g of sodium acetate (Fluka, cat. no. 71183) in 800 mL of H$_2$O. Adjust pH to 5.2 with glacial acetic acid, the vol to 1 L, and autoclave at 121°C.

h. Blocking buffer: 2% bovine serum albumin (BSA) (Fluka, cat. no. 05491) in PBS, pH 7.4.

i. Phage elution buffer: 0.1 M HCl, pH 2.2, adjusted with solid glycine (Fluka, cat. no. 50049), 1% BSA.

8. Electroporation apparatus, cuvettes, and electroporation protocols (Bio-Rad, Hercules, CA).

9. Polyethylene glycol (PEG)-NaCl solution: to 200 g PEG 6000 (Fluka, cat. no. 81260) and 146.1 g NaCl, add ddH$_2$O to 1 L, autoclave at 121°C.

10. For bacterial plasmid minipreps, we use Miniprep Express Matrix (BIO 101, Vista, CA, cat. no. 2000-200).

11. Ampicillin (Amp) (Fluka, cat. no. 10047) stock solution 100 mg/mL in H$_2$O. Sterilize by filtration and store in aliquots at –20°C.

12. Tetracyclin (Tc) (Fluka, cat. no. 87130) stock solution 12.5 mg/mL in ethanol-H$_2$O (50% v/v). Store at –20°C in the dark.

13. Kanamycin (Kan) (Fluka, cat. no. 60615) stock solution 25 mg/mL in H$_2$O. Sterilize by filtration and store in aliquots at –20°C.

3. Methods

3.1. Vector Preparation

1. Spread *E. coli* XL1-Blue cells harboring pJuFo II on LB agar plates containing 100 µg/mL Amp and 12.5 µg/mL Tc (LB^{Amp,Tc}) to obtain single colonies.

2. Pick a single colony with a sterile loop, transfer to a sterile tube containing 5 mL LB^{Amp,Tc} medium and grow overnight.

3. Inoculate 500 mL LB^{Amp,Tc} medium with the overnight culture from **step 2** and grow at 37°C, 220 rpm, until culture is stationary.

4. Collect bacteria by centrifugation (10 min at 4000g at 4°C).

5. Lyse cells and isolate plasmid DNA according to the Qiagen MAXIprep protocols. Determine DNA concentration and purity by gel electrophoresis and store at –20°C until use.

6. Set up restriction digests as follows:

Purified pJuFo II DNA	10 µg
10X Restriction buffer	10 µL
*Eco*RI	30 U
*Xho*I	30 U
ddH$_2$O to	100 µL
CIP	0.5 U

Digest at 37°C for 3 h. Add another 3 U of CIP and incubate for 10 min at 37°C.

7. Run restricted DNA on 1.0% agarose gel (2 µg DNA/lane at 80 V), excise linearized pJuFo II vector (approx 4.3 kb), and purify the band using a commercial silica based kit (*see* **Note 1**) or by electroelution *(25)*.

8. Ethanol precipitate the eluted DNA by adding 1/10 vol 3 M NaAc, pH 5.2, and 2.5 vol EtOH, store at least 2 h at –20°C.

9. Collect DNA by centrifugation in a microcentrifuge (30 min, >10,000g at 4°C), air-dry pellet, and resuspend linearized DNA in 40 μL TE buffer, store at –20°C until use.
10. To test the vector self-ligation, set up a ligation with the linearized vector (0.1 μL vector, 1 μL T4 DNA ligase, 1 μL 10X ligation buffer, ddH$_2$O to a total volume of 10 μL and incubate 4–14 h at room temperature). Pulse 100 μL competent *E. coli* XL1-Blue cells *(26)* (*see* **Note 2**) at 2.5 kV, 25 μF, and 200 Ohm. Flush cuvette immediately after electroporation with 1 mL prewarmed (37°C) SB medium and incubate 1 h in a microcentrifuge tube at 37°C with shaking (200 rpm). Spread 10 and 100 μL of the transformants on LB^{Amp,Tc} plates, incubate at 37°C overnight, and determine the number of transformants (*see* **Note 3**).

3.2. Preparation of cDNA Inserts

Alternatively to the following protocol, premade commercially available λ-ZAP libraries (Stratagene) can be used to prepare *Eco*RI/*Xho*I-restricted cDNA inserts *(16)*.

1. Isolate poly(A)$^+$ mRNA from a source of choice using a standard procedure. Reliable results are obtained with the Qiagen Oligotex direct mRNA kit (*see* **Note 4**).
2. Generate cDNA using the Stratagene cDNA synthesis kit.
3. Digest the cDNA with *Eco*RI and *Xho*I and run the restriction mixture on a 1.0% agarose gel. Cut out inserts exceeding 500 bp in length (*see* **Note 5**).
4. Purify size-selected cDNAs according to **Subheading 3.1.**, **steps 7–9**.

3.3. Generation of cDNA Phage Surface Display Libraries

1. Set up test ligations to determine the best vector/cDNA insert ratio:

Linearized pJuFo II DNA	250 ng
10× ligase buffer	1 μL
T4 DNA ligase (>1 Weiss U)	1 μL
cDNA inserts	10, 50, and 100 ng
Water to an end volume of	20 μL

   Ligate for at least 4 h at room temperature or overnight.

2. Electroporate ligation mixture into competent *E. coli* XL1-Blue cells (pulse at 2.5 kV, 0.2-cm gap cuvettes, 25 μF, and 200 Ohms) flush the electroporation cuvette immediately with 1 mL prewarmed SB medium, and plate on LB^{Amp,Tc} plates to determine the optimal vector:insert ratio.
3. For construction of a library, scale up ligation using the optimal vector:insert ratio determined in **step 2** to obtain a library size of >10^7 colony forming units (cfus).
4. Heat-inactivate the ligation mixture (10 min at 60°C). Precipitate DNA by adding 1/10 vol of 3 *M* NaAc (pH 5.2), 1/200 vol of glycogen (20 mg/mL), and 2.5 vol. of EtOH (>2 h at –20°C). Collect DNA by centrifugation (30 min at >10,000g), air-dry the pellets, and resuspend in 20 mL ddH$_2$O.
5. Electroporate *E. coli* cells as in **step 2**. Flush the electroporation cuvette immediately with 1 mL prewarmed SB medium, then with additional 2 mL SB medium, transfer to a 50-mL conical tube, and incubate immediately for exactly 60 min at 37°C with shaking at 200 rpm (*see* **Note 6**).
6. Add SB medium (Amp 20 μg/mL) prewarmed at 37°C to a vol of 10 mL and immediately titer the transformants by plating 10, 1, 0.1, and 0.01 μL on LB^{Amp,Tc} plates. Incubate plates overnight at 37°C and calculate the primary size of the library from the number of transformants obtained from the dilutions plated.
7. Incubate the 10-mL culture for 60 min at 37°C at 250 rpm.

8. Add Amp to a final concentration of 50 µg/mL and incubate for an additional 1 h at 37°C at 250 rpm.
9. Add helper phage VCSM13 (10^{12} plaque forming units [pfu]) and adsorb 5 min at 37°C (*see* **Note 7**).
10. Adjust the culture volume to 100 mL with SB medium (Amp 100 µg/mL, Tc 12.5 µg/mL) prewarmed at 37°C and incubate on a shaker (250 rpm at 37°C for 2 h).
11. Add Kan to a final concentration of 70 µg/mL and incubate overnight (250 rpm) at 30° or 37°C (*see* **Note 7**).
12. Spin down cells (5000*g* at 4°C for 15 min). Place the supernatant in a clean bottle and add 1/5 vol cold PEG-NaCl solution.
13. Place the bottle in an ice-water bath for 60 min to precipitate phage.
14. Spin down (8000*g* at 4°C for 20 min) and discard supernatant. Allow the bottle to drain on a paper towel (5 min) to remove as much PEG as possible.
15. Resuspend pellet in 2 mL TBS containing 1% BSA, clean by microcentrifugation (2 min at >10,000*g*), transfer the supernatant to a new tube, and store at 4°C; for long-term storage, store at –20°C.
16. Titer the phage suspension by infecting 200-µL aliquots of freshly grown *E. coli* XL1-Blue cells ($A_{600 nm} = 0.5$) with 1 µL of 10^{-8}, 10^{-9}, 10^{-10}, and 10^{-11} phage dilutions in LB medium for 15 min at room temperature and plate on LB^{Amp,Tc}. Incubate plates overnight at 37°C to visualize transfectants.

3.4. Testing the Complexity of the Library

1. To test the successful generation of a library, pick 20 colonies from the titering plates, grow in 3 mL LB$^{Amp, Tc}$ medium, and make plasmid minipreps using the Miniprep Express Matrix kit.
2. Digest minipreps with *Eco*RI/*Xho*I and visualize the size distribution of the inserts on agarose gels. Inserts should be of variable size (*see* **Note 8**).
3. Alternatively, presence and distribution of inserts can be tested by polymerase chain reaction (PCR), starting directly from colonies picked from the titering plates (*see* **Note 9**).

3.5. Enrichment of Phage with Affinity for a Ligand

This section deals with antigens or antibodies used as ligands to enrich phage by affinity interaction. For the use of other ligands *see* **Note 10**.

1. Coat 2 to 3 wells of a 96-well enzyme-linked immunosorbent assay (ELISA) plate with 150 µL antigen or antibody solution (10 µg/mL). Choose a coating buffer suitable for your protein. Use PBS, pH 8.0, if working with monoclonal antibodies. Coating may be performed at 37°C for 1 h or at 4°C overnight.
2. Wash twice with PBS, pH 7.4. Block by filling the wells completely with blocking buffer, seal the plate, and incubate for 1 h at 37°C, then wash twice with PBS.
3. Add 10^{11} cfus of phage from the cDNA library (generally 50 µL) to each well, seal the plate, and incubate for 2 h at 37°C.
4. Remove phage, wash each well once with water. Wash 10 times with TBS containing 0.5% Tween-20, 10 times with TBS, and twice with distilled water at room temperature. Pipet up and down several times to remove nonspecified phage, fill completely with washing buffer (*see* **Note 11**). Elute phage with 150 µL phage elution buffer. Incubate for 10 min at room temperature.
5. Remove the eluate by pipetting up and down vigorously and transfer immediately to a tube containing 18 µL 1 *M* Tris (pH 8.0) per 150 µL eluate to neutralize.

6. For further rounds of panning, use eluted phage to infect 4 mL of freshly grown *E. coli* XL1-Blue cells ($A_{600\ nm}$ = 0.5) for 15 min at 37°C.

7. Add prewarmed (37°C) SB medium (Amp 20 µg/mL, Tc 12.5 µg/mL) to 10 mL. Immediately plate 20, 10, and 1 µL on LB agar to determine the number of packaged phagemids. Incubate plates overnight at 37°C and count colonies.

8. Incubate the culture for 1 h at 37°C, 250 rpm, and repeat **Subheading 3.3., steps 7–16**.

9. Perform successive rounds of affinity enrichment, repeating **steps 1–6**. If phage that are able to interact with the ligand are present in the library, the number of phage eluted from the wells (**step 5**) will increase with the rounds of panning. Determine the percentage yield of phage (% yield = [Number of phage eluted/Number of phage applied] × 100) for each round of panning. If no enrichment is seen in the following 5 rounds of panning, the desired clone is not likely to be isolated from the library, and the search should be concluded (*see* **Note 12**).

10. Enrichment of specific cDNAs can also be detected by PCR amplification directly from phage samples obtained after each round of panning (*see* **Note 13**).

11. If phage enrichment is successful proceed with subcloning of the inserts into a suitable vector to produce recombinant protein or with sequence determination to identify the encoded products (*see* **Note 14**).

4. Notes

1. In our hands, Geneclean and Nucleotrap glassmilk kits from Bio 101 and Machery-Nagel (Dürren, Germany) work efficiently for the isolation of cDNA inserts from agarose gels.

2. We normally produce elecrocompetent cells in our laboratory. However, the transformation efficiency for construction of cDNA libraries should be >1 × 10^8 cfus/µg of transformed plasmid DNA and needs to be tested for each new batch of cells. Commercially available competent cells are warranted in terms of transformation efficiency.

3. A low vector background is extremely important for the construction of cDNA libraries displayed on phage surface, because empty phagemids generated during superinfection with helper phage can overgrow specific phage during successive rounds of panning. The background of the vector can be reduced by treatment of the restricted DNA with alkaline phosphatase. Dephosphorylation of the vector may reduce the efficiency of ligation but is, however, strongly recommended. More than two to three colonies/plate for 100 µL transformation mixture cannot be tolerated.

4. Any cDNA library depends primarily on the quality of the mRNA used for cDNA synthesis. Make sure that methods or kits used for mRNA isolation work properly. Directional *Eco*RI/*Xho*I cloning of cDNAs generated by the ZAP cDNA synthesis kit into the modified pJuFo II vector takes advantage of the methylation of internal restriction sites in the cDNAs, enhancing the quality of the libraries.

5. Separation of small cDNAs from larger cDNA inserts on agarose gels is always incomplete. Therefore, inserts of >500 bp will be present in the library in spite of the cutoff.

6. The electroporation step is crucial for the quality of the library and should be done very carefully. The ligation mixture can be transformed in several portions to reduce the risk related to electroporation. Sometimes electroporation is not successful due to accidental discharging of the condensor, which results in the well known electrotransformation spark and the complete killing of the cells.

7. We use the Kan-resistant helper phage VCSM13 for superinfection. However, helper phage R408 works as well as VCSM13. If R408 helper phage is used, addition of Kan should be omitted. To generate a good phage surface displayed library, *E. coli* cells should be superinfected with helper phage at a multiplicity of infection of 10.

8. The diversity of a library can be defined as the percentage of different clones contained in the library. Therefore, a high percentage of inserts differing in size, together with the absence of empty clones, warrant a good quality of the library. Since the expected diversity is in the range of 10^6 to 10^9 independent clones, this experiment gives only a glance of the diversity of the library. Screening of a statistically significant number of colonies for an exact determination of the diversity is limited by practical reasons.

9. PCR amplification primers and conditions. To directly scan single clones and/or the whole phage repertoire the following PCR primer set allow visualization of the insert distribution and direct subcloning into the high level expression vector pQE30 (Qiagen, Hilden). The two primers are flanking the cDNA construct, priming at the 5′ end on the Fos gene and at the 3′ end on the vector. The 3′ primer contains a *Hind*III restriction side for direct subcloning of the PCR product into pQE30 as *Bgl*II/*Hind*III fragment.

    5′ JuFo: 5′-AAAGAAAAGCTGGAGTTCATCCTGGC-3′
    3′ JuFo: 5′-GGCCAGTGAATTGTAATACGACCCCAAGCTTGGG-3′

    PCRs are carried out at 94°C for 60 s, 55°C for 60 s, and 72°C for 60 s, for 30–35 cycles in the PCR amplification mixture containing 2.5 mM Mg^{2+}. As target for PCR amplifications, it is possible to directly transfer bacteria from single colonies to tubes containing the PCR mixture. However, to obtain reproducible results, we recommend to use 1 µL of 1/400 diluted plasmid minipreps as PCR targets.

10. Many different ligands have been successfully used to screen cDNA libraries displayed on phage surface *(13)*. In addition to mass law considerations *(27)*, two important points have to be taken into account: *(1)* the cDNA molecules obtained after reverse transcription are thought to contain the information for all those proteins that were expressed in the source at the time of mRNA isolation. It is, therefore, crucial to ensure that the desired gene product is actually expressed in the source chosen for mRNA extraction; and *(2)* we draw attention to the fact that many ligands, when immobilized directly to solid phase, might become altered loosing their biological functionality. A ligand used to screen phage libraries should be tagged or immobilized to solid phase in such a way that allows retention of its ability to interact with the wanted target.

11. To avoid contamination of pipets and subsequent cross-contaminations, we always use aerosol-resistant tips for all steps in phage handling, especially for the washing procedure. For an excellent theoretical review about the factors influencing the probability of selecting phage able to interact with the ligand, including washing steps, *see* **ref. 27**.

12. Enrichment depends on many factors inclucing the quality of the library, the affinity of the ligands for the target, and the number of specific ligands available for a specific target. Therefore, enrichment of libraries with biologically active homogeneous ligands are likely to be successful if the interacting gene product is present in the library. When using complex ligands like human serum, success will depends on the percentage of the specific ligand (e.g., antibody) present in the serum used. It is almost impossible to select specific phagemids from background if binding of a wanted gene product to its specific ligand is limiting compared to the unspecific binding *(27)*.

13. Phage PCR is carried out with the same primer set and conditions as described in **Note 6**, but with a diluted phage solution (approx 10^{11} cfu) as target. To visualize the enrichment, 10 µL of the PCR products are loaded on a 1.2% agarose gel. The size distribution of the PCR mixture is condensing from a smear in early rounds to more intense discrete bands in later rounds of successful enrichment. Prominent PCR bands are likely to represent enriched genes and can be directly excised from the gel for subcloning, sequencing, and protein production.

14. Low amounts of protein derived from inserts of putative clones can be produced as secreted protein by isopropyl-β-D-thiogalactopyranoside (IPTG) (2 mM) induction of the *Lac*

promoter controlling Fos-expression in absence of helper phage. Alternatively, phage derived from single clones can be directly used in phage ELISA experiments or Western blot analyses as described previously *(15)*. If large amounts of protein are required, subcloning in high-level expression vectors is required.

Acknowledgments

We are grateful to Prof. Dr. K. Blaser and Prof. Dr. M. Breitenbach for their continuous encouragement, helpful discussions, and critical comments of the manuscript. This work was supported by the Swiss National Science Foundation (grant no. 31.63381.00).

References

1. Smith, G. P. (1985) Filamentous fusion phage: novel expression vectors that display cloned antigens on the surface of the virion. *Science* **228,** 1315–1317.
2. Scott, J. K. and Smith, G. P. (1990) Searching for peptide ligands with an epitope library. *Science* **249,** 386–390.
3. Smith, G. P. (1991) Surface presentation of protein epitopes using bacteriophage expression systems. *Curr. Opin. Biotechnol.* **2,** 668–673.
4. Barbas III, C. F. and Lerner, R. A. (1991) Combinatorial immunoglobulin libraries on the surface of phage (Phabs): rapid selection of antigen-specific Fabs. *Compan. Meth. Enzymol.* **2,** 119–124.
5. Hoogenboom, H. R. and Winter, G. (1992) By-passing immunisation. Human antibodies from synthetic repertoires of germline VH-gene segments rearranged *in vitro. J. Mol. Biol.* **227,** 381–388.
6. Barbas III, C. F., Bain, J. D., Hoekstra, D. M., and Lerner, R. A. (1992) Semisynthetic combinatorial antibody libraries: a chemical solution to the diversity problem. *Proc. Natl. Acad. Sci. USA* **89,** 4457–4461.
7. Janda, K. D., Lo, C. H. L., Li, T., Barbas III, C. F., Wirshing, P., and Lerner, R. A. (1994) Direct selection for a catalytic mechanism from combinatorial antibody libraries. *Proc. Natl. Acad. Sci. USA* **91,** 2532–2536.
8. Devlin, J. J., Panganiban, L. C., and Devlin, P. E. (1990) Random peptide libraries: a source of specific protein binding molecules. *Science* **249,** 404–406.
9. Bass, S., Greene, R., and Wells, J. A. (1990) Hormone phage: an enrichment method for variant proteins with altered binding properties. *Proteins* **8,** 309–314.
10. Rebar, E. J. and Pabo, C. O. (1994) Zinc finger phage: affinity selection of fingers with new DNA-binding specificities. *Science* **263,** 671–673.
11. Crameri, R. (1998) Recombinant *Aspergillus fumigatus* allergens: from the nucleotide sequences to clinical applications. *Int. Arch. Allergy Immunol.* **115,** 99–114.
12. Borrebaeck C. A. K. (1998) Tapping the potential of molecular libraries in functional genomics. *Immunol. Today* **19,** 524–527.
13. Crameri, R. and Walter, G. (1999) Selective enrichment and high-throughput screening of phage surface-displayed cDNA libraries from complex allergenic systems. *Combin. Chem. High Throughput Screen* **2,** 63–72.
14. Kay, B. K., Winter, J., and McCafferty, J. (eds.), (1996) *Phage display of peptides and proteins. A laboratory manual.* Academic Press, San Diego.
15. Crameri, R. and Suter, M. (1993) Display of biologically active proteins on the surface of filamentous phages: a cDNA cloning system for selection of functional gene products linked to the genetic information responsible for their production. *Gene* **137,** 69–75.

16. Crameri, R. (1997) pJuFo: a phage surface display system for cloning genes based on protein-ligand interaction, in *Gene cloning and analysis. Current innovations* (Schaefer, B. C., et. al., eds.), Horizon Scientific Press, Whymondham, UK, pp. 29–42.

17. Jaspers, L., Messens, J., de Keyser, A., Eeckhout, D., van den Brande, I., Gansemans, Y, et al. (1995) Surface expression and ligand-based selection of cDNAs fused to filamentous phage gene VI. *BioTechnology* **13**, 378–382.

18. Crameri, R., Jaussi, R., Menz, G., and Blaser, K. (1994) Display of expression products of cDNA libraries on phage surfaces. A versatile screening system for selective isolation of genes by specific gene-product/ligand interaction. *Eur. J. Biochem.* **222**, 53–58.

19. Pernelle, C., Clerc, F. F., Dureuil, C., Bracco, L., and Torque, B. (1993) An efficient screening assay for the rapid and precise determination of affinities between leucine zipper domains. *Biochemistry* **32**, 11,682–11,687.

20. Hottinger, M., Gramatikoff, K., Georgiev, O., Chaponnier, C., Schaffner, W., and Hübscher, U. (1995) The large subunit of HIV-1 reverse transcriptase interacts with β-actin. *Nucleic Acids Res.* **23**, 736–741.

21. Palzkill, T., Huang, W., and Weinstock, G. M. (1998) Mapping protein-ligand interactions using whole genome phage display libraries. *Gene* **221**, 79–83.

22. Grob, P., Baumann, S., Ackermann, M., and Suter, M. (1998) A system for stable indirect immobilisation of multimeric recombinant proteins. *Immunotechnology* **4**, 155–163.

23. Kola, A., Baensch, M., Bautsch, W., Klos, A., and Köhl, J. (1999) Analysis of the C5a anaphylatoxin core domain using a C5a phage library selected on differentiated U937 cells. *Mol. Immunol.* **36**, 145–152.

24. Hennecke, M., Kola, A., Baensch, M., Wrede, A., Klos, A., Bautsch, W., and Köhl, J. (1997) A selection system to study C5a-C5a-receptor interactions: phage display of a novel C5a anaphylatoxin, Fos-C5a[ala27]. *Gene* **184**, 263–272.

25. Sambrook. J., Fritsch, E. F., and Maniatis, T. (eds.), (1989) *Molecular cloning: a laboratory manual.* CSH Laboratory Press, Cold Spring Harbor, New York.

26. Aushubel, F. (ed.), (1987) *Current protocols in molecular biology.* Wiley & Sons, New York.

27. Levitan, B. (1998) Stochastic modeling and optimization of phage display. *J. Mol. Biol.* **277**, 893–916.

35

Screening for Protein–Protein Interactions in the Yeast Two-Hybrid System

R. Daniel Gietz and Robin A. Woods

1. Introduction

The two-hybrid system (THS) (*1*) is a molecular genetic screen that detects protein–protein interactions. The protein specified by the yeast *GAL4* gene activates the transcription of genes involved in galactose metabolism. It has two functional domains, a DNA binding domain, $Gal4_{BD}$, and a transcriptional activating domain, $Gal4_{AD}$, which interact with DNA sequences in the promoter regions of *GAL1*, *GAL2*, and *GAL7* to stimulate transcription. The screen involves two plasmids; one carries the $GAL4_{BD}$ sequence fused, in-frame, to a sequence coding for a "bait" protein, and the other carries $GAL4_{AD}$ sequences, fused to "prey" sequences from a cDNA library. The two plasmids are introduced, typically by transformation, into a yeast strain carrying a reporter gene coupled to a *GAL1*, *GAL2*, or *GAL7* promoter. If the proteins encoded by the bait and prey sequences interact to allow correct positioning of the $Gal4_{AD}$ and $GaL4_{BD}$ moieties, the reporter gene is activated. Transformants are plated on medium that allows the detection of reporter gene activation. The plasmid carrying the $GAL4_{AD}:cDNA$ plasmid can be recovered, and the positive cDNA isolated and characterized.

Numerous modifications of the THS system have been developed, many of them employing positive selection of reporter gene activation. In our laboratory, we use yeast strains with several reporter genes (*see* **Table 1**). Primary selection is for one or more nutritional markers fused to a galactose gene promoter, e.g., *GAL1-HIS3* or *GAL2-ADE2* or both, and positive transformants are then tested for the activation of *GAL1-lacZ* or *GAL7-lacZ*. Specific plasmid–yeast stain combinations allow the detection of protein–protein interactions that require phosphorylation (*2,3*). Various applications of the THS have been recently reviewed by Vidal and Legrain (*4*), Fashena et al. (*5*), and Pandey and Mann (*6*) and Pirson et al. (*7*).

Large numbers of transformants are required to screen a mammalian cDNA library by the yeast THS, and this necessitates efficient transformation protocols. We have developed the LiAc/SS-DNA/PEG transformation protocol (*8*) to generate the numbers of transformants required for such screens. A recent study reports the use of the yeast mating system to combine the bait and prey plasmids in a single diploid cell (*9*).

From: *Methods in Molecular Biology, vol. 185: Embryonic Stem Cells: Methods and Protocols*
Edited by: K. Turksen © Humana Press Inc., Totowa, NJ

Table 1
Two-Hybrid Yeast Strains

Yeast Strain	Genotype	Reporter Genes	Plasmid Selection	Reference
AH109	*MATa, trp1-901, leu2-3, 112, ura3-52, his3-200, gal4Δ, gal80Δ, URA3 : : MEL1* $_{UAS}$ *-MEL1* $_{TATA}$ *–lacZ, LYS2 : : GAL1* $_{UAS}$ *-GAL1* $_{TATA}$ *-HIS3, GAL2* $_{UAS}$ *-GAL2* $_{TATA}$ *-ADE2,*	*MEL1-lacZ* *GAL1-HIS3* *GAL2-ADE2* *MEL1-MEL1*	*TRP1* *LEU2*	ClonTech Laboratories
Y190	*MATa, ade2-101, gal4Δ, gal80Δ, his3Δ-200, leu2-3,112 trp1Δ-901, ura3-52, URA3 : : GAL1-lacZ, lys2 : : GAL1-HIS3, cyh^r*	*GAL1-lacZ* *GAL1-HIS3*	*TRP1* *LEU2* *LYS2*	*(10)*
PJ69-4A	*MATa, ade2Δ, gal4Δ, gal80Δ, his3Δ-200, leu2-3,112 trp1Δ-901, ura3-52, met1 : : GAL7-lacZ, ade2 : : GAL2-ADE2 lys2 : : LEU2 GAL1-HIS3,*	*GAL7-lacZ* *GAL2-ADE2* *GAL1-HIS3*	*TRP1* *URA3* *LYS2*	*(11)*
KGY37	*MATa, ade2-101, gal4Δ, gal80Δ, his3Δ-200, leu2Δ-inv pUC18, trp1Δ-901, ura3Δ-inv : : GAL1-lacZ,* lys2Δ-inv : : *GAL1-HIS3,*	*GAL1-lacZ* *GAL1-HIS3*	*TRP1* *LEU2* *URA3* *LYS2*	*(12)*

Transformations were carried out in microtiter plates—a protocol for this procedure can be found in Gietz and Woods *(8)*.

A THS screen involves the following steps:

1. Preparation of the *GAL4$_{BD}$* bait plasmid carrying the gene of interest.
2. Transformation of the bait plasmid into a reporter yeast strain and checking for auto-activation of the reporter gene(s).
3. Preparation and amplification of the cDNA library in the *GAL4$_{AD}$* prey plasmid.
4. Transformation of the reporter yeast strain carrying the bait plasmid with the prey plasmid and screening for positive transformants.
5. Identification of true positive interactions and isolation of the appropriate prey plasmids.

2. Materials

1. β-Mercaptoethanol (β-ME), (Sigma, St. Louis, MO, cat. no. M6250)
2. Glass beads (425–600 μm), (Sigma, cat. no. G-8722).
3. Lithium acetate dihydrate, (Sigma, cat. no. L-6883).
4. Micotiter plates (96 well), (Fisher Scientific [http://www.fisher1.com], cat. nos. 07-200-104 and 07-200-376).
5. Microtiter plate replicator (96 well), (Fisher Scientific, cat. no. 05-450-9).
6. N,N-dimethyl formamide, (Sigma, cat. no. D8654.)
7. ONPG (o-nitrophenyl-β-D-galactopyranoside), (Sigma, cat. no. M1127).

Table 2
Two-Hybrid Plasmid Vectors

Plasmid	Binding Sequence[a]	Restriction Sites	Selected Marker	Hemagluttinin Tag	References
pAS1	$GAL4_{BD}$	*Sal*1, *Bam*H1, *Sma*1, *Nco*1, *Sfi*1, *Nde*1, *Eco*R1	*TRP1*	Yes	*(10)*
pAS2	$GAL4_{BD}$	*Nde*1, *Nco*1, *Sfi*1, *Sma*1, *Bam*H1, *Sal*1	*TRP1*	Yes	*(13)*
pGBT9	$GAL4_{BD}$	*Eco*R1, *Sma*1, *Bam*H1, *Sal*1, *Pst*1	*TRP*	No	*(14)*
pACT1	$GAL4_{AD}$	*Bgl*II, *Eco*R1, *Bam*H1, *Xho*1, *Bgl*II	*LEU2*	Yes	*(10)*
pACT2	$GAL4_{AD}$	*Nde*1, *Nco*1, *Sfi*1, *Sma*1, *Bam*H1, *Sac*1, *Xho*1, *Bgl*II	*LEU2*	Yes	*(13)*
pGAD10	$GAL4_{AD}$	*Bgl*II, *Xho*1, *Bam*H1, *Eco*R1, *Bgl*II	*LEU2*	No	*(14)*
pGAD424	$GAL4_{AD}$	*Eco*R1, *Sma*1, *Bam*H1, *Sal*1, *Pst*1, *Bgl*II	*LEU2*	No	*(14)*

[a]The $GAL4_{BD}$ codes for amino acids 1–147 of Gal4, and the $GAL4_{AD}$ codes for amino acids 768–881 of Gal4.

8. Polyethylene glycol (PEG) 3350, (Sigma, cat. no. P-3640).
9. Salmon sperm DNA, (Sigma, cat. no. D-1626).
10. Tris-EDTA (TE) buffer: 10 mM Tris-HCl, 1 mM Na$_2$ EDTA, pH 8.0.
11. X-GAL (5-bromo-4-chloro-3-indolyl-β-D-galactopyranoside), (Sigma, cat. no. B 9146).
12. Yeast cracking buffer: 10 mM Tris-HCl, pH 8.0, 100 mM NaCl, 1 mM EDTA, 2% (v/v) Triton X-100, 1% (w/v) sodium dodecyl sulfate (SDS).

2.1. Yeast Strains

The genotypes, reporter genes, and nutritional markers used for plasmid selection of five yeast strains commonly used for THS screens are listed in **Table 1**.

2.2. Two-Hybrid Plasmid Vectors

The properties of a number of plasmids used for THS screens are listed in **Table 2**. All of the $GAL4_{BD}$ plasmids are selected using the *TRP1* marker and all the $GAL4_{AD}$ plasmids using *LEU2*. Several plasmids of both types contain the hemaglutinnin (HA) tag, which allows for the immunological detection of the fusion protein with appropriate antibodies.

2.3. Bacterial Strains

Plasmids are routinely amplified in and purified from the *Escherichia coli* strain DH5α *(F-/ endA1 hsdR17 glnV44 thi-1 recA1 gyrA relA1 Δ[lacIZYA-argF]U169 deoR[φ80dlacΔ(lacZ)M15])*. Positive prey plasmids are recovered into *E. coli* strain KC8 *(hsdR leuB600 trpC9830 pyrF*::Tn5 hisB463 lacΔX74 strA galU galK) and then electroporated *(15)* into DH5α for plasmid DNA preparation. Procedures, media, and solutions for bacteriological techniques and plasmid manipulation can be found in Ausubel et al. *(16)* or Sambrooke et al. *(17)*.

2.4. Yeast Growth Media

2.4.1. Yeast Extract-Peptone-Adenine-Dextrose Medium

Yeast strains are routinely grown on or in yeast extract-peptone-adenine-dextrose (YPAD) medium; the adenine is added to decrease the selective advantage of *ade2* to *ADE2* reversions. Double-strength YPAD, 2X YPAD broth, reduces the doubling time of yeast strains and increases transformation efficiency.

Component	YPAD Agar	2X YPAD Broth
Difco Bacto Yeast Extract	6 g	12 g
Difco Bacto Peptone	12 g	24 g
Glucose	12 g	24 g
Adenine hemisulphate	60 mg	60 mg
Difco Bacto Agar	10 g	—
Distilled–deionized water	600 mL	600 mL

Place a 1.0-L medium bottle or Erlenmeyer flask containing 600 mL water and a magnetic stir bar on a stir plate, add the ingredients, except agar, and mix until dissolved. Add the agar for YPAD agar medium and autoclave for 20 min. After autoclaving, swirl the bottles of YPAD agar to ensure even distribution of the agar. Equilibrate the bottles of YPAD agar to 55°C in a water bath and use each bottle to pour twenty 100 × 15 mm standard Petri plates. Allow the poured plates to dry overnight, and then store them in plastic sleeves in a refrigerator or cold room. The 2X YPAD broth should also be stored in the cold. Yeast extract-peptone-dextrose (YEPD) agar and broth media can be purchased from Becton Dickinson Microbiology Systems, Cockeysville, MD 21030, USA (BBL YEPD agar and YEPD broth). Adenine hemisulfate should be added to these media to make YPAD agar and 2X YPAD broth.

2.4.2. Synthetic Complete Medium

The plasmids used in THS carry one or more of the following selectable markers: *URA3* (uracil requirement), *TRP1* (tryptophan requirement), *HIS3* (histidine requirement), *LEU2* (leucine requirement), and *ADE2* (adenine requirement). Synthetic complete (SC) selection medium is based on Difco Yeast Nitrogen Base (without amino acids) with the addition of a mixture of amino acids, purines, pyrimidines, and vitamins *(18)*. Selection for particular genetic markers is achieved by the omission of these specific components from the mixture (*see* **Subheading 2.4.3.**).

Ingredient	SC Selection Medium
Difco Yeast Nitrogen Base without amino acids	4.0 g
Amino acid mixture	1.2 g
Glucose	12.0 g
Difco Bacto agar (agar is omitted to make liquid SC selection medium)	10.0 g
Distilled–deionized H$_2$O	600.0 mL

Place a 1.0-L medium bottle or Erlenmeyer flask containing 600 mL water and a magnetic stir bar on a stir plate, add the ingredients and mix until dissolved. Adjust the pH to 5.6 with 1.0 N NaOH. Add the agar for SC agar medium and autoclave for

15 min. After autoclaving, swirl the bottles of agar medium to ensure even distribution of the agar. Equilibrate the bottles of agar medium to 55°C in a water bath and use each one to pour twenty 100×15 mm standard Petri plates. One bottle of SC selection medium will pour 7–9 of the 150×15 mm standard Petri plates used for the screening transformation. Since SC selection medium is light sensitive, the plates should be dried in the dark at room temperature overnight and then stored in sealed bags in the dark at 4°C.

2.4.3. Amino Acid Mixture

Add the following ingredients *(18)* to a polypropylene 250-mL bottle and mix by thorough shaking with several glass marbles.

Adenine SO4	**0.5 g**	Methionine	2.0 g
Arginine	2.0 g	Phenylalanine	2.0 g
Aspartic acid	2.0 g	Serine	2.0 g
Glutamic acid	2.0 g	Threonine	2.0 g
Histidine HCl	**2.0 g**	**Tryptophan**	**2.0 g**
Inositol	2.0 g	Tyrosine	2.0 g
Isoleucine	2.0 g	**Uracil**	**2.0 g**
Leucine	**4.0 g**	Valine	2.0 g
Lysine HCl	2.0 g	p-aminobenzoic acid	0.2 g

Omit the ingredients in bold type to select for specific plasmid markers (*ADE2*, adenine SO$_4$; *HIS3, histidine HCl; LEU2*, leucine; *TRP1,* tryptophan; *URA3*, uracil).

2.5. Solutions

2.5.1. Lithium Acetate (1.0 M)

Add 5.1 g of lithium acetate dihydrate to 50 mL of water in a 100-mL medium bottle and stir on a magnetic stir plate until dissolved. Sterilize by autoclaving for 15 min and store at room temperature.

2.5.2. PEG MW 3350 (50% w/v)

Add 50 g of PEG 3350 to 30 mL of distilled–deionized water in a 150-mL beaker and mix on a stirring hot plate with medium heat until dissolved. Allow the solution to cool to room temperature and make the volume up to 100 mL in a 100-mL measuring cylinder. Seal the cylinder with Parafilm™ and mix thoroughly by inversion. Transfer the solution to a glass medium bottle and autoclave for 15 min. Store securely capped at room temperature. Evaporation of water from the solution will increase the concentration of PEG in the transformation reaction and severely reduce the yield.

2.5.3. Single-Stranded Carrier DNA (2.0 mg/mL)

Add 200 mg of salmon sperm DNA to 100 mL of TE buffer in a beaker and stir at 4°C for 1 to 2 h. Store at –20°C in 1.0-mL samples. The carrier DNA must be denatured in a boiling water bath for 5 min before use and immediately chilled in ice-water. It can be boiled 3 or 4 times without the loss of activity.

2.5.4. Z Buffer

Dissolve 13.79 g of $NaH_2PO_4 \cdot H_2O$, 750 mg of KCl, and 246.0 mg of $MgSO_4 \cdot 7 H_2O$ in 1000 mL of ddH_2O, and titrate to a pH of 7.0 with 10 N NaOH. Before use, add 270 mL of β-ME to 100 mL of Z buffer.

2.5.5. X-Gal (20 mg/mL)

Dissolve 1.0 g of X-Gal in 50 mL of N,N-dimethyl formamide and store at –20°C.

2.5.6. Z Buffer/β-Me/X-Gal

Add 270 μL of β-ME and 1.67 mL of X-gal solution to 100 mL of Z buffer immediately before use.

2.5.7. ONPG (4 mg/mL)

Dissolve 200 mg of ONPG in 50 mL sterile double-distilled water. Store at –20°C in aliquots.

3. Methods

3.1. Preparation of the GAL4$_{BD}$: GOI Fusion Plasmid

Your "Gene of Interest" (*GOI*) encoding the bait protein must be cloned into a suitable THS vector in-frame with the binding domain fragment of the Gal4 protein. **Table 2** contains a list of THS *GAL4$_{BD}$* vectors for making this fusion plasmid. The DNA sequences and fusion reading frames of the multicloning sites (MCS) in these vectors are given in **Fig. 1**.

A number of cloning strategies can be used to clone YFG into a *GAL4$_{BD}$* vector:

1. Polymerase chain reaction (PCR) amplification of the *GOI* open reading frame (ORF) with the addition of unique *GAL4$_{BD}$* vector restriction sites to the ends of the primers.
2. Using existing *GAL4$_{BD}$* vector MCS restriction sites within *GOI* ORF.
3. Blunt-end ligation of a restriction fragment containing the *GOI* ORF into the appropriate frame in the plasmid.

If the *GOI* has identifiable protein domains or motifs, these can be fused to the *GAL4$_{BD}$*, especially if the *GOI* is relatively large. The most important aspect of this cloning is to ensure that the *GOI* is in-frame with the *GAL4$_{BD}$*, so that a fusion protein can be produced. Finally, the fusion junction of any plasmid constructed using the blunt-end ligation strategy should be sequenced prior to performing any screen to confirm the ORF fusion. The *GAL4$_{BD}$* sequencing primer, 5′-TCA TCG GAA GAG AGT AG-3′, can be used for *GAL4$_{BD}$* vectors such as pGBT9, pAS1, and pAS2. More details concerning cloning strategies can be found in Gietz et al. (*19*) and Parchaliuk et al. (*20*).

3.2. Transformation of the GAL4$_{BD}$: GOI Fusion Plasmid into Yeast

The first step in a THS screen is to transform the constructed *GAL4$_{BD}$*:GOI fusion plasmid into the appropriate yeast strain. The rapid LiAc/SS carrier DNA/PEG transformation protocol yields sufficient transformants for the isolation of single plasmid transformants. Although actively growing yeast cells give the highest yields,

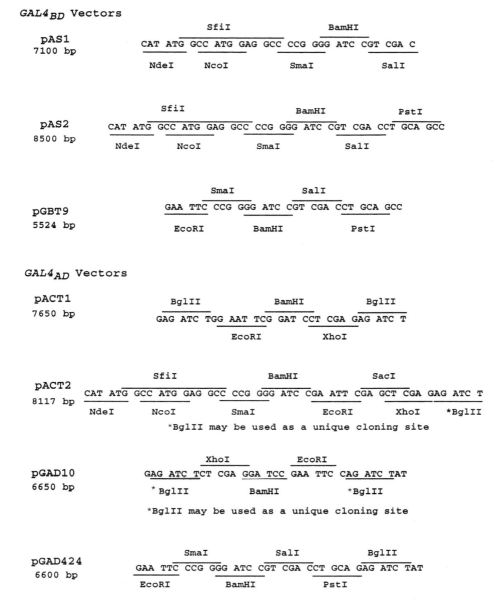

Fig. 1. Reading Frames of the Multi-cloning Sites in THS Vectors

cultures that are several days or even weeks old can be used. You should set up separate transformations for each $GAL4_{BD}:GOI$ fusion construct you have prepared.

1. Inoculate the yeast strain onto a YPAD plate and incubate overnight at 30°C.
2. Scrape a 20-µL sample of yeast from the plate and resuspend the cells in 1.0 mL of sterile water.
3. Pellet the cells for 30 s at maximum speed (~15,000g) in a microcentrifuge.

4. Boil a sample of carrier DNA for 5 min and immediately chill in an ice-water bath.
5. Add the following reagents to the cell pellet in descending order.

PEG 3350 (50% w/v)	240 μL
LiAc 1.0 *M*	36 μL
Carrier DNA (2 mg/mL)	50 μL
Sterile double-distilled water	30 μL
$GAL4_{BD}$:*GOI* plasmid DNA (1 μg)	5 μL

6. Vortex mix vigorously to suspend the cell pellet.
7. Heat shock in a water bath at 42°C for 40 min (*see* **Note 1**).
8. Microcentrifuge at ~15,000*g* for 30 s.
9. Remove the supernatant and resuspend the cells in 1 mL of sterile water.
10. Plate 200 μL of the suspension onto each of two SC-Trp plates.
11. Incubate the SC-Trp plates for 3 to 4 d at 30°C (*see* **Note 2**).
12. Isolate several $GAL4_{BD}$:*GOI* transformants and test them for auto-activation of the reporter genes.

3.3.1. Testing for Auto-Activation of the GAL1-HIS3 Reporter Gene

The *HIS3* reporter gene is leaky in some yeast strains; this phenotype is suppressed by adding 3-aminotriazole (3-AT) to the medium. An appropriate dilution of yeast cells containing the $GAL4_{BD}$:*GOI* plasmid should be plated onto Petri plates of SC-His medium containing 1, 5, 10, and 25 m*M* 3-AT to give about 500 cells/plate. The suspension should be plated onto two SC-Trp medium plates as a control for growth. Auto-activation of *GAL1-HIS3* will give colony growth on medium containing 25 m*M* 3-AT after 1 wk of incubation. The concentration of 3-AT needed to eliminate background growth is plasmid- and strain-dependent. Higher levels of 3-AT may be required to suppress background *GAL1-HIS3* expression when screening plasmid constructs based on pAS1 or pAS2, or when using the strain Y190 (*see* **Note 3**).

3.3.2. Testing for Auto-Activation of the lacZ Reporter Gene

You can use the two control plates of SC-Trp from the previous test.

1. Place a sterile 75-mm circle of Whatman no. 1 filter paper on top of the colonies on SC-Trp medium and ensure that the paper makes good contact with the colonies. Punch through the filter in an asymmetric pattern with an 18-gauge needle to mark the orientation of the paper.
2. Remove the paper and immerse it in liquid nitrogen for 10–15 s.
3. Remove the filter and allow it to thaw on a sheet of plastic wrap, colony side up. Freeze and thaw the paper twice more.
4. Soak another circle of filter paper in 1.25 mL of Z buffer/β-ME/X-gal in a Petri dish.
5. Place the first filter, colony side up, onto the filter soaked with Z buffer/β-ME/X-gal.
6. Put the lid on the Petri dish, seal it in a plastic bag and incubate it at 37°C for 120 min.
7. Place the filters on filter paper soaked in 1.0 *M* Na$_2$CO$_3$ and incubate for 5 min (*21*). Strong auto-activation of *lacZ* will give a blue color within 1 to 2 h.

3.4. Expression of the Fusion Protein

The steady state expression of the $GAL4_{BD}$:*GOI* fusion protein should be assayed by Western blotting. This can be done with a specific antibody for the product of the

GOI. The vectors pAS1, pAS2, pACT1, and pACT2 contain the HA tag *(11)*, which is recognized by the 12CA5 mAb available from Pharmacia (http://www.apbiotech.com) or equivalent, (Babco, http://www.babco.com/epitope.shtml). A Gal4$_{BD}$ antibody from Santa Cruz Biotechnology (http://www.scbt.com) can also be used.

3.5. Library Transformation Efficiency Test

Before embarking on a large-scale THS screen, you should perform a $GAL4_{AD}$:$cDNA$ plasmid library transformation efficiency test. This will ensure efficient use of the library plasmid DNA and allow you to plan on a specific number of transformants for THS screening. Too much library DNA will result in multiple $GAL4_{AD}$:$cDNA$ library plasmids in a single yeast cell; this will complicate the analysis of two-hybrid positives.

The high efficiency LiAc/SS carrier DNA/PEG transformation protocol can be used to transform 0.1, 0.5, 1, and 2 µg, of the $GAL4_{AD}$:$cDNA$ library plasmid DNA into the THS yeast strain containing the $GAL4_{BD}$:GOI plasmid.

3.5.1. Day 1

1. Inoculate the yeast strain carrying the $GAL4_{BD}$:GOI plasmid into 25 mL of SC selection medium and incubate at 30°C on a rotary shaker at 200 rpm overnight. The culture should reach a titer of 1 to 2×10^7 cells/mL.
2. Incubate a bottle of 2X YPAD broth and a 250-mL flask at 30°C.

3.5.2. Day 2

1. Determine the titer of the overnight culture. This can be done by measuring the optical density at 545 or 600 nm of a 10^{-2} dilution of the culture in sterile water. For most yeast strains, a suspension containing 10^6 cells/mL has an OD_{545} of 0.1. Alternatively, you can count the number of cells with a hemocytometer. For accurate determination of the cell titer, you should determine the relationship between OD or hemocytometer count and colony counts for your yeast strain.
2. Dispense 50 mL of prewarmed 2X YPAD into the prewarmed 250-mL flask and return it to the 30°C incubator.
3. Calculate the volume of suspension that contains 2.5×10^8 cells and transfer to a 50-mL centrifuge tube. Pellet the cells at 3000*g* for 5 min. Resuspend the cells in 10 mL of the prewarmed 2X YPAD and transfer into the flask. The starting titer will be 5×10^6 cells/mL.
4. Incubate the culture at 30°C on a rotary shaker at 200 rpm for 4 to 5 h and determine the cell titer. When the cells have divided at least twice (cell titer $\geq 2 \times 10^7$/mL) harvest the cells by centrifugation at 3000*g* for 5 min in a 50-mL centrifuge tube (*see* **Note 4**).
5. Boil a sample of carrier DNA for 5 min and chill in an ice-water bath.
6. Wash the cells in 25 mL of sterile water and resuspend them in 1 mL of sterile water.
7. Transfer the suspension to a 1.5-mL microcentrifuge tube, centrifuge at ~15,000*g* again, and discard the supernatant.
8. Add water to a final volume of 1 mL and vortex mix vigorously to resuspend the cells.
9. Pipet 100-µL samples (approx 10^8 cells) into four individual 1.5-mL microfuge tubes, one for each plasmid concentration, centrifuge at ~15,000*g* for 20 s, and remove the supernatant.
10. Make up sufficient transformation mixture, lacking plasmid DNA and water, for the planned number of transformations plus one extra. For four transformations, make

sufficient for five: 1200 μL of PEG, 180 μL of 1 *M* LiAc, and 250 μL of boiled carrier DNA. Keep the transformation mixture minus DNA in ice-water.

11. Add 326 μL of transformation mixture minus plasmid DNA to each transformation tube.
12. Add 0.1, 0.5, 1.0, and 2.0 μg of the $GAL4_{AD}$:cDNAcf:L, to the transformation tubes, and resuspend the cells by vortex mixing vigorously.
13. Incubate the transformation tubes at 42°C for 40 min.
14. Microcentrifuge at ~15,000g for 20 s and remove the transformation mixture.
15. Pipet 1 mL of sterile water into each tube. Loosen the pellet by stirring with a micropipet tip, and then vortex mix vigorously.
16. Dilute the suspensions 10^{-2} (10 μL into 1 mL), vortex mix thoroughly, and plate duplicate 10- and 100-μL samples onto plates of appropriate SC selection medium. The 10-μL samples should be pipetted into 100-μL puddles of sterile water.
17. Incubate the plates at 30°C for 3 to 4 d, and count the number of transformants on the plates.

Calculate the transformation efficiency (transformants/1 μg plasmid/10^8 cells) and transformant yield (total number of transformants/transformation) (*see* **Note 5**).

Scaling up the volumes of transformation mixture 60-fold and using 1 μg plasmid DNA per "unit transformation" should result in an overall yield of 120×10^6 transformants ($2 \times 10^6 \times 60$). This should be sufficient for the most demanding two-hybrid screen.

3.6. The THS Library Screen

The High Efficiency LiAc/SS Carrier DNA/PEG Transformation Protocol can be scaled up to ensure that an adequate number of transformants is screened.

3.6.1. Day 1

1. Inoculate the yeast strain carrying the bait plasmid *(GAL4$_{BD}$:GOI)* into SC selection medium. The volumes of medium and flask sizes for 30-, 60-, or 120-fold scale-up are listed below:

	Scale-up		
	30X	60X	120X
SC selection medium	50 mL in 250 mL	100 mL in 500 mL	200 mL in 1 L

2. Incubate at 30°C on a rotary shaker at 200 rpm overnight.
3. Warm an appropriate vol of 2× YPAD and a culture flask(s) at 30°C overnight (*see* **Subheading 3.6.2., step 1**).

3.6.2. Day 2

1. Determine the titer of the overnight culture and calculate the volume required for regrowth from a starting titer of 5×10^6 cells/mL. The numbers of cells, volumes of overnight culture, and the vol of 2X YPAD and flask sizes for regrowth are as follows:

	Scale-up		
	30X	60X	120X
Cells required	7.5×10^8	1.5×10^9	3.0×10^9
Volume of SC culture (approx)	40 mL	80 mL	160 mL
2X YPAD for regrowth	150 mL	300 mL	600 mL
Flask size for regrowth	1000 mL	2×1000 mL	3×1000 mL

(A larger number of flasks can be used. The volume of medium should be no more than one fifth of the flask volume.)

2. Use a sterile pipet or measuring cylinder to measure the appropriate volume of overnight culture, and transfer it to an appropriate number of sterile centrifuge tubes.
3. Pellet the cells at 3000*g* for 5 min and discard the supernatant.
4. Resuspend the pellet in warm 2X YPAD and dispense into the flask(s) for regrowth. Make up to the volume(s) indicated in **step 1**.
5. Incubate the flasks at 30°C on a rotary shaker at 200 rpm until the cell titer reaches 2×10^7/mL. This may take 4 to 5 h.
6. Boil sufficient carrier DNA (*see* **step 10**) for 5 min and chill in ice-water until required.
7. Harvest the cells by centrifugation at 3000*g* and discard the supernatant.
8. Wash the cells twice in half the regrowth culture volume of sterile water and transfer the suspension to a single 50-mL centrifuge tube (30X and 60X scale-up) or divide it between two tubes (120X scale-up).
9. Centrifuge and discard the supernatant.
10. Prepare the transformation mixture for the appropriate scale-up:

	Scale-up		
	30X	60X	120X
PEG 50%	7.20 mL	14.40 mL	28.40 mL
LiAc 1 M	1.08 mL	2.16 mL	4.32 mL
Carrier DNA 2 mg/mL	1.50 mL	3.00 mL	6.00 mL
Plasmid DNA plus sterile water	1.02 mL	2.04 mL	4.08 mL
Total volume	10.80 mL	21.60 mL	42.80 mL

11. Add the transformation mixture to the cell pellet(s) and vortex mix vigorously until the pellet is completely resuspended.
12. Incubate the tubes at 42°C for 45 min. Mix by inversion at 5-min intervals to ensure temperature equilibration.
13. Centrifuge at 3000*g* for 5 min and remove the supernatant.
14. Add sterile water to the cell pellet(s) (30X, 20 mL; 60X, 40 mL; 120X, 40 mL) and resuspend the cells by pipetting up and down and then vortex mixing vigorously.
15. Spread 400-µL samples of the cell suspension onto 150-mm plates of appropriate SC selection medium. You will need 50 plates for a 30X scale-up and 100 plates for 60X and 120X.
16. Incubate the plates at 30°C for 3–10 d until colonies have grown.
17. Score the plates for colonies that show an interaction between the proteins specified by the $GAL4_{BD}$:*GOI* and $GAL4_{AD}$:cDNA plasmids.

Check the plates and subculture positives to fresh SC –Trp –Leu –His + 3-AT after 4 d. Continue checking and subculturing for a maximum of 2 wk.

3.7. Assay for lacZ Reporter Gene Activity

Once positives have been subcultured and duplicates set aside, *lacZ* gene activation can be assayed. The positives should be maintained on the appropriate SC selection medium and tested as set out in **Subheading 3.3.2.**

3.8. Storage of the THS Positives

Positives that activate the *lacZ* reporter should be stored frozen. Streak them onto fresh SC-Trp-Leu-His + 3-AT plates and incubate at 30°C for 24–48 h. Suspend a

20-µL sample of cells in 1 mL of sterile 20% glycerol in a 1.5-mL microcentrifuge tube or cryotube and store at –70°C. Large numbers of positives can be inoculated in a grid pattern onto 150-mm SC-Trp-Leu-His + 3-AT plates and stored frozen in microtiter plates. Sterilize a 96-well microtiter plate replicator by flaming in ethanol and then allow the replicator to cool. Gently place the sterile replicator onto a plate of SC selection medium and press gently to mark the surface of the agar. Patch the positive colonies onto the imprints and incubate the plate overnight. The next day, flame and cool the replicator, and lower onto the plate of patched colonies so that the prongs make contact with each inoculum. Move the replicator gently to transfer cells to the prongs. Remove the replicator, and lower into a sterile microtiter plate containing 150 µL of sterile 20% glycerol in each well. Jiggle the replicator using a gentle rotating action to suspend the cells in the glycerol. Remove the replicator, replace the lid on the microtiter plate, and store it, sealed in a plastic bag, at –70°C.

3.9. Characterizing THS Positives

The $GAL4_{AD}$:cDNA plasmids in the THS positives that activate the reporter genes can be isolated by the method of Hoffman and Winston (22).

1. Inoculate the THS positives onto SC-Trp-Leu-His + 3-AT plates in 2-cm^2 patches and incubate at 30°C overnight.
2. Scrape 50-µL samples of cells from the plate and resuspend them in 500 µL of sterile water in 1.5-µL microcentrifuge tubes, pellet at top speed in a microcentrifuge for 1 min, and discard the supernatant.
3. Add 200 µL of Yeast cracking buffer, and gently resuspend the cell pellet using a micropipet tip.
4. Add 200 µL of 425 to 600-µm glass beads and 200 µL of buffer saturated phenol: chloroform.
5. Vortex mix each sample vigorously for 30 s and then place on ice. Repeat twice, leaving the samples on ice for 30 s between treatments.
6. Microcentrifuge at ~15,000g for 1 min.
7. Remove the aqueous phase to a fresh tube and precipitate the nucleic acids with 20 µL of 3.0 M sodium acetate (pH 6.0) and 500 µL of 95% ethanol. Incubate at –20°C for 30 min, and then microcentrifuge at ~15,000g for 5 min at 4°C. Wash the pellet of plasmid DNA with 100 µL of 70% ethanol and dry it for 5 min at room temperature. Dissolve the pellet of plasmid DNA in 25 µL of TE (10 mM Tris-HCl, pH 8.0, 1 mM EDTA).

The plasmids should be electroporated (15) into an *E. coli* strain containing a *leuB* mutation, such as KC8 or HB101. This will allow for the direct selection of the yeast *LEU2* gene on the $GAL4_{AD}$:cDNA plasmid.

3.10. Reconstruction of THS Positives

It is essential to reconstruct each THS positive before proceeding with further analysis. $GAL4_{AD}$:cDNA plasmid DNA isolated from each THS positive strain should be transformed back into the THS yeast strain containing the $GAL4_{BD}$:*GOI* plasmid by the protocol described in **Subheading 3.2.** Transformants can then tested for the activation of the *HIS3* reporter gene by plating onto SC -Trp-Leu-His + 3-AT medium, and for *lacZ* activation by the procedure described in **Subheading 3.3.2.**

3.10.1. Failure of THS Positives to Reconstruct

A single yeast cell can be simultaneously transformed by several $GAL4_{AD}$:cDNA plasmids. At high plasmid concentrations, the frequency of such co-transformation can be as high as 30% (our unpublished observations). If the original THS positive isolate contained several $GAL4_{AD}$:cDNA plasmids, only one of which was responsible for reporter gene activation, it is likely that a failure to reconstruct the positive phenotype is because the plasmid used is not the one responsible for reporter gene activation. Isolate an additional 20 plasmids from original THS positives and characterize them by digestion with *Eco*RI. Repeat the reconstruction with each distinct plasmid type that you are able to identify. In some cases, a THS positive can result from deletions between direct repeats within the bait gene, giving rise to an auto-activating $GAL4_{BD}$:*GOI* bait plasmid *(23)*. For this reason, the $GAL4_{BD}$:*GOI* bait plasmid isolated from the original THS positive should be tested for auto-activation.

3.11. Characterizing the Strength of the THS Positive Interaction

The activation of the *lacZ* reporter can be quantified using the liquid ONPG assay *(24)* modified for application to yeast.

1. Inoculate the THS positives onto SC-Trp-Leu-His + 3-AT plates in 2-cm^2 patches and incubate at 30°C overnight. Scrape 50-µL samples of cells from the plate and resuspend them in 1 mL of sterile water in 1.5-mL microcentrifuge tubes. Dilute 5 µL of suspension into 1 mL of water and measure the OD_{600}. Pellet the remainder of the suspension at ~15,000*g* in a microcentrifuge for 1 min and discard the supernatant.
2. Resuspend the cell pellet in 100 µL of Z buffer and add glass beads to one half the volume of the resuspended cells.
3. Chill on ice and vortex mix for 30 s followed by a 30-s incubation on ice. Repeat 4 times.
4. Mix duplicate samples of 50 µL of extract with 450 µL with Z buffer/β-ME, and equilibrate in a 37°C water bath for 5 min.
5. Add 160 µL of ONPG (4 mg/mL) and incubate at 37°C. As soon as a yellow color is evident, add 0.4 mL of 1.0 *M* Na_2CO_3 to terminate the reaction, and note the elapsed time. Incubation times can range from 1–60 min.
6. Centrifuge at ~15,000*g* for 1 min and measure the absorbance at 420 nm. The activity of β-galactosidase in Miller units is calculated using the formula:

$$\text{Units} = (A_{420} \times 1000) / (t \times V \times OD_{600})$$

where t = elapsed time (min), V = volume of culture used (1 mL), and OD_{600} = optical density of cell suspension.

3.12. Deletion Mapping of Interacting Domains

The identification of protein motifs responsible for the interaction can be accomplished by deletions of the *GOI* and library cDNA genes using restriction sites or with a PCR strategy. Begin by deleting the 3′ ends of both genes, to preserve the fusion junctions of the translated proteins. Altering the fusion junction can dramatically affect the steady state level of fusion protein expression. When deleting the 5′ end of a gene in a bait or prey plasmid, it is recommended that the steady state level of expression of the fusion protein expression be determined by Western blot to ensure accurate interpretation of the *lacZ* assay.

3.13. False Positives

A successful THS screen may result in the isolation of several hundred apparent positives. Many of these will be true positives, in which activation of the reporter genes requires an interaction between the proteins encoded by the *GOI* and the *cDNA* sequence, but some will almost certainly be false positives. Three classes of false positives can occur in a THS screen. They can be tested for at the same time that you carry out the reconstruction of THS positives. Type 1 false positives arise when the $GAL4_{AD}:cDNA$ library plasmid activates reporter genes without interacting with the $GAL4_{BD}$ plasmid. They can be identified by transforming the $GAL4_{AD}:cDNA$ plasmid into the THS yeast strain lacking the $GAL4_{BD}$ plasmid. If activation occurs, it is the result of auto-activation by the $GAL4_{AD}:cDNA$ plasmid. Type 2 false positives occur when the $GAL4_{AD}:cDNA$ plasmid activates the reporter genes in the presence of any $GAL4_{BD}$ plasmid. They can be identified by co-transformation of the THS yeast strain with the $GAL4_{AD}:cDNA$ plasmid and an unrelated $GAL4_{BD}$ plasmid. Growth on SC-Trp-Leu-His + 3-AT indicates a Type 2 false positive. Type 3 false positives result from the interaction of a $GAL4_{AD}:cDNA$ plasmid with an "empty" $GAL4_{BD}$ plasmid. They can be identified by co-transformation of the THS yeast strain with the $GAL4_{AD}:cDNA$ plasmid and an empty $GAL4_{BD}$ plasmid. Growth on SC-Trp-Leu-His + 3-AT indicates a Type 3 false positive. More detailed discussions of false positives can be found in Bartel et al. *(25)*, Gietz et al. *(19)*, and Parchaliuk et al. *(20)*.

4. Notes

1. Extending the duration of incubation at 42°C to 180 min increases the number of transformants to approx 1×10^6/µg plasmid/10^8 cells with some strains.
2. If you use the wrong amino acid mixture for your yeast strain–plasmid combination, you will obtain:
   a. No transformants if your yeast strain requires a nutrient that is absent from the amino acid mixture.
   b. Confluent growth if your yeast strain does not require the component missing from the amino acid mixture.
3. Auto-activation by the $GAL4_{BD}:GOI$ construct can usually be overcome by cloning your bait gene into a different vector, such as pGBT9, if pAS1 or pAS2 were used previously. Alternatively, the construct can be modified by deletion to remove the region responsible for the auto-activation. An alternative approach to dampen the auto-activation of specific $GAL4_{BD}:GOI$ constructs uses the SSB24 repressor sequence fused to the $GAL4_{BD}$ in pGBT9 *(26)*.
4. Completion of two divisions is required for high transformation efficiency. Yields remain high for at least another cell doubling. High efficiencies are also obtained if the cells are grown overnight from a small inoculum, 2×10^4 cells/mL, and harvested at mid to late log phase, $>1.0 \times 10^8$ cells/mL.
5. For example, if the transformation reaction containing 1.0 µg of plasmid gave an average of 200 colonies on the two 10-µL sample plates, the calculation is:

$$200 \times 100 \ (10^{-2} \text{ dilution factor}) \times 100 \ (10 \text{ µL plating factor}) \times 1 \ (\text{plasmid factor})$$

$$\text{Transformation Efficiency} = 2 \times 10^6/\text{µg plasmid DNA}/10^8 \text{ cells}$$

References

1. Fields, S. and Song, O. (1989) A novel genetic system to detect protein-protein interactions. *Nature* **340**, 245–246.

2. Keegan, K. and Cooper, J. A. (1996) Use of the two hybrid system to detect the association of the protein-tyrosine-phosphatase, SHPTP2, with another SH2-containing protein, Grb7. *Oncogene* **12**, 1537–1544.

3. Osborne, M. A, Dalton, S., and Kochan, J. P. (1995) The yeast tribrid system—genetic detection of trans-phosphorylated ITAM-SH2-interactions. *Biotechnology* **13**, 1474–1478.

4. Vidal, M. and Legrain, P. (1999) Yeast forward and reverse 'n'-hybrid systems. *Nucleic Acids Res.* **27**, 919–929.

5. Fashena, S. J., Serebriiskii, I., and Golemis, E. A. (2000) The continued evolution of two-hybrid screening approaches in yeast: how to outwit different preys with different baits. *Gene* **250**, 1–14.

6. Pandey, A. and Mann, M. (2000) Proteomics to study genes and genomes. *Nature* **405**, 837–846.

7. Pirson, I., Jacobs, C., Vandenbroere, I., El Housni, H., Dumont, J. E., and Perez-Morga, D. (1999) Use of two-hybrid methodology for identifying proteins of interest in endocrinology. *Mol. Cell. Endocrinol.* **151**, 137–141.

8. Gietz, R. D. and Woods, R. A. (1998) Transformation of yeast by the lithium acetate/single-stranded carrier DNA/PEG method, in *Methods in Microbiology* (Brown, A. J. P. and Tuite, M. F., eds.), Academic Press, New York, pp. 53–66.

9. Uetz, P., Giot, L., Cagney, G., Mansfield, T. A., Judson, R. S., Knight, J. R., et al. (2000) A comprehensive analysis of protein-protein interactions in *Saccharomyces cerevisiae*. *Nature* **403**, 623–627.

10. Durfee, T., Becherer, K., Chen, R.-I., Yeh, S. H., Yang, Y., Kilburn, A. K., et al. (1993) The retinoblastoma protein associates with the protein phosphatase type 1 catalytic subunit. *Genes Dev.* **7**, 555–569.

11. James, P., Halladay, J., and Craig, E. A. (1996) Genomic libraries and a host strain for highly efficient two-hybrid selection in yeast. *Genetics* **144**, 1425–1436.

12. Graham, K. C. (1996) *Production of two S. cerevisiae strains designed to enhance utilization of the yeast two-hybrid system.* M.Sc. Thesis, University of Manitoba, Winnipeg Manitoba, Canada.

13. Harper, J. W., Adami, G., Wei, N., Keyomarsi, K., and Elledge, S. J. (1993) The p21 Cdk-interacting protein Cip1 is a potent inhibitor of G1 cyclin-dependent kinases. *Cell* **75**, 805–816.

14. Bartel, P. L., Chien, C.-T., Sternglanz, R., and Fields, S. (1993) Using the two-hybrid system to detect protein-protein interactions, in *Cellular Interactions in Development: A Practical Approach* (Hartley, D. A., ed.), Oxford University Press, Oxford, pp. 153–179.

15. Dower, W. J., Miller, J. F., and Ragsdale, C. W. (1988) High efficiency transformation of *E. coli* by high voltage electroporation. *Nucleic Acids Res.* **16**, 6127–6145.

16. Ausubel, F. M., Brent, R., Kingston, R. E., Moore, D. D., Seidman, J. G., Smith J. A. and Stuhl, K. (1989) *Current Protocols in Molecular Biology.* John Wiley & Sons, New York.

17. Sambrooke, J., Fritsch, E. F., and Maniatis, T. (1989) *Molecular Cloning: A Laboratory Manual.* 2nd ed. CSH Laboratory Press, Cold Spring Harbor, N.Y.

18. Rose, M. D. (1987) Isolation of genes by complementation in yeast. *Methods Enzymol.* **152**, 481–504.

19. Gietz, R. D., Triggs-Raine B., Robbins A., Graham K. C., and Woods, R. A. (1997) Identification of proteins that interact with a protein of interest: applications of the yeast two-hybrid system. *Mol. Cell. Biochem.* **172**, 67–79.

20. Parchaliuk, D. L., Kirkpatrick, R. D., Simon, S. L., Agatep, R., and Gietz, R. D. (1999) Yeast two-hybrid system: part A: screen preparation. *Technical Tips Online* (http://tto.trends.com) **1**:66:P01616

21. Tanahashi, H. and Tabira, T. (2000) Alkaline treatment after X-gal staining reaction for *Escherichia coli* β-galactosidase enhances sensitivity. *Anal. Biochem.* **279,** 122–123.

22. Hoffman, C. S. and Winston, F. (1987) A ten-minute DNA preparation from yeast efficiently releases autonomous plasmids for transformation of *Escherichia coli. Gene* **57,** 267–272.

23. El Hounsni, H., Vandenbroere, I., Perez-Morga, D., Christophe, D., and Pirson, I. (1998) A rare case of false positive in a yeast two-hybrid screening: the selection of rearranged bait constructs that produce a functional Gal4 activity. *Anal. Biochem.* **262,** 94–96.

24. Miller, J. H. (1972) *Experiments in Molecular Genetics*. CHS Laboratory Press, Cold Spring Harbor, N.Y.

25. Bartel, P. L., Chien, C.-T., Sternglanz, R., and Fields, S. (1993) Elimination of false positives that arise in using the two-hybrid system. *Biotechniques* **14,** 920–924.

26. Cormack, R. S. and Somssich, I. E. (1997) Dampening of bait proteins in the two-hybrid system. *Anal. Biochem.* **248,** 184–186.

Index

A

Activin, embryonic stem cell differentiation marker, 22

Adipocyte,
induction from embryonic stem cells,
cell maintenance, 109–111, 113, 114
embryoid body differentiation, 111, 112, 115
materials, 109, 110, 113, 114
Oil Red O staining, 109, 110, 112, 113
requirements for lineage commitment, 107, 108
reverse transcription-polymerase chain reaction analysis of gene expression during differentiation,
annealing temperature, 113, 114
primers, 110
RNA preparation, 110, 113, 115
X-Gal staining, 109, 110, 112, 113

Antibody,
phage display (see Phage display)
single chain Fv,
applications, 433, 434
intracellular antibody capture technology, 434
isolation of intracellular fragments binding to specific antigens,
β–galactosidase filter assay, 437, 442
materials, 435–438
overview, 435
phage library screening with antigen, 435, 436, 438, 439, 445
plasmid extraction from yeast, 438, 442, 443
polymerase chain reaction, 438, 443, 444
sublibrary preparation, 436, 439, 440, 445

transformation of bacteria, 436, 441
Western blot, 438, 444
yeast antigen–antibody interaction screening, 436–438, 441–443, 445
yeast expression library, 436, 440, 445

Apoptosis,
detection during embryonic stem cell differentiation, 42, 44
DNA fragmentation analysis, 42

B

Bcr-Abl fusion,
chronic myelogenous leukemia role, 83
immortalization of cells, 83
tetracycline-regulated gene expression analysis of embryonic stem cell hematopoiesis,
Bcr-Abl clone establishment, 89, 90, 93
differentiation induction, 86, 90–93
embryonic stem cell maintenance, 85–88, 93
materials, 85, 86, 93
OP9 stromal cell line maintenance, 86, 90, 93
overview, 83–85
Tet-off parental cell line generation, 88, 89
vectors, 86, 87

BMPs (see Bone morphogenetic proteins)

Bone morphogenetic proteins (BMPs),
embryonic stem cell differentiation marker, 23
neural induction in ectoderm, 217, 218

BrdU (see Bromodeoxyuridine)

Bromodeoxyuridine (BrdU), cell cycle analysis, 31